MOLECULAR AND CELL BIOLOGY
VOLUME 31

Non-Neuronal Cells of the Nervous System: Function and Dysfunction

MOLECULAR AND CELL BIOLOGY

VOLUME 31

Non-Neuronal Cells of the Nervous System: Function and Dysfunction

Part II: Biochemistry, Physiology and Pharmacology

Series Editor:

E. Edward Bittar
University of Wisconsin - Madison
Madison, Wisconsin,
USA

Volume Editor:

Leif Hertz,
Gilmour, Ontario,
Canada

2004

ELSEVIER

Amsterdam - Boston - Heidelberg - London - New York - Oxford
Paris - San Diego - San Francisco - Singapore - Sydney - Tokyo

ELSEVIER B.V.	ELSEVIER Inc.	ELSEVIER Ltd	ELSEVIER Ltd
Sara Burgerhartstraat 25	525 B Street, Suite 1900	The Boulevard, Langford Lane	84 Theobalds Road
P.O. Box 211, 1000 AE	San Diego, CA 92101-4495	Kidlington, Oxford OX5 1GB	London WC1X 8RR
Amsterdam, The Netherlands	USA	UK	UK

First edition 2004

Library of Congress Cataloging in Publication Data
A catalog record is available from the Library of Congress.

British Library Cataloguing in Publication Data
A catalogue record is available from the British Library.

ISBN: 0-444-51451-1 (set of three volumes)
ISSN: 1569-2558 (series)

∞ The paper used in this publication meets the requirements of ANSI/NISO Z39.48-1992 (Permanence of Paper).
Printed in The Netherlands.

TABLE OF CONTENTS

Volume II

Biochemistry, Physiology and Pharmacology

Chapter 17
A role for lactate released from astrocytes in energy production during neural activity? ... 391
Eugene L. Roberts Jr. and Ching-Ping Chih

Chapter 18
Principles of the measurement of neuro-glial metabolism using in vivo ^{13}C NMR spectroscopy ... 409
Rolf Gruetter

Chapter 19
Ion, transmitter and drug effects on energy metabolism in astrocytes ... 435
Leif Hertz, Liang Peng, Christel C. Kjeldsen,
Brona S. O'Dowd and Gerald A. Dienel

Chapter 20
Role of astrocytes in homeostasis of glutamate and GABA during physiological and pathophysiological conditions ... 461
Arne Schousboe and Helle S. Waagepetersen

Chapter 21
Astrocytic receptors and second messenger systems ... 475
Elisabeth Hansson and Lars Rönnbäck

Chapter 22
Transactivation in astrocytes as a novel mechanism of neuroprotection ... 503
Liang Peng

Chapter 23
Roles of glia cells in cholesterol homeostasis in the brain ... 519
Jin-ichi Ito and Shinji Yokoyama

Chapter 24
Non-neuronal cells in the nervous system: sources and targets of neuroactive steroids ... 535
Roberto C. Melcangi, Iñigo Azcoitia, Mariarita Galbiati,
Valerio Magnaghi, Daniel Garcia-Ovejero and Luis M. Garcia-Segura

Chapter 25
Expression of neurotrophic factors and cytokines and their receptors on astrocytes in vivo ... 561
Takao Nakagawa and Joan P. Schwartz

Chapter 26
The nitric oxide/cyclic GMP pathway in CNS glial cells 575
Agustina García and María Antonia Baltrons

Chapter 27
Potassium homeostasis in the brain at the organ and cell level 595
Wolfgang Walz

Chapter 28
Potassium and glia-derived slow potential shifts in relation to behaviour 611
Peter R. Laming

Chapter 29
Regulation of Ca^{2+} stores in glial cells 635
Giovanni Scapagnini, Thomas J. Nelson and Daniel L. Alkon

Chapter 30
Decoding calcium wave signaling 661
A.H. Cornell-Bell, P. Jung and V. Trinkaus-Randall

Chapter 31
Mathematical modeling of intracellular and intercellular calcium signaling 689
Jian-Wei Shuai, Suhita Nadkarni, Peter Jung, Ann Cornell-Bell
and Vickery Trinkaus-Randall

Chapter 32
pH regulation in non-neuronal brain cells and interstitial fluid 707
Suzanne D. McAlear and Mark O. Bevensee

Chapter 33
AVP effects and water channels in non-neuronal CNS cells 747
Ye Chen and Maria Spatz

Advances in
Molecular and
Cell Biology

A role for lactate released from astrocytes in energy production during neural activity?

Eugene L. Roberts Jr.[a,b,*] and Ching-Ping Chih[b]

[a]*Department of Neurology, University of Miami School of Medicine, Miami, FL 33136, USA*
[b]*Geriatric Research, Education, and Clinical Center, Miami VA Medical Center, Miami FL 33125, USA*
[*]*Correspondence address: Department of Neurology, University of Miami School of Medicine, P.O. Box 016960, Miami, FL 33101, USA*
E-mail: e.roberts@miami.edu

Contents

1. Introduction
2. Major assertions of the ANLSH
   2.1. Assertions
   2.2. Do active neurons prefer lactate to glucose?
   2.3. Convincing evidence showing that active neurons prefer lactate to glucose is still lacking
   2.4. Does activity-induced glucose uptake occur predominantly in astrocytes?
   2.5. Are neurons lactate consumers, and astrocytes lactate producers?
3. Role of lactate as an energy substrate in the brain
4. Concluding remarks

Glucose has long been viewed as the main energy source for brain cells, while lactate served as a secondary energy source. Recently, the astrocyte–neuron lactate shuttle hypothesis (ANLSH) has proposed that, when neurons are activated, they preferentially consume lactate generated by astrocytes to meet their energetic needs. In this chapter, we examine the theoretical basis and current evidence for the ANLSH. From this examination, we find little support for the ANLSH and much credibility for the conventional view of glucose utilization.

1. Introduction

Glucose has long been viewed as the primary energy source for brain cells. Lactate, a glucose metabolite, has been thought of as a secondary energy source utilized by brain cells when its levels are high (Nemoto et al., 1974) or when glucose levels are low

Advances in Molecular and Cell Biology, Vol. 31, pages 391–407

ISBN: 0-444-51451-1

(Thurston et al., 1983). Recently however, a new model of brain glucose utilization termed the astrocyte–neuron lactate shuttle hypothesis (ANLSH) (see Pellerin et al., 1998a; Sibson et al., 1998; Magistretti and Pellerin, 1999; Magistretti et al., 1999; Rothman et al., 1999; Magistretti, 2000) has drastically redefined the role of lactate in brain energy metabolism. According to the ANLSH, lactate released from astrocytes is the primary energy source for active neurons. Neuronal activity is thought to stimulate astrocytic production of lactate indirectly by releasing glutamate, which is taken up by astrocytes via high affinity Na^+-dependent transporters. Increases in glutamate and Na^+ in glia activate glutamine synthetase and Na^+-K^+ ATPase, respectively. The latter stimulates astrocytic ATP consumption, leading to activation of anaerobic glycolysis (conversion of glucose to lactate) in astrocytes. The resulting lactate is envisaged to be transported out of astrocytes and into neurons, where it is oxidized to fuel the activity-related energy needs of neurons.

The conventional (Fig. 1) and ANLSH (Fig. 2) models of glucose metabolism in brain tissue differ most noticeably in where and how glucose is utilized. In the conventional model, neural activation stimulates Na^+-K^+ ATPase activity, causing increased ATP utilization. ATP consumption activates glycolysis (the conversion of glucose to pyruvate) and increases glucose oxidation in both neurons and astrocytes. In contrast, the ANLSH compartmentalizes glucose and lactate consumption resulting from neural activity. According to the ANLSH, neural activation stimulates glucose consumption in astrocytes, but not in neurons, via anaerobic glycolysis. Instead of metabolizing glucose during neural activity, neurons consume lactate released from astrocytes.

The ANLSH has received widespread attention since it represents a significant departure from the classical view of glucose utilization in the brain. If true, the ANLSH would require a major change in how we think about brain energy metabolism. In this chapter, we examine the theoretical background and critically review the experimental evidence regarding the ANLSH to see whether such a change in thought is warranted.

2. Major assertions of the ANLSH

2.1. *Assertions*

Studies supporting the ANLSH assert that (i) active neurons preferentially use lactate instead of glucose (Magistretti and Pellerin, 1996, 1999; Tsacopoulos and Magistretti, 1996; Magistretti, 1999; Magistretti et al., 1999), (ii) activity-induced glucose utilization occurs predominantly, if not exclusively, in astrocytes (Sibson et al., 1998; Magistretti, 1999), and (iii) lactate is produced by astrocytes and shuttled to neurons during activity (Magistretti et al., 1999; Magistretti, 2000). In Section 2.2, we discuss the theoretical background and experimental evidence regarding these assertions.

2.2. *Do active neurons prefer lactate to glucose?*

The brain has similar concentrations of glucose and lactate (about 1 mM; Kuhr and Korf, 1988; Prichard et al., 1991; Harada et al., 1992; Silver and Erecinska, 1994; Fray et al., 1996; Hu and Wilson, 1997; Lowry et al., 1998). Several questions deserve

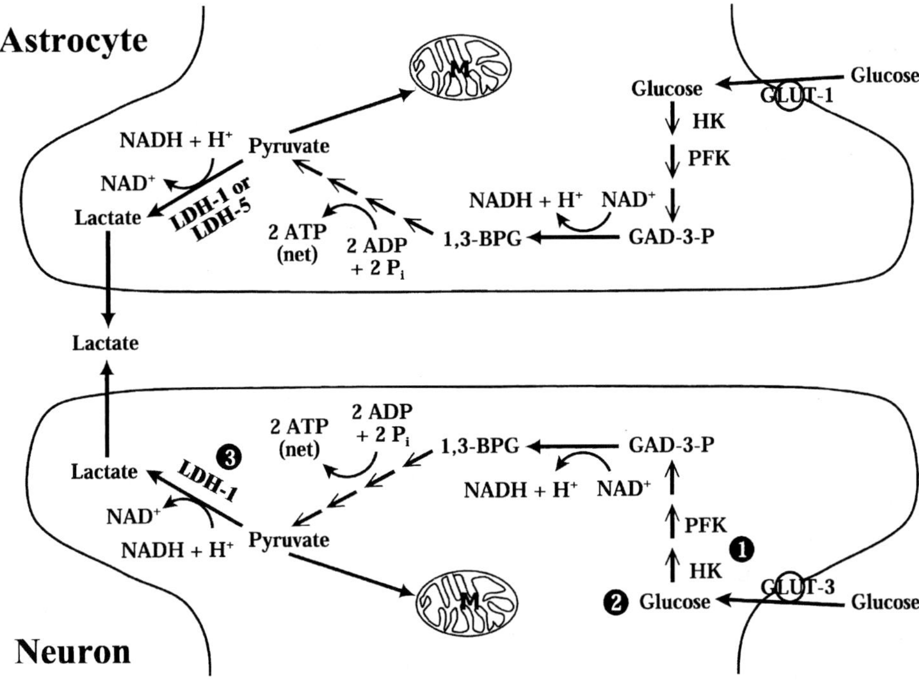

Fig. 1. Conventional view of glucose metabolism in astrocytes and neurons when neural activity increases. Increased Na^+, K^+-ATPase activity boosts ATP consumption in both cell types, because decreasing ATP levels leads to rapid activation of the glycolytic enzymes hexokinase (HK) and phosphofructokinase (PFK). Thus, increased neural activity leads to increased consumption of glucose in both neurons and astrocytes via either aerobic or anaerobic pathways. Glucose has the following advantages over lactate as an energy substrate during neural activity (see numbers on figure): (1) HK and PFK are very sensitive to declines in ATP levels and activate rapidly when energy demand increases. Also, conversion of glucose to pyruvate is more favorable energetically than conversion of lactate to pyruvate (see Section 2.2.3). (2) Glucose is readily available to both neurons and astrocytes. In particular, glucose concentrations in brain cells exceed greatly the K_m for HK (see Section 2.2.1). Also, GLUT3 transports glucose seven times faster than GLUT1. (3) When glucose is metabolized during neural activity, both NADH shuttles (see Section 2.2.4) and lactate production are available to maintain the cytoplasmic redox balance. These advantages are in contrast to the limitations of the ANLSH pointed out in Fig. 2. Abbreviations: GLUT1 and GLUT3: glucose transporters 1 and 3, respectively; GAD-3-P: glyceraldehyde-3-phosphate; 1,3-BPG: 1,3-bisphosphoglycerate; LDH-1 and LDH-5: lactate dehydrogenases 1 and 5, respectively; M: mitochondrion.

consideration when comparing these substrates as energy sources for active neurons: (1) Are neurons equipped to use these substrates? (2) Which metabolic pathway (glucose or lactate utilization) responds faster to increases in energy demand? (3) Which metabolic pathway is energetically more favorable? (4) Which pathway makes it easiest to maintain the cytoplasmic redox balance? and (5) Is glycolysis under some conditions advantageous for local synthesis of ATP in neurons? In the following analysis, we discuss whether lactate can be the principal energy substrate of active brain cells, as proposed by the ANLSH.

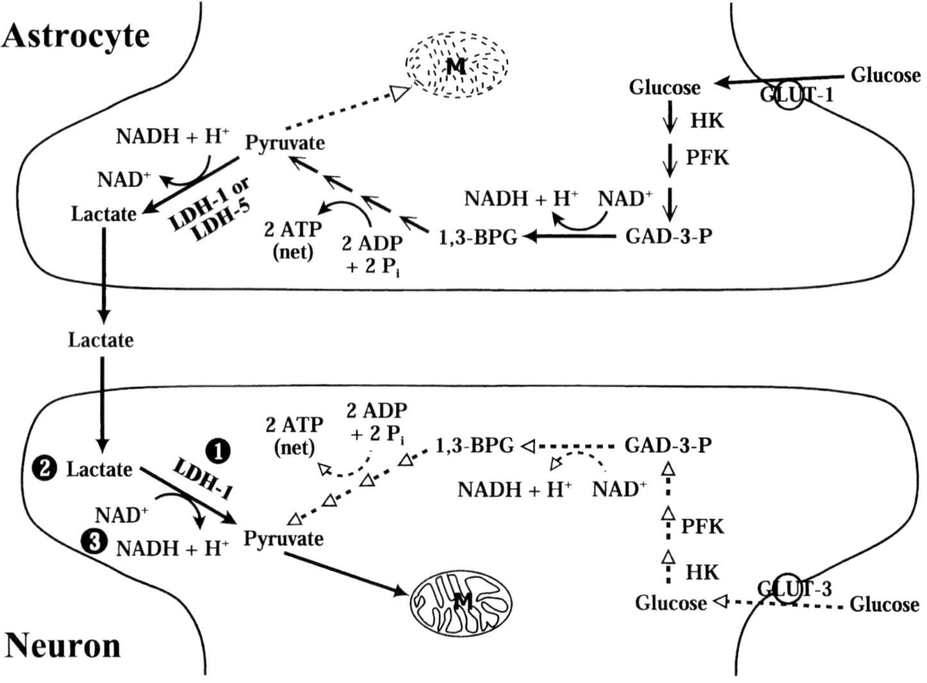

Fig. 2. Neuronal and astrocytic glucose metabolism during neural activity according to the astrocyte–neuron lactate shuttle hypothesis (ANLSH). The ANLSH requires that astrocytes do not consume glucose oxidatively during neural activity, but instead metabolize all glucose to lactate. The ANLSH also requires that active neurons do not metabolize glucose but instead metabolize lactate produced by astrocytes. In the figure, dashed lines and open arrow heads show metabolic pathways not available to neurons (the glycolytic pathway) and to astrocytes (the oxidative phosphorylation pathway), according to the ANLSH. Limitations in the ANLSH's view of energy metabolism pointed out in the figure are: (1) In contrast to HK and PFK, LDH is not regulated by changes in ATP levels, so LDH responds slowly to increased cellular energy demands. Also, conversion of lactate to pyruvate may not be favored thermodynamically during neural activity (see Section 2.2.3). (2) Increases in extracellular lactate levels lag well behind the start of neural activity. This tends to limit lactate utilization as a choice for meeting the energy demands of neural activity since increases in lactate oxidation depend heavily upon an increase in available lactate (see Section 2.2.2). (3) If lactate is the only available substrate for neurons, then the cytoplasmic redox balance in neurons will depend entirely upon NADH shuttles (see Section 2.2.4). Also, LDH must compete with glyceraldehyde-3-phosphate-dehydrogenase (the enzyme converting GAD-3-P to 1,3-BPG) for NAD^+. This competition can limit lactate utilization. These limitations in the ANLSH model contrast with the advantages of the conventional model (Fig. 1).

2.2.1. Neurons and astrocytes are well equipped to utilize both glucose and lactate

Both neurons and astrocytes have high activities of glycolytic enzymes (Cimino et al., 1998; Lai et al., 1999). High specific activities of glycolytic enzymes are found within synaptosomal membranes and nerve endings (Wilson, 1972; Knull, 1978; Kao-Jen and Wilson, 1980; Lim et al., 1983). This high glycolytic capacity in neurons and astrocytes is coupled to high capacity glucose transporters (Vannucci et al., 1997). The dominant glucose transporter in neurons (GLUT3) transports glucose seven times faster than the dominant glucose transporter in astrocytes (GLUT1) (Vannucci et al., 1997). Much evidence exists for both neurons and astrocytes increasing their glycolytic rates in response to increased energy demands (e.g., Walz and Mukerji, 1988; Peng et al., 1994; Sokoloff, 1999).

A high concentration of blood glucose and efficient glucose transporters in the brain help maintain brain glucose levels between 0.35 and 2.6 mM (Silver and Erecinska, 1994; Fray et al., 1996; Hu and Wilson, 1997; Lowry et al., 1998). Glucose is evenly distributed between the intra- and extracellular compartments of the brain (Pfeuffer et al., 2000). Increased blood flow accompanying brain stimulation (Fellows and Boutelle, 1993) may increase glucose delivery to the brain. Even with this possible increased delivery of glucose, brain glucose levels have been found to either increase (Fellows and Boutelle, 1993) or decrease (up to 20%; see data of Fray et al., 1996; Hu and Wilson, 1997) transiently during neural stimulation. However, glucose levels were well above the K_m for hexokinase (0.04 mM; Lowry and Passonneau, 1964) in these studies. Thus, although neuronal processes are not directly coupled to the circulation as astrocytic processes are, the high glucose concentrations in the brain and the presence of GLUT3 suggest that neuronal glucose supplies are adequate.

Lactate dehydrogenase (LDH) and lactate transporters are also present in both neurons and astrocytes (Bittar et al., 1996; Pellerin et al., 1998b). In vitro studies have shown that neurons and astrocytes can oxidize lactate at similar rates (Peng et al., 1994).

2.2.2. *Activation of glycolysis occurs rapidly during increased energy demand*

During neural activity, oxygen utilization goes up rapidly (Malonek and Grinvald, 1996; Malonek et al., 1997), so rapid increases in the supply of pyruvate to mitochondria are necessary to maintain oxidative metabolism. Both glucose and lactate can supply pyruvate to mitochondria. However, the glycolytic pathway, which converts glucose to pyruvate, can be activated much faster than lactate dehydrogenation, which converts lactate to pyruvate.

The activities of key glycolytic enzymes are tightly regulated by energy demand (Clarke et al., 1989). Thus, when energy demand is low, glycolysis is inhibited due to down-regulation of the key glycolytic enzymes hexokinase and phosphofructokinase by high ATP levels (Clarke et al., 1989). For example, brain hexokinase is 97% inhibited under resting conditions (Clarke et al., 1989). This basal inhibition gives cells the potential to respond quickly to increased energy demands. This is particularly important for neurons because they shift energy demands drastically between their active and inactive states (Ames, 2000). During neural activity, increased Na^+–K^+ ATPase activity in neurons should augment their ATP utilization. The resulting changes in adenine nucleotide and phosphate levels then activate several key regulatory enzymes in glycolysis and oxidative phosphorylation. Thus, rapid increases in the rates of glucose utilization can occur without increases in glucose concentrations. For example, in cultured cerebellar neurons, neuronal glucose utilization increased two-fold without an increase in glucose concentration during heightened energy demand (Peng et al., 1994).

In contrast to glycolytic enzymes, the activity of LDH is not coupled to energy demand, and it is high enough that the reaction will be close to its thermodynamic equilibrium. Thus, for lactate dehydrogenation to increase, either lactate levels must go up, or pyruvate levels must decrease. An increase in lactate levels may be the only viable choice since pyruvate levels go up during increased neural activity (Goldberg et al., 1966; Ferrendelli and McDougal, 1971). For example, in cultured cerebellar neurons exposed to heightened energy demand and constant levels of either glucose, pyruvate, or lactate, the rates of both glucose and pyruvate oxidation increased while the rate of lactate oxidation was not significantly altered (Peng et al., 1994). Thus, for lactate to be a major substrate for neuronal oxidative metabolism during neural activity, its extracellular concentrations must increase, and do so quickly.

However, instead of going up rapidly with neural activity, increases in extracellular lactate in situ lag behind neural activity (Fellows et al., 1993; Fray et al., 1996; Hu and Wilson, 1997). For example, in rats undergoing 5 min of continuous tail pinch stimulation, lactate showed no measurable increases for the first 2.5 min of stimulation, and peaked (55–80% greater than control) 5 min after cessation of stimulation (Fellows et al., 1993). Sometimes extracellular lactate does not increase until after neural activity has ended. For example, in the electrically stimulated rat brain, lactate levels began increasing 10–12 s after termination of a 5 s stimulation, and peaked (40–100% above control) after another 50–60 s (see Hu and Wilson, 1997). These increases in extracellular lactate appear far too slow to meet the energy demands of neural activity.

Some investigators have also argued that lactate may become a significant energy source for neurons during prolonged activation when lactate levels increase (Hu and Wilson, 1997). However, increases in extracellular lactate concentrations during prolonged neural activity are relatively modest (35–135%; Fellows et al., 1993; Fray et al., 1996; Hu and Wilson, 1997), and may not create conditions thermodynamically favorable for driving the LDH-catalyzed reaction in neurons toward lactate oxidation. Also, such a utilization of lactate would come at the expense of glucose utilization since both lactate dehydrogenation and glycolysis need NAD^+. This means lactate must compete with glucose for utilization. Early during neural activity, as glycolytic rates increase, pyruvate (Goldberg et al., 1966; Ferrendelli and McDougal, 1971) and H^+ (Hochachka and Mommsen, 1983; Chesler and Kaila, 1992) levels go up, and the cytosolic NAD^+/NADH ratio decreases (Clarke et al., 1989). These changes favor the production of lactate via the LDH-catalyzed reaction rather than its utilization.

Even if conditions support lactate oxidation in neurons during activity, LDH must compete with glyceraldehyde-3-phosphate dehydrogenase (GAPDH) for NAD^+ (Fig. 2). Both LDH and GAPDH are present in high concentrations in brain cells (McIlwain and Bachelard, 1985). Heightened glyceraldehyde-3-P levels resulting from increased glycolytic activity boost the activity of GAPDH (Williamson, 1965). Also, increases in pyruvate levels caused by greater glycolytic activity may limit lactate oxidation in neurons since LDH-1 is highly sensitive to pyruvate product inhibition (Stambaugh and Post, 1966, also see Section 2.3.2). Thus, high glycolytic rates may preclude lactate use in neurons.

2.2.3. Glucose oxidation is energetically more favorable than lactate oxidation

Glucose conversion to pyruvate is more advantageous energetically than conversion of lactate to pyruvate. The reason for this is that the HK- and PFK-catalyzed reactions of glycolysis are very favorable thermodynamically and are essentially irreversible under intracellular conditions (Lehninger et al., 1993; Chih et al., 2001a). In contrast, thermodynamic conditions are highly unfavorable for the conversion of lactate to pyruvate during neural activity (also see Section 2.2.2). In addition, conversion of glucose to pyruvate yields two net ATPs, while conversion of lactate to pyruvate yields no ATP. These factors favor glucose utilization during increased ATP consumption. Thus, neurons gain an energetic advantage by oxidizing glucose instead of lactate once increased energy demand has activated key glycolytic enzymes.

2.2.4. Maintaining the cytoplasmic redox balance may be more difficult with lactate as the principal energy substrate

Both glycolysis and lactate dehydrogenation decrease the cytoplasmic redox ratio (NAD^+/NADH). This means that NAD^+ must be regenerated faster to accommodate higher rates of glycolysis or lactate dehydrogenation during increased energy demand. When glucose is used as a substrate, the cytoplasmic redox balance can be maintained by NADH shuttles (oxidative metabolism) and lactate production (anaerobic glycolysis). The fact that lactate is produced during neural activity suggests that activation of NADH shuttles in the brain may lag behind increases in energy demand. Glycerol-3-P dehydrogenase (glycerol-3 phosphate shuttle) and malate dehydrogenase (malate–aspartate shuttle) (Cheeseman and Clark, 1988; McKenna et al., 1993) are involved in reoxidizing NADH generated from glycolysis via GAPDH. Each NADH shuttle system depends on the availability of carrier metabolites (Berry et al., 1992). The carrier for the glycerol-3-P shuttle, dihydroxyacetone-3-P, is also a glycolytic intermediate. When the glycolytic rate increases, dihydroxyacetone-3-P levels also go up (Williamson, 1965), so the rate of NAD^+ regeneration can increase (Berry et al., 1992) to maintain a high glycolytic rate.

When lactate is used as substrate, the cytoplasmic redox balance can rely only on NADH shuttles. Also, without increases in glycolysis, dihydroxyacetone-3-P levels will not go up with increased energy demand, eliminating a driving force for the regeneration of NAD^+ via one shuttle mechanism. Moreover, the accessibility of LDH to the glycerol-3-P shuttle may differ from that of GAPDH. For example, in hepatocytes, NADH coming from lactate dehydrogenation failed to access the glycerol-3-P shuttle, possibly because the enzymes for glycolysis are segregated from those for gluconeogenesis (Berry et al., 1992). Thus, the NADH shuttle may be a limiting factor for lactate oxidation. Indeed, Peng and Hertz (Peng et al., 1994) found that, in cultured cerebellar granule neurons exposed to either glucose, pyruvate, or lactate and to high K^+ concentrations, the rates of glucose and pyruvate oxidation doubled, while the rate of lactate oxidation showed no statistically significant change. The fact that pyruvate oxidation, but not lactate oxidation, can increase with elevated energy demand shows that lactate oxidation is limited at the conversion of lactate to pyruvate, possibly due to a limited supply of NAD^+. Thus, the ability of neurons to increase lactate oxidation during heightened energy demand is limited even in the absence of glucose.

2.2.5. Glycolysis may generate ATP locally in both neurons and astrocytes for ion pumping and other processes

Anaerobic glycolysis may be an essential source of energy for localized areas of brain cells (Wu et al., 1997). This point is particularly true for synapses and for the most distal and finest astrocytic processes (see chapters by Derouiche and by Hertz, Peng et al.). For example, mitochondria occur with low frequency in axonal varicosities, and are indeed absent in 50% of synaptic boutons in the CA3 to CA1 projection in the hippocampus (Shepherd and Harris, 1998). Also, mitochondria are rarely observed in dendritic spine heads in the cerebral cortex (Wu et al., 1997). However, glycolytic enzymes are bound to postsynaptic densities (PSDs; Wu et al., 1997), which are proteinaceous structures attached to the surface of dendritic spine heads containing neurotransmitter receptors, ion channels, and protein kinases. These protein kinases consume ATP generated in PSDs via glycolysis (Wu et al., 1997). The presence of glycolytic enzymes, LDH (Wu et al., 1997), and the monocarboxylate transporter (MCT) (Bergersen et al., 2001) in PSDs, and the rarity of mitochondria in dendritic spine heads in the cerebral cortex (Wu et al., 1997), suggest that at least some glucose may be used anaerobically and that the lactate produced via glycolysis in PSDs may be released via the MCT. This segregation of glycolytic enzymes from mitochondria may provide a reasonable explanation for the increase in lactate production during heightened neural activity even when there is no shortage of oxygen.

Also, energy generated from glycolysis has been linked to ion transport via Na^+–K^+ ATPase. Studies of erythrocytes (Proverbio and Hoffman, 1977), smooth muscle (Paul et al., 1979), and cardiac muscle (Weiss and Hiltbrand, 1985) have provided compelling evidence that glycolytically generated ATP is linked to Na^+–K^+ ATPase. Although no direct evidence for a similar link has been identified yet in the brain, both glycolytic enzymes and Na^+–K^+ ATPase are bound at high specific activity within synaptosomal membranes (Knull, 1978; Lim et al., 1983). Also, several studies have shown that ion transport is less robust when glucose is replaced either by lactate or by pyruvate (Lipton and Robacker, 1983; Raffin et al., 1992; Roberts, 1993; Silver et al., 1997). The association of glycolytic enzymes with Na^+–K^+ ATPase in synaptic membranes (Knull, 1978; Lim et al., 1983) suggests that glucose, rather than lactate, is used as the substrate to provide energy for the Na^+–K^+ pump in neurons.

2.3. Convincing evidence showing that active neurons prefer lactate to glucose is still lacking

Several studies have been considered as strong evidence for the concept that active neurons use lactate instead of glucose. In the following sections, we discuss the uncertainties of these studies.

2.3.1. Substrate utilization in the isolated retinal preparation

Poitry-Yamate et al. (1995) reported that Müller cells fed ^{14}C-labeled glucose released 70% less radioactive ^{14}C-lactate to the bathing medium when photoreceptors were included in the cell suspension. They concluded that lactate released from retinal glia (Müller cells) was used preferentially by photoreceptors even in the presence of glucose. However, because the mixed neuronal–glial samples contained only about half the number of glial cells per mg of protein (Poitry-Yamate et al., 1995), the radioactivity of bath lactate in the mixed photoreceptor-Müller cell cultures had to be half that of Müller cells alone (see Chih et al., 2001a). This dilution of radioactivity, and not an increased lactate utilization by photoreceptors, would account for most of the decrease in radiolabeled lactate/mg protein in the bath. Thus, results from this study do not support the assertion that neurons prefer to metabolize lactate from astrocytes instead of ambient glucose.

2.3.2. Distributions of LDH-1 and LDH-5

The dominant presence of the LDH isoform LDH-1 in neurons has also been cited as support for the ANLSH. ANLSH proponents suggest that LDH-1 drives the LDH-catalyzed reaction toward pyruvate production (Bittar et al., 1996). However, the direction of a reaction is determined by the relative concentrations of its substrates and products, and is not influenced by the enzyme catalyzing the reaction (Lehninger et al., 1993). Compared with LDH-5, LDH-1 has a lower K_m for pyruvate or lactate, a lower V_{max}, and is more easily inhibited by its substrates and products (Cahn et al., 1962; Stambaugh and Post, 1966; Everse and Kaplan, 1973; Nitisewojo and Hultin, 1976). The fact that LDH-1 is more easily inhibited by its substrates and products may have important functional consequences (Cahn et al., 1962). For example, LDH-1 has a higher sensitivity to pyruvate substrate inhibition and lactate product inhibition, which may help direct pyruvate generated from glycolysis toward the TCA cycle instead of toward lactate production (Cahn et al., 1962). This would allow full oxidation of glucose in cells where LDH-1 is the dominant isoform, such as neurons and cardiac muscle cells (Cahn et al., 1962). Also, LDH-1 has a high sensitivity to pyruvate product inhibition ($K_i = 0.18$ mM) (Stambaugh and Post, 1966). Since physiological pyruvate levels are between 0.1 and 0.2 mM (McIlwain and Bachelard, 1985), any slight increase in pyruvate levels during heightened glycolysis may help inhibit LDH-1 and prevent lactate oxidation. Thus, the dominant presence of LDH-1 does not help lactate compete with glucose as an energy substrate in neurons.

2.3.3. Substrate utilization in isolated chicken embryo ganglia

A study of substrate utilization in chicken embryo ganglia (Larrabee, 1995) has also been cited as evidence that brain tissue preferentially uses lactate instead of glucose (Tsacopoulos and Magistretti, 1996). Briefly, this study showed that isolated chicken embryo ganglia produced more CO_2 from lactate than from glucose when both substrates were present. However, as stated earlier, the rate of glycolysis is tightly regulated by energy status and is largely inhibited under resting conditions. In contrast, lactate utilization can be increased by increasing the concentration gradient for lactate. In Larrabee's (1995) study, lactate concentrations were about five times greater than physiological lactate levels. Since glycolysis is down-regulated in resting brain tissue, lactate may become the dominant substrate. An analogous situation occurs in resting red muscle (Pearce and Connett, 1980), where lactate oxidation can account for 70% of energy production when ambient lactate levels are 8 mM, which is high enough to suppress both glucose and glycerol oxidation. However, this result does not mean that lactate is the preferred substrate for red muscle cells during activity. It means only that lactate levels are high enough to inhibit utilization of other substrates when cellular energy demand is low. Since the ANLSH concerns energy utilization by active neurons, the study by Larrabee (1995) does not apply to the ANLSH. It is also important to note that in this study, the cell type(s) using lactate was (were) not identified. As will be discussed below, astrocytes are equally as able to use lactate as neurons (Peng et al., 1994).

2.3.4. *In vivo measurements of lactate levels during neural activity*

In another study thought to support the ANLSH (Hu and Wilson, 1997), transient changes in extracellular lactate and glucose levels were measured in the rat hippocampus in response to repetitive trains of electrical stimulation. After each stimulus train, glucose and lactate declined transiently, and the data were interpreted as showing that lactate declined more in later stimulus trains after mean lactate levels had gone up (Hu and Wilson, 1997). However, results from only one experiment were reported, and no statistical analysis of the data was provided. Also, each repetitive stimulation was given when lactate was sharply declining and glucose was gradually increasing. This complicated interpretation of the actual changes in glucose and lactate caused by stimulation. Finally, this kind of experiment provides no details regarding what cell types were using glucose or lactate.

2.3.5. *Utilization of lactate under stress*

Hippocampal slices subjected to hypoxic (Schurr et al., 1997) or excitotoxic (Schurr et al., 1999) stress fail to recover synaptic transmission after these stress conditions, even in the presence of sufficient glucose, if the monocarboxylate transport inhibitor 4-CIN (alpha-cyano-4-hydroxycinnamate) is present (Schurr et al., 1999). These studies have been viewed as strong support for the ANLSH (e.g., Magistretti et al., 1999), because they suggest that neurons need lactate during recovery from stress, and that glucose alone is insufficient for such recovery. However, 4-CIN blocks pyruvate and lactate entry into isolated mitochondria (Halestrap and Denton, 1974; Cox et al., 1985) besides blocking lactate transport. Also, 4-CIN compromises normal oxidative metabolism and blocks neural activity (e.g., Amos and Richards, 1996; Chih et al., 2001a). These effects of 4-CIN on mitochondrial function, rather than inhibition of lactate transport, may explain why hippocampal slices do not recover from hypoxia or excitotoxic stress in these experiments. Moreover, these studies apply to pathological stress, and not to stress resulting from normal physiological activity.

2.4. *Does activity-induced glucose uptake occur predominantly in astrocytes?*

A study by Shulman and coworkers is thought by proponents of the ANLSH to provide the most conclusive evidence that activity-induced glucose utilization occurs mainly in astrocytes (Sibson et al., 1998). Briefly, Sibson et al. (1998) found a 1:1 ratio between total oxidative glucose consumption and astrocytic glutamate cycling in the rat brain. They concluded that activity-induced glucose utilization is driven by astrocytic glutamate cycling, and that their results were consistent with the ANLSH.

However, the 1:1 ratio obtained from this study can vary significantly depending on the assumptions used to calculate this ratio. For example, if the rate of glutamine synthesis via the anaplerotic pathway during normal activity is 30% of total glutamine synthesis, as others have observed (Künnecke et al., 1993; Lapidot and Haber, 2000), instead of being unchanged, as assumed by Sibson et al., then the stoichiometry between oxidative glucose consumption and glutamate cycling would be roughly 1.5:1 rather than 1:1 (estimated from data in Sibson et al., 1998). It is consistent with this conclusion that Gruetter and coworkers have found a much higher rate of glucose oxidation than of glutamate cycling (see chapter by Gruetter). Also, because the mode of astrocytic glucose consumption was not identified, it is impossible to determine the percentage of glucose being used by astrocytes or neurons (Chih et al., 2001b) even if the 1:1 stoichiometry holds true. If as little as 6% of glucose is metabolized aerobically in astrocytes, then 50% of glucose could have been metabolized by neurons and still satisfy the 1:1 stoichiometry. In this context it is interesting that Rothman and coworkers recently have determined by aid of the astrocyte-specific precursor, acetate, that oxidative metabolism in astrocytes accounts for 15% of oxidative metabolism in the brain in vivo (Lebon et al., 2002). The ANLSH assumes that activity-induced glucose consumption in astrocytes is entirely anaerobic. However, many studies (Swanson, 1992; Eriksson et al., 1995; Hertz et al., 1998) have shown that oxidative metabolism can support astrocytic glutamate uptake. Thus, the reported 1:1 ratio between oxidative glucose consumption and astrocytic glutamate cycling does not necessarily support the ANLSH.

Activity-induced 2-DG uptake occurs primarily in the neuropil (Kadekaro et al., 1985), a region enriched in axon terminals, dendrites, and synapses. Since the membranes of synaptic terminals contain a high density of the glucose transporter GLUT3 (McCall et al., 1994; Leino et al., 1997), little apparent reason exists to speculate that neurons do not take up glucose as the ANLSH suggests. Critical experiments determining the actual locus of glucose uptake during activity in situ are in their infancy, but evidence has recently been obtained that both cell types may contribute about equally to glucose utilization (Wittendorp-Rechenmann et al., 2001).

2.5. Are neurons lactate consumers, and astrocytes lactate producers?

A key point of the ANLSH is that, during neural activity, lactate is produced by astrocytes and then shuttled to and used by neurons. Although astrocytes produce lactate (Pellerin and Magistretti, 1994; Demestre et al., 1997), no evidence exists showing that neurons use lactate released from astrocytes during neural activity. In fact, in vitro studies have shown that neurons also produce lactate when energy demands increase (Walz and Mukerji, 1988; Peng et al., 1994).

The fact that neuronal cultures produce less lactate than cultures of astrocytes has been used as indirect evidence that neurons are primarily lactate consumers, and that astrocytes are mainly lactate producers (Schousboe et al., 1997). However, less neuronal lactate production may simply mean that neurons oxidize glucose more fully (also see Section 2.2.2). As a result, neuronal cultures may not need to consume as much glucose as astrocytic cultures (Lopes-Cardozo et al., 1986). It should be stressed that less lactate production by neurons at rest does not mean neurons have a smaller glycolytic capacity than astrocytes, because when oxidative metabolism is blocked, neurons produce as much lactate as astrocytes (Walz and Mukerji, 1988; Peng and Hertz, 2002).

The fact that LDH-1, the dominant LDH isoform in neurons, has a lower K_m for lactate than LDH-5 has also been used as evidence that neurons may be better suited to use lactate than astrocytes (Bittar et al., 1996), which have both LDH-5 and LDH-1. However, the lower V_{max} of LDH-1 may offset the effect of a lower K_m. For example, LDH-1 catalyzes a lower rate of lactate production than LDH-5 despite its lower K_m for pyruvate (Pesce et al., 1967). Also, in vitro studies have shown that astrocytes oxidize lactate at approximately the same rate as neurons (Peng et al., 1994). Thus, the distribution of LDH apparently does not play a significant role in affecting the rate of lactate utilization in neurons and astrocytes.

The heterogeneous distribution of MCT subtypes between neurons and glia has also been suggested as being consistent with the ANLSH (Broer et al., 1997; Pellerin et al., 1998b). However, the kinetic characteristics of the MCT isoforms currently provide little support for the ANLSH. Indeed, the initial rates of lactate uptake by cultured astrocytes and neurons are similar for lactate concentrations of 1–5 mM (Dienel and Hertz, 2001), suggesting that differences in MCT isoforms do not play a significant role in lactate transport under normal conditions. Also, at physiological lactate concentrations, metabolically driven lactate uptake in both neurons and astrocytes corresponds to at most 25% of the resting rate of glucose oxidation in the brain (Dienel and Hertz, 2001). Thus, lactate probably does not replace glucose as the principal metabolic substrate in neurons.

3. Role of lactate as an energy substrate in the brain

The above discussion does not rule out the possibility that lactate is used as a substrate in the brain. Because of the relatively slow transport of lactate across the blood–brain barrier (McIlwain and Bachelard, 1985), most of the lactate produced during neural activity probably has to be utilized by brain cells. Also, lactate can become an important energy substrate when blood lactate levels increase above normal (Ames, 2000). Besides, many in vitro studies have shown that brain tissue can use lactate to maintain neural function (Schurr et al., 1988). Thus, lactate can clearly serve as an alternative energy substrate in the brain. The major difference between the conventional hypothesis and the ANLSH concerns how lactate is produced and used in the brain during neural activity. The conventional hypothesis contends that glucose is the major substrate for both neurons and astrocytes during normal neural activity. Lactate production occurs during neural activity when increases in the glycolytic flux surpass increases in the rate of oxidative phosphorylation, when oxygen is temporarily in short supply, or in areas where mitochondria are rare, such as dendritic spine heads (Wu et al., 1997). Lactate produced by active neurons and astrocytes is then used by inactive brain cells and by those cells with little glycolytic capacity, such as oligodendrocytes (Dringen et al., 1993a). Lactate may also be used by astrocytes for glycogen synthesis (Dringen et al., 1993b). The ANLSH contends that lactate is produced by astrocytes and used by neurons during neural activity. So far, there is no clear evidence that shuttling of lactate between astrocytes and neurons actually occurs in situ during neural activity. Also, as discussed above, the kinetic properties of enzymes and transporters involved in the metabolism of glucose and lactate do not support the ANLSH. Indeed, the lack of mechanisms for activating LDH in response to increases in energy demand, and the slow increases in lactate levels during neural activity, suggest that lactate is not intended for use as the major energy substrate during neural activity. Instead, lactate is used by inactive brain cells to keep brain lactate levels low.

4. Concluding remarks

Both neurons and astrocytes metabolize glucose and lactate. The ANLSH compartmentalizes the use of glucose and lactate during neural activity by identifying neurons as exclusive lactate consumers and astrocytes as exclusive glucose consumers and lactate producers. However, the ANLSH model of substrate utilization during neural activity is not well supported by current evidence. From a theoretical point of view, glucose is a better substrate for both neurons and astrocytes during neural activity: First, glycolytic enzymes are activated rapidly in response to increased energy demands while LDH is not regulated by energy demand. Second, increases in lactate levels, which drive lactate oxidation, lag well behind neural activity. Third, the conversion of glucose to pyruvate is thermodynamically more favorable than the conversion of lactate to pyruvate. There is no apparent reason why neural activity should only activate astrocytic glycolysis but not neuronal glycolysis, as suggested by the ANLSH, since neurons have a high glycolytic capacity and glucose is readily available to neurons. Experimental evidence also does not support the major assertions of the ANLSH (1) that active neurons prefer lactate to glucose, (2) that activity-induced glucose uptake occurs predominantly in astrocytes, and (3) that astrocytes are the lactate producers and neurons are the lactate consumers. In particular, convincing evidence for shuttling of lactate between astrocytes and neurons during neural activity is still lacking. Thus, both theoretical considerations and experimental evidence provide little support for the ANLSH and much credibility for the conventional view of glucose utilization. Experiments are needed to determine where glucose uptake takes place during neural activity so that the issues raised by the ANLSH may be resolved.

Acknowledgements

We would like to thank Dr Peter Lipton (Department of Physiology, University of Wisconsin School of Medicine) for his invaluable advice and ideas regarding the topics covered in this chapter. This work was supported by an award from the American Heart Association, Florida/Puerto Rico Affiliate.

References

Ames, A., 2000. CNS energy metabolism as related to function. Brain Res. Rev. 34, 42–68.

Amos, B.J., Richards, C.D., 1996. Intrinsic hydrogen ion buffering in rat CNS neurones maintained in culture. Exp. Physiol. 81, 261–271.

Bergersen, L., Warhaug, O., Helm, J., Thomas, M., Laake, P., Davies, A.J., Wilson, M.C., Halestrap, A.P., Ottersen, O.P., 2001. A novel postsynaptic density protein: the monocarboxylate transporter MCT2 is co-localized with glutamate receptors in postsynaptic densities of parallel fiber-Purkinje cell synapses. Exp. Brain Res. 136, 523–534.

Berry, M.N., Phillips, J.W., Grivell, A.R., 1992. Interactions between mitochondria and cytoplasm in isolated hepatocytes. Curr. Top. Cell. Regul. 33, 309–328.

Bittar, P.G., Charnay, Y., Pellerin, L., Bouras, C., Magistretti, P.J., 1996. Selective distribution of lactate dehydrogenase isoenzymes in neurons and astrocytes of human brain. J. Cereb. Blood Flow Metab. 16, 1079–1089.

Broer, S., Rahman, B., Pellegri, G., Pellerin, L., Martin, J.L., Verleysdonk, S., Hamprecht, B., Magistretti, P.J., 1997. Comparison of lactate transport in astroglial cells and monocarboxylate transporter 1 (MCT 1) expressing *Xenopus laevis* oocytes. Expression of two different monocarboxylate transporters in astroglial cells and neurons. J. Biol. Chem. 272, 30096–30102.

Cahn, R.D., Kaplan, N.O., Levine, L., Zwilling, L.E., 1962. Nature and development of lactic dehydrogenases. Science 136, 962–969.

Cheeseman, A.J., Clark, J.B., 1988. Influence of the malate–aspartate shuttle on oxidative metabolism in synaptosomes. J. Neurochem. 50, 1559–1565.

Chesler, M., Kaila, K., 1992. Modulation of pH by neuronal activity. Trends Neurosci. 15, 396–402.

Chih, C.P., He, J., Sly, T.S., Roberts, E.L. Jr., 2001a. Comparison of glucose and lactate as substrates during NMDA-induced activation of hippocampal slices. Brain Res. 893, 143–154.

Chih, C.P., Lipton, P., Roberts, E.L. Jr., 2001b. Do active cerebral neurons really use lactate rather than glucose? Trends Neurosci. 24, 573–578.

Cimino, M., Balduini, W., Marini, P., Cattabeni, F., Court, J.A., Bianchi, M., Magnani, M., 1998. Expression of hexokinase mRNA in human hippocampus. Mol. Brain Res. 53, 297–300.

Clarke, D.D., Lajtha, A.L., Maker, H.S., 1989. Intermediary metabolism, 4th ed, In: Basic Neurochemistry. Raven Press, New York, pp. 542–550.

Cox, D.D.W.G., Drower, J., Bachelard, H.S., 1985. Effects of metabolic inhibitors on evoked activity and the energy state of hippocampal slices superfused in vitro. Exp. Brain Res. 57, 464–470.

Demestre, M., Boutelle, M., Fillenz, M., 1997. Stimulated release of lactate in freely moving rats is dependent on the uptake of glutamate. J. Physiol. (Lond.) 499, 825–832.

Dienel, G.A., Hertz, L., 2001. Glucose and lactate metabolism during brain activation. J. Neurosci. Res. 66, 824–838.

Dringen, R., Gebhardt, R., Hamprecht, B., 1993a. Glycogen in astrocytes: possible function as lactate supply for neighboring cells. Brain Res. 623, 208–214.

Dringen, R., Schmoll, D., Cesar, M., Hamprecht, B., 1993b. Incorporation of radioactivity from [^{14}C]lactate into the glycogen of cultured mouse astroglial cells: evidence for gluconeogenesis in brain cells. Biol. Chem. 374, 343–347.

Eriksson, G., Peterson, A., Iverfeldt, K., Walum, E., 1995. Sodium-dependent glutamate uptake as an activator of oxidative metabolism in primary astrocyte cultures from newborn rat. Glia 15, 152–156.

Everse, J., Kaplan, N.O., 1973. Lactate dehydrogenases: structure and function. Adv. Enzymol. 37, 61–133.

Fellows, L.K., Boutelle, M.G., 1993. Rapid changes in extracellular glucose levels and blood flow in the striatum of the freely moving rat. Brain Res. 604, 225–231.

Fellows, L.K., Boutelle, M.G., Fillenz, M., 1993. Physiological stimulation increases nonoxidative glucose metabolism in the brain of the freely moving rat. J. Neurochem. 60, 1258–1263.

Ferrendelli, J.A., McDougal, D.B. Jr., 1971. The effect of audiogenic seizures on regional CNS energy reserves, glycolysis and citric acid cycle flux. J. Neurochem. 18, 1207–1220.

Fray, A.E., Forsyth, R.J., Boutelle, M.G., Fillenz, M., 1996. The mechanisms controlling physiologically stimulated changes in rat brain glucose and lactate: a microdialysis study. J. Physiol. (Lond.) 496, 49–57.

Goldberg, N.D., Passonneau, J.V., Lowry, O.H., 1966. Effects of changes in brain metabolism on the levels of citric acid cycle intermediates. J. Biol. Chem. 241, 3997–4003.

Halestrap, A.P., Denton, R.M., 1974. Specific inhibition of pyruvate transport in rat liver mitochondria and human erythrocytes by cyano-4-hydroxycinnamate. Biochem. J. 138, 313–316.

Harada, M., Okuda, C., Sawa, T., Murakami, T., 1992. Cerebral extracellular glucose and lactate concentrations during and after moderate hypoxia in glucose-infused and saline-infused rats. Anesthesiology 77, 728–734.

Hertz, L., Swanson, R.A., Newman, G.C., Marrif, H., Juurlink, B.H.J., Peng, L.A., 1998. Can experimental conditions explain the discrepancy over glutamate stimulation of aerobic glycolysis. Dev. Neurosci. 20, 339–347.

Hochachka, P.W., Mommsen, T.P., 1983. Protons and anaerobiosis. Science 219, 1391–1397.

Hu, Y.B., Wilson, G.S., 1997. A temporary local energy pool coupled to neuronal activity: fluctuations of extracellular lactate levels in rat brain monitored with rapid-response enzyme-based sensor. J. Neurochem. 69, 1484–1490.

Kadekaro, M., Crane, A.M., Sokoloff, L., 1985. Differential effects of electrical stimulation of sciatic nerve on metabolic activity in spinal cord and dorsal root ganglion in the rat. Proc. Natl Acad. Sci. USA 82, 6010–6013.

Kao-Jen, J., Wilson, J.E., 1980. Localization of hexokinase in neural tissue: electron microscopic studies of rat cerebellar cortex. J. Neurochem. 35, 667–678.

Knull, H.R., 1978. Association of glycolytic enzymes with particulate fractions from nerve endings. Biochim. zBiophys. Acta 522, 1–9.

Kuhr, W.G., Korf, J., 1988. Extracellular lactic acid as an indicator of brain metabolism: continuous on-line measurement in conscious, freely moving rats with intrastriatal dialysis. J. Cereb. Blood Flow Metab. 8, 130–137.

Künnecke, B., Cerdan, S., Seelig, J., 1993. Cerebral metabolism of [1,2-^{13}C] glucose and [U-^{13}C] 3-hydroxybutyrate in rat brain as detected by ^{13}C NMR spectroscopy. NMR Biomed. 6, 264–277.

Lai, J.C., Behar, K.L., Liang, B.B., Hertz, L., 1999. Hexokinase in astrocytes: kinetic and regulatory properties. Metab. Brain Dis. 14, 125–133.

Lapidot, A., Haber, S., 2000. Effect of acute insulin-induced hypoglycemia on fetal versus adult brain fuel utilization, assessed by ^{13}C MRS isotopomer analysis of [U-^{13}C] glucose metabolites. Dev. Neurosci. 22, 444–455.

Larrabee, M.G., 1995. Lactate metabolism and its effects on glucose metabolism in an excised neural tissue. J. Neurochem. 64, 1734–1741.

Lebon, V., Petersen, K.F., Cline, G.W., Shen, J., Mason, G.F., Dufour, S., Behar, K.L., Shulman, G.I., Rothman, D.L., 2002. Astroglial contribution to brain energy metabolism in humans revealed by ^{13}C nuclear magnetic resonance spectroscopy: elucidation of the dominant pathway for neurotransmitter glutamate repletion and measurement of astrocytic oxidative metabolism. J. Neurosci. 22, 1523–1531.

Lehninger, A.L., Nelson, D.L., Cox, M.M., 1993. Principles of Biochemistry, 2nd ed. Worth Publishers, New York.

Leino, R.L., Gerhart, D.Z., Vanbueren, A.M., McCall, A.L., Drewes, L.R., 1997. Ultrastructural localization of GLUT 1 and GLUT 3 glucose transporters in rat brain. J. Neurosci. Res. 49, 617–626.

Lim, L., Hall, C., Leung, T., Mahadevan, L., Whatley, S., 1983. Neurone-specific enolase and creatine phosphokinase are protein components of rat brain synaptic plasma membrane. J. Neurochem. 41, 1177–1182.

Lipton, P., Robacker, K., 1983. Glycolysis and brain function: $[K^+]_o$ stimulation of protein synthesis and K^+ uptake require glycolysis. Fed. Proc. 42, 2875–2880.

Lopes-Cardozo, M., Larsson, O.M., Schousboe, A., 1986. Acetoacetate and glucose as lipid precursors and energy substrates in primary cultures of astrocytes and neurons from mouse cerebral cortex. J. Neurochem. 46, 773–778.

Lowry, J.P., O'Neill, R.D., Boutelle, M.G., Fillenz, M., 1998. Continuous monitoring of extracellular glucose concentrations in the striatum of freely moving rats with an implanted glucose biosensor. J. Neurochem. 70, 391–396.

Lowry, O.H., Passonneau, J.V., 1964. The relationships between substrates and enzymes of glycolysis in brain. J. Biol. Chem. 239, 31–42.

Magistretti, P.J., 1999. Brain energy metabolism. In: Zigmond, M.J., Bloom, F.E., Landis, S.C., Roberts, J.L., Squire, L.R., (Eds.), Fundamental Neuroscience. Academic Press, New York, pp. 389–413.

Magistretti, P.J., 2000. Cellular bases of functional brain imaging: insights from neuron–glia metabolic coupling. Brain Res. 886, 108–112.

Magistretti, P.J., Pellerin, L., 1996. Cellular bases of brain energy metabolism and their relevance to functional brain imaging: evidence for a prominent role of astrocytes. Cerebral Cortex 6, 50–61.

Magistretti, P.J., Pellerin, L., 1999. Cellular mechanisms of brain energy metabolism and their relevance to functional brain imaging. Philos. Trans. R. Soc. Lond. B: Biol. Sci. 354, 1155–1163.

Magistretti, P.J., Pellerin, L., Rothman, D.L., Shulman, R.G., 1999. Energy on demand. Science 283, 496–497.

Malonek, D., Dirnagl, U., Lindauer, U., Yamada, K., Kanno, I., Grinvald, A., 1997. Vascular imprints of neuronal activity: relationships between the dynamics of cortical blood flow, oxygenation, and volume changes following sensory stimulation. Proc. Natl Acad. Sci. USA 94, 14826–14831.

Malonek, D., Grinvald, A., 1996. Interactions between electrical activity and cortical microcirculation revealed by imaging spectroscopy: implications for functional brain mapping. Science 272, 551–554.

McCall, A.L., Van Bueren, A.M., Moholt-Siebert, M., Cherry, N.J., Woodward, W.R., 1994. Immunohistochemical localization of the neuron-specific glucose transporter (GLUT3) to neuropil in adult rat brain. Brain Res. 659, 292–297.

McIlwain, H., Bachelard, H.S., 1985. Biochemistry and the Central Nervous System, 5th ed. Churchill Livingstone, New York.

McKenna, M.C., Tildon, J.T., Stevenson, J.H., Boatright, R., Huang, S., 1993. Regulation of energy metabolism in synaptic terminals and cultured rat brain astrocytes: differences revealed using aminooxyacetate. Dev. Neurosci. 15, 320–329.

Nemoto, E.M., Hoff, J.T., Severinghaus, J.W., 1974. Lactate uptake and metabolism by brain during hyperlactatemia and hypoglycemia. Stroke 5, 48–53.

Nitisewojo, P., Hultin, H.O., 1976. A comparison of some kinetic properties of soluble and bound lactate dehydrogenase isoenzymes at different temperatures. Eur. J. Biochem. 67, 87–94.

Paul, R.J., Bauer, W., Pease, W., 1979. Vascular smooth muscle: aerobic glycolysis linked to sodium and potassium transport proceses. Science 206, 1414–1416.

Pearce, F.J., Connett, R.J., 1980. Effect of lactate and palmitate on substrate utilization of isolated rat soleus. Am. J. Physiol. 238, C149–C159.

Pellerin, L., Magistretti, P.J., 1994. Glutamate uptake into astrocytes stimulates aerobic glycolysis: a mechanism coupling neuronal activity to glucose utilization. Proc. Natl Acad. Sci. USA 91, 10625–10629.

Pellerin, L., Pellegri, G., Bittar, P.G., Charnay, Y., Bouras, C., Martin, J.L., Stella, N., Magistretti, P.J., 1998a. Evidence supporting the existence of an activity-dependent astrocyte–neuron lactate shuttle. Dev. Neurosci. 20, 291–299.

Pellerin, L., Pellegri, G., Martin, J.L., Magistretti, P.J., 1998b. Expression of monocarboxylate transporter mRNAs in mouse brain: support for a distinct role of lactate as an energy substrate for the neonatal vs. adult brain. Proc. Natl Acad. Sci. USA 95, 3990–3995.

Peng, L., Hertz, L., 2002. Amobarbital inhibits K^+-stimulated glucose oxidation in cerebellar granule neurons by two mechanisms. Eur. J. Pharmacol. 446, 53–61.

Peng, L., Zhang, X.H., Hertz, L., 1994. High extracellular potassium concentrations stimulate oxidative metabolism in a glutamatergic neuronal culture and glycolysis in cultured astrocytes but have no stimulatory effect in a GABAergic neuronal culture. Brain Res. 663, 168–172.

Pesce, A., Fondt, T.P., Stolzenbach, F., Castillo, F., Kaplan, N.O., 1967. The comparative enzymology of lactic dehydrogenases. J. Biol. Chem. 242, 2151–2167.

Pfeuffer, J., Tkac, I., Gruetter, R., 2000. Extracellular–intracellular distribution of glucose and lactate in the rat brain assessed noninvasively by diffusion-weighted ^{1}H nuclear magnetic resonance spectroscopy *in vivo*. J. Cereb. Blood Flow Metab. 20, 736–746.

Poitry-Yamate, C.L., Poitry, S., Tsacopoulos, M., 1995. Lactate released by Muller glial cells is metabolized by photoreceptors from mammalian retina. J. Neurosci. 15, 5179–5191.

Prichard, J., Rothman, D., Novotny, E., Petroff, O., Kuwabara, T., Avison, M., Howseman, A., Hanstock, C., Shulman, R., 1991. Lactate rise detected by ^{1}H NMR in human visual cortex during physiologic stimulation. Proc. Natl Acad. Sci. USA 88, 5829–5831.

Proverbio, F., Hoffman, J.F., 1977. Membrane compartmentalized ATP and its preferential use by Na,K-ATPase of human red cell ghosts. J. Gen. Physiol. 69, 605–632.

Raffin, C.N., Rosenthal, M., Busto, R., Sick, T.J., 1992. Glycolysis, oxidative metabolism, and brain potassium ion clearance. J. Cereb. Blood Flow Metab. 12, 34–42.

Roberts, E.L. Jr., 1993. Glycolysis and recovery of potassium ion homeostasis and synaptic transmission in hippocampal slices after anoxia or stimulated potassium release. Brain Res. 620, 251–258.

Rothman, D.L., Sibson, N.R., Hyder, F., Shen, J., Behar, K.L., Shulman, R.G., 1999. *In vivo* nuclear magnetic resonance spectroscopy studies of the relationship between the glutamate–glutamine neurotransmitter cycle and functional neuroenergetics. Philos. Trans. R. Soc. Lond. B: Biol. Sci. 354, 1165–1177.

Schousboe, A., Westergaard, N., Waagepetersen, H.S., Larsson, O.M., Bakken, I.J., Sonnewald, U., 1997. Trafficking between glia and neurons of TCA cycle intermediates and related metabolites. Glia 21, 99–105.

Schurr, A., Miller, J.J., Payne, R.S., Rigor, B.M., 1999. An increase in lactate output by brain tissue serves to meet the energy needs of glutamate-activated neurons. J. Neurosci. 19, 34–39.

Schurr, A., Payne, R.S., Miller, J.J., Rigor, B.M., 1997. Brain lactate is an obligatory aerobic energy substrate for functional recovery after hypoxia: further in vitro validation. J. Neurochem. 69, 423–426.

Schurr, A., West, C.A., Rigor, B.M., 1988. Lactate-supported synaptic function in the rat hippocampal slice preparation. Science 240, 1326–1328.

Shepherd, G.M.G., Harris, K.M., 1998. Three-dimensional structure and composition of CA3–CA1 axons in rat hippocampal slices: implications for presynaptic connectivity and compartmentalization. J. Neurosci. 18, 8300–8310.

Sibson, N.R., Dhankhar, A., Mason, G.F., Rothman, D.L., Behar, K.L., Shulman, R.G., 1998. Stoichiometric coupling of brain glucose metabolism and glutamatergic neuronal activity. Proc. Natl Acad. Sci. USA 95, 316–321.

Silver, I.A., Deas, J., Erecinska, M., 1997. Ion homeostasis in brain cells—differences in intracellular ion responses to energy limitation between cultured neurons and glial cells. Neuroscience 78, 589–601.

Silver, I.A., Erecinska, M., 1994. Extracellular glucose concentration in mammalian brain: continuous monitoring of changes during increased neuronal activity and upon limitation in oxygen supply in normo-, hypo-, and hyperglycemic animals. J. Neurosci. 14, 5068–5076.

Sokoloff, L., 1999. Energetics of functional activation in neural tissues. Neurochem. Res. 24, 321–329.

Stambaugh, R., Post, D., 1966. Substrate and product inhibition of rabbit muscle lactic dehydrogenase heart (H_4) and muscle (M_4) isozymes. J. Biol. Chem. 241, 1462–1467.

Swanson, R.A., 1992. Astrocyte glutamate uptake during chemical hypoxia in vitro. Neurosci. Lett. 147, 143–146.

Thurston, J.H., Hauhart, R.E., Schiro, J.A., 1983. Lactate reverses insulin-induced hypoglycemic stupor in suckling–weanling mice: biochemical correlates in blood, liver, and brain. J. Cereb. Blood Flow Metab. 3, 498–506.

Tsacopoulos, M., Magistretti, P.J., 1996. Metabolic coupling between glia and neurons. J. Neurosci. 16, 877–885.

Vannucci, S.J., Maher, F., Simpson, I.A., 1997. Glucose transporter proteins in brain: delivery of glucose to neurons and glia. Glia 21, 2–21.

Walz, W., Mukerji, S., 1988. Lactate release from cultured astrocytes and neurons: a comparison. Glia 1, 366–370.

Weiss, J., Hiltbrand, B., 1985. Functional compartmentation of glycolytic versus oxidative metabolism in isolated rabbit heart. J. Clin. Invest. 75, 436–447.

Williamson, J.R., 1965. Metabolic control in the perfused rat heart. In: Chance, B., Estabrook, R.W., Williamson, J.R. (Eds.), Control of Energy Metabolism. Academic Press, New York, pp. 333–355.

Wilson, J.E., 1972. The localization of latent brain hexokinase on synaptosomal mitochondria. Arch. Biochem. Biophys. 150, 96–104.

Wittendorp-Rechenmann, E., Lam, C.D., Nehlig, A., 2001. First in vivo demonstration of the uptake of $[^{14}C]$deoxyglucose by astrocytes and neurons: a microautoradiographic study. J. Cereb. Blood Flow Metab. 21(Suppl. 1), S321.

Wu, K., Aoki, C., Elste, A., Rogalski-Wilk, A.A., Siekevitz, P., 1997. The synthesis of ATP by glycolytic enzymes in the postsynaptic density and the effect of endogenously generated nitric oxide. Proc. Natl Acad. Sci. USA 94, 13273–13278.

Advances in
Molecular and
Cell Biology

Principles of the measurement of neuro-glial metabolism using in vivo ^{13}C NMR spectroscopy

Rolf Gruetter

Departments of Radiology and Neuroscience, University of Minnesota, Minneapolis, MN, USA
Correspondence address: Center for MR Research, 2021 6th Street SE, Minneapolis, MN 55455, USA.
Tel.: +1-612-625-6582; fax: +1-612-626-2004. E-mail: gruetter@cmrr.umn.edu(R.G.)

Contents

1. Introduction
2. Key elements of ^{13}C tracer methodology measured by in vivo ^{13}C NMR spectroscopy
   2.1. 'Tracer methods'
   2.2. Dynamic isotopomer analysis
   2.3. ^{13}C NMR methodological aspects
3. Glial metabolism I: brain glycogen
   3.1. Brain glycogen, an endogenous store of fuel
   3.2. Human brain glycogen metabolism in vivo
4. Glial metabolism II: the glutamate–glutamine cycle
   4.1. Glutamate turnover: neuronal oxygen metabolism and the malate–aspartate shuttle
   4.2. Glutamine turnover: the hallmark of glial metabolism
   4.3. Anaplerosis and the astroglial TCA cycle
5. Concluding remarks

This chapter reviews some recent achievements and insights obtained by ^{13}C NMR in the brain in rats and humans in vivo. The studies discussed include (i) the demonstration that brain glycogen is an important store of glucose equivalents in the brain, providing significant fuel during hypoglycemia; (ii) the demonstration of slow brain glycogen metabolism in non-activated awake brain; (iii) the presence of significant anaplerosis (pyruvate carboxylase activity) in brain in vivo; (iv) the measurement of the energy metabolism of neurons and glia and the metabolic trafficking of glutamate between these two major metabolic compartments; (v) the measurement of a major regulatory metabolic element of oxidative metabolism in the brain, the malate–aspartate shuttle; and (vi) the finding that brain glycogen metabolism is deranged following hypoglycemic episodes,

Advances in Molecular and Cell Biology, Vol. 31, pages 409–433

ISBN: 0-444-51451-1

suggesting an involvement in the hypoglycemia unawareness syndrome clinically observed in diabetes.

1. Introduction

The propagation of electrical impulses between brain cells is accomplished by chemical transmission, achieved by releasing signaling molecules (neurotransmitters) from the presynaptic bouton that interact with receptors on the postsynaptic neuron and thus mediate the transmission of electrical signals from one neuron to the next. It is becoming increasingly clear that normal brain function not only involves the function of neurons on both sides of the synaptic cleft. In addition to the pre- and the postsynaptic neuron, the astroglial compartment has recently gained increased attention by virtue of its importance in maintaining the functionality of the synapse (Schousboe et al., 1993a; Hansson and Ronnback, 1994; Vernadakis, 1996; Magistretti and Pellerin, 1999). Among the many neurotransmitter systems, that of glutamate is probably most abundantly distributed in the central nervous system. The accepted mechanism for the action of glutamate (Shank and Aprison, 1979; Westergaard et al., 1995; Daikhin and Yudkoff, 2000; Lieth et al., 2001), is a prime example for the importance of the interplay of electrical events and metabolism in the action of this important neurotransmitter, as illustrated in the scheme of Fig. 1.

Glutamate uptake from the synaptic cleft is characterized by concurrent electrical events due to electrogenic glutamate transporters which are critical in maintaining a low extracellular glutamate concentration in order to avoid excitotoxicity and to maintain

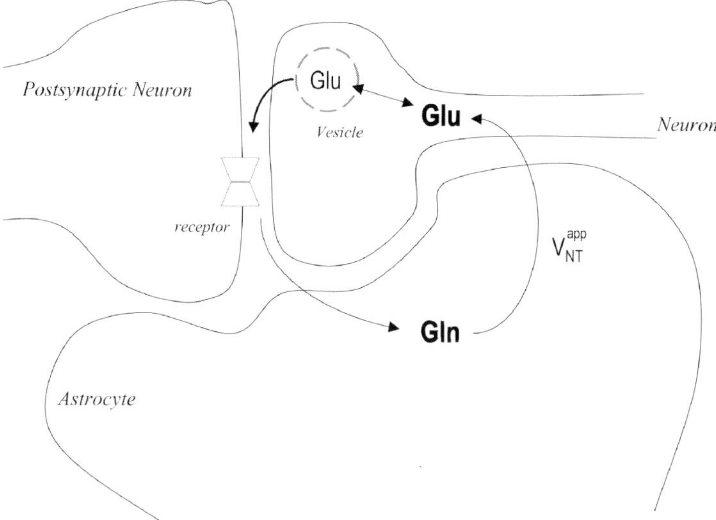

Fig. 1. Scheme depicting the metabolism of neurotransmitter glutamate, which forms the basis for the concept of the glutamate–glutamine cycle, with the rate VNT. Uptake of glutamate in astrocytes from the synaptic cleft (not shown as a separate step) is followed by conversion into electrophysiologically inactive glutamine.

postsynaptic excitability. Most of the synaptic glutamate is cleared into the astrocytes surrounding the synaptic cleft (Yudkoff et al., 1993; Bergles et al., 1997; Zigmond, 1999; see also the chapter by Schousboe and Waagepetersen). Following uptake into the glial cell, glutamate is converted by glutamine synthetase into the electrophysiologically inactive glutamine, which is transported back to the neuron and converted to glutamate, thereby maintaining the nerve terminal concentration of neurotransmitter glutamate, in a 'glutamate–glutamine cycle' (Yudkoff et al., 1988; Kanamori and Ross, 1995; Westergaard et al., 1995; Brand et al., 1997; Rothman et al., 1999; Daikhin and Yudkoff, 2000). It is therefore quite clear that glutamatergic transmission involves metabolism (through the glutamate–glutamine cycle), as well as energy metabolism of the astroglial compartment due to the requirements for the restoration of the ion balance and for glutamine synthesis (Magistretti and Pellerin, 1996; Attwell and Laughlin, 2001). Indeed, it has been reported that the uptake of Glu into the astrocytes was associated with increased glucose metabolism in the astrocyte (Magistretti et al., 1993; Eriksson et al., 1995), thereby linking stimulated energy metabolism between the astroglial and neuronal compartments during neurotransmission. Because of the mostly neuronal localization of glutamate and the exclusively astroglial localization of glutamine synthesis, the measurement of label transfer from glutamate to glutamine, uniquely possible using ^{13}C NMR spectroscopy (Gruetter, 1993; Gruetter et al., 1994; Hassel et al., 1997; Rothman et al., 1999; Chhina et al., 2001; Bluml et al., 2002; Gruetter, 2002), in principle could serve as an indicator of the rate of the glutamine–glutamate cycle, thought to reflect the rate of glutamatergic action (Yudkoff et al., 1988; Gruetter et al., 1998a; Rothman et al., 1999). While conceptually very simple in its formulation, the interpretation of the labeling of glutamate and glutamine needs to take into account many additional reactions, whose activity cannot be neglected in vivo, such as the rate of glial oxidative metabolism and pyruvate carboxylase, and, potentially, glycogen metabolism. This chapter deals with the requirements on the modeling that are necessary to understand in order to use in vivo NMR spectroscopy for the assessment of glutamatergic metabolism in the intact brain in vivo, but it does not deal with the technical requirements to implement a successful ^{13}C NMR spectroscopy program and NMR methodology.

2. Key elements of ^{13}C tracer methodology measured by in vivo ^{13}C NMR spectroscopy

NMR spectroscopy is a non-destructive method that allows the measurement of signals from several compounds and distinct positions within the molecule. The information content of detecting ^{13}C label by NMR is illustrated in Fig. 2. While detection of the ^{1}H NMR signal of hydrogen nuclei adjacent to ^{13}C nuclei is clearly most sensitive (Fig. 2B), the detection of label by directly measuring the signal of ^{13}C provides more information (Fig. 2A). A full review of the methodology involved is beyond the scope of this paper, however, it is important to recognize that many methodological difficulties need to be overcome in measuring ^{13}C label in vivo and this field is far from being fully developed as illustrated by some examples are provided at the end of this section.

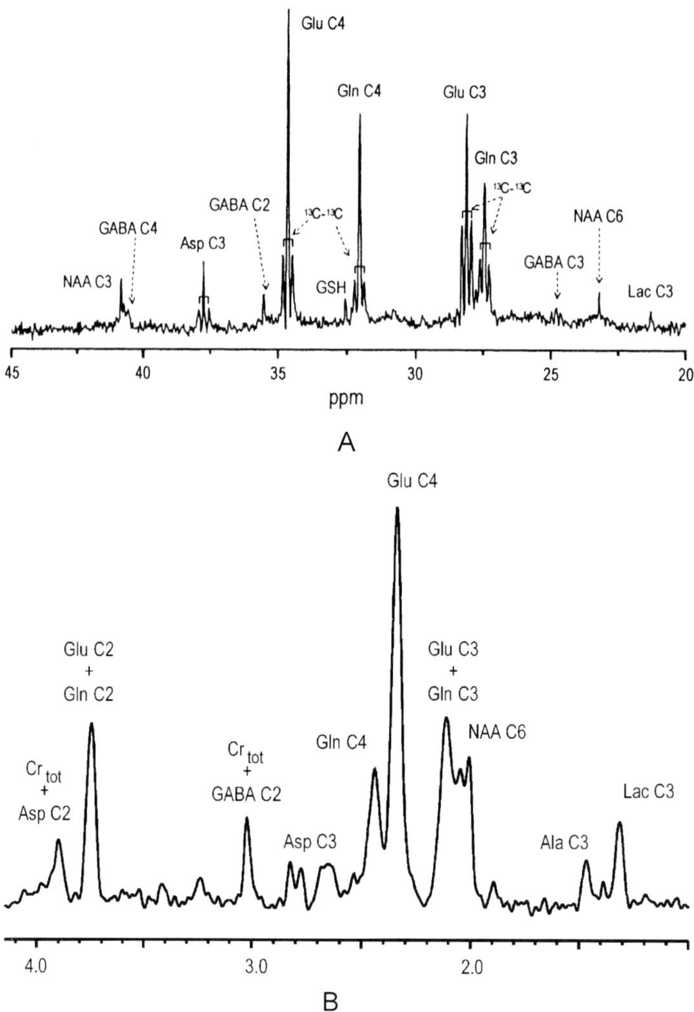

Fig. 2. In vivo NMR detection of ^{13}C label. (A) illustrates a spectral region depicting the detection of ^{13}C label for the C3 and C4 region of amino acids in a 400 μl volume of rat brain at 9.4 Tesla. Reproduced from (Choi et al., 2000). (B) shows the spectral region of the ^{1}H spectrum covering the ^{13}C label in all compounds but glucose in a 120 μl volume. Reproduced from Pfeuffer et al. (1998). (NAA - N-acetyl-aspartate; Asp - aspartate; Glu - glutamate; Gln - glutamine; Lac - lactate; Ala - alanine; Crtot - phosphocreatine + creatine).

2.1. 'Tracer methods'

The administration of a tracer, whether stable or radioactive, and the ability to follow its metabolism non-invasively opens unique opportunities to study the brain in action. When the highly sensitive radiotracers, label in different metabolic pools cannot be distinguished, unless the measurement of the incorporated radioactivity is performed for each compound separately which is only possible using postmortem analysis or if one uses non-metabolizable analogs, such as deoxyglucose, which is trapped following phosphorylation by hexokinase (Sokoloff et al., 1977).

The fundamental mathematical principle that underlies the modeling of tracer turnover curves is in principle the same, regardless of the type of tracer used: In all cases, the rate of label appearance in the product pool P is given by the sum of metabolic fluxes from any substrate multiplied by the probability that this particular substrate was labeled, ${}^{13}S/S$, resulting in Eq. (1) which is from (Gruetter, 2002):

$$\frac{\mathrm{d}^{13}P}{\mathrm{d}t}(t) = \sum_i V_i^{(\mathrm{in})} \frac{{}^{13}S_i(t)}{S_i} - \sum_j V_j^{(\mathrm{out})} \frac{{}^{13}P(t)}{P} \qquad (1)$$

For example, the Sokoloff method measures the tissue radioactivity 45 min after administering a measured bolus of labeled glucose, when the radioactivity from non-phosphorylated sources is negligible and when dephosphorylation is not significant. A further extension of the Sokoloff method is the measurement of tissue radioactivity as a function of time, to which a suitable model of the tracer compartments is fitted. The elegance of the Sokoloff method is its operation in the true tracer mode, i.e., when the kinetics of the product buildup do neither affect the tracer kinetics nor the biochemical reaction, leading in principle to a simplified mathematical problem, as indicated by the schematic representation of label incorporation (Fig. 3).

Label incorporation into a product from a precursor, such as into glutamate from glucose, is based on the same fundamental mathematical principles as, e.g., the Sokoloff method, yet several quite profound differences must be discussed. In contrast to the radiotracer method, the signal is detected in a naturally occurring compound, which, because of the inherently lower sensitivity of NMR, must be highly concentrated and enriched in order to be measurable, and inherently includes an upper limit of the measurable label incorporation in tissue. These aspects typically lead to 'tracer' curves, i.e., label incorporation curves, similar to what is shown in the middle in Fig. 3, with the label in a specific compound (such as the highly concentrated glutamate) reaching a steady-state value after some time.

2.2. *Dynamic isotopomer analysis*

The true power of modeling label incorporation into tissue pools, as measured by NMR, however, is harnessed by taking into account the ability of NMR to measure the rate of label incorporation not only into different molecules, but also into different positions in a given molecule (Cerdan et al., 1990; Badar-Goffer et al., 1992; Mason et al., 1992; Schousboe et al., 1993b; Shank et al., 1993; Gruetter et al., 1998a), such as the C2, C3 and C4 of glutamate (Fig. 3). The measurement of label incorporation into multiple positions in a molecule in effect is very similar to the measurement of the label distribution in a molecule, traditionally dubbed isotopomer analysis (Malloy et al., 1990; Jeffrey et al., 1991). It has been shown that the *time-resolved* measurement of label incorporation into the glutamate C3 and C4 is equivalent to the dynamic measurement of the simultaneous, but separate measurement of [4-^{13}C] glutamate and [3,4-${}^{13}C_2$] glutamate from the ^{13}C–^{13}C singlet and doublet signals (isotopomers) at the C4 position (Jeffrey et al., 1999).

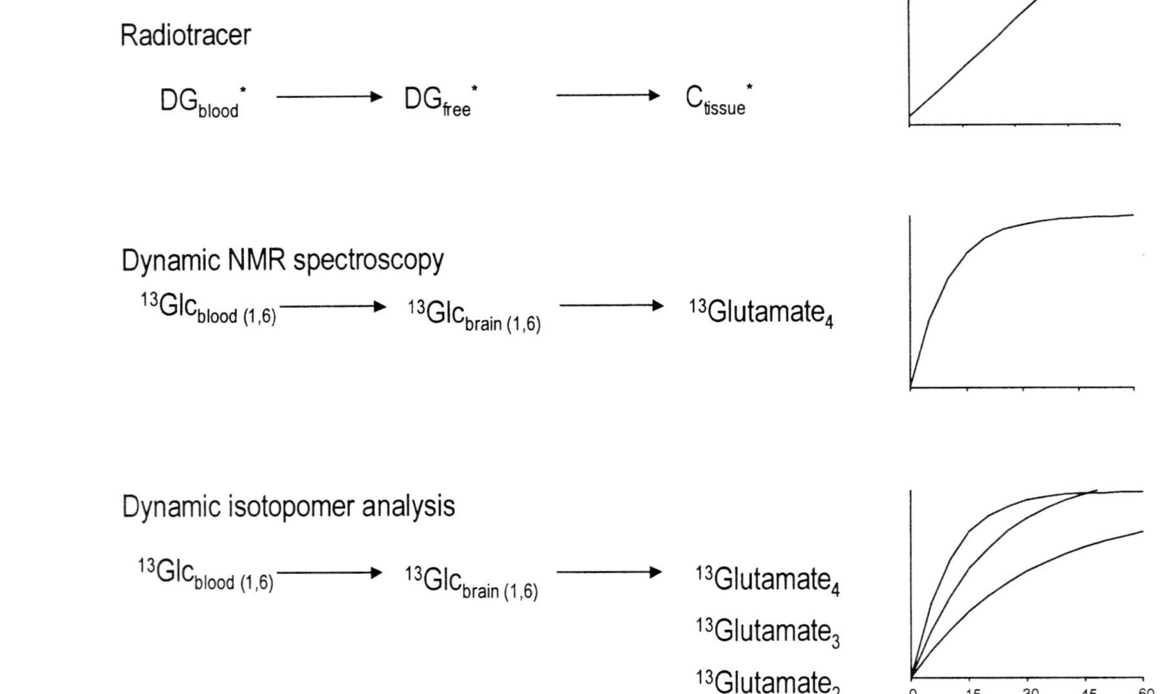

Fig. 3. Measurement of labeling kinetics using radiotracer methods in comparison to in vivo NMR methods. (Top) Radiotracer methods such as the Sokoloff method use a non-metabolizable analog such as deoxyglucose (DG) to measure the accumulation in the metabolic products (C^{*}_{tissue}), mostly due to deoxy-glucose-6-P, leading to simplified first-order tracer kinetics (Sokoloff et al., 1977). (Middle) The traditional method of measuring metabolic fluxes is based on dynamic NMR spectroscopy, mostly focusing on the measurement of label incorporation from a precursor, such as glucose (Glc) into the C4 of a molecule, such as glutamate ($^{13}Glu_4$) (Malloy et al., 1987; Fitzpatrick et al., 1990). (bottom) The full power of NMR spectroscopy is exploited by dynamic isotopomer analysis, which in its generality measures the label incorporation into multiple positions of the same molecule, leading to measurement of multiple time courses of label incorporation into the C2, C3 and C4 of glutamate.

It is therefore proposed to name the time-resolved measurement of label incorporation into multiple positions of a given molecule 'dynamic isotopomer analysis' (bottom in Fig. 3).

When fitting a model of compartmentalized cerebral metabolism to such a model, it is important to recognize that the cost function will be evaluated for all fitted time courses simultaneously. In strictly mathematical terms, this is an extension of the one-dimensional case applicable to the measurement of label incorporation into the glutamate C4 for instance.

In this context it may be confusing that some models seem to be governed by many more differential equations than parameters that are fitted or number of time courses that are measured. However, in practice, even with so-called simpler models involving only a few explicit differential equations, many more differential equations are in fact involved. In those cases they do not enter the model explicitly because it has been assumed that the small pool size compared to the metabolic rate leads to a negligible effect on the measured metabolic rate. In other words, in the case of a series of chemical reactions involving at least three pools of metabolites, when assuming that the second pool is small compared to the third pool and compared to metabolic flux, the labeling rate of the third pool is not likely to be substantially affected whether that of the second pool was explicitly calculated or not. For metabolic branching points, such as 2-oxoglutarate, however, it is necessary to explicitly include the calculation of the rate of labeling of 2-oxoglutarate, even though the pool size of 2-oxoglutarate itself is unlikely to affect the label turnover curve. Hence, the argument that the number of differential equations is too large needs to take into account whether this implies more reactions with a significant impact on the labeling curves or just a more explicit mathematical formulation of reality.

Lastly, an important difference between NMR measurements using isotopes and radiotracer methods lies in the fact that the degree of isotopic labeling in the NMR studies typically is very high, such that small fluctuations of the isotopic enrichment of the precursor pool are not likely to affect the outcome of the analysis.

In any of these methodologies it is clear that the rate of label incorporated *as a function of time* can in principle be related to the metabolic rate and thus allows measuring *absolute* metabolic fluxes. Another interesting approach consists in measuring the relative amount of label in different molecules or even different positions between different molecules when metabolic steady-state has been achieved (Gruetter et al., 1998a; Gruetter et al., 2001; Bluml et al., 2002; Lebon et al., 2002), and some relative fluxes can be derived using equations such as the following:

$$\frac{^{13}P}{P} = \frac{V_n^{(\mathrm{in})}}{\sum_j V_j^{(\mathrm{out})}} \frac{^{13}S_n}{S_n} \tag{2}$$

Eq. (2) is derived from Eq. (1) assuming steady-state and only one substrate S_n leading to label incorporation into the product P. Such steady-state analysis can lead to a simplified analysis and has also been used for the measurement of differential labeling in the C4 following acetate labeling (Lebon et al., 2002). In the case of acetate labeling, the scrambling of label into many molecules needs to be taken into account. For example, in the case of the glutamate–glutamine cycle, not only are there four metabolic pools to be considered (glutamate and glutamine in the neuronal and glial compartments) that can be labeled, but also the magnitude of the fluxes between the mitochondrial Krebs cycle intermediate and the cytosolic glutamate, V_x, is expected to have an effect on the calculated relative metabolic rates. That the derived labeling can depend on this exchange can be appreciated from the following 'Gedankenexperiment': Consider a case where the glutamate does not have a significant exchange with 2-oxoglutarate in the neuron (small V_x). In this case the relative labeling of neuronal glutamate ($^{13}Glu^{(n)}/Glu^{(n)}$) will be identical to that of glial glutamine. However, as the exchange rate increases, so does the contribution of unlabeled carbon from the neuronal Krebs cycle to neuronal glutamate, leading to increasingly different labeling in glutamate relative to glutamine, which may very well affect the interpretation of the relative quantitative magnitude of the glutamate–glutamine cycle. Therefore, for the calculation of relative rates of the glutamate–glutamine cycle, at least 6 pools into which label is accumulated, need to be considered even for the case of labeling from acetate or [2-^{13}C] glucose.

2.3. ^{13}C NMR methodological aspects

Unfortunately, the technical development of ^{13}C NMR spectroscopy in vivo has been limited to a handful sites worldwide (Gruetter et al., 1998a; Bluml, 1999; Rothman et al., 1999; Chen et al., 2001; Chhina et al., 2001; Henry et al., 2002) and largely requires further development. The technical underpinnings of ^{13}C NMR in vivo is not the focus of this chapter, but shall be briefly illustrated by reviewing a few key advances in this field.

In 1992, two important advances in MR methodology were introduced to in vivo ^{13}C NMR spectroscopy: First, the use of automated shimming (i.e., in vivo optimization of the main static magnetic field, B_0, such that it becomes largely independent of the spatial coordinates) dramatically improved sensitivity by narrowing linewidths by almost an order of magnitude compared to what was reported at the time (Gruetter and Boesch, 1992; Gruetter, 1993). Second, the introduction of three-dimensional localization allowed for well-defined volumes to be measured (Gruetter et al., 1992a,b). Two improvements were immediately realizable, namely (i) the complete elimination of the intense scalp lipid signals from the extra-cerebral tissue, which overlap with numerous signals from amino acids, and (ii) the collection of signals from a well-defined volume, which together with excellent shimming improved the spectral resolution. The increases in sensitivity were demonstrated with the rather surprising observation that *natural abundance* signals from brain metabolites can be observed in vivo, such as those from *myo*-inositol (Gruetter et al., 1992b). These methodological advances lead to the landmark discovery that labeling of glutamine can be detected in the in vivo brain (Gruetter, 1993; Gruetter et al., 1994), which is now recognized as a window to study cerebral metabolic compartmentation (Sonnewald et al., 1994; Bachelard, 1998; Cruz and Cerdan, 1999; Magistretti et al., 1999; Rothman et al., 1999) and provides a unique window on the brain. Localization has also proven critical in the ability of NMR to detect and measure the signals of brain glycogen, because the several-fold higher concentration of muscle glycogen requires dedicated efforts to eliminate any non-cerebral source of glycogen signal (Choi et al., 1999; Choi et al., 2000; Oz et al., 2003).

3. Glial metabolism I: brain glycogen

The brain relies on a continuous supply of glucose from the blood for maintaining normal brain function, yet the brain maintains a significant level of brain glycogen making it the largest endogenous carbohydrate reserve in the brain. The concentration of brain glycogen has been estimated at a few μmol/g (Choi and Gruetter, 2003) and references therein, however, several recent studies have suggested that due to the rapid postmortem glycogenolysis (Lowry et al., 1964; Swanson et al., 1989; Choi et al., 1999), as well as glycogen breakdown during the assay itself, these brain glycogen concentrations may have been underestimated (Cruz and Dienel, 2002; Kong et al., 2002). The problems with the biochemical determination of brain glycogen clearly point to the need of a non-invasive method for its measurement. In vivo NMR spectroscopy has unique capabilities in that regard, as shall be illustrated below.

3.1. Brain glycogen, an endogenous store of fuel

Brain glycogen metabolism and concentrations obviously can be measured by NMR when using suitable methodology (Choi et al., 1999, 2000). Pulse-chase experiments demonstrated that glycogen breakdown in the α-chloralose anesthetized rat was very slow with a turnover rate on the order of 0.5 μmol/g/h during glucose infusions (Choi et al., 1999). The study further precluded label turnover as the only mechanism by which label was transferred to the brain glycogen pool and it was concluded that net brain glycogen synthesis must have occurred (Choi et al., 1999). In a follow-up study, the effect of insulin was measured by measuring the effect of hyperinsulinemia on label incorporation at a controlled plasma glucose level fixed close to normoglycemia (Choi et al., 2003). The results showed that in vivo at mild hyperglycemia, plasma insulin has a profound effect on brain glycogen metabolism and leads to a net accumulation. These findings were in agreement with data from cell cultures showing an effect of high insulin concentrations on culture glycogen concentrations (Nelson et al., 1968; Dringen and Hamprecht, 1992; Sorg and Magistretti, 1992; Swanson and Choi, 1993). The study extends previous studies suggesting an effect of insulin on brain glycogen content at supraphysiologic hyperglycemia (Daniel et al., 1977).

It is quite plausible that brain glycogen may serve as a reservoir of glucosyl units that are mobilized whenever demand for glucose is in excess of supply and such a neuroprotective role for brain glycogen can be implied from, e.g., the mechanism of glutamate neurotransmission (Fig. 4), where glycogen is able to provide energy during hypoglycemia to maintain a low extracellular glutamate concentration. Such a neuroprotective role for brain glycogen has been suggested on the basis of preloaded astrocytes in co-culture with neurons (Swanson and Choi, 1993) and for axonal survival during glucodeprivation (Wender et al., 2000). A recent study demonstrated that degradation of brain glycogen

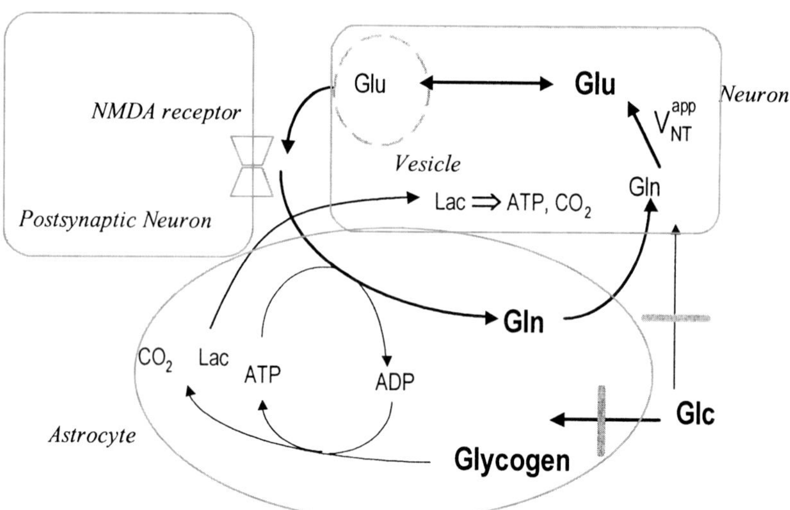

Fig. 4. Potential mechanism of a neuroprotective role of brain glycogen during hypoglycemia. In this scheme, an impaired supply of glucose to the brain (as indicated by the shaded bars) leads to an energy deficit, that ultimately can result in glutamate excitotoxicity, as the mechanism of the glutamate neurotransmission involves glial energy metabolism. However, glycogenolysis can produce the extra fuel to maintain glial function and thus a low extracellular glutamate concentration. Some of the glycogen is metabolized oxidatively to CO_2 and some of it can be exported to the neuron for oxidative metabolism (Magistretti and Pellerin, 1996).

initiated by hypoglycemia started when the brain glucose concentration approached zero, which is the point at which glucose transport became rate-limiting for metabolism (Choi et al., 2003). Interestingly, at this point cerebral blood flow (CBF) was increased abruptly (Fig. 5A), indicating an attempt by the brain to increase supply, by decreasing the arterio-venous gradient for glucose (Choi et al., 2001). The textbook literature implied that brain glycogen must be a limited storage form for glucose due to its low content and, thus, the role of brain glycogen as a glucose reservoir has been generally dismissed in the literature. Nonetheless, during hypoglycemia, glycogen need only account for part of the deficit in glucose supply and hence can survive longer periods of sustained hypoglycemia. Indeed, a preliminary estimate indicated conservatively that brain glycogenolysis accounted for a majority of the deficit in glucose supply, supporting the quantitative importance of brain glycogen in hypoglycemia (Choi et al., 2003). Measurements of brain glycogen during hypoglycemia indicated that brain glycogen degradation occurred at a rate during hypoglycemia that resulted in brain glycogen concentrations to be substantial even after 2 h of moderate hypoglycemia (Fig. 5B). It is therefore likely that the brain tries to defend itself against moderate hypoglycemia by using brain glycogen and by increasing CBF and that these defenses are triggered by the point at which glucose transport becomes rate-limiting for metabolism, or by the point at which the brain glucose concentrations become rate-limiting for metabolism.

Because brain glycogen is an insulin-sensitive glucose reservoir it is interesting to explore whether brain glycogen metabolism is deranged following hypoglycemia. Indeed,

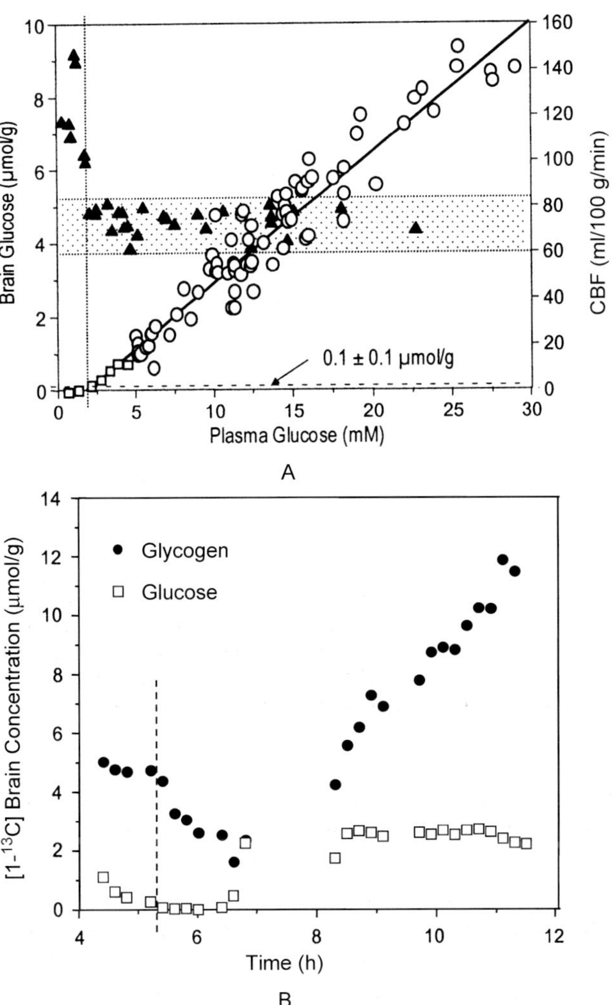

Fig. 5. Effect of hypoglycemia on cerebral blood flow, brain glucose concentrations and brain glycogen metabolism as measured by NMR. (A) Measurements of brain glucose concentration (left scale) as a function of plasma glucose concentration. The solid line is the fit of the reversible Michaelis–Menten model to the eu- and hyperglycemic brain glucose (open circles). The open squares indicate brain glucose measurements below 4.5 mM plasma glucose. When brain glucose approaches zero (dotted vertical line), the measurement of cerebral blood flow (solid triangles, right scale) indicate CBF values above the 95% confidence interval (shaded area) and this is also the point where brain glycogenolysis started (arrow in A and vertical dotted line in B). (B) Measurement of the effect of insulin-induced hypoglycemia at 4 h on brain glycogen metabolism and glucose concentrations. When the brain glucose concentrations (open squares) approached zero, brain glycogenolysis started (dotted vertical line) at a rate that sustained brain glycogen concentrations (solid circles) for at least 2 h. Restoration of brain glucose concentrations at $t = 7$ h typically resulted in a brain glycogen rebound (supercompensation). Modified from (Choi et al., 2001) and (Choi et al., 2003).

data suggest that brain glycogen concentrations following hypoglycemia increase substantially above normal (Fig. 5B) leading to increased neuroprotection. It is therefore reasonable to conclude that brain glycogen metabolism may play a role in the development of defective recognition of hypoglycemia (hypoglycemia unawareness) by the brain as proposed recently (Choi et al., 2003).

3.2. Human brain glycogen metabolism in vivo

The studies mentioned in Section 3.1 have been performed in anesthetized animals and, consequently, the legitimate question arises as to whether the slow brain glycogen turnover observed also translates to the awake brain. This is of interest, since several studies suggesting an involvement of brain glycogen in brain activation have done these measurements in the conscious animal (Swanson, 1992; Dienel et al., 2002). Our results thus far suggest that brain glycogen is only utilized when supply is insufficient to cover demands in metabolism, possibly only when brain glucose concentrations become so low that they significantly limit the rate of glucose phosphorylation. Some of the reported increases in brain glucose metabolism observed during focal activation imply increased usage of carbohydrates other than blood glucose because the reported increases (Hyder et al., 1997) exceed the transport capacity of the blood-brain barrier by several-fold (Choi et al., 2001). One such source of glucose equivalents is brain glycogen, which is present in sizable amounts in brain (Sagar et al., 1987; Choi et al., 1999; Cruz and Dienel, 2002; Kong et al., 2002). It is possible that parts of the glycogen molecule may undergo rather rapid metabolism. However, in line with our results in the α-chloralose anesthetized rat brain (Choi et al., 1999), we found that metabolism of bulk brain glycogen was also very slow in the awake rat brain, with a turnover time on the order of that of NAA (Choi and Gruetter, 2003) and a total brain glycogen concentration of ~ 3 μmol/g wet weight in line with the literature.

However, the important question arises how these measurements relate to human brain glycogen metabolism, which has never been measured in vivo. The question remained as to whether in the conscious human, brain glycogen metabolism is also slow. We have adapted our previously developed methods to the measurement of brain glycogen in humans and measured the rate of label incorporation into brain glycogen during administration of [1-^{13}C] glucose in humans (Oz et al., 2003). The results indicated a much slower rate of label incorporation in the human than in the rat (Fig. 6) with a turnover rate of approximately 0.15 μmol/g/h. Such a slow rate of turnover certainly does not suggest an involvement of brain glycogen metabolism in the background activity of the conscious human brain. Nonetheless, it does not preclude the activation of the reservoir in conditions of extreme metabolic demand. Instead, it favors the overall influence of the sleep–wake cycle on brain glycogen metabolism as reported (Kong et al., 2002), and supported by altered gene expression (Petit et al., 2002): For example, it is quite conceivable that small bursts of brain activity will lead to transient mismatches in glucose supply and demand causing, e.g., small decreases in brain glycogen that can accrue over time during the day. Such a slow rate of turnover of glycogen in humans suggests that turnover of the glucosyl units in brain glycogen may require days and that altered brain glycogen concentrations

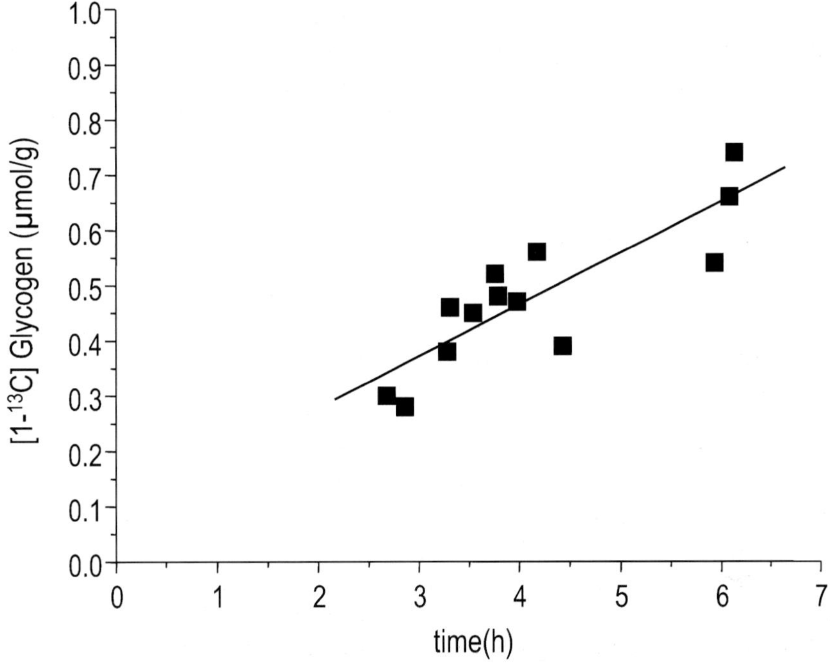

Fig. 6. Measurement of the label incorporation into human brain glycogen C1. The solid squares represent the measurement of ^{13}C label in glycogen C1 following administration of 80 g of 1-^{13}C glucose in three humans begun at $t = 0$ min. The solid line is the result of a linear regression of the measurements in the first 5 h of the study and indicates a rate of label incorporation consistent with a very slow glycogen turnover rate on the order of 0.1–0.2 μmol/g/h. From Oz et al. (2003).

(such as a super-compensation following a hypoglycemic episode) may require time on the order of a week to be restored to normal. This time scale of brain glycogen metabolism is consistent with the time scale it takes to revert the syndrome of hypoglycemia unawareness and is consistent with the proposed involvement of brain glycogen in the pathogenesis of impaired recognition of hypoglycemia.

4. Glial metabolism II: the glutamate–glutamine cycle

Because of the ever increasing importance of functional MRI, a mechanism of which is the activation-dependent change in the venous concentration of deoxyhemoglobin, the question whether there is tight coupling between glucose and oxygen consumption in the brain has become of paramount importance. The landmark study by Fox and Raichle in the late 1980s suggested that there are indeed large increases in glucose metabolism and CBF that exceed the changes in oxygen metabolism (Fox et al., 1988). In principle, NMR provides the unique capability to measure cerebral concentration changes of brain glucose and lactate, both of which are key components in addressing this question, and increases in lactate and glucose concentration have been reported (Prichard et al., 1991; Sappey-Marinier et al., 1992; Frahm et al., 1996). In addition, from the incorporation of label from a suitable precursor such as glucose, into glutamate, the cerebral oxygen consumption could be computed. A majority of studies in the brain have focused on measuring glutamate turnover (Rothman et al., 1992; Hyder et al., 1997; Lukkarinen et al., 1997; Pan et al., 2000), which was motivated by the fact that glutamate turnover is linked to the metabolism of the Krebs cycle (Mason et al., 1992; Chatham et al., 1995; Mason et al., 1995; Yu et al., 1997; Gruetter et al., 1998a, 2001; Cruz and Cerdan, 1999), and that the glutamate C4 resonance, which is labeled in the first turn of the Krebs cycle, presents a readily detectable NMR signal due to the high concentration of glutamate. Using this methodology, one such study compared the rate of label incorporation and found a significant difference between the activated and the resting visual cortex, indicating that the cerebral oxygen consumption increased at most by 30%, which is approximately half of the blood flow increase measured using this stimulation paradigm (Chen et al., 2001). This study supported the idea that oxygen consumption increases are less than the associated blood flow increases, leading to a net decrease in deoxy-hemoglobin content during focal activation, which forms the basis of blood-oxygen-level-dependent functional MRI (Ogawa et al., 1998).

Perhaps the major advantage of in vivo NMR is not to provide neuroscientists with an alternative alternative method to measure CMR_{O2} and CMR_{glc} (although this may be very useful as indicated above), but to shed light on metabolic processes not accessible by any other method, one of which (glycogen) was addressed above and some of which will be discussed below.

4.1. Glutamate turnover: neuronal oxygen metabolism and the malate–aspartate shuttle

The measurement of cerebral oxygen consumption from turnover of glutamate (as referred to in the previous paragraph) assumes a direct stoichiometric relationship between that measurement and the rate of oxygen consumption. Unfortunately, this relationship is not directly inferred, as the brain is intricately compartmentalized, which shall be discussed further below. In addition, most of the glutamate signal that is observed is in the cytosol, whereas the labeling occurs in the mitochondrion and hence label has to be transported across the charged inner mitochondrial membrane (Fig. 7), which has been shown to be the rate limiting step in many tissues, such as the heart (Chatham et al., 1995; Yu et al., 1997; Sherry et al., 1998) and the liver (Garcia-Martin et al., 2002).

Initially it was thought that the exchange between 2-oxoglutarate and glutamate, V_x, is very fast in the brain in vivo. However, many studies in heart, liver and muscle have indicated the opposite in these tissues (Chatham et al., 1995; Yu et al., 1997; Sherry et al., 1998; Garcia-Martin et al., 2002). More recent evidence now suggests that in the brain V_x is on the order of the flux through pyruvate dehydrogenase, V_{PDH} (Gruetter et al., 2001; Choi et al., 2002), which may vary in pathologic conditions (Henry et al., 2002). The observation that V_x was comparable to the flux through pyruvate dehydrogenase implied that the malate–aspartate shuttle may be a major mechanism mediating the

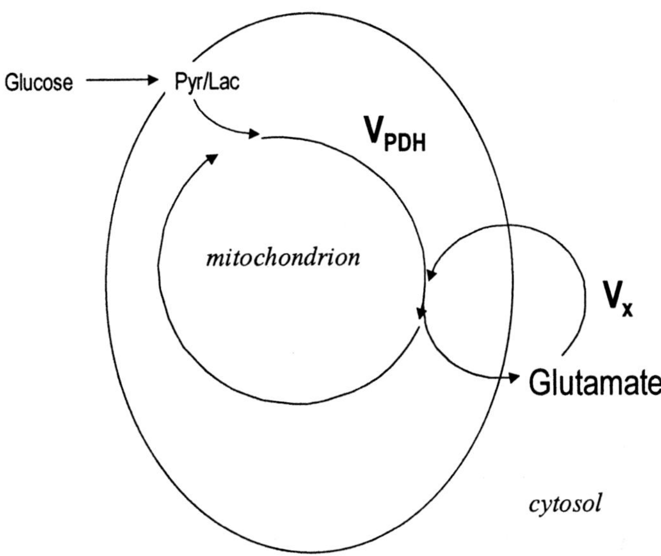

Fig. 7. Measurement of oxidative glucose consumption from the flow of label from glucose to glutamate. The label in glucose (or any other precursor, such as pyruvate or acetyl-CoA for that matter) is metabolized in the mitochondrion and then transferred to glutamate, most of which is in the cytosol. The flow of label into glutamate thus is in principle a combined effect of Krebs cycle flux (V_{PDH}) and label exchange across the mitochondrial membrane (V_x).

exchange of label across the mitochondrial membrane (LaNoue and Tischler, 1974; Yu et al., 1997; Gruetter, 2002). The assumption that V_x is very fast will affect the modeling results depending on the pool sizes that participate in this exchange (Gruetter et al., 2001; Gruetter, 2002).

Furthermore, it is important to recognize that because of the mostly neuronal localization of glutamate, the measurement of glutamate turnover alone, as done in several previous studies (Rothman et al., 1999; Pan et al., 2000; Sibson et al., 2001), mainly measures neuronal metabolism. The astroglial compartment does contain significant oxidative capacity for metabolism, as pointed out recently (Gruetter et al., 2001), despite some previous assumptions to the contrary (Sibson et al., 1998). This shall be discussed in Section 4.2.

4.2. Glutamine turnover: the hallmark of glial metabolism

It is well known that brain metabolism is characterized by at least two major compartments with a large neuronal and a small glial glutamate pool associated with the Krebs cycle. As pointed out previously, these two pools are metabolically linked by the glutamate–glutamine cycle. The compartmentation of brain metabolism is based on that of several enzymes. In addition to those of glycogen (see above), glutamine synthetase (Martinez-Hernandez et al., 1976) and pyruvate carboxylase (Shank et al., 1985) are almost exclusively in the glial compartment, as summarized in more detail elsewhere

(Bachelard, 1998; Cruz and Cerdan, 1999; Gruetter et al., 2001; Gruetter, 2002). Compartmentation furthermore extends to metabolites, such as glutamate (neuronal) and glutamine (glial) (Ottersen et al., 1992; Shupliakov et al., 1997), as well as to mitochondria and other systems (Schousboe et al., 1993b).

It is of interest to note that the exclusively glial localization of glutamine synthetase implies that the observation of glutamine synthesis in vivo, first achieved in human brain (Gruetter, 1993; Gruetter et al., 1994), is a direct manifestation of glial metabolism, whereas the observation of label incorporation into glutamate implies a mainly neuronal event. Clearly, glutamine synthesis can be measured non-invasively by NMR, as shown in Fig. 2, and therefore demonstrates the ability of NMR to study cerebral compartmentation non-invasively in intact brain.

The mechanism of inactivation of glutamate by uptake into the astroglial compartment implies a much more active role for astrocytes than is conventionally assumed, due to the imperative involvement of glial energy metabolism (Eriksson et al., 1995; Magistretti and Pellerin, 1996; Silver and Erecinska, 1997). The neuron-astrocyte triade thus has to be considered the functional unit intimately involved in achieving chemical transmission (Magistretti et al., 1993, 1999; Tsacopoulos and Magistretti, 1996). The link between astrocytes and neurons is generally accepted from a metabolic as well as from a neurophysiological standpoint (Bergles et al., 1997), yet differences exist as to the precise coupling and the specific energetics involved.

The simplest scheme for measuring glutamate neurotransmission in vivo is shown in Fig. 8. This model assumes very rapid exchange V_x and negligible glial Krebs cycle rate, as well as negligible anaplerosis. Based on this simple and elegant scheme, it was proposed that the rate of glutamate/glutamine inter-conversion (the glutamate–glutamine cycle), identified in the scheme in Fig. 8 by V_{NT}^{app}, is equal to the glucose consumption rate (Sibson et al., 1998; Rothman, 2001; Shulman et al., 2001a,b; Shen and Rothman, 2002; Rothman et al., 2003). This elegant, but perhaps oversimplified model assumed that the two ATP produced by glycolysis were almost completely consumed by glutamine synthesis and restoration of the ion balance through the Na/K ATPase with a negligible oxidative metabolism in the glial compartment. Under these circumstances it was postulated that glucose metabolism must be directly linked to glutamate neurotransmission with a 1:1 stoichiometry. The proposal put forth by Shulman and coworkers (Sibson et al., 1998; Rothman et al., 1999), that the glial ATP production needed to maintain neuronal glutamate is solely provided by glycolysis pathway is intriguing as it emphasizes the coupling between neurons and glia at the level of energy metabolism. However, only a few percent of pyruvate molecules need to be diverted to the Krebs cycle to generate as many ATP as are formed in the absence of oxidative metabolism of glucose in the astrocytes.

Furthermore, a recent study (Choi et al., 2002) measured brain glucose and glycogen metabolism in deep pentobarbital anesthesia under conditions similar to what was used (Sibson et al., 1998) and what had been shown to result in isoelectric coma (Contreras et al., 1999). In that study it was shown that the brain glucose concentration changed only slightly despite a drastic reduction in electrical activity and that a substantial gradient in brain glucose concentration relative to that in plasma persisted, as illustrated in Fig. 9.

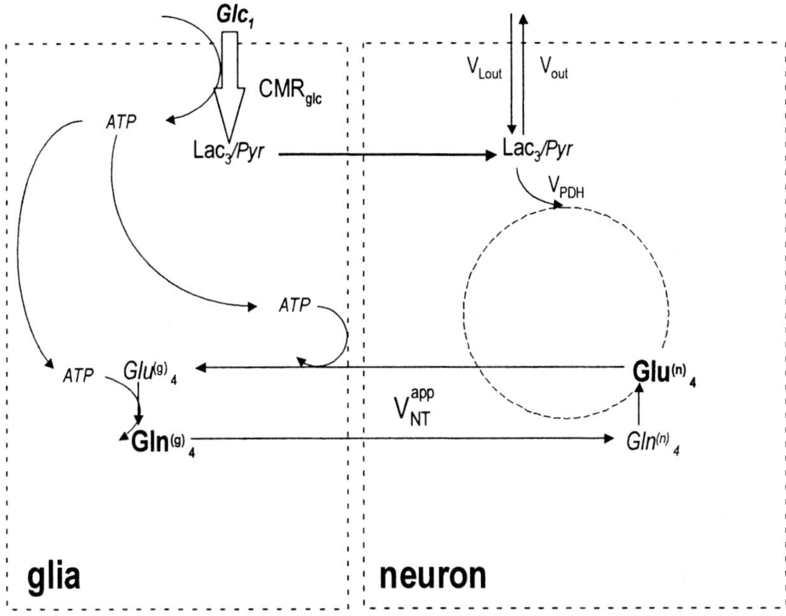

Fig. 8. Proposed link between glucose metabolism and glutamate neurotransmission. Adapted according to Sibson et al. (1998). The proposed stoichiometric coupling between glucose metabolism and glutamate neurotransmission is based on the assumption that all of the glucose is metabolized in the glia and that the two ATP produced are consumed by the need to restore ion potential following glutamate uptake and by glutamine synthetase. Oxidative metabolism of glucose and anaplerosis (pyruvate carboxylation) are neglected in this simple, yet elegant model.

In addition, that study indicated that when using the simplified scheme shown in Fig. 8 (Choi et al., 2002), similar metabolic rates as those reported by (Sibson et al., 1998) were obtained, however, the rate of label incorporation into glutamate C4 and C3 was inconsistent with that observed (dashed line in Fig. 10). Thus, the study reiterated the importance of minimizing the number of assumptions made in the modeling, which was also emphasized by two other independent studies (Gruetter et al., 2001; Henry et al., 2002). One additional assumption of the study by Sibson et al. (1998) was that the magnitude of glutamine synthesis not related to neurotransmitter cycling was constant over a large range of electrical activity. A surprising observation of our study was that under deep pentobarbital anesthesia, astrocyte metabolism was as significant as was neuronal metabolism with approximately equal magnitude per volume brain tissue. This observation is consistent with results from previous studies in culture, suggesting an effect of barbiturates on neuronal metabolism that is different in magnitude from that on astrocytes (Yu et al., 1983; Hertz et al., 1986; Swanson and Seid, 1998; Qu et al., 2000).

Oxygen consumption has been reported to increase in cultured astrocytes when exposed to extracellular glutamate (Eriksson et al., 1995) and large increases in oxygen consumption have been reported in brain during functional activity (Hyder et al., 1996, 1997), which support the idea that oxygen metabolism in astrocytes is stimulated during focal

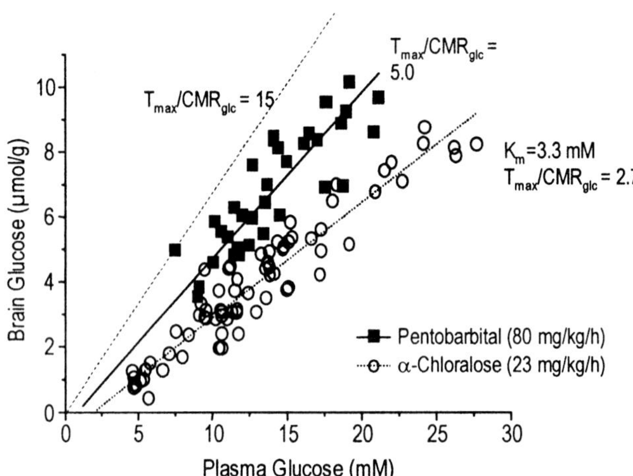

Fig. 9. Effect of pentobarbital anesthesia on brain glucose content in the rat brain. Shown is a comparison of the brain glucose concentration between light α-chloralose anesthesia and deep pentobarbital anesthesia. Modeling of brain glucose transport according to previous studies (Gruetter et al., 1998b; Choi et al., 2001; Seaquist et al., 2001) indicated a decreased rate of glucose metabolism (CMR_{glc}) relative to the apparent maximal rate of glucose transport (T_{max}). Even under deep pentobarbital anesthesia, brain glucose concentrations were significantly lower than expected if glucose metabolism was abolished (as indicated by the dashed line).

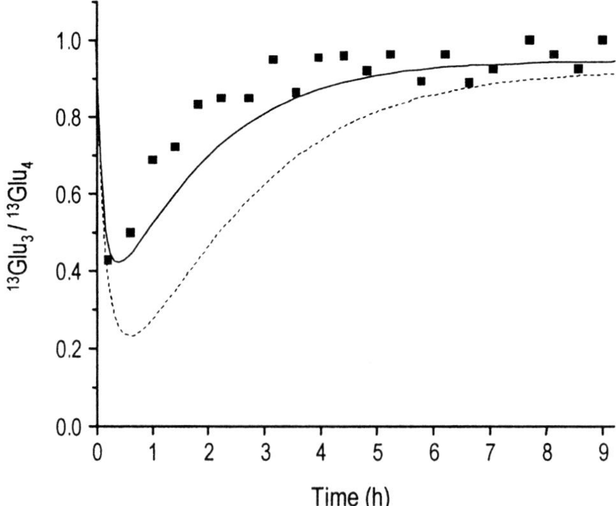

Fig. 10. Effect of the exchange rate between 2-oxoglutarate and glutamate, V_x, on the relative labeling of glutamate C3 and C4 during deep pentobarbital anesthesia. When assuming $V_x = 0.57$ μmol/g/min and fitting to the measured label incorporated into the C4 of glutamate only, an oxidative glucose consumption rate similar to that reported by Sibson et al. (1998) was obtained ($V_{PDH} = 0.15$ μmol/g/min), however, the label incorporation into the C3 of glutamate relative to that into the C4 (solid squares) was not very well reproduced (dashed line). In contrast using the scheme in (Gruetter et al., 1998; Gruetter et al., 2001; Gruetter, 2002), lead to a much better approximation (solid line), indicating that V_x is on the order of V_{PDH} also in deep pentobarbital anesthesia, and thus brain activity dependent.

activation. Glucose metabolism at rest is likely to be oxidative in glia, as judged from the well-known incorporation of acetate label into glutamine (Van den Berg, 1973; Hassel et al., 1997; Waniewski and Martin, 1998; Dienel et al., 1999), and proposed from ^{13}C studies in vivo by NMR (Brand et al., 1997; Hassel et al., 1997; Bluml et al., 2002; Lebon et al., 2002). The previously put forth argument that the majority of glial metabolism in resting brain is probably oxidative (Gruetter, 2002) thus appears valid and there is now general consensus that glial energy metabolism has a significant oxidative component on the order of 0.2–0.1 of that in the glutamatergic compartment, which we were the first to show in vivo (Gruetter et al., 2001).

The presence of dominant oxidative metabolism in the astrocyte does not disprove the hypothesis that lactate produced in astrocytes is also a fuel for oxidative metabolism in neurons. The results obtained first by our laboratory and then confirmed by others suggest that in the human brain approximately one sixth of the ATP production from glucose measured by NMR is in the glial compartment. This leaves at least five sixth of the lactate for export to neurons, if assuming the extreme case that phosphorylation of glucose is an exclusively glial process.

4.3. Anaplerosis and the astroglial TCA cycle

Astrocytes thus clearly have oxygen metabolism at rest and during activation. Assuming, as implied by the scheme in Fig. 8, that the glutamate–glutamine cycle is the sole contributor to flux through glutamine synthetase, the labeling of the carbon backbone of glutamate and glutamine must be identical at isotopic and metabolic steady-state. However, early studies in rat brain extracts (Lapidot and Gopher, 1994), and in human brain (Gruetter et al., 1998a, 2001) have reported that this is not the case. In this context, the inequality of the label distribution between brain glutamine and glutamate does depend on the relative rate of the glutamate–glutamine cycle relative to other reactions as can be deduced from Eq. (2). Furthermore, all potential contributions of label must be taken into account and thus the effect of a variable V_x must be accounted for. We demonstrated that the inequality of label in glutamate and glutamine implied significant contribution of pyruvate carboxylase to the flux through glutamine synthetase (Gruetter et al., 1998a, 2001). Therefore, other metabolic reactions must contribute substantially to the labeling of glial glutamate, and eventually glial glutamine.

One reaction that can lead to a differential labeling of glutamine and glutamate at the different positions of the molecule is the glial enzyme pyruvate carboxylase, which can label the C2 more than the C3 when administering glucose labeled at the 1 and/or 6 position. Pyruvate carboxylase activity is significant in vivo (Lapidot and Gopher, 1994; Gruetter et al., 1998a, 2001; Shen et al., 1999). Although differences exist as to the magnitude of the flux through pyruvate carboxylase, the relative amount of label incorporation into glutamine differs from that into glutamate (Martin et al., 1993; Lapidot and Gopher, 1994; Gruetter et al., 1998a, 2001). As pointed out previously (Gruetter, 2002), even the lowest reported value of 0.04 μmol/g/min (Shen et al., 1999) results in a rate of ATP generation that amounts to $\sim 2/3$ of the ATP needed for glutamate uptake and conversion to glutamine. A recent study measured the labeling of glutamate and glutamine from 2-^{13}C glucose and concluded that in the rat brain pyruvate carboxylase contributes approximately 30% to the flux of glutamine synthetase (Sibson et al., 2001). In this context it is important to note that using 2-^{13}C glucose labels the C3 of glutamate directly through pyruvate carboxylase and indirectly through the pentose phosphate shunt.

The fact that pyruvate carboxylase activity is now accepted as a substantial and significant metabolic flux in astrocytes and that astrocytes thus have substantial oxidative energy metabolism calls into question the underlying mechanism for the proposed 1:1 stoichiometric relationship between glutamate neurotransmission and oxidative glucose metabolism (summarized in Fig. 8), but it does not rule out the proposed predominantly astrocytic location of incremental glucose metabolism during activation as suggested (Magistretti and Pellerin, 1996), which remains an intriguing hypothesis. In fact, the observation that during hypoglycemia, astrocytic glycogen accounts for a majority of the metabolic deficit (see above) implicitly supports the presence of this mechanism.

5. Concluding remarks

The new non-invasive method ^{13}C NMR has come a long way: Increases in sensitivity and methodology have paved the way for many new measurements that are now feasible in the live and intact brain, leading to unique insights in anaplerosis, glial and neuronal energy metabolism, metabolic trafficking, brain glycogen metabolism and the regulation of oxidative energy metabolism. It is concluded that considerable care must be exercised when attempting to interpret and model the measured rates of label incorporation.

Acknowledgements

Supported in part by grants from the US Public Health Service, NIH R21DK58004 (RG), R21NS451119 (RG), R01NS42005 (RG), R01NS38672 (RG), P41RR08079, M01RR00400, and the Whitaker Foundation (RG) and Juvenile Diabetes Research Foundation International (RG). The encouragement and support from colleagues at the Center for MR Research is appreciated, in particular Drs Kamil Ugurbil, Elizabeth R, Seaquist, Wei Chen, Xiao-Hong Zhu. Special thanks to the members of my group, Drs In-Young Choi, Ivan Tkac, Josef Pfeuffer, Gulin Oz, Pierre-Gilles Henry and Melissa J. Terpstra for their tireless efforts and hard work, as well as to Hong-Xia Lei, Sarah L. Crawford, Dee M. Koski and Tian-Wen Yue for their assistance in these experiments.

References

Attwell, D., Laughlin, S.B., 2001. An energy budget for signaling in the grey matter of the brain. J. Cereb. Blood Flow Metab. 21, 1133–1145.

Bachelard, H., 1998. Landmarks in the application of ^{13}C-magnetic resonance spectroscopy to studies of neuronal/glial relationships. Dev. Neurosci. 20, 277–288.

Badar-Goffer, R.S., Ben-Yoseph, O., Bachelard, H.S., Morris, P.G., 1992. Neuronal–glial metabolism under depolarizing conditions. A ^{13}C NMR study. Biochem. J. 282, 225–230.

Bergles, D.E., Dzubay, J.A., Jahr, C.E., 1997. Glutamate transporter currents in Bergmann glial cells follow the time course of extrasynaptic glutamate. Proc. Natl Acad. Sci. USA 94, 14821–14825.

Bluml, S., 1999. In vivo quantitation of cerebral metabolite concentrations using natural abundance ^{13}C MRS at 1.5 T. J. Magn. Reson. 136, 219–225.

Bluml, S., Moreno-Torres, A., Shic, F., Nguy, C.H., Ross, B.D., 2002. Tricarboxylic acid cycle of glia in the in vivo human brain. NMR Biomed. 15, 1–5.

Brand, A., Richter-Landsberg, C., Leibfritz, D., 1997. Metabolism of acetate in rat brain neurons, astrocytes and cocultures: metabolic interactions between neurons and glia cells, monitored by NMR spectroscopy. Cell Mol. Biol. (Noisy-le-grand) 43, 645–657.

Cerdan, S., Kunnecke, B., Seelig, J., 1990. Cerebral metabolism of [1,2-$^{13}C_2$]acetate as detected by in vivo and in vitro ^{13}C NMR. J. Biol. Chem. 265, 12916–12926.

Chatham, J.C., Forder, J.R., Glickson, J.D., Chance, E.M., 1995. Calculation of absolute metabolic flux and the elucidation of the pathways of glutamate labeling in perfused rat heart by ^{13}C NMR spectroscopy and nonlinear least squares analysis. J. Biol. Chem. 270, 7999–8008.

Chen, W., Zhu, X., Gruetter, R., Seaquist, E.R., Adriany, G., Ugurbil, K., 2001. Study of tricarboxylic acid cycle flux changes in human visual cortex during hemifield visual stimulation using $^{(1)}H$-[$^{(13)}C$] MRS and fMRI. Magn. Reson. Med. 45, 349–355.

Chhina, N., Kuestermann, E., Halliday, J., Simpson, L.J., Macdonald, I.A., Bachelard, H.S., Morris, P.G., 2001. Measurement of human tricarboxylic acid cycle rates during visual activation by $^{(13)}C$ magnetic resonance spectroscopy. J. Neurosci. Res. 66, 737–746.

Choi, I.Y., Tkac, I., Ugurbil, K., Gruetter, R., 1999. Noninvasive measurements of [1-$^{(13)}C$]glycogen concentrations and metabolism in rat brain in vivo. J. Neurochem. 73, 1300–1308.

Choi, I.Y., Tkac, I., Gruetter, R., 2000. Single-shot, three-dimensional "non-echo" localization method for in vivo NMR spectroscopy. Magn. Reson. Med. 44, 387–394.

Choi, I.Y., Lee, S.P., Kim, S.G., Gruetter, R., 2001. In vivo measurements of brain glucose transport using the reversible Michaelis–Menten model and simultaneous measurements of cerebral blood flow changes during hypoglycemia. J. Cereb. Blood Flow Metab. 21, 653–663.

Choi, I.Y., Lei, H., Gruetter, R., 2002. Effect of deep pentobarbital anesthesia on neurotransmitter metabolism in vivo: on the correlation of total glucose consumption with glutamatergic action. J. Cereb. Blood Flow Metab. 22, 1343–1351.

Choi, I.Y., Gruetter, R., 2003. In vivo ^{13}C NMR assessment of brain glycogen concentration and turnover in the awake rat. Neurochem. Int. 43, 317–322.

Choi, I.Y., Seaquist, E.R., Gruetter, R., 2003. Effect of hypoglycemia on brain glycogen metabolism in vivo. J. Neurosci. Res. 72, 25–32.

Contreras, M.A., Chang, M.C., Kirkby, D., Bell, J.M., Rapoport, S.I., 1999. Reduced palmitate turnover in brain phospholipids of pentobarbital-anesthetized rats. Neurochem. Res. 24, 833–841.

Cruz, F., Cerdan, S., 1999. Quantitative ^{13}C NMR studies of metabolic compartmentation in the adult mammalian brain. NMR Biomed. 12, 451–462.

Cruz, N.F., Dienel, G.A., 2002. High glycogen levels in brains of rats with minimal environmental stimuli: implications for metabolic contributions of working astrocytes. J. Cereb. Blood Flow Metab. 22, 1476–1489.

Daikhin, Y., Yudkoff, M., 2000. Compartmentation of brain glutamate metabolism in neurons and glia. J. Nutr. 130, 1026S–1031S.

Daniel, P.M., Love, E.R., Pratt, O.E., 1977. The influence of insulin upon the metabolism of glucose by the brain. Proc. R. Soc. Lond. B, Biol. Sci. 196, 85–104.

Dienel, G.A., Liu, K., Popp, D., Cruz, N.F., 1999. Enhanced acetate and glucose utilization during graded photic stimulation. Neuronal–glial interactions in vivo. Ann. N. Y. Acad. Sci. 893, 279–281.

Dienel, G.A., Wang, R.Y., Cruz, N.F., 2002. Generalized sensory stimulation of conscious rats increases labeling of oxidative pathways of glucose metabolism when the brain glucose–oxygen uptake ratio rises. J. Cereb. Blood Flow Metab. 22, 1490–1502.

Dringen, R., Hamprecht, B., 1992. Glucose, insulin, and insulin-like growth factor I regulate the glycogen content of astroglia-rich primary cultures. J. Neurochem. 58, 511–517.

Eriksson, G., Peterson, A., Iverfeldt, K., Walum, E., 1995. Sodium-dependent glutamate uptake as an activator of oxidative metabolism in primary astrocyte cultures from newborn rat. Glia 15, 152–156.

Fitzpatrick, S.M., Hetherington, H.P., Behar, K.L., Shulman, R.G., 1990. The flux from glucose to glutamate in the rat brain in vivo as determined by ^{1}H-observed, ^{13}C-edited NMR spectroscopy. J. Cereb. Blood Flow Metab. 10, 170–179.

Fox, P.T., Raichle, M.E., Mintun, M.A., Dence, C., 1988. Nonoxidative glucose consumption during focal physiologic neural activity. Science 241, 462–464.

Frahm, J., Kruger, G., Merboldt, K.D., Kleinschmidt, A., 1996. Dynamic uncoupling and recoupling of perfusion and oxidative metabolism during focal brain activation in man. Magn. Reson. Med. 35, 143–148.

Garcia-Martin, M.L., Garcia-Espinosa, M.A., Ballesteros, P., Bruix, M., Cerdan, S., 2002. Hydrogen turnover and subcellular compartmentation of hepatic [2-$^{(13)}$C]glutamate and [3-$^{(13)}$C]aspartate as detected by $^{(13)}$C NMR. J. Biol. Chem. 277, 7799–7807.

Gruetter, R., 1993. Automatic, localized in vivo adjustment of all first- and second-order shim coils. Magn. Reson. Med. 29, 804–811.

Gruetter, R., 2002. In vivo ^{13}C NMR studies of compartmentalized cerebral carbohydrate metabolism. Neurochem. Int. 41, 143–154.

Gruetter, R., Boesch, C., 1992. Fast, non-iterative shimming on spatially localized signals: in vivo analysis of the magnetic field along axes. J. Magn. Reson. 96, 323–334.

Gruetter, R., Novotny, E.J., Boulware, S.D., Mason, G.F., Rothman, D.L., Prichard, J.W., Shulman, R.G., 1994. Localized ^{13}C NMR spectroscopy of amino acid labeling from [1-^{13}C] D-glucose in the human brain. J. Neurochem. 63, 1377–1385.

Gruetter, R., Novotny, E.J., Boulware, S.D., Rothman, D.L., Mason, G.F., Shulman, G.I., Shulman, R.G., Tamborlane, W.V., 1992b. Direct measurement of brain glucose concentrations in humans by ^{13}C NMR spectorscopy. Proc. Natl Acad. Sci. USA 89, 1109–1112.

Gruetter, R., Seaquist, E.R., Ugurbil, K., 2001. A mathematical model of compartmentalized neurotransmitter metabolism in the human brain. Am. J. Physiol. Endocrinol. Metab. 281, E100–E112.

Gruetter, R., Rothman, D.L., Novotny, E.J., Shulman, R.G., 1992a. Localized ^{13}C NMR spectroscopy of *myo*-inositol in the human brain in vivo. Magn. Reson. Med. 25, 204–210.

Gruetter, R., Seaquist, E., Kim, S.-W., Ugurbil, K., 1998b. Localized *in vivo* ^{13}C NMR of glutamate metabolism. Initial results at 4 Tesla. Dev. Neurosci. 20, 380–388.

Gruetter, R., Ugurbil, K., Seaquist, E.R., 1998a. Steady-state cerebral glucose concentrations and transport in the human brain. J. Neurochem. 70, 397–408.

Hansson, E., Ronnback, L., 1994. Astroglial modulation of synaptic transmission. Perspect. Dev. Neurobiol. 2, 217–223.

Hassel, B., Bachelard, H., Jones, P., Fonnum, F., Sonnewald, U., 1997. Trafficking of amino acids between neurons and glia in vivo. Effects of inhibition of glial metabolism by fluoroacetate. J. Cereb. Blood Flow Metab. 17, 1230–1238.

Henry, P.G., Lebon, V., Vaufrey, F., Brouillet, E., Hantraye, P., Bloch, G., 2002. Decreased TCA cycle rate in the rat brain after acute 3-NP treatment measured by in vivo ^{1}H-[^{13}C] NMR spectroscopy. J. Neurochem. 82, 857–866.

Hertz, E., Shargool, M., Hertz, L., 1986. Effects of barbiturates on energy metabolism by cultured astrocytes and neurons in the presence of normal and elevated concentrations of potassium. Neuropharmacology 25, 533–539.

Hyder, F., Chase, J.R., Behar, K.L., Mason, G.F., Siddeek, M., Rothman, D.L., Shulman, R.G., 1996. Increased tricarboxylic acid cycle flux in rat brain during forepaw stimulation detected with H-1 [C-13] NMR. Proc. Natl Acad. Sci. USA 93, 7612–7617.

Hyder, F., Rothman, D.L., Mason, G.F., Rangarajan, A., Behar, K.L., Shulman, R.G., 1997. Oxidative glucose metabolism in rat brain during single forepaw stimulation: a spatially localized ^{1}H[^{13}C] nuclear magnetic resonance study. J. Cereb. Blood Flow Metab. 17, 1040–1047.

Jeffrey, F.M., Rajagopal, A., Malloy, C.R., Sherry, A.D., 1991. ^{13}C-NMR: a simple yet comprehensive method for analysis of intermediary metabolism. Trends Biochem. Sci. 16, 5–10.

Jeffrey, F.M., Reshetov, A., Storey, C.J., Carvalho, R.A., Sherry, A.D., Malloy, C.R., 1999. Use of a single $^{(13)}$C NMR resonance of glutamate for measuring oxygen consumption in tissue. Am. J. Physiol. 277, E1111–E1121.

Kanamori, K., Ross, B.D., 1995. Steady-state in vivo glutamate dehydrogenase activity in rat brain measured by ^{15}N NMR. J. Biol. Chem. 270, 24805–24809.

Kong, J., Shepel, P.N., Holden, C.P., Mackiewicz, M., Pack, A.I., Geiger, J.D., 2002. Brain glycogen decreases with increased periods of wakefulness: implications for homeostatic drive to sleep. J. Neurosci. 22, 5581–5587.

LaNoue, K.F., Tischler, M.E., 1974. Electrogenic characteristics of the mitochondrial glutamate–aspartate antiporter. J. Biol. Chem. 249, 7522–7528.

Lapidot, A., Gopher, A., 1994. Cerebral metabolic compertmentation. Estimation of glucose flux via pyruvate carboxylase/pyruvate dehydrogenase by ^{13}C NMR isotopomer analysis of [U-^{13}C] D-glucose metabolites. J. Biol. Chem. 269, 27198–27208.

Lebon, V., Petersen, K.F., Cline, G.W., Shen, J., Mason, G.F., Dufour, S., Behar, K.L., Shulman, G.I., Rothman, D.L., 2002. Astroglial contribution to brain energy metabolism in humans revealed by ^{13}C nuclear magnetic resonance spectroscopy: elucidation of the dominant pathway for neurotransmitter glutamate repletion and measurement of astrocytic oxidative metabolism. J. Neurosci. 22, 1523–1531.

Lieth, E., LaNoue, K.F., Berkich, D.A., Xu, B., Ratz, M., Taylor, C., Hutson, S.M., 2001. Nitrogen shuttling between neurons and glial cells during glutamate synthesis. J. Neurochem. 76, 1712–1723.

Lowry, O., Passonneau, J., Hasselberger, F., Schulz, D., 1964. Effect of ischemia on known substrates and cofactors of the glycolytic pathway in brain. J. Biol. Chem. 239, 18–30.

Lukkarinen, J., Oja, J.M., Turunen, M., Kauppinen, R.A., 1997. Quantitative determination of glutamate turnover by ^{1}H-observed, ^{13}C-edited nuclear magnetic resonance spectroscopy in the cerebral cortex ex vivo: interrelationships with oxygen consumption. Neurochem. Int. 31, 95–104.

Magistretti, P.J., Sorg, O., Yu, N., Martin, J.L., Pellerin, L., 1993. Neurotransmitters regulate energy metabolism in astrocytes: implications for the metabolic trafficking between neural cells. Dev. Neurosci. 15, 306–312.

Magistretti, P., Pellerin, L., 1996. Cellular mechanisms of brain energy metabolism. Relevance to functional brain imaging and to neurodegenerative disorders. Ann. N. Y. Acad. Sci., 777.

Magistretti, P.J., Pellerin, L., 1999. Cellular mechanisms of brain energy metabolism and their relevance to functional brain imaging. Philos. Trans. R. Soc. Lond. B, Biol. Sci. 354, 1155–1163.

Magistretti, P.J., Pellerin, L., Rothman, D.L., Shulman, R.G., 1999. Energy on demand. Science 283, 496–497.

Malloy, C.R., Sherry, A.D., Jeffrey, F.M., 1987. Carbon flux through citric acid cycle pathways in perfused heart by ^{13}C NMR spectroscopy. FEBS Lett. 212, 58–62.

Malloy, C., Sherry, A., Jeffrey, F., 1990. Analysis of tricarboxylic acid cycle of the heart using ^{13}C isotope isomers. Am. J. Physiol. 259, H987–H995.

Martin, M., Portais, J.C., Labouesse, J., Canioni, P., Merle, M., 1993. [1-^{13}C]glucose metabolism in rat cerebellar granule cells and astrocytes in primary culture. Evaluation of flux parameters by ^{13}C- and ^{1}H-NMR spectroscopy. Eur. J. Biochem. 217, 617–625.

Martinez-Hernandez, A., Bell, K.P., Norenberg, M.D., 1976. Glutamine synthetase: glial localization in brain. Science 195, 1356–1358.

Mason, G.F., Rothman, D.L., Behar, K.L., Shulman, R.G., 1992. NMR determination of the TCA cycle rate and alpha-ketoglutarate/glutamate exchange rate in rat brain. J. Cereb. Blood Flow Metab. 12, 434–447.

Mason, G.F., Gruetter, R., Rothman, D.L., Behar, K.L., Shulman, R.G., Novotny, E.J., 1995. Simultaneous determination of the rates of the TCA cycle, glucose utilization, α-ketoglutarate/glutamate exchange, and glutamine synthesis in human brain by NMR. J. Cereb. Blood Flow Metab. 15, 12–25.

Nelson, S.R., Schulz, D.W., Passonneau, J.V., Lowry, O.H., 1968. Control of glycogen levels in brain. J. Neurochem. 15, 1271–1279.

Ogawa, S., Menon, R.S., Kim, S.G., Ugurbil, K., 1998. On the characteristics of functional magnetic resonance imaging of the brain. Annu. Rev. Biophys. Biomol. Struct. 27, 447–474.

Ottersen, O., Zhang, N., Walberg, F., 1992. Metabolic compartmentation of glutamate and glutamine: morpholocial evidence obtained by quantitative immunocytochemistry in rat cerebellum. Neuroscience 46, 519–534.

Oz, G., Henry, P.G., Seaquist, E.R., Gruetter, R., 2003. Direct, noninvasive measurement of brain glycogen metabolism in humans. Neurochem. Int. 43, 323–329.

Pan, J.W., Stein, D.T., Telang, F., Lee, J.H., Shen, J., Brown, P., Cline, G., Mason, G.F., Shulman, G.I., Rothman, D.L., Hetherington, H.P., 2000. Spectroscopic imaging of glutamate C4 turnover in human brain. Magn. Reson. Med. 44, 673–679.

Petit, J.M., Tobler, I., Allaman, I., Borbely, A.A., Magistretti, P.J., 2002. Sleep deprivation modulates brain mRNAs encoding genes of glycogen metabolism. Eur. J. Neurosci. 16, 1163–1167.

Pfeuffer, J., Tkac, I., Choi, I.-Y., Merkle, H., Ugurbil, K., Garwood, M., Gruetter, R., 1999. Localized in vivo ^{1}H NMR detection of neurotransmitter labeling in rat brain during infusion of [1-^{13}C] D-glucose. Magn. Reson. Med. 41, 1077–1083.

Prichard, J., Rothman, D., Novotny, E., Petroff, O., Kuwabara, T., Avison, M., Howseman, A., Hanstock, C., Shulman, R., 1991. Lactate rise detected by ^{1}H NMR in human visual cortex during physiologic stimulation. Proc. Natl Acad. Sci. USA 88, 5829–5831.

Qu, H., Waagepetersen, H.S., van Hengel, M., Wolt, S., Dale, O., Unsgard, G., Sletvold, O., Schousboe, A., Sonnewald, U., 2000. Effects of thiopental on transport and metabolism of glutamate in cultured cerebellar granule neurons. Neurochem. Int. 37, 207–215.

Rothman, D.L., Novotny, E.J., Shulman, G.I., Howseman, A.M., Petroff, O.A., Mason, G., Nixon, T., Hanstock, C.C., Prichard, J.W., Shulman, R.G., 1992. ^{1}H-[^{13}C] NMR measurements of [4-^{13}C]glutamate turnover in human brain. Proc. Natl Acad. Sci. USA 89, 9603–9606.

Rothman, D.L., Sibson, N.R., Hyder, F., Shen, J., Behar, K.L., Shulman, R.G., 1999. In vivo nuclear magnetic resonance spectroscopy studies of the relationship between the glutamate-glutamine neurotransmitter cycle and functional neuroenergetics. Philos. Trans. R. Soc. Lond. B, Biol. Sci. 354, 1165–1177.

Rothman, D.L., 2001. Studies of metabolic compartmentation and glucose transport using in vivo MRS. NMR Biomed. 14, 149–160.

Rothman, D.L., Behar, K.L., Hyder, F., Shulman, R.G., 2003. In vivo NMR studies of the glutamate neurotransmitter flux and neuroenergetics: implications for brain function. Annu. Rev. Physiol. 65, 401–427.

Sagar, S.M., Sharp, F.R., Swanson, R.A., 1987. The regional distribution of glycogen in rat brain fixed by microwave irradiation. Brain Res. 417, 172–174.

Sappey-Marinier, D., Calabrese, G., Fein, G., Hugg, J.W., Biggins, C., Weiner, M.W., 1992. Effect of photic stimulation on human visual cortex lactate and phosphates using ^{1}H and ^{31}P magnetic resonance spectroscopy. J. Cereb. Blood Flow Metab. 12, 584–592.

Schousboe, A., Westergaard, N., Hertz, L., 1993a. Neuronal–astrocytic interactions in glutamate metabolism. Biochem. Soc. Trans. 21, 49–53.

Schousboe, A., Westergaard, N., Sonnewald, U., Petersen, S.B., Huang, R., Peng, L., Hertz, L., 1993b. Glutamate and glutamine metabolism and compartmentation in astrocytes. Dev. Neurosci. 15, 359–366.

Seaquist, E.R., Damberg, G.S., Tkac, I., Gruetter, R., 2001. The effect of insulin on in vivo cerebral glucose concentrations and rates of glucose transport/metabolism in humans. Diabetes 50, 2203–2209.

Shank, R., Aprison, M., 1979. Biochemical aspects of the neurotransmitter function of glutamate. In: al, F.L.e. (Ed.), Glutamic Acid: Advances in Biochemistry and Physiology. Raven Press, New York, pp. 139–150.

Shank, R.P., Bennett, G.S., Freytag, S.O., Campbell, G.L., 1985. Pyruvate carboxylase: an astrocyte-specific enzyme implicated in the replenishment of amino acid neurotransmitter pools. Brain Res. 329, 364–367.

Shank, R.P., Leo, G.C., Zielke, H.R., 1993. Cerebral metabolic compartmentation as revealed by nuclear magnetic resonance analysis of D-[1-^{13}C]glucose metabolism. J. Neurochem. 61, 315–323.

Shen, J., Rothman, D.L., 2002. Magnetic resonance spectroscopic approaches to studying neuronal: glial interactions. Biol. Psychiatry 52, 694–700.

Shen, J., Petersen, K.F., Behar, K.L., Brown, P., Nixon, T.W., Mason, G.F., Petroff, O.A., Shulman, G.I., Shulman, R.G., Rothman, D.L., 1999. Determination of the rate of the glutamate/glutamine cycle in the human brain by in vivo ^{13}C NMR. Proc. Natl Acad. Sci. USA 96, 8235–8240.

Sherry, A.D., Zhao, P., Wiethoff, A.J., Jeffrey, F.M., Malloy, C.R., 1998. Effects of aminooxyacetate on glutamate compartmentation and TCA cycle kinetics in rat hearts. Am. J. Physiol. 274, H591–H599.

Shulman, R.G., Hyder, F., Rothman, D.L., 2001a. Lactate efflux and the neuroenergetic basis of brain function. NMR Biomed. 14, 389–396.

Shulman, R.G., Hyder, F., Rothman, D.L., 2001b. Cerebral energetics and the glycogen shunt: neurochemical basis of functional imaging. Proc. Natl Acad. Sci. USA 98, 6417–6422.

Shupliakov, O., Ottersen, O.P., Stormmathisen, J., Brodin, L., 1997. Glial and neuronal glutamine pools at glutamatergic synapses with distinct properties. Neuroscience 77, 1201–1212.

Sibson, N.R., Dhankhar, A., Mason, G.F., Rothman, D.L., Behar, K.L., Shulman, R.G., 1998. Stoichiometric coupling of brain glucose metabolism and glutamatergic neuronal activity. Proc. Natl Acad. Sci. USA 95, 316–321.

Sibson, N.R., Mason, G.F., Shen, J., Cline, G.W., Herskovits, A.Z., Wall, J.E., Behar, K.L., Rothman, D.L., Shulman, R.G., 2001. In vivo $^{(13)}$C NMR measurement of neurotransmitter glutamate cycling, anaplerosis and TCA cycle flux in rat brain during. J. Neurochem. 76, 975–989.
Silver, I.A., Erecinska, M., 1997. Energetic demands of the Na^+/K^+ ATPase in mammalian astrocytes. Glia 21, 35–45.
Sokoloff, L., Reivich, M., Kennedy, C., Des Rosiers, M.H., Patlak, C.S., Pettigrew, K.D., Sakurada, O., Shinohara, M., 1977. The [^{14}C]deoxyglucose method for the measurement of local cerebral glucose utilization: theory, procedure, and normal values in the conscious and anesthetized albino rat. J. Neurochem. 28, 897–916.
Sonnewald, U., Gribbestad, I.S., Westergaard, N., Nilsen, G., Unsgard, G., Schousboe, A., Petersen, S.B., 1994. Nuclear magnetic resonance spectroscopy: biochemical evaluation of brain function in vivo and in vitro. Neurotoxicology 15, 579–590.
Sorg, O., Magistretti, P.J., 1992. Vasoactive intestinal peptide and noradrenaline exert long-term control on glycogen levels in astrocytes: blockade by protein synthesis inhibition. J. Neurosci. 12, 4923–4931.
Swanson, R.A., 1992. Physiologic coupling of glial glycogen metabolism to neuronal activity in brain. Can. J. Physiol. Pharmacol. 70, S138–S144.
Swanson, R.A., Choi, D.W., 1993. Glial glycogen stores affect neuronal survival during glucose deprivation in vitro. J. Cereb. Blood Flow Metab. 13, 162–169.
Swanson, R.A., Sagar, S.M., Sharp, F.R., 1989. Regional brain glycogen stores and metabolism during complete global ischaemia. Neurol. Res. 11, 24–28.
Swanson, R.A., Seid, L.L., 1998. Barbiturates impair astrocyte glutamate uptake. Glia 24, 365–371.
Tsacopoulos, M., Magistretti, P., 1996. Metabolic coupling between glia and neurons. J. Neurosci. 16, 877–885.
Van den Berg, C., 1973. A model of compartmentation in mouse brain based on glucose and acetate metabolism. In: Balazs, E., Cremer, J. (Eds.), Metabolic compartmentation in the brain. MacMillan, London, pp. 137–166.
Vernadakis, A., 1996. Glia–neuron intercommunications and synaptic plasticity. Prog. Neurobiol. 49, 185–214.
Waniewski, R.A., Martin, D.L., 1998. Preferential utilization of acetate by astrocytes is attributable to transport. J. Neurosci. 18, 5225–5233.
Wender, R., Brown, A.M., Fern, R., Swanson, R.A., Farrell, K., Ransom, B.R., 2000. Astrocytic glycogen influences axon function and survival during glucose deprivation in central white matter. J. Neurosci. 20, 6804–6810.
Westergaard, N., Sonnewald, U., Schousboe, A., 1995. Metabolic trafficking between neurons and astrocytes: the glutamate/glutamine cycle revisited. Dev. Neurosci. 17, 203–211.
Yu, X., Alpert, N.M., Lewandowski, E.D., 1997. Modeling enrichment kinetics from dynamic ^{13}C NMR spectra: theoretical analysis and practical considerations. Am. J. Physiol. 41, C2037–C2048.
Yu, A.C., Hertz, E., Hertz, L., 1983. Effects of barbiturates on energy and intermediary metabolism in cultured astrocytes. Prog. Neuropsychopharmacol. Biol. Psychiatry 7, 691–696.
Yudkoff, M., Nissim, I., Daikhin, Y., Lin, Z., Nelson, D., Pleasure, D., Erecinska, M., 1993. Brain glutamate metabolism: neuronal–astroglial relationships. Dev. Neurosci. 15, 343–350.
Yudkoff, M., Nissim, I., Pleasure, D., 1988. Astrocyte metabolism of [^{15}N]glutamine: implications for the glutamine–glutamate cycle. J. Neurochem. 51, 843–850.
Zigmond, M.J., 1999. Fundamental Neuroscience. Academic Press, San Diego, pp. 402–412.

Advances in
Molecular and
Cell Biology

Ion, transmitter and drug effects on energy metabolism in astrocytes

Leif Hertz,[a,*] Liang Peng,[a] Christel C. Kjeldsen,[b]
Brona S. O'Dowd[c] and Gerald A. Dienel[d]

[a]*College of Basic Medical Sciences China Medical University, Heping District, Shenyang, 110001, P.R. China*
[*]*Correspondence address: RR 2, Box 245, Gilmour, Ont., Canada K0L 1W0*
E-mail: lhertz@northcom.net
[b]*Centre of Psychiatry, Copenhagen County, Denmark*
[c]*Centre for Magnetic Resonance, University of Queensland, Gehrmann Laboratories, St. Lucia, Queensland 4072, Australia*
[d]*Department of Neurology, University of Arkansas for Medical Sciences, Little Rock, AR 72205, USA*

Contents

1. Introduction
2. Mechanisms of metabolic stimulation
   2.1. Increase in ADP/ATP ratio
   2.2. Increase in intramitochondrial Ca^{2+} concentration
   2.3. Glycogen as a re-chargeable energy reservoir
3. Ion effects on metabolism
   3.1. The potassium ion (K^+)
   3.2. The sodium ion (Na^+)
4. Transmitter effects on metabolism
   4.1. Second messenger systems
   4.2. Individual transmitters
5. Drug effects on metabolism
   5.1. Interaction with ion effects
   5.2. Interaction with transmitter effects
6. Concluding remarks

Accumulation of K^+ in astrocytes by stimulation of the extracellular K^+-sensitive site of the Na^+, K^+-ATPase and by activation of the Na^+, K^+, $2Cl^-$ cotransporter plays a major role in ion homeostasis and therefore in energy metabolism in the CNS. Na^+-driven uptake of glutamate also contributes to stimulation of astrocytic energy metabolism by activating the intracellular Na^+-dependent site of the Na^+, K^+-ATPase, but the energy expenditure for glutamate uptake is probably quantitatively less important, and it may be

Advances in Molecular and Cell Biology, Vol. 31, pages 435–460

ISBN: 0-444-51451-1

met by oxidative degradation of accumulated glutamate. Several transmitters affect energy metabolism in astrocytes, by Ca^{2+}-mediated stimulation of mitochondrial dehydrogenases and glutaminase and by activation of the breakdown of glycogen, likely to serve as a rechargeable energy substrate in peripheral processess of the astrocytes, which are too thin to contain any mitochondria. Certain drugs classically assumed to act solely on neurons owe a substantial part of their effect to interactions with effects of ions or transmitters on astrocytes, thereby altering energy metabolism in astrocytes and thus in the brain.

1. Introduction

With the recent establishment that the metabolism of cortical astrocytes accounts for a sizeable fraction of energy metabolism, including oxidative metabolism, in the brain in situ (see chapters by Gruetter and by Håberg and Sonnewald) comes the question to what extent astrocyte metabolism is regulated during brain activity. As of yet, this has not been determined in the brain in situ, although it is known that oxidative metabolism in astrocytes in the rat brain is substantially increased during spreading depression (Dienel et al., 2001), a situation in which the extracellular potassium concentration ($[K^+]_e$) is greatly increased (Vyskocil et al., 1972; Somjen, 1979). This correlation is probably not co-incidental, since there is now overwhelming evidence that regulation of $[K^+]_e$ homeostasis in the brain depends heavily on an initial active, and thus energy-consuming, uptake of K^+ into astrocytes (see chapter by Walz). In addition, it is well established that transmitter glutamate is predominantly accumulated by astrocytes, in co-transport with sodium ions (Na^+), which subsequently must be extruded by active transport (see chapter by Schousboe and Waagepetersen). This represents another, although quantitatively probably less prominent, stimulus of energy metabolism. Evidence has also been obtained in cell culture experiments that transmitters like noradrenaline increase glycolysis (Subbarao and Hertz, 1991; Magistretti et al., 1993) and oxidative metabolism (Subbarao and Hertz, 1991) in astrocytes, but have no corresponding effect in neurons (Subbarao and Hertz, 1990a). The stimulation of oxidative metabolism is probably secondary to an increase in intramitochondrial calcium (Ca^{2+}), which is secondary to a transmitter-induced increase in free cytosolic Ca^{2+} ($[Ca^{2+}]_i$). That a similar stimulation may occur also in the brain in situ is suggested by the repeated observation that administration of α-adrenergic antagonists reduce glucose metabolism, measured by the autoradiographic 2-deoxy-D-[^{14}C]glucose technique, in most brain areas of the rat brain (Savaki et al., 1982; Inoue et al., 1991; French et al., 1995). It is also well established that many transmitters as well as elevated $[K^+]_e$ stimulate breakdown of glycogen (glycogenolysis) in brain and in astrocytes (Hof et al., 1988; Subbarao and Hertz, 1990b; Subbarao et al., 1995; Waagepetersen et al., 2000). In the present chapter we will describe effects of individual ions and transmitters known to stimulate energy metabolism in astrocytes, after we have discussed the general mechanisms involved and the potential role of glycogen metabolism during brain activation. Effects of some drugs, which may interact with ion and transmitter effects on brain metabolism will also be included.

2. Mechanisms of metabolic stimulation

2.1. Increase in ADP/ATP ratio

In neurons an elevated intracellular Na^+ concentration following excitation is a major stimulus of Na^+,K^+-ATPase activity and thus of glucose metabolism due to the resulting decrease in ATP and concomitant increase in ADP and AMP, which stimulate oxidative phosphorylation and phosphofructokinase (PFK) activity and thus glycolysis (see chapter by Roberts and Chih; Hertz and Dienel, 2002). Since normal astrocytes are non-excitable cells, it can a priori be excluded that a similar excitation-induced increase in intracellular Na^+ can provide a metabolic stimulus. This does, however, not exclude that post-excitatory stimulation of Na^+,K^+-ATPase activity might enhance energy metabolism in astrocytes, since the extracellular K^+ concentration ($[K^+]_e$) in the brain increases as a result of brain activation (see chapter by Walz). Increased $[K^+]_e$ can stimulate the extracellular, K^+-sensitive site of the Na^+,K^+-ATPase, because the affinity of the enzyme for K^+ is low enough that maximum activity is not achieved at resting $[K^+]_e$. Moreover, the intracellular Na^+ concentration in astrocytes increases not only during Na^+-coupled uptake of glutamate (and other amino acids) but also during K^+ uptake by a co-transporter, mediating coupled uptake of K^+, Na^+ and $2Cl^-$ (see chapter by Walz). Thus, as illustrated in Fig. 1, elevated $[K^+]_e$ stimulates energy metabolism (oxidative metabolism as well as glycolysis) via an increase in ADP/ATP ratio, both by a direct stimulation of the extracellular K^+-sensitive site of the Na^+,K^+-ATPase and by enhancing co-transporter activity and thus uptake of Na^+, which secondarily stimulates the Na^+,K^+-ATPase at its intracellular the Na^+-sensitive site. This figure also shows that glutamate-induced uptake of glutamate together with Na^+ similarly stimulates energy metabolism by an effect of the accumulated Na^+ on the intracellular site of the Na^+,K^+-ATPase.

In addition, ADP-mediated stimulation of energy metabolism can be evoked by transmitters which either activate Na^+,K^+-ATPase activity (Mercado and Hernandez, 1992; Hajek et al., 1996) or stimulate an energy-requiring process, such as cellular uptake of glutamate in conjunction with Na^+ (Hansson and Rönnbäck, 1992; Alexander et al., 1997), stimulating the Na^+,K^+-ATPase at its intracellular Na^+-stimulated site. Transmitters have long been known to stimulate glycogenolysis in brain, and since virtually all glycogen in brain is located in astrocytes (Ibrahim, 1975), glycogenolysis must also be an astrocytic effect (Fig. 1). The functional importance of stimulation of glycogenolysis is indicated by the observation that glycogen turnover, i.e., both glycogenolysis and re-synthesis of glycogen are enhanced during brain activity (Swanson et al., 1992; Dienel and Cruz, 2003). Thus, glycogenolysis and subsequent re-synthesis of glycogen may possibly account for a sizeable fraction of energy metabolism in brain during and after enhanced activity. This conclusion raises the question what role glycogen turnover plays during brain activation.

2.2. Increase in intramitochondrial Ca^{2+} concentration

A decrease in ATP and increase in ADP levels are not the only stimuli regulating energy metabolism in mammalian cells. During the last 10–15 years, studies in other

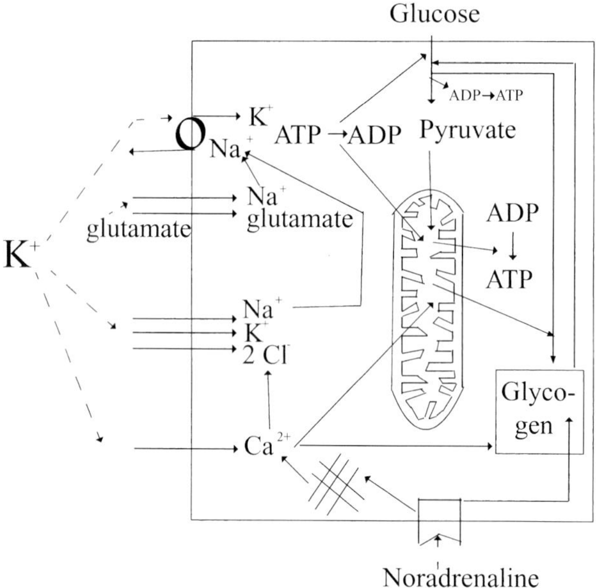

Fig. 1. Schematic illustration of energy-requiring and energy-yielding reactions in astrocytes. Even slightly elevated extracellular K^+ concentrations ($[K^+]_e$) stimulate the extracellular, K^+-sensitive site of the Na^+,K^+-ATPase, stimulating active uptake of K^+, as well as of the co-transporter, accumulating jointly Na^+,K^+ and $2Cl^-$ and at slightly higher concentrations they open a voltage sensitive L-channel for Ca^{2+}, allowing Ca^{2+} entry into the cell. Extracellular glutamate stimulates uptake of glutamate, co-transported with Na^+, which subsequently stimulates the intracellular, the Na^+-sensitive site of the Na^+,K^+-ATPase, as does Na^+ accumulated by the co-transporter. Stimulation of Na^+,K^+-ATPase activity leads to conversion of ATP to ADP, and the altered ATP/ADP ratio stimulates both cytosolic glycolysis and mitochondrial oxidative metabolism. In addition increased intracellular Ca^{2+} ($[Ca^{2+}]_i$) stimulates oxidative metabolism as well as glycogenolysis. Simultaneously or slightly later glycogen is re-synthesized from glucose, partly fueled by oxidatively derived energy. The transmitter noradrenaline stimulates glycogenolysis by a β-adrenergic effect and glycolysis by an α-adrenergic effect (not shown). It also causes a release of bound Ca^{2+} from the endoplasmic reticulum by an α-adrenergic effect, leading to an increase in $[Ca^{2+}]_i$ and subsequent increase in glycogenolysis. Thus CNS activation, including the establishment of memory, which is characterized by increases in $[K^+]_e$ and extracellular glutamate and the release of noradrenaline evokes a concerted increase in energy utilization and energy production by glycolysis, oxidative metabolism and glycogenolysis in astrocytes.

tissues than brain have shown that an increase in free intramitochondrial Ca^{2+}, secondary to a rise in free cytosolic Ca^{2+} concentration, $[Ca^{2+}]_i$, within seconds causes a direct stimulation of the mitochondrial dehydrogenases, pyruvate dehydrogenase, isocitrate dehydrogenase, and α-ketoglutarate dehydrogenase (McCormack and Denton, 1990; Rutter et al., 1996), as well as of glutaminase activity (Halestrap, 1989) and of oxidative phosphorylation (Robb-Gaspers et al., 1998). It is highly likely that a similar response occurs in astrocytes, since astrocytic $[Ca^{2+}]_i$ is increased by a multitude of transmitters (see chapter by Hansson and Rönnbäck), including ATP (Peuchen et al., 1996), and noradrenaline (Fig. 1), which cause a release of Ca^{2+} from the endoplasmic reticulum. It is in agreement with this concept that noradrenaline stimulates pyruvate dehydrogenation (Hertz and Peng, 1992; Chen and Hertz, 1999), α-ketoglutarate dehydrogenation

(Subbarao and Hertz, 1991) and hydrolysis of glutamine to glutamate (Huang and Hertz, 1995) in mouse astrocytes in primary cultures, and that ATP depolarizes the mitochondrial membrane potential, which may reflect entry of Ca^{2+} into the mitochondria (Peuchen et al., 1996).

Entry of extracellular Ca^{2+} can also lead to an increase in $[Ca^{2+}]_i$ in astrocytes during exposure to elevated $[K^+]_e$ (Fig. 1), an effect which is due to opening of voltage-sensitive Ca^{2+} channels (Hertz et al., 1989; Duffy and MacVicar, 1994; Zhao et al., 1996; Thorlin et al., 1998). An influx of Ca^{2+} occurs also during mechanical stimulation, probably due to opening of stretch-activated Ca^{2+} channels (e.g., Peuchen et al., 1996). Although it cannot a priori be assumed that increases in $[Ca^{2+}]_i$ evoked by Ca^{2+} entry and by release from the endoplasmic reticulum exert similar effect (since the subcellular localization differ—see also chapters by Cornell-Bell et al. and by Scapagnini et al.), mechanical stimulation of astrocytes in primary cultures has been shown to cause mitochondrial depolarization (Peuchen et al., 1996). This observation is analogous to the finding that opening of voltage-sensitive Ca^{2+} channels triggers an increase in free intramitochondrial Ca^{2+} in epithelial cells (Lawrie et al., 1996). Opening of voltage-activated Ca^{2+} channels by exposure to elevated $[K^+]_e$ within the range occurring during neuronal stimulation (Subbarao et al., 1995) also causes glycogenolysis, accentuating the question what the functional role of glycogenolysis is in the activated brain.

2.3. Glycogen as a re-chargeable energy reservoir

Glycogenolysis may rapidly provide a large amount of energy. Recent experiments have shown that glycogen is present in the resting brain at a higher concentration than previously realized, i.e., ~10 μmol/glucose equivalent per g wet wt (Cruz and Dienel, 2002; Kong et al., 2002b; Dienel and Cruz, 2003). It is often believed that glycogen storage serves the purpose of providing the brain with glucose equivalents for use during failure of glucose delivery. In the rat the 'resting' rate of glucose utilization is 0.7 μmol/min per g wet wt (Sokoloff et al., 1977), meaning that all brain glycogen would be depleted within 15 min, if it were to replace glucose (see, however, chapter by Gruetter for a different conclusion). Moreover, in most cases glucose deprivation is combined with deprivation of O_2, increasing the rate of glucose utilization in order to provide sufficient energy by glycolysis alone, so that glycogen stores would last even shorter. However, recent experiments have indicated that glycogen metabolism is much more dynamic than would be expected from an emergency store, as reflected by the observation that both glycogenolysis and glycogen re-synthesis are stimulated during brain activation (Swanson et al., 1992; Dienel et al., 2002; Dienel and Cruz, 2003). Glycogen is also broken down rapidly and subsequently re-synthesized at specific stages of a one-trial avoidance learning task in day-old chicks, i.e., immediately after training (Fig. 2) and again around 55 min after training (O'Dowd et al., 1994). Memory consolidation in this animal model is an energy requiring process, characterized by release of K^+, glutamate and noradrenaline at specific time points (Hertz et al., 1996). As illustrated in the figure, the degradation of glycogen can occur very rapidly, probably exceeding 1 μmol/min per g wet wt of the brain or, with astrocytes constituting less than 20% of brain cortical volume (see chapter by

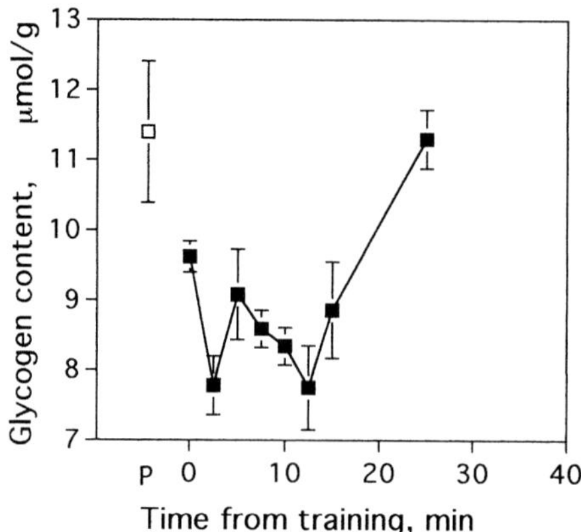

Fig. 2. Content of glycogen in the forebrain of day old chicks after pretraining, involving pecking at a water-coated bead (open square) and representative of the non-aversively trained animal, and after one-trial aversive training, involving pecking at a red aversively tasting bead and a blue neutral bead (filled-in squares, with the training performed seconds before the first point). The aversive training induces a rapid decline in glycogen content (~4 µmol/g wet wt within 2.5 min), which is statistically significant between 0 and 2.5 min, followed by a maintained reduction in glycogen level between 2.5 and 12.5 min (possibly with some oscillation of the level) and complete restoration of aversive-training level of glycogen between 12.5 and 25 min. Not shown in the figure is that there is a second decline in glycogen level around 55 min post-training, likewise followed by complete recovery during a 10 min period.

Wolff and Chao), more than 5 µmol/min per g wet wt of astrocytes. If this value applies to the rat brain, and if astrocytes metabolize glucose at the same rate as the average brain, this value is seven times higher than the rate of glucose utilization (0.7 µmol/min per g wet wt in the rat brain).

Storage of glucose as glycogen is an energy requiring process. Synthesis of glycogen requires three high energy phosphate equivalents (one ATP, one UTP, one from cleavage of pyrophosphate) for each glucose equivalent incorporated. However, glycogenolysis provides the advantage that less ATP is required to 'prime' glucose equivalents in glycogen than glucose for further metabolism, when there is an abrupt demand for energy. Metabolism of glucose to pyruvate via the glycolytic pathway requires an 'investment' of two ATP (to form glucose-6-phosphate and fructose-1,6-bisphosphate), and it yields four ATP for a net production of two ATP. Degradation of glycogen does not require the initial ATP to convert glucose to glucose-6-P, so the net, immediately available energy yield is three ATP, disregarding the ATP consumed in the synthesis of glycogen. If the ATP consumption required for synthesis of glycogen is also taken into account, there is an added cost to degrading glucose via glycogen, which is subsequently re-synthesized. It is therefore likely that activity-induced turnover of glycogen rather than direct utilization of glucose as a metabolic fuel provides an unknown metabolic advantage to astrocytes. Evidence will be presented, suggesting that this advantage may be that glycogen can serve at a re-chargeable energy store in a spatio-temporal manner, i.e., provide energy at specific locations (where oxidatively derived energy may be less readily available) at specific times (when the demand for energy is high), and then subsequently be 're-charged' when there is no urgent demand for energy. Due to the rapidity of the breakdown of glycogen and the fact that three molecules of ATP are generated per molecule glucose equivalent degraded glycolytically, compared to two molecules of ATP per molecule glucose, glycogenolysis may be able to supply more ATP more rapidly.

The specific location where glycogen is most likely to be of special importance as an energy reservoir is in the most peripheral parts of the astrocytic processes (or filopodia and lamellopodia), which form a specific functional compartment, which accounts for 80% of the total cell surface (see chapters by Derouiche and by Wolff and Chao). Ultrastructural analysis has demonstrated a widespread distribution of glycogen in astrocytic processes throughout the neuropil, and glycogen particles are small enough, with diameters ranging between 10 and 40 nm, depending on staining method (Cataldo and Broadwell, 1986; Wender et al., 2000) to reside within the 50–100 nm wide lamellae of even the most peripheral astrocytic processes (Peters et al., 1991). These domains show extensive immunostaining for Cx43, the major astrocytic gap junction protein, and the abundant localization of Cx43 partly reflects the formation of autaptic gap junctions onto other fine processes or onto major branches of the same astrocyte (Rohlmann and Wolff, 1996; Wolff et al., 1998—see also chapter by Wolff and Chao). It is likely that glycogenolysis can spread both intra- and intercellularly through gap junctions (Cruz et al., 1999; Dienel and Cruz, 2003), since gap junctional communication has been found to coordinate vasopressin-induced glycogenolysis in rat hepatocytes (Eugenin et al., 1998). Immunoreactivity for the glycogen degrading enzyme glycogen phosphorylase has similarly shown that the enzyme is localized mainly in astrocytes, with diffuse staining throughout the cytoplasm and processes ensheathing capillaries, and in the fine processes and lamellae adjacent to synaptic structures (Richter et al., 1996). Since mitochondria are far too large to be located here, oxidatively generated energy can only reach the peripheral astrocytic regions by diffusion of ATP and/or phosphocreatine. This process may be too slow to keep pace with rapidly developing energy demand, e.g., for uptake of neuronally released K^+, which accordingly may be responded to by immediate glycogenolysis, as will be discussed below.

Subsequently, the energy generated by complete oxidative degradation of pyruvate may partly be utilized for regeneration of glycogen at times when the demand for glucose degradation is less acute. For efficient operation of such a 'glycogen shunt' it would be a prerequisite to fuel at least part of the energy demands for glycogen synthesis by oxidatively derived energy. Type I hexokinase, the enzyme converting glucose to glucose-6-phosphate (the first metabolic reaction in both glycolysis and glycogen synthesis), has characteristics consistent with this requirement, since it binds reversibly to mitochondria and, as illustrated in Fig. 1, selectively utilizes oxidatively generated ATP to fuel the phosphorylation of glucose (BeltrandelRio and Wilson, 1992; de Cerqueira Cesar and Wilson, 1995).

3. Ion effects on metabolism

3.1. The potassium ion (K^+)

3.1.1. Carrier-mediated effects

Neuronal excitation is accompanied by K^+ release, and $[K^+]_e$ in brain increases from its resting level (3 mM) during neuronal excitation, up to a maximum of 12 mM during seizures and intense stimulation (see chapter by Walz). An even larger, and completely reversibly increase in $[K^+]_e$ (to >50 mM) occurs during energy deprivation and 'spreading depression', a peculiar phenomenon during which repetitive waves of inhibition of electrical activity, triggered by local application of glutamate, high $[K^+]_e$, or by strong mechanical stimulation, spreads across the cerebral cortex (Somjen, 2001). As discussed above, excess K^+ is subsequently cleared from the extracellular space, initially mainly into astrocytes, partly by active, Na^+,K^+-ATPase-mediated K^+ uptake into astrocytes, triggered by the increased $[K^+]_e$, and partly by activation of the Na^+, K^+, Cl^- co-transporter (Fig. 1)

Any elevation of the concentration of $[K^+]_e$ above its resting level causes a stimulation of Na^+,K^+-ATPase activity at its extracellular K^+-sensitive site in cultured astrocytes and in astrocytes isolated from mature brain by gradient centrifugation, but not in corresponding preparations of neurons (e.g., Henn et al., 1972; Grisar et al., 1979; Hajek et al., 1996). In cultured astrocytes maximum Na^+,K^+-ATPase activity is reached at a $[K^+]_e$ of ~ 12 mM, the enzyme activity follows Michaelis–Menten kinetics with a K_m of 1.9 mM for K^+, and the maximum activity (V_{max}) is higher than in neurons (Hajek et al., 1996). In cultured neurons and synaptosomes (Kimelberg et al., 1978) the enzyme has a 3- to 5-fold higher affinity for K^+ compared to astrocytes, and it is therefore operating at maximal velocity (substrate level $\gg K_m$) at resting $[K^+]_e$, and it is accordingly not stimulated by above-normal $[K^+]_e$. Therefore, at elevated $[K^+]_e$, K^+ uptake will mainly occur in astrocytes, whereas neuronal re-accumulation may be favored at normal or lowered $[K^+]_e$ due to the higher affinity of the neuronal Na^+,K^+-ATPase. In addition, neuronal K^+ accumulation will obviously be activated by stimulation of Na^+,K^+-ATPase activity during extrusion of excess Na^+, accumulated as a result of the action potential, but this process may be slower than the re-establishment of a resting $[K^+]_e$.

Consistent with the stimulation of Na^+,K^+-ATPase in astrocytes only by elevated $[K^+]_e$ (as long as the elevation is not high enough to lead to neuronal excitation) an increase of $[K^+]_e$ from 5 to 12 mM increases glucose phosphorylation in mouse astrocytes in primary cultures and in neuronal-astrocytic co-cultures from the rat by 25–50% (Fig. 3), whereas 12 mM $[K^+]_e$ is not sufficient to elicit an action potential and does not enhance glucose metabolism in neurons in primary cultures (Peng et al., 1994, 1996; Huang et al., 1994; Honneger and Pardo, 1999). The metabolic stimulation in astrocytes is almost completely inhibited by ouabain, an inhibitor of Na^+,K^+-ATPase activity.

Since K^+ release during neuronal stimulation causes a much larger relative increase in $[K^+]_e$ than in intracellular K^+ concentration (due to the smaller volume and the lower concentration), and neuronal activity is sensitive to the membrane potential, which is mainly determined by the ratio between extra- and intracellular K^+ (see chapter by Walz), an intense astrocytic K^+ accumulation may primarily serve the purpose of restoring $[K]_e$, and it may be most active at the most peripheral processes in the neuropil. In accordance with the concept that the energy supply for urgent metabolic processes in distant astrocytic processes partly may be met by glycogenolysis, it has been shown by Raffin et al. (1992) that iodoacetate, an inhibitor of glycolysis and glycogenolysis, affects the clearance of $[K^+]_e$ resulting from neuronal stimulation in the rat brain in vivo. As can be seen from Fig. 4, the halflife for the first half of K^+ clearance from its peak level towards baseline was only slightly, and non-significantly, inhibited by hypoxia, whereas it

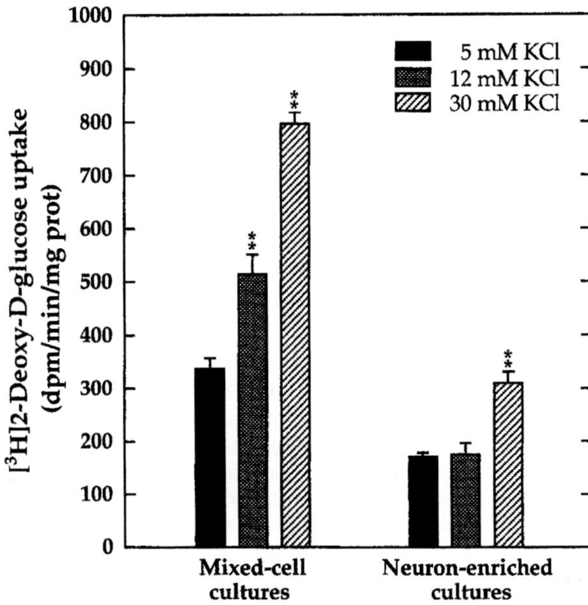

Fig. 3. Effect of elevated extracellular K^+ level ($[K^+]_e$) on deoxyglucose (DG) phosphorylation in mixed neuronal-astrocytic aggregate cultures and in neuron-enriched aggregate cultures from rat brain. The rates of DG phosphorylation were measured during a 30 min incubation in tissue culture medium with 5.5 mM glucose and a fixed concentration of [^{3}H]DG. Note that both 12 and 30 mM $[K^+]_e$ stimulate the DG phosphorylation rate in the mixed neuronal-astrocytic cultures above the rate obtained with 5 mM $[K^+]_e$. On the other hand, only 30 mM $[K^+]_e$ has a stimulatory effect in the neuron-enriched cultures, presumably secondary to $[K^+]_e$-induced excitation. The absence of effect by 12 mM $[K^+]_e$ in the neuronal aggregates suggests an effect on astrocytes, which is consistent with the stimulatory effect of 12 mM $[K^+]_e$ on DG phosphorylation in astrocyte cultures consistently reported by Peng et al. (1994, 1996) and Huang et al. (1994). Vertical bars denote SD. The stimulatory effects of 12 and 30 mM $[K^+]_e$ in mixed-cell cultures and of 30 mM $[K^+]_e$ in neuron-enriched aggregates are statistically significant, as is the difference between the effects of 12 and 30 mM $[K^+]_e$ in the mixed-cell cultures ($P < 0.05$ or better). From Honneger and Pardo (1999).

was almost doubled by iodoacetate, indicating that the rate of K^+ removal was reduced by almost one half; in contrast, the second half of K^+ clearance to the resting level of $[K^+]_e$ was virtually unaffected by iodoacetate but substantially inhibited by hypoxia. The first conclusion of these observations is that K^+ homeostasis at the cellular level of the brain is active and energy-requiring. The second conclusion is that the initial uptake in astrocytes (and perhaps also in some parts of neurons) mainly is fueled by glycolytic metabolism, probably to a large extent of the local glycogen stores and to a minor extent of the glucose present. In contrast, neuronal re-accumulation at lower $[K^+]_e$ as well as the later part of the astrocytic accumulation (depending upon the rate with which mitochondrially formed ATP and phosphocreatine can reach the uptake sites) is fueled by oxidatively generated energy. Since iodoacetate also abolishes subsequent oxidative metabolism of glucose-derived pyruvate, a third conclusion is that either metabolism is slow enough that some pyruvate has remained, or different compounds are oxidized, probably primarily glutamate (see chapter by Schousboe and Waagepetersen).

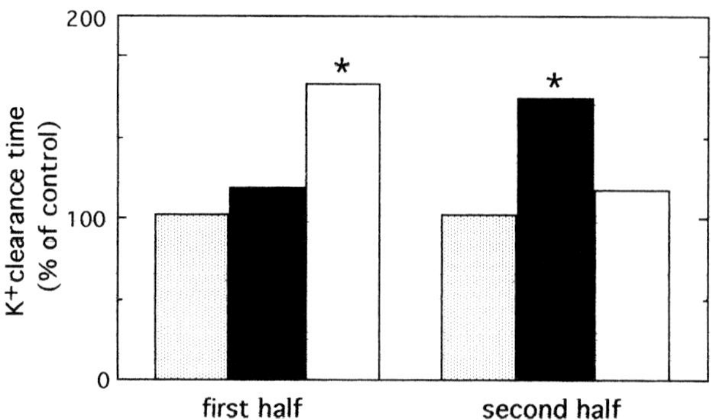

Fig. 4. Effect of hypoxia and of inhibition of glycolysis on K^+ clearance from the extracellular space in rat brain. Hypoxia was induced by administration of inspired air with a severely reduced oxygen content, and glycolysis was inhibited by application of a 5 mM solution of iodoacetate (IOA) to the surface of the brain. $[K^+]_e$ was measured with an extracellular K^+-sensitive electrode implanted approximately 100 μm below the cortical surface. After direct cortical stimulation $[K^+]_e$ initially increased to a peak (1/1 in inset in upper left corner) and then decreased to reach its resting level (0 in inset). The time to reach one half of the peak increase (indicated by arrow at $\frac{1}{2}$) was considered as the first one half of K^+ clearance and the time from $\frac{1}{2}$ to 0 as the second one half. The times required for $[K^+]_e$ to decline from 1/1 to $\frac{1}{2}$ and from $\frac{1}{2}$ to 0 were determined under control conditions (gray bars), indicated as 100% in the graph (which should not suggest that they were similar), and compared with the corresponding times during hypoxia (black bars) and during inhibition of glycolysis (white bars). Hypoxia had no significant effect on the first one half of K^+ clearance but almost doubled the time needed for the second half of K^+ clearance. In contrast, inhibition of glycolysis by aid of iodoacetate almost doubled the first one half of K^+ clearance, but had no significant effect on the time needed for the second half of K^+ clearance. Significant differences from control conditions are indicated by asterisks.

3.1.2. Channel-mediated effects

The $[K^+]_e$ which is required to cause Ca^{2+} channel-mediated glycogenolysis and stimulate glycogen phosphorylase in brain tissue (Ververken et al., 1982; Cambray-Deakin et al., 1988; Hof et al., 1988) is within the range occurring in the brain in vivo (i.e., well below 12 mM). Spreading depression with its very high $[K^+]_e$ is accompanied by a reduction of glycogen level by one third and complete restoration after 10 min (Lauritzen et al., 1990). However, after 5 min of sensory stimulation, when the glycogen level fell by one quarter, the glycogen level had not been restored 15 min after cessation of stimulation (Cruz and Dienel, 2002).

Elevation of $[K^+]_e$ also stimulates glycogenolysis in well differentiated astrocytes in primary cultures (Subbarao et al., 1995). This is illustrated in Table 1, where glycogenolysis is indicated as the percentage release of previously incorporated $[^{14}C]$glucose during 1–10 min of exposure to 10–30 mM $[K^+]_e$. It can be seen that 40–45% of the label is released after at least 1 min exposure to a $[K^+]_e$ of 10 mM or more, and that this is about the maximum amount releasable by elevated $[K^+]_e$. The release is fast, as shown by the fact that slightly more than one half is completed after exposure for only $\frac{1}{2}$ min; however, since recently incorporated radioactivity may be preferentially released, it is not possible to calculate the amount released, based on the pool size. More than 75% of the response was abolished by Ca^{2+} depletion (even though no Ca^{2+}-chelating agent had been added) and a similar reduction was evoked by 100 nM

Table 1
K^+-stimulated glycogenolysis in primary cultures of mouse astrocytes, expressed as percentage release of radioactivity previously incorporated from [^{3}H]glucose, measured under control conditions and after medium modification

$[K^+]_e$ (mM)	Time (min)	Condition	Glycogenolysis control	Medium modification	Glycogenolysis modified
10	1	Control + K^+	42.5 ± 8.3	$-Ca^{2+}$ + K^+	9.3 ± 1.6
20	10	Control + K^+	43.8 ± 7.7	Nifedipine, 100 nM + K^+	12.5 ± 7.8
30	$\frac{1}{2}$	Control + K^+	26.5 ± 2.2	Diazepam, 20 nM + K^+	41.9 ± 3.1

Individual cultures were incubated with tissue culture medium containing [6-^{3}H]glucose (20 μCi/ml) at a glucose concentration of 3 mM during a 30 min period as described by Quach et al. (1982) using cultures from the same batch (containing virtually the same amount of cell protein) for the experimental conditions to be compared, i.e., especially those in the same horizontal row. After medium removal and thorough wash the cultures were re-fed with similar, non-labeled medium or similar, non-labeled medium modified as indicated in the table, both containing the indicated K^+ concentration. After the indicated time the medium was removed and the cultures washed with ice-cold isotonic NaCl solution and harvested in a similar solution. Glycogen was isolated on a filterpaper, and after removal of non-glycogen label by washing it was dissolved in boiling distilled water, and radioactivity was determined and expressed relative to that in similarly treated cultures from the same batches, which had not been exposed to an elevated K^+ concentration and in which no glycogenolysis had occurred. All values are means ± SEM values. Data are from Subbarao et al. (1995).

nifedipine (Subbarao et al., 1995), a blocker of the L-channel for Ca^{2+}. Thus, the glycogenolytic effect is secondary to opening of voltage-dependent L-channels.

3.1.3. Co-transporter-mediated effects

An increase of $[K^+]_e$ in the range occurring in the brain in vivo also stimulates the activity of a co-transporter jointly accumulating K^+, Na^+ and Cl^-. Simultaneous uptake of K^+, Na^+ and Cl^- by the K^+, Na^+, $2Cl^-$ co-transporter is metabolically driven by the electrochemical gradient of Na^+ (and to a minor extent of Cl^-). It therefore must lead to stimulation of the Na^+,K^+-ATPase at its intracellular Na^+-sensitive site (Fig. 1). This results in subsequent exchange of intracellular Na^+ with extracellular K^+, i.e., the co-transporter mediates in essence a net uptake of K^+ and Cl^-, which for osmotic reasons is accompanied by water uptake, i.e., cell swelling, when the cell membrane is permeable for water (see chapter by Chen and Spatz). In cultured mouse astrocytes co-transporter activity is stimulated by high $[K^+]_e$ (Walz and Hertz, 1984). This stimulation is probably a result of K^+-stimulated Ca^{2+} entry through L-channels (Fig. 1), since co-transporter activity is dependent upon the presence of Ca^{2+} and is inhibited by 0.5 μM nifedipine (Su et al., 2000). Thus, stimulation of the rate of oxygen consumption in microdissected or cultured astrocytes by elevated $[K^+]_e$ (Hertz, 1966; Hertz et al., 1973; Hertz and Hertz, 1979) may at least partly reflect stimulation of co-transporter activity. This conclusion is supported by the observation that the stimulation of oxidative metabolism, is inhibited by furosemide, a co-transporter inhibitor (Hertz, 1986).

Stimulation of co-transporter activity by high $[K^+]_e$ is probably also the principal reason for a K^+-induced increase in O_2 consumption in brain slices, which is

Ca^{2+}-dependent (Hertz and Schou, 1962) and rather insensitive to ouabain (suggesting that a Na^+ gradient, necessary to drive the co-transport, persists for a considerable amount of time after inhibition of Na^+ extrusion, although its magnitude becomes gradually reduced). This stimulation of O_2 consumption in brain slices is accompanied by astrocytic cell swelling (Hertz and Kjeldsen, 1985; Hertz and Dienel, 2002), and it is abolished by ethacrynic acid, another co-transporter inhibitor. This inhibition as well as the resistance to ouabain is illustrated in Table 2, which shows a rate of O_2 consumption slightly above 100 μmol/g wet wt during incubation in a 'balanced salt solution' (Hertz and Kjeldsen, 1985) with 6 mM glucose. Under control conditions an increase in the K^+ concentration to 50 mM causes a 70% increase in rate of O_2 uptake (Ashford and Dixon, 1935; Dickens and Greville, 1935; Hertz and Schou, 1962), but in the presence of 0.5 mM ethacrynic acid this stimulation is abolished, whereas the rate of O_2 consumption at normal K^+ concentration is unaffected. Ouabain (10^{-5} and 10^{-4} M) reduces the K^+-induced stimulation, but it does not abolish it. These findings suggest that the K^+-induced stimulation of O_2 consumption is mainly an astrocytic phenomenon, probably due to inactivation of neuronal Na^+ channels during maintained depolarization. In contrast, electrical stimulation of brain slices, as developed by McIlwain (1951) is likely to stimulate the neuronal population (due to the pulsatile rather than maintained depolarization), and it is inhibited by tetrodotoxin and several other drugs that have little effect on the K^+-mediated stimulation of oxygen uptake (for review, see Hertz, 1977).

As illustrated in Fig. 5, spreading depression, which is accompanied by very high $[K^+]_e$, and must represent one of the strongest possible demands for K^+ clearance, is accompanied by a 40% stimulation of oxidative metabolism of acetate (Dienel et al., 2001; Dienel and Cruz, 2003), a substrate that is exclusively taken up and therefore oxidized by astrocytes (O'Dowd, 1995; Waniewski and Martin, 1998; Hertz and Dienel, 2002).

Table 2
Rates of oxygen consumption (μmol/g wet wt per h) in rat brain slices incubated in a balanced salt solution with 6 mM glucose under control conditions, in the presence 0.5 mM ethacrynic acid, an inhibitor of the Na^+, K^+, $2Cl^-$ co-transporter, and in the presence of ouabain (10^{-5} and 10^{-4} M)

Condition	Rate of O_2 consumption balanced medium	Rate of O_2 consumption 50 mM K^+	Percent stimulation
Control	111.1 ± 4.04	188.7 ± 4.47	50.7
0.5 mM ethacrynic acid	114.0 ± 1.20	114.0 ± 3.47	0.1
10^{-5} M ouabain	137.0 ± 5.26	174.3 ± 3.62	27.2
10^{-4} M ouabain	86.7 ± 4.40	112.4 ± 6.35	29.6

Brain cortex slices of 0.5 mm thickness (first slices only) were prepared from adult Wistar rats as described by Hertz and Kjeldsen (1985) and incubated in balanced bicarbonate-buffered saline medium, containing 6 mM glucose and 5 mM KCl and prepared with or without the drugs indicated and equilibrated with a CO_2/O_2 mixture (5%/95%) in a closed chamber equipped with a Clark oxygen electrode and completely filled with medium. The oxygen tension was recorded, and the rate of O_2 consumption, expressed per g initial wet wt, calculated from the total O_2 content in the fluid volume and the rate of decline in O_2 tension during the experiment, which was found to be constant for at least 30 min. After a stable decline in O_2 tension had been recorded for 10–15 min, a concentrated KCl solution was added to a final concentration of 50 mM, and the decline in O_2 tension followed for another 10 – 15 min. Results are means ±SEM.

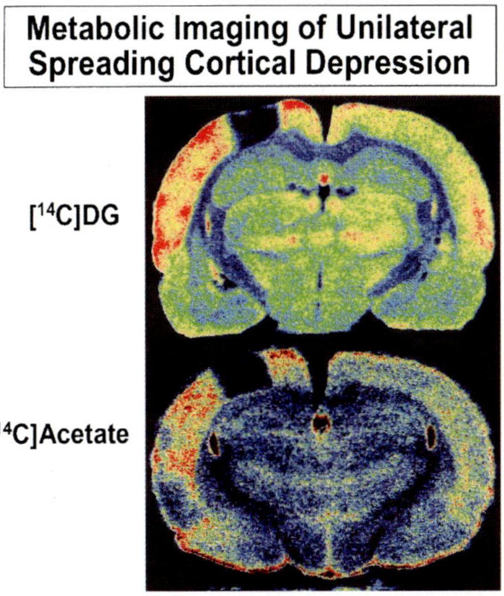

Metabolic Imaging of Unilateral Spreading Cortical Depression

[14C]DG

[14C]Acetate

Fig. 5. Metabolic imaging of unilateral spreading cortical depression. An intravenous pulse of $[^{14}C]$ tracer was injected at 20 min after induction of unilateral spreading depression by topical application of KCl to the dura of left cerebral cortex of the conscious rat. The labeling period was 5 min for both tracers, and autoradiographs were prepared from serial coronal sections. Spreading depression caused heterogeneous increases in labeling of the left compared to the untreated right cerebral cortex with both $[^{14}C]$DG and $[1\text{-}^{14}C]$acetate; red indicates high metabolic rate, with progressively lower rates represented by yellow, green, blue, and black. The dark area in the left cortex is the KCl application site; the cortical tissue below the KCl site had very low uptake of all tracers, presumably due to the very high KCl levels. Labeling with DG and acetate was highest near the KCl application site, and tended to be higher than average in the most dorsal and most ventral layers of left cerebral cortex. Note that labeling by acetate was heterogeneous in gray matter in both hemispheres, and corpus callosum (white matter) had lower levels compared to gray matter structures. Modified from Hertz and Dienel (2002).

Blockade of voltage-dependent Ca^{2+} channels does not prevent the initial wave of spreading depression, but after application of blockers of either the L-, N- or P/Q-type of voltage-dependent Ca^{2+} channels, application of KCl to the cortical surface elicited one, or at most a few, waves of spreading depression, rather than the usual repetitive waves (Richter et al., 2002). This inhibition was interpreted as indicating an influence of voltage-gated Ca^{2+} channels on cortical excitability. However, inhibition of restorative processes mediated by the co-transporter, and restoring intracellular K^+ would be in perfect agreement with the unaffected initial response and secondary development of refractoriness. It would be of interest to establish the effect of Ca^{2+} channel blockers on glucose and acetate metabolism during spreading depression.

3.1.4. K^+-induced stimulation of enzymes involved in glucose metabolism

Several enzymes involved in glucose metabolism are stimulated by elevated K^+ concentrations (see, e.g., Hertz and Dienel, 2002), including pyruvate carboxylase (Ruiz-Amil et al., 1965; McClure et al., 1971). Accordingly pyruvate carboxylation increases with a rise in the extracellular K^+ concentration from 2 to 25 mM in cultured astrocytes (Kaufman and Driscoll, 1992).

3.2. The sodium ion (Na^+)

3.2.1. Carrier-mediated effects

The Na^+, K^+, $2Cl^-$ co-transporter is not the only carrier that is metabolically driven by the Na^+ gradient across the astrocytic cell membrane. Many amino acids are accumulated by a Na^+-driven active uptake, none more important for energy metabolism than glutamate, which is avidly accumulated by the astrocyte-specific EAAT1 and EAAT2 (see chapter by Schousboe and Waagepetersen). Administration of glutamate accordingly stimulates the rate of O_2 consumption in cultured astrocytes (Eriksson et al., 1995; Hertz and Hertz, 2003). Since glutamate is oxidatively degraded in astrocytes (Yu et al., 1982; McKenna et al., 1996; Hertz and Hertz, 2003), the increase in O_2 consumption may reflect oxidation of glutamate as an alternate metabolic substrate. It is in agreement with this conclusion that it repeatedly has been observed in several different laboratories that deoxyglucose phosphorylation is unaltered or slightly reduced in the presence of glutamate (Hertz et al., 1998; Peng et al., 2001; Chen and Liao, 2001; Qu et al., 2001). However, a glutamate-mediated stimulation of glucose phosphorylation in primary cultures of astrocytes was reported by Pellerin and Magistretti (1994) and by Sokoloff et al. (1996), possibly reflecting a lack of ability of the cultures used by these authors to degrade glutamate oxidatively. It is in agreement with this interpretation that Peng et al. (2001), who found no glutamate-induced stimulation of glucose phosphorylation, did observe such a stimulation after administration of the non-metabolizeable D-aspartate, which utilizes the same carrier as glutamate.

3.2.2. Channel-mediated effects

Exchange between intracellular H^+ and extracellular Na^+, brought about by monensin, stimulates not only glucose phosphorylation (Yarowsky et al., 1986), but also glucose oxidation in astrocytes (Peng et al., 2001), presumably by activation of the intracellular Na^+-sensitive site of the Na^+,K^+-ATPase. The evoked alkalosis might also stimulate glucose metabolism, but the stimulation is inhibited by ouabain, indicating that it is secondary to enhanced Na^+,K^+-ATPase activity (Yarowsky et al., 1986; Peng et al., 1994).

4. Transmitter effects on metabolism

4.1. Second messenger systems

4.1.1. The phosphoinositide second messenger system

Phospholipase C catalyzes formation of inositoltrisphosphate (IP_3) and diacylglycerol (DAG) from the membrane lipid phosphatidylinositide 4,5-bisphosphate (PIP_2) (see chapter by Hertz, Chen et al.). IP_3 stimulates protein kinase C, activating downstream signal transduction, and DAG causes release of Ca^{2+} from intracellular stores on the endoplasmic reticulum, leading to an increase in $[Ca^{2+}]_i$ (see chapter by Scapagnini et al.), which subsequently can lead to activation of glycogen phosphorylase, the Na^+, K^+, $2Cl^-$ co-transporter and/or direct stimulation of energy metabolism, caused by an elevation of intramitochondrial Ca^{2+}, as illustrated in Fig. 1.

4.1.2. The adenylyl cyclase second messenger system

The activated and released α subunit of the G-protein G_s stimulates adenylyl cyclase, leading to the formation of cAMP from ATP. In turn, cAMP binds to the regulatory subunits of inactive protein kinase A, releasing the free catalytic subunits, which are then able to phosphorylate and activate their target proteins, which include phosphorylase kinase. The activated, phosphorylase kinase converts an inactive form of glycogen phosphorylase, phosphorylase b to its active form, phosphorylase a. The free catalytic subunits of protein kinase C are also able to enter the cell nucleus, where they phosphorylate the transcription factor CRE-binding protein (CREB), leading to the activation of cAMP-inducible genes containing the regulatory sequence cAMP responsive element (CRE) and thus to synthesis of specific proteins.

4.2. Individual transmitters

4.2.1. Noradrenaline

Adrenergic receptors are expressed on cerebral astrocytes (see chapter by Hansson and Rönnbäck), indicating that astrocytes represent a major target for activation of locus coeruleus, the nucleus of origin for noradrenergic fibers to the brain (Stone and Ariano, 1989; Stone et al., 1992). Rather than releasing transmitter from conventional synapses many of these fibers release noradrenaline from varicosities from which it can diffuse to all adjacent cells, including astrocytes and endothelial cells.

The 'classical' adrenergic receptor activating the phosphoinositide second messenger system is the α_1-adrenergic receptor, but astrocytes also express phospholipase C-linked α_2-adrenergic receptors. Both the α_1-adrenergic receptor agonist phenylephrine and the α_2-adrenergic receptor agonist clonidine stimulate formation of labeled CO_2 from 1-[^{14}C]glutamate, indicating activation of the α-ketoglutarate dehydrogenase complex (αKGDH), as summarized in Table 3 (Subbarao and Hertz, 1991). The same two subtype agonists also stimulate hydrolysis of glutamine to glutamate (Huang and Hertz, 1995). However, only the α_2-adrenergic agonist clonidine was found to significantly stimulate the pyruvate dehydrogenase complex (PDH) (Table 3), as indicated by its stimulation of the formation rate of labeled CO_2 from 1-[^{14}C]pyruvate (Chen and Hertz, 1999). Thus, an increase in $[Ca^{2+}]_i$ is a necessary, but not sufficient requirement for stimulation of mitochondrial dehydrogenases by stimulation of the phophoinositide second messenger system.

Na^+, K^+-ATPase activity in astrocytes is also increased by clonidine, but not by phenylephrine (Hajek et al., 1996). Stimulation of glycolysis (Subbarao and Hertz, 1991) and of glutamate uptake (Hansson and Rönnbäck, 1992; Alexander et al., 1997) is, in contrast, dependent upon stimulation of α_1-adrenergic receptors. The reason for dissimilar effects of different phospholipase C-linked transmitter agonists may be differences in subcellular localization of increased $[Ca^{2+}]_i$ and/or differences in downstream signaling; such differences do exist, as indicated by an ability of the α_2-adrenergic agonist dexmedetomidine, but not of the α_1-adrenergic agonist phenylephrine to induce phosphorylation at the MAP kinases ERK1/2 (see chapter by Peng).

Stimulation of glycogenolysis by noradrenaline in non-cultured cerebral cortical tissue (Magistretti, 1988) and in retina (Ghazi and Osborne, 1989) has indicated a metabolic

Table 3
Transmitters affecting glucose and glutamine metabolism in primary cultures of mouse astrocytes

Transmitter	Subtype	Messenger	PDH	αKGDH	PAG	Glycolysis	Glycogenolysis
Noradrenaline	β	cAMP		0	0	0	X
Noradrenaline	α_1	Ca^{2+}	0	X	X	X	0
Noradrenaline	α_2	Ca^{2+}	X	X	X	0	X
Serotonin	5-HT_{2A}	Ca^{2+}	0				X
Serotonin	5-HT_{2B}	Ca^{2+}	0				X
Dopamine	D(1)-like	cAMP					X
Histamine	H1	Ca^{2+}					X
Histamine	H2	cAMP					X
Adenosine	A2B	cAMP					X
ATP	P2Y	Ca^{2+}					X
VIP	VPAC	cAMP					X
Secretin	–	cAMP					X
Substance P	NK	Ca^{2+}					X
Vasopressin	V1b/V3	Ca^{2+}	0				

Stimulation of the PDH was measured by determination of rate of $^{14}CO_2$ production from [1-^{14}C]pyruvate; stimulation of the α-ketoglutarate dehydrogenase complex (αKGDH) by determination of rate of $^{14}CO_2$ production from [1-^{14}C]glutamate; stimulation of phosphate-activated glutaminase (PAG) by incubation with [U-^{14}C]glutamine followed by HPLC analysis and counting of radioactivity in the glutamate and glutamine fractions; stimulation of glycolysis by measurement of lactate formation; and stimulation of glycogenolysis in our own experiments by release of radioactivity from preloaded glycogen, as described in the legend of Table 1 and in experiments by other investigators as indicated in the respective publications. X indicates significant stimulation and 0 that no stimulation was observed. Information for messenger of subtype-specific receptors are from chapter by Hansson and Rönnbäck; from Jin et al. (2001) (D(1)-like receptor); from Dickenson and Hill (1994) (H1 receptor); from Dartt (1989) (VPAC receptor); from Olde et al. (1998) (secretin receptor); and from Ueda (1999) (NK receptor).

effect on astrocytic energy metabolism in non-cultured tissue, and the activation of glycogen phosphorylase was found to be mediated by β-adrenergic receptors (Ververken et al., 1982; Table 3). The glycogenolytic effect at specific stages of imprinting in day-old chicks is also a β-adrenergic effect, as indicated by its inhibition by propranolol (O'Dowd, 1995). Studies using cultured astrocytes have confirmed a β-adrenergic stimulation of glycogenolysis and indicated an additional effect of α_2-adrenergic stimulation by showing that glycogenolysis can be stimulated by isoproterenol and clonidine, but not by phenylephrine (Subbarao and Hertz, 1990b). The effects can be explained by a stimulatory effect of Ca^{2+} on glycogen phosphorylase and by the well known activation of phosphorylase by protein kinase A.

Since glycogen is not only rapidly degraded but also re-synthesized during and after brain activation and at specific stages of learning (Fig. 2), it becomes of importance to establish whether noradrenaline also enhances glycogen synthesis. Magistretti and coworkers have demonstrated that noradrenaline exerts long-term control of glycogen levels in astrocytes and within 9 h leads to a large, protein synthesis-dependent induction of glycogen synthase and increase in glycogen, an effect mediated by protein kinase A and its phosphorylation of CREB (Sorg and Magistretti, 1992; Pellegri et al., 1996). However, such a long-term effect is unable to explain re-synthesis of glycogen concurrently with or relatively soon after glycogenolysis. We have therefore measured the effect of noradrenaline, phenylephrine, clonidine and isoproterenol on incorporation of label from [U-^{14}C]glucose into glycogen, realizing that this is not a 'clean' experiment, because of simultaneous stimulation of glycogenolysis by most of the agonists. The results, as percentage of incorporation in the absence of any adrenergic agonist (control), are shown in Table 4. It can be seen that noradrenaline and the β-adrenergic agonist, isoproterenol exert an inhibition, which is statistically significant in the case of isoproterenol, and can be explained by simultaneous glycogenolysis. This inhibition is counteracted by the β-adrenergic antagonist, alprenolol. In spite of the glycogenolytic activity by the a_2-adrenergic agonist clonidine, clonidine tended to cause an increase in incorporation of labeled glucose into glycogen, and the a_2-adrenergic antagonist yohimbine significantly decreased this incorporation, further supporting the stimulation of glycogen synthesis by a_2-adrenergic agonists. Phenylephrine had no effect (Table 4).

4.2.2. Serotonin (5-HT)

Serotonergic fibers spread from the raphe nuclei across the entire brain in much the same way as noradrenergic fibers spread from locus coeruleus. Like noradrenaline, serotonin increases glycogenolysis in brain tissue and in retina (Magistretti, 1988; Ghazi and Osborne, 1989). It has been reported that glycolysis in retina is 5-HT_1 receptor-mediated, but glycogenolysis in astrocytes can be stimulated by activation of simultaneously expressed high-affinity 5-HT_{2B} (Kong et al., 2002a) and lower affinity

Table 4
Transmitter effects on incorporation of [6-^{3}H]glucose (3 mM) for 30 min into glycogen

Transmitter	Percentage incorporation
Control	100 ± 3.0
Noradrenaline (NA)	85.9 ± 6.8
Phenylephrine	91.0 ± 8.5
Clonidine	111.3 ± 15.0
Isoproterenol	62.8 ± 10.7
NA + Alprenolol	97.2 ± 3.9
NA + Yohimbine	65.5 ± 5.7

Individual cultures were incubated with tissue culture medium containing [6-^{3}H]glucose (20 μCi/ml) at a glucose concentration of 3 mM during a 60-min period in tissue culture medium with or without the noradrenergic agonists (0.1 μM) and antagonists (1 μM) indicated, using cultures from the same batch (containing virtually the same amount of cell protein) for all experimental conditions. After the indicated time the medium was removed and the cultures washed with ice-cold isotonic NaCl solution and harvested in a similar solution. Glycogen was isolated on a filter paper, and after removal of non-glycogen label by washing, it was dissolved in boiling distilled water, and radioactivity per culture was determined and expressed relative to that in similarly treated cultures which had not been exposed to an elevated K^+ concentration. All values are means ± SEM values.

5-HT_{2A} receptors (Table 3), and the concentration dependence of the glycogenolytic effect is similar to that for a biphasic increase in $[Ca^{2+}]_i$ (Chen et al., 1995). In contrast to clonidine, but similar to phenylephrine, serotonin does not stimulate pyruvate dehydrogenation (Table 3; Chen, 1996).

4.2.3. Dopamine

In primary cultures of mouse cerebrocortical astrocytes 10 μM dopamine causes glycogenolysis (Table 3), which is more marked than that evoked by the corresponding concentration of serotonin (L. Peng, unpublished experiments). Intraventricularly injected dopamine increases glycogenolysis in the mouse brain (Leonard, 1975). However, it does not influence glycogenolysis in retina (Ghazi and Osborne, 1989).

4.2.4. Histamine

Histamine resembles noradrenaline and serotonin by stimulating glycogenolysis in brain tissue (Magistretti, 1988) and in primary cultures of astrocytes (Arbones et al., 1990). As in the case of noradrenaline, both a G_s-coupled, cAMP-linked receptor subtype, the H_2 receptor, and a phospholipase C-linked, $G_{i/o}$-coupled receptor, the H_1 receptor, stimulate glycogenolysis (Table 3). No data are available whether histamine stimulates mitochondrial dehydrogenases.

4.2.5. Purinergic receptors

Both P1 receptors, activated by adenosine, and phospholipase C-linked P2Y ATP receptor mediate glycogenolysis in primary cultures of astrocytes (Magistretti, 1988; Sorg et al., 1995). The G_s-coupled, cAMP-linked A2b receptor mimics its β-adrenergic counterpart by promoting long-term glycogen synthesis in astrocytes (Allaman et al., 2002).

4.2.6. Peptidergic receptors

Vasoactive intestinal peptide (VIP) was among the very first agents shown to promote glycogenolysis in cultured astrocytes, and it was suggested that this effect was exerted via stimulation of adenylyl cyclase activity (Magistretti et al., 1983). VIP also induces long-term glycogen re-synthesis in a similar manner as noradrenaline (Sorg and Magistretti, 1992), and it stimulates glycogenolysis in non-cultured cerebrocortical tissue (Magistretti, 1988) and in retina (Ghazi and Osborne, 1989). Other neuropeptides which stimulate glycogenolysis include substance P (Medrano et al., 1994) and secretin (Sorg and Magistretti, 1992). It is unknown whether vasopressin (AVP) stimulates glycogenolysis in astrocytes, but it does not enhance the activity of mitochondrial dehydrogenases (Table 3).

5. Drug effects on metabolism

5.1. Interaction with ion effects

5.1.1. Barbiturates

Barbiturates inhibit K^+ uptake into cultured astrocytes (Hertz, 1979), apparently especially at elevated $[K^+]_e$ (Hertz, 1986), and they depress K^+-stimulated oxidative metabolism in astrocytes quite potently, whereas they have only little effect on O_2 consumption and production of labeled CO_2 from [U-^{14}C]glucose at normal $[K^+]_e$ (Hertz et al., 1986). Since they also inhibit K^+-stimulated respiration in brain slices (for review, see Hertz, 1977), it is likely that they interfere with co-transporter-mediated K^+ uptake. These observations should not lead to the impression that barbiturate effects are mainly exerted on astrocytes, since they cause a very large decrease in production of labeled CO_2 from [U-^{14}C]glucose in cerebellar granule neurons, a glutamatergic preparation (Peng and Hertz, 2002), and since in the brain in vivo they mainly affect neuronal metabolism (see chapter by Gruetter).

5.1.2. Benzodiazepines

Astrocytes express the so-called mitochondrial-type benzodiazepine binding sites (see chapter by Bélanger et al.), which are important for the synthesis of neurosteroids from cholesterol (see chapter by Melcangi et al.). The pharmacological profile of these sites is quite different from that of neuronal benzodiazepine receptors. For example the neuronal benzodiazepine antagonist flumazenil has only low affinity for these sites, whereas PK11195, which binds with very low affinity to neurons, has a high affinity for the mitochondrial-type benzodiazepine binding sites (Bender and Hertz, 1987).

In addition to these sites astrocytes also express benzodiazepine receptors, which partly mimic the neuronal benzodiazepine receptors linked to $GABA_A$ receptors (Backus et al., 1988). However, in astrocytes activation of these receptors causes depolarization, rather than hyperpolarization (see chapter by Hansson and Rönnbäck). An interaction between Ca^{2+} channel ligands and benzodiazepines in astrocytes had long been anticipated (Bender and Hertz, 1985), when it was shown that clinically used benzodiazepines like diazepam increase depolarization-induced Ca^{2+} uptake through L-channels into cultured astrocytes during exposure to K^+ concentrations too low to exert maximum channel opening on their own (Zhao et al., 1996), probably reflecting an additional benzodiazepine-mediated partial depolarization. A benzodiazepine-mediated increase in $[Ca^{2+}]_i$ has been confirmed in freshly isolated astrocytes (Fraser et al., 1995), and membrane-associated peripheral-type benzodiazepine receptors have been described in heart, liver, adrenal gland, testis, hemopoietic cells, and, not least, cells of the cardiovascular system (Bolger, 1993; Woods and Williams, 1996). Because glycogenolysis is stimulated by Ca^{2+} entry through voltage-gated channels, the glycogenolytic effects of submaximally effective K^+ concentrations is also enhanced by benzodiazepines (Subbarao et al., 1995). This effect is illustrated in the bottom line of Table 1, showing that glycogenolysis, which after $\frac{1}{2}$ min of exposure to 30 mM $[K^+]_e$ under control conditions amounts to 27% of previously incorporated label, in the presence of 20 nM diazepam is increased to 42% of the incorporated radioactivity.

5.2. Interaction with transmitter effects

5.2.1. Therapeutics

It is likely that many clinically used drugs, that are receptor agonists or antagonists exert effects on astrocytes. An example is the anti-depressant drug fluoxetine (Prozac), which is a specific serotonin re-uptake inhibitor (SSRI), but has sufficient activity for at least the 5-HT_{2B} receptor that clinically relevant concentrations of fluoxetine cause a direct stimulation of glycogenolysis in cultured astrocytes, which do not express the serotonin transporter (Chen et al., 1995; Kong et al., 2002a). In spite of its promotion of glycogenolysis, fluoxetine causes an inhibition of the production of labeled CO_2 from U-[^{14}C]glucose (Chen, 1996).

Other serotonin-specific reuptake inhibitors, e.g., paroxetine (Paxol) have a lower affinity for the 5-HT_2 receptor, and do not by themselves stimulate glycogenolysis at usual therapeutic concentrations. Chronic treatment of cultured astrocytes with fluoxetine can nevertheless be used as a model of the effect of a drug specifically stimulating 5-HT_2 receptors, which in any case will be stimulated during in vivo treatment with any serotonin-specific reuptake inhibitor (due to inhibition of neuronal serotonin transporters). Chronic treatment with 1 μM fluoxetine initially downregulates 5-HT_2 receptor-mediated glycogenolysis in cultured astrocytes (Chen et al., 1995), but subsequently upregulates it (Kong et al., 2002a). The upregulation appears after 2–3 weeks of treatment, which is similar to the lag time for the onset of fluoxetine's therapeutic effect. Like fluoxetine, the anti-migraine drug methysergide displaces serotonin from its binding site on astrocytes in primary cultures (Hertz et al., 1979), but it remains to be demonstrated whether it has any effect on glycogenolysis.

5.2.2. Drugs of abuse

Both amphetamine (Nowak, 1988) and 3,4-methylenedioxymethamphetamine ('ecstasy') (Darvesh et al., 2002) cause an enhancement of glycogenolysis in the rodent brain, which was suggested at least partly to result from a concomitant increase in body temperature. A single subcutaneous dose of 10–40 mg/kg of 3,4-methylenedioxymethamphetamine caused a dose-dependent reduction of glycogen, which reached 40%, lasted for more than 1 h, and could be inhibited by a 5-HT_2 antagonist. However, 3,4-methylenedioxymethamphetamine does cause a temperature-independent effect on glycogen breakdown in astrocytes, as shown by the observation that active glycogen phosphorylase in primary cultures of astrocytes is increased by 70% in the presence of 5 μM 3,4-methylenedioxymethamphetamine (Poblete and Azmitia, 1995).

Metabolic effects have been observed after chronic administration of cocaine (1 and 3 μM) to immature mouse astrocyte cultures (simulating human drug exposure during the last third of gestation), in which the usual stimulatory effect on α-ketoglutarate dehydrogenase activity in response to noradrenaline became abolished (Peng and Hertz, 1992). After treatment for 21 days this effect persisted during 'withdrawal', i.e., discontinuation of drug treatment, for the entire period investigated (more than one month from the time of withdrawal). The implications of this finding are scary, considering the wide-spread abuse of cocaine by pregnant women.

6. Concluding remarks

The demonstration by Su et al. (2000) that the Na^+, K^+, $2Cl^-$ co-transporter is activated by opening of L-channels for Ca^{2+} (Fig. 1) has been crucial for the understanding of effects on oxidative metabolism by $[K^+]_e$ of the magnitude found extracellularly during brain activity It can now be concluded that activation of this transporter requires both high enough $[K^+]_e$ to activate voltage-gated Ca^{2+} as well as Na^+,K^+-ATPase activity to extrude accumulated Na^+(Fig. 1). There is also no longer any doubt that K^+ accumulation in astrocytes by this mechanism and by stimulation of the extracellular K^+-sensitive site of the Na^+,K^+-ATPase plays a major role in ion homeostasis (see chapter by Walz) and therefore in energy metabolism in the CNS. Any claim that K^+ is passively distributed in astrocytes should by now be regarded as obsolete. Although other processes, probably especially Na^+-driven uptake of glutamate also contribute to Na^+,K^+-ATPase-mediated stimulation of astrocytic energy metabolism by activating the intracellular Na^+-dependent site of the Na^+,K^+-ATPase, they are probably of quantitatively minor importance compared to the regulation of K^+ homeostasis.

In addition to ion-stimulated energy metabolism, emerging evidence indicates an important role of transmitter effects on energy metabolism in astrocytes, both by stimulation of mitochondrial dehydrogenases and glutaminase (glutamine is also an energy substrate for brain [see chapter by Schousboe and Waagepetersen]) and by activating glycogenolysis. Although it has been realized for some time that glycogen stores turn over dynamically during neuronal activation (Swanson et al., 1992; O'Dowd et al., 1994), it is only recently that attention has been drawn to the possibility that glycogen may serve as a re-chargeable energy substrate in peripheral processes of the astrocytes, which are too thin to contain any mitochondria (Dienel and Cruz, 2003).

With the increasing realization of the importance of astrocytes in brain energy metabolism comes emerging evidence that drugs classically assumed to act solely on neurons may owe a substantial amount of their effect to actions on astrocytes, including astrocytic energy metabolism. The list of such drugs is likely to grow in the future. Many of them may interact with effects of ions or transmitters and thereby affect energy metabolism in astrocytes and thus in the brain.

References

Alexander, G.M., Grothusen, J.R., Gordon, S.W., Schwartzman, R.J., 1997. Intracerebral microdialysis study of glutamate reuptake in awake, behaving rats. Brain Res. 766, 1–10.

Allaman, I., Lengacher, S., Magistretti, P.J., Pellerin, L., 2002. A2B receptor activation promotes glycogen synthesis in astrocytes through modulation of gene expression. Am. J. Physiol. Cell Physiol. (Nov. 6, published ahead of print).

Arbones, L., Picatoste, F., Garcia, A., 1990. Histamine stimulates glycogen breakdown and increases $^{45}Ca^{2+}$ permeability in rat astrocytes in primary culture. Mol. Pharmacol. 37, 921–927.

Ashford, C.A., Dixon, K.C., 1935. The effect of potassium on the glycolysis of brain tissue with reference to the Pasteur effect. Biochem. J. 29, 157–168.

Backus, K.H., Kettenmann, H., Schachner, M., 1988. Effect of benzodiazepines and pentobarbital on the GABA-induced depolarization in cultured astrocytes. Glia 1, 132–140.

BeltrandelRio, H., Wilson, J.E., 1992. Coordinated regulation of cerebral glycolytic and oxidative metabolism, mediated by mitochondrially bound hexokinase dependent on intramitochondrially generated ATP. Arch. Biochem. Biophys. 296, 667–677.

Bender, A., Hertz, L., 1985. Pharmacological evidence that the non-neuronal diazepam binding site in primary cultures of glial cells is associated with a calcium channel. Eur. J. Pharmacol. 110, 287–288.

Bender, A.S., Hertz, L., 1987. Pharmacological characterization of diazepam receptors in neurons and astrocytes in primary cultures. J. Neurosci. Res. 18, 366–372.

Bolger, G.T., 1993. Cardiovascular actions of peripheral benzodiazepines. In: Giesen-Crouse, E. (Ed.) Peripheral Benzodiazepine Receptors. Academic Press, New York, pp. 153–167.

Cambray-Deakin, M., Pearce, B., Morrow, C., Murphy, S., 1988. Effects of extracellular potassium on glycogen stores of astrocytes in vitro. J. Neurochem. 51, 1846–1851.

Cataldo, A.M., Broadwell, R.D., 1986. Cytochemical identification of cerebral glycogen and glucose-6-phosphatase activity under normal and experimental conditions. II. Choroid plexus and ependymal epithelia, endothelia and pericytes. J. Neurocytol. 15, 511–524.

de Cerqueira Cesar, M., Wilson, J.E., 1995. Application of a double isotopic labeling method to a study of the interaction of mitochondrially bound rat brain hexokinase with intramitochondrial compartments of ATP generated by oxidative phosphorylation. Arch. Biochem. Biophys. 324, 9–14.

Chen, Y., 1996. Effects of selected transmitters on free cytosolic calcium concentration and pyruvate dehydrogenation in primary cultures of mouse astrocytes. PhD Thesis. University of Saskatchewan, Saskatoon, Saskatchewan, Canada.

Chen, Y., Hertz, L., 1999. Noradrenaline effects on pyruvate decarboxylation—correlation with calcium signaling. J. Neurosci. Res. 58, 599–606.

Chen, C.-J., Liao, S.-L., 2001. Effects of glutamate on glucose utilization and lactate output in cultured rat cortical astrocytes. J. Cereb. Blood Flow Metab. 21(Suppl. 1), S254.

Chen, Y., Peng, L., Zhang, X., Stolzenburg, J.-U., Hertz, L., 1995. Further evidence that fluoxetine interacts with a 5-HT_{1C} receptor. Brain Res. Bull. 38, 153–159.

Cruz, N.F., Adachi, K., Dienel, G.A., 1999. Metabolite trafficking during K^+-induced spreading cortical depression: rapid efflux of lactate from cerebral cortex. J. Cereb. Blood Flow Metab. 19, 380–392.

Cruz, N.F., Dienel, G.A., 2002. High glycogen levels in brains of rats with minimal environmental stimuli implications for metabolic contributions of working astrocytes. J. Cereb. Blood Flow Metab. 22, 1476–1489.

Dartt, D.A., 1989. Signal transduction and control of lacrimal gland protein secretion: a review. Curr. Eye Res. 8, 619–636.

Darvesh, A.S., Shankaran, M., Gudelsky, G.A., 2002. 3,4-Methylenedioxymethamphetamine produces glycogenolysis and increases the extracellular concentration of glucose in rat brain. J. Pharmacol. Exp. Ther. 301, 138–144.

Dickens, F., Greville, G.D., 1935. The metabolism of normal and tumour tissue (XIII). Neutral salt effects. Biochem. J. 29, 1468–1483.

Dickenson, J.M., Hill, S.J., 1994. Interactions between adenosine A1- and histamine H1-receptors. Int. J. Biochem. 26, 959–969.

Dienel, G.A., Cruz, N.F., 2003. Neighborly interactions of metabolically-activated astrocytes in vivo. Neurochem. Int. 43, 339–354.

Dienel, G.A., Liu, K., Cruz, N.F., 2001. Local uptake of ^{14}C-labeled acetate and butyrate in rat brain in vivo during spreading cortical depression. J. Neurosci. Res. 66, 812–820.

Dienel, G.A., Wang, R.Y., Cruz, N.F., 2002. Generalized sensory stimulation of conscious rats increases oxidative metabolism of glucose when the brain glucose/oxygen uptake ratio rises. J. Cereb. Blood Flow Metab. 22, 1490–1502.

Duffy, S., MacVicar, B.A., 1994. Potassium-dependent calcium influx in acutely isolated hippocampal astrocytes. Neuroscience 61, 51–61.

Eriksson, G., Peterson, A., Iverfeldt, K., Walum, E., 1995. Sodium-dependent glutamate uptake as an activator of oxidative metabolism in primary astrocyte cultures from newborn rat. Glia 15, 152–156.

Eugenin, E.A., Gonzalez, H., Saez, C.G., Saez, J.C., 1998. Gap junctional communication coordinates vasopressin-induced glycogenolysis in rat hepatocytes. Am. J. Physiol. 274, G1109–G1116.

Fraser, D.D., Duffy, S., Angelides, K.J., Perez-Velazquez, J.L., Kettenmann, H., MacVicar, B.A., 1995. $GABA_A$/benzodiazepine receptors in acutely isolated hippocampal astrocytes. J. Neurosci. 15, 2720–2732.

French, N., Lalies, M.D., Nutt, D.J., Pratt, J.A., 1995. Idazoxan-induced reductions in cortical glucose use are accompanied by an increase in noradrenaline release; complementary [^{14}C]2-deoxyglucose and microdialysis studies. Neuropharmacology 34, 605–613.

Ghazi, H., Osborne, N.N., 1989. Agonist-induced glycogenolysis in rabbit retinal slices and cultures. Br. J. Pharmacol. 96, 895–905.

Grisar, T., Frere, J.M., Franck, G., 1979. Effects of K^+ ions on kinetic properties of the (Na^+–K^+)-ATPase (EC 3.6.1.3.) in bulk isolated glia cells, perikarya and synaptosomes from rabbit brain cortex. Brain Res. 165, 87–103.

Hajek, I., Subbarao, K.V., Hertz, L., 1996. Stimulation of Na^+,K^+-ATPase activity in astrocytes and neurons by K^+ and/or noradrenaline. Neurochem. Int. 28, 335–342.

Halestrap, A.P., 1989. The regulation of the matrix volume of mammalian mitochondria in vivo and in vitro and its role in the control of mitochondrial metabolism. Biochim. Biophys. Acta 973, 355–382.

Hansson, E., Rönnbäck, L., 1992. Adrenergic receptor regulation of amino acid neurotransmitter uptake in astrocytes. Brain Res. Bull. 29, 297–301.

Henn, F.A., Haljamae, H., Hamberger, A., 1972. Glial cell function: active control of extracellular K^+ concentration. Brain Res. 43, 437–443.

Hertz, L., 1966. Neuroglial localization of potassium and sodium effects on respiration. J. Neurochem. 13, 1373–1387.

Hertz, L., 1977. Drug-induced alterations of ion distribution at the cellular level of the central nervous system. Pharmacol. Rev. 29, 35–65.

Hertz, L., 1979. Inhibition by barbiturates of an intense net uptake of potassium into astrocytes. Neuropharmacology 18, 629–633.

Hertz, L., 1986. Potassium as a signal in metabolic interactions between neurons and astrocytes. In: Grisar, T., Franck, G., Hertz, L., Norton, W.Y., Sensenbrenner, M., Woodbury, D.M. (Eds.), Dynamic Properties of Glia Cells. Pergamon Press, Oxford, pp. 215–224.

Hertz, L., Baldwin, F., Schousboe, A., 1979. Serotonin receptors on astrocytes in primary cultures: effects of methysergide and fluoxetine. Can. J. Physiol. Pharmacol. 57, 223–226.

Hertz, L., Bender, A.S., Woodbury, D., White, S.A., 1989. Potassium induced calcium uptake in astrocytes and its potent inhibition by a calcium channel blocker. J. Neurosci. Res. 22, 209–215.

Hertz, L., Dienel, G.A., 2002. Energy metabolism in the brain. Int. Rev. Neurobiol. 51, 1–101.

Hertz, L., Dittmann, L., Mandel, P., 1973. K^+ induced stimulation of oxygen uptake in cultured cerebral glial cells. Brain Res. 60, 517–520.

Hertz, L., Gibbs, M.E., O'Dowd, B.S., Sedman, G.L., Robinson, S.R., Peng, L., Huang, R., Hertz, E., Hajek, I., Sykova, I., Ng, K.T., 1996. Astrocyte–neuron interaction during one-trial aversive learning in the neonate chick. Neurosci. Biobehav. Rev. 20, 537–551.

Hertz, E., Hertz, L., 1979. Polarographic measurement of oxygen uptake by astrocytes in primary cultures using the tissue culture flask as the respirometer chamber. In Vitro 15, 429–436.

Hertz, E., Hertz, L., 2003. Cataplerotic TCA cycle flux determined as glutamate-sustained oxygen consumption in primary cultures of astrocytes. Neurochem. Int. 43, 355–361.

Hertz, L., Kjeldsen, C.S., 1985. Functional role of the potassium-induced stimulation of oxygen uptake in brain slices studied with cesium as a probe. J. Neurosci. Res. 14, 83–93.

Hertz, L., Peng, L., 1992. Energy metabolism at the cellular level of the CNS. Can. J. Physiol. & Pharmacol. 70, S145–S157.

Hertz, L., Schou, M., 1962. Univalent cations and the respiration of brain cortex slices. Biochem. J. 85, 93–104.

Hertz, E., Shargool, M., Hertz, L., 1986. Effects of barbiturates on energy metabolism by cultured astrocytes and neurons in the presence of normal and elevated potassium concentrations. Neuropharmacology 25, 533–539.

Hertz, L., Swanson, R.A., Newman, G.C., Mariff, H., Juurlink, B.H.J., Peng, L., 1998. Can experimental conditions explain the discrepancy whether or not glutamate stimulates aerobic glycolysis? Dev. Neurosci. 20, 339–347.

Hof, P.R., Pascale, E., Magistretti, P.J., 1988. K^+ at concentrations reached in the extracellular space during neuronal activity promotes a Ca^{2+}-dependent glycogen hydrolysis in mouse cerebral cortex. J. Neurosci. 8, 1922–1928.

Honneger, P., Pardo, B., 1999. Separate neuronal and glial Na^+,K^+-ATPase isoforms regulate glucose utilization in response to membrane depolarization and elevated extracellular potassium. J. Cereb. Blood Flow Metab. 19, 1051–1059.
Huang, R., Hertz, L., 1995. Noradrenaline induced stimulation of glutamine metabolism in primary cultures of astrocytes. J. Neurosci. Res. 41, 677–683.
Huang, R., Peng, L., Chen, Y., Hajek, I., Zhao, Z., Hertz, L., 1994. Signalling effect of monoamines and of elevated potassium concentrations on brain energy metabolism at the cellular level. Dev. Neurosci. 16, 337–351.
Ibrahim, H.Z.M., 1975. Glycogen and its related enzymes of metabolism in the central nervous system. Adv. Anat. Cell Biol. 52, 5–87.
Inoue, M., McHugh, M., Pappius, H.M., 1991. The effect of alpha-adrenergic receptor blockers prazosin and yohimbine on cerebral metabolism and biogenic amine content of traumatized brain. J. Cereb. Blood Flow Metab. 11, 242–252.
Jin, L.Q., Wang, H.Y., Friedman, E., 2001. Stimulated D(1) dopamine receptors couple to multiple Galpha proteins in different brain regions. J. Neurochem. 78, 981–990.
Kaufman, E.E., Driscoll, B.F., 1992. Carbon dioxide fixation in neuronal and astroglial cells in culture. J. Neurochem. 58, 258–262.
Kimelberg, H.K., Biddlecome, S., Narumi, S., Bourke, R.S., 1978. ATPase and carbonic anhydrase activities of bulk-isolated neuron, glia and synaptosome fractions from rat brain. Brain Res. 141, 305–323.
Kong, E.K., Peng, L., Chen, Y., Yu, A.C., Hertz, L., 2002a. Up-regulation of 5-HT_{2B} receptor density and receptor-mediated glycogenolysis in mouse astrocytes by long-term fluoxetine administration. Neurochem. Res. 27, 113–120.
Kong, J., Shepel, P.N., Holden, C.P., Mackiewicz, M., Pack, A.I., Geiger, J.D., 2002b. Brain glycogen decreases with increased periods of wakefulness: implications for homeostatic drive to sleep. J. Neurosci. 22, 5581–5587.
Lauritzen, M., Hansen, A.J., Kronborg, D., Wieloch, T., 1990. Cortical spreading depression is associated with arachidonic acid accumulation and preservation of energy charge. J. Cereb. Blood Flow Metab. 10, 115–122.
Lawrie, A.M., Rizzuto, R., Pozzan, T., Simpson, A.W., 1996. A role for calcium influx in the regulation of mitochondrial calcium in endothelial cells. J. Biol. Chem. 271, 10753–10759.
Leonard, B.E., 1975. A study of the neurohumoral control of glycolysis in the mouse brain in vivo: role of noradrenaline and dopamine. Z. Naturforsch. 30, 385–391.
Magistretti, P.J., 1988. Regulation of glycogenolysis by neurotransmitters in the central nervous system. Diab. Metab. 14, 237–246.
Magistretti, P.J., Manthorpe, M., Bloom, F.E., Varon, S., 1983. Functional receptors for vasoactive intestinal polypeptide in cultured astroglia from neonatal rat brain. Regul. Pept. 6, 71–80.
Magistretti, P.J., Sorg, O., Yu, N., Martin, J.L., Pellerin, L., 1993. Neurotransmitters regulate energy metabolism in astrocytes: implications for the metabolic trafficking between neural cells. Dev. Neurosci. 15, 306–312.
McClure, W.R., Lardy, H.A., Kneifel, H.P., 1971. Rat liver pyruvate carboxylase. I. Preparation, properties and cation specificity. J. Biol. Chem. 246, 3569–3578.
McCormack, J.G., Denton, R.M., 1990. The role of mitochondrial Ca^{2+} transport and matrix Ca^{2+} in signal transduction in mammalian tissues. Biochim. Biophys. Acta 1018, 287–291.
McIlwain, H., 1951. Metabolic response in vitro to electrical stimulation of sections of mammalian brain. Biochem. J. 49, 382–393.
McKenna, M.C., Sonnewald, U., Huang, X., Stevenson, J., Zielke, H.R., 1996. Exogenous glutamate concentration regulates the metabolic fate of glutamate in astrocytes. J. Neurochem. 66, 386–393.
Medrano, S., Gruenstein, E., Dimlich, R.V., 1994. Substance P receptors on human astrocytoma cells are linked to glycogen breakdown. Neurosci. Lett. 167, 14–18.
Mercado, R., Hernandez, J., 1992. Regulatory role of a neurotransmitter (5-HT) on glial Na^+,K^+-ATPase in the rat brain. Neurochem. Int. 21, 119–127.
Nowak, T.S. Jr., 1988. Effects of amphetamine on protein synthesis and energy metabolism in mouse brain: role of drug-induced hyperthermis. J. Neurochem. 50, 285–294.
O'Dowd, B.S., 1995. Metabolic activity in astrocytes is essential for the consolidation of one-trial passive avoidance memory in day-old chick. PhD Thesis. La Trobe University, Bundoora, Victoria, Australia.
O'Dowd, B.S., Gibbs, M.E., Ng, K.T., Hertz, E., Hertz, L., 1994. Astrocytic glycogenolysis energizes memory processes in neonate chicks. Dev. Brain Res. 78, 137–141.

Olde, B., Sabirsh, A., Owman, C., 1998. Molecular mapping of epitopes involved in ligand activation of the human receptor for the neuropeptide, VIP, based on hybrids with the human secretin receptor. J. Mol. Neurosci. 11, 127–134.

Pellegri, G., Rossier, C., Magistretti, P.J., Martin, J.L., 1996. Cloning, localization and induction of mouse brain glycogen synthase. Mol. Brain Res. 38, 191–1999.

Pellerin, L., Magistretti, P.J., 1994. Glutamate uptake into astrocytes stimulates aerobic glycolysis: a mechanism coupling neuronal activity to glucose utilization. Proc. Natl Acad. Sci. USA 91, 10625–10629.

Peng, L., Hertz, L., 1992. Long-lasting abolishment of noradrenaline induced stimulation of oxidative metabolism after chronic exposure of developing mouse astrocytes to cocaine. Brain Res. 581, 334–338.

Peng, L., Hertz, L., 2002. Amobarbital inhibits K(+)-stimulated glucose oxidation in cerebellar granule neurons by two mechanisms. Eur. J. Pharmacol. 446, 53–61.

Peng, L., Juurlink, B.H.J., Hertz, L., 1996. Pharmacological and developmental evidence that the potassium-induced stimulation of deoxyglucose uptake in astrocytes is a metabolic manifestation of increased Na^+-K^+-ATPase activity. Dev. Neurosci. 18, 353–359.

Peng, L., Swanson, R.A., Hertz, L., 2001. Effects of L-glutamate, D-aspartate and monensin on glycolytic and oxidative glucose metabolism in mouse astrocyte cultures. Neurochem. Int. 38, 437–443.

Peng, L., Zhang, X., Hertz, L., 1994. High extracellular potassium concentrations stimulate oxidative metabolism in a glutamatergic neuronal culture and glycolysis in cultured astrocytes, but have no stimulatory effect in a GABAergic neuronal culture. Brain Res. 663, 168–172.

Peters, A., Palay, S.L., Webster, deF. H., 1991. The Fine Structure of the Nervous System: Neurons and their Supporting Cells, 3rd ed. Oxford University Press, New York, pp. 276–295.

Peuchen, S., Duchen, M.R., Clark, J.B., 1996. Energy metabolism of adult astrocytes in vitro. Neuroscience 71, 855–870.

Poblete, J.C., Azmitia, E.C., 1995. Activation of glycogen phosphorylase by serotonin and 3,4-methylene-dioxymethamphetamine in astroglial-rich primary cultures: involvement of the 5-HT_{2A} receptor. Brain Res. 680, 9–15.

Raffin, C.N., Rosenthal, M., Busto, R., Sick, T.J., 1992. Glycolysis, oxidative metabolism, and brain potassium clearance. J. Cereb. Blood Flow Metab. 12, 34–42.

Qu, H., Eloqayli, H., Unsgård, G., Sonnewald, U., 2001. Glutamate decreases pyruvate carboxylase activity and spares glucose as energy substrate in cultured cerebellar astrocytes. J. Neurosci. Res. 66, 1127–1132.

Quach, T.T., Rose, C., Duchemin, A.M., Schwartz, J.C., 1982. Glycogenolysis induced by serotonin in brain: identification of a new class of receptor. Nature 298, 373–375.

Richter, F., Ebersberger, A., Schaible, H.G., 2002. Blockade of voltage-gated calcium channels in rat inhibits repetitive cortical spreading depression. Neurosci. Lett. 334, 123–126.

Richter, K., Hamprecht, B., Scheich, H., 1996. Ultrastructural localization of glycogen phosphorylase predominantly in astrocytes of the gerbil brain. Glia 17, 263–273.

Robb-Gaspers, L.D., Burnett, P., Rutter, G.A., Denton, R.M., Rissuto, R., Thomas, A.P., 1998. Integrating cytosolic calcium signals into mitochondrial metabolic responses. EMBO J. 17, 4987–5000.

Rohlmann, A., Wolff, J.R., 1996. Subcellular topography and plasticity of gap junction distribution in astrocytes. In: Spray, D.C., Dermietzel, R. (Eds.), Gap Junctions in the Nervous System. Landes, Austin, Texas, pp. 175–192.

Ruiz-Amil, M., De Torrontegui, G., Palacian, E., Catalina, L., Losada, M., 1965. Properties and function of yeast pyruvate carboxylase. J. Biol. Chem. 240, 3485–3492.

Rutter, G.A., Burnett, P., Rizzuto, R., Brini, M., Murgia, M., Pozza, T., Tavare, J.M., Denton, R.M., 1996. Subcellular imaging of intramitochondrial Ca^{2+} with recombinant targeted aequorin: significance for the regulation of pyruvate dehydrogenase activity. Proc. Natl Acad. Sci. USA 93, 5489–5494.

Savaki, H.E., Kadekaro, M., McCulloch, J., Sokoloff, L., 1982. The central noradrenergic system in the rat: metabolic mapping with alpha-adrenergic blocking agents. Brain Res. 234, 65–79.

Sokoloff, L., Reivich, M., Kennedy, C., Des Rosiers, M.H., Patlak, C.S., Pettigrew, K.D., Sakurada, O., Shinohara, M., 1977. The [^{14}C]deoxyglucose method for the measurement of local cerebral glucose utilization: theory, procedure, and normal values in the conscious and anesthetized albino rat. J. Neurochem. 28, 897–916.

Sokoloff, L., Takahashi, S., Gotoh, J., Driscoll, B.F., Law, M.J., 1996. Contribution of astroglia to functionally activated energy metabolism. Dev. Neurosci. 18, 343–352.

Somjen, G.G., 1979. Extracellular potassium in the mammalian central nervous system. Ann. Rev. Physiol. 41, 159–177.

Somjen, G.G., 2001. Mechanisms of spreading depression and hypoxic spreading depression-like depolarization. Physiol. Rev. 81, 1065–1096.

Sorg, O., Magistretti, P.J., 1992. Vasoactive intestinal peptide and noradrenaline exert long-term control on glycogen levels in astrocytes: blockade by protein synthesis inhibition. J. Neurosci. 12, 4923–4931.

Sorg, O., Pellerin, L., Stolz, M., Beggah, S., Magistretti, P.J., 1995. Adenosine triphosphate and arachidonic acid stimulate glycogenolysis in primary cultures of mouse cerebral astrocytes. Neurosci. Lett. 188, 109–112.

Stone, E.A., Ariano, M.A., 1989. Are glial cells target of the central noradrenergic system? A review of the evidence. Brain Res. Rev. 14, 297–309.

Stone, E.A., John, S.M., Zhang, Y., 1992. Studies of the cellular localization of biochemical responses to catecholamines in the brain. Brain Res. Bull. 29, 285–288.

Su, G., Haworth, R.A., Dempsey, R.J., Sun, D., 2000. Regulation of Na(+)–K(+)–Cl(−) co-transporter in primary astrocytes by dibutyryl cAMP and high $[K(+)]_0$. Am. J. Physiol. Cell. Physiol. 279, 1710–1721.

Subbarao, K.V., Hertz, L., 1990a. Noradrenaline induced stimulation of oxidative metabolism in astrocytes but not in neurons in primary cultures. Brain Res. 527, 346–349.

Subbarao, K.V., Hertz, L., 1990b. Effects of adrenergic agonists on glycogenolysis in primary cultures of astrocytes. Brain Res. 527, 346–349.

Subbarao, K.V., Hertz, L., 1991. Stimulation of energy metabolism in astrocytes by adrenergic agonists. J. Neurosci. Res. 28, 399–405.

Subbarao, K.V., Stolzenburg, J.-U., Hertz, L., 1995. Pharmacological characteristics of potassium-induced glycogenolysis in astrocytes. Neurosci. Lett. 196, 45–48.

Swanson, R.A., Morton, M.M., Sagar, S.M., Sharp, F.R., 1992. Sensory stimulation induces local cerebral glycogenolysis: demonstration by autoradiography. Neuroscience 51, 451–461.

Thorlin, T., Eriksson, P.S., Rönnbäck, L., Hansson, E., 1998. Receptor-activated Ca^{2+} increases in vibrodissociated cortical astrocytes: a nonenzymatic method for acute isolation of astrocytes. J. Neurosci. Res. 54, 390–401.

Ueda, H., 1999. In vivo molecular signal transduction of peripheral mechanisms of pain. Jpn. J. Pharmacol. 79, 263–268.

Ververken, D., Van Veldhoven, P., Proost, C., Carton, H., De Wulf, H., 1982. On the role of calcium ions in the regulation of glycogenolysis in brain cortical slices. J. Neurochem. 38, 1286–1295.

Vyskocil, F., Kriz, N., Bures, J., 1972. Potassium-selective microelectrodes used for measuring the extracellular brain potassium during spreading depression and anoxic depolarization in rats. Brain Res. 39, 255–259.

Waagepetersen, H.S., Westergaard, N., Schousboe, A., 2000. The effects of isofagomine, a potent glycogen phosphorylase inhibitor, on glycogen metabolism in cultured mouse cortical astrocytes. Neurochem. Int. 36, 435–440.

Walz, W., Hertz, L., 1984. Intense furosemide-sensitive potassium accumulation into astrocytes in the presence of pathologically high extracellular potassium levels. J. Cereb. Blood Flow Metab. 4, 301–304.

Waniewski, R.A., Martin, D.L., 1998. Preferential utilization of acetate by astrocytes is attributable to transport. J. Neurosci. 18, 5225–5233.

Wender, R., Brown, A.M., Fern, R., Swanson, R.A., Farrell, K., Ransom, B.R., 2000. Astrocytic glycogen influences axon function and survival during glucose deprivation in central white matter. J. Neurosci. 20, 6804–6810.

Wolff, J.R., Stuke, K., Missler, M., Tytko, H., Schwarz, P., Rohlmann, A., Chao, T.I., 1998. Autocellular coupling by gap junctions in cultured astrocytes: a new view on cellular autoregulation during process formation. Glia 24, 121–140.

Woods, M.J., Williams, D.C., 1996. Multiple forms and locations for the peripheral-type benzodiazepine receptor. Biochem. Pharmacol. 52, 1805–1814.

Yarowsky, P., Boyne, A.F., Wierwille, R., Brookes, N., 1986. Effect of monensin on deoxyglucose uptake in cultured astrocytes: energy metabolism is coupled to sodium entry. J. Neurosci. 6, 859–866.

Yu, A.C.H., Schousboe, A., Hertz, L., 1982. Metabolic fate of (^{14}C)-labelled glutamate in astrocytes. J. Neurochem. 39, 954–966.

Zhao, Z., Hertz, L., Code, W.E., 1996. Effects of benzodiazepines on potassium induced increase in free cytosolic calcium concentration in astrocytes and neurons in primary cultures. Can. J. Physiol. Pharmacol. 74, 273–277.

Advances in
Molecular and
Cell Biology

Role of astrocytes in homeostasis of glutamate and GABA during physiological and pathophysiological conditions

Arne Schousboe* and Helle S. Waagepetersen

Department of Pharmacology, Royal Danish School of Pharmacy, Universitetsparken 2, DK-2100 Copenhagen, Denmark.
**Correspondence address: Tel.: +45-3530-6330; fax: +45-3530-6021.*
E-mail: as@mail.dfh.d(A.S.)

Contents

1. Introduction
2. Glutamate
   2.1. Release
   2.2. Uptake
   2.3. Metabolism
3. GABA
   3.1. Release
   3.2. Uptake
   3.3. Metabolism
4. Concluding remarks

Glutamatergic and GABAergic neurotransmission is terminated by high affinity uptake of the released neurotransmitters into the neuronal or astroglial entities of the synapses. In case of glutamate astroglial transport prevails over neuronal transport, whereas for GABA the opposite is the case. Regardless of this, the astroglial contribution to removal of the transmitter is of crucial importance for function. Malfunction, which may occur, e.g., as a result of energy failure, therefore has serious consequences, such as neuronal degeneration. In this context, the metabolism of the transmitters is also of interest, since astroglial cells control the availability of precursors for biosynthesis of glutamate and GABA in glutamatergic and GABAergic neurons, respectively. The energy balance plays an important role in these processes as well.

Advances in Molecular and Cell Biology, Vol. 31, pages 461–474

ISBN: 0-444-51451-1

1. Introduction

Generally the homeostatic mechanisms operating in relation to neurotransmission consist of enzymatic processes regulating biosynthesis and biodegradation of the neurotransmitters as well as plasma membrane transporters, which conduct uptake of the transmitters into nerve endings. In the case of excitatory and inhibitory neurotransmission, mediated by the amino acids glutamate and γ-aminobutyrate (GABA), respectively, the homeostasis is more complicated, since astrocytes play a prominent role in glutamatergic and GABAergic neurotransmission with regard to both membrane transporters and the metabolic machinery involved (Hertz et al., 1999; Waagepetersen et al., 1999b; Schousboe, 2003). It is thought provoking that the role of astrocytes is particularly prominent in the neurotransmission processes utilizing glutamate and GABA, since neurotransmission mediated by these two transmitters account for at least 90% of all synaptic neurotransmission in the central nervous system.

In this context it may be of interest to note important quantitative and functional differences between glutamatergic and GABAergic neurotransmission. From studies of rates of release of neurotransmitter glutamate and GABA it may be concluded that the amount of released glutamate exceeds that of GABA by at least a factor of 10 (Gram et al., 1988; Palaiologos et al., 1988; Waagepetersen et al., 2001b). In keeping with this, the capacity of glutamate transporters in the central nervous system or cultured neural cells is far higher than that of GABA transporters (Balcar and Johnston, 1973; Schousboe et al., 1977a,b; Larsson et al., 1981; Drejer et al., 1982). Another important difference with regard to the transporters is their distribution between neurons and astrocytes, where astrocytic glutamate transporters by far outnumber neuronal glutamate transporters (Hertz, 1979; Schousboe, 1981; Hertz et al., 1999; Danbolt, 2001), whereas in the case of GABA neuronal transporters dominate (Schousboe and Kanner, 2002; Schousboe, 2003). These aspects of glutamate and GABA homeostasis are schematically presented in Fig. 1 (adapted from Hertz and Schousboe, 1987), which shows preferential uptake of released glutamate into glia and preferential uptake of released GABA into neurons. A mainly neuronal re-uptake of GABA, but not of glutamate, results in the ability of GABA synapses to operate primarily by recycling of the neurotransmitter, whereas glutamatergic neurotransmission is dependent upon continuous de novo synthesis of neurotransmitter from either astrocytically accumulated glutamate or from glucose. Due to the fact that key enzymes in these processes (glutamine synthetase and pyruvate carboxylase) are expressed only in astrocytes (see Hertz et al., 1999), glutamatergic neurons are totally dependent on metabolic interactions with surrounding astrocytes for supply of precursors for synthesis of transmitter glutamate (Westergaard et al., 1995; Hertz et al., 1999). Interruption of the glial components of these interactions almost immediately abolishes glutamatergic neurotransmission (Keyser and Pellmar, 1994).

The present review is aimed at a discussion of these aspects of glutamate and GABA homeostasis during normal function of the CNS and during conditions of energy failure induced by hypoxia, hypoglycemia or ischemia. An additional discussion focusing on the beneficial or harmful consequences of interactions between neurons and astrocytes during focal ischemia is provided in the chapter by Håberg and Sonnewald.

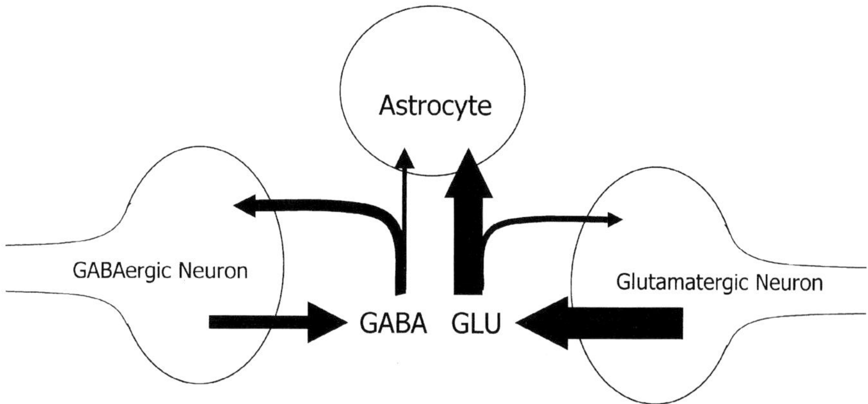

Fig. 1. Illustration of evoked release and uptake of glutamate and GABA in GABAergic or glutamatergic neurons and astrocytes. The sizes of the *arrows* provide an estimate of the relative magnitudes of the respective fluxes. It can be seen that neuronally released glutamate to a major extent is accumulated into astrocytes, whereas most of the released GABA is reaccumulated into neurons.

2. Glutamate

2.1. Release

Depolarization of glutamatergic neurons, typically by activation of different types of glutamate receptors, leads to influx of Ca^{2+} and subsequent release of vesicularly stored glutamate (McMahon and Nicholls, 1991). Depending upon the depolarizing signal the release may under certain conditions consist not only of vesicular glutamate but also to a considerable extent of glutamate released by reversal of the glutamate transporters, i.e., originating from the cytoplasmic, 'metabolic' pool of glutamate (Belhage et al., 1992; Bernath, 1992; Jensen et al., 2000; Waagepetersen et al., 2001a; Bak et al., 2003). Vesicular release is energy dependent and is therefore likely to be decreased during energy failure (Nicholls and Attwell, 1990). On the contrary, energy failure will lead to a massive release of glutamate via a reversal of the glutamate carriers, as indeed observed during ischemia and hypoglycemia (Benveniste et al., 1984; Hagberg et al., 1985; Sandberg et al., 1986; Phillis et al., 2000). This is because the plasma membrane glutamate transporters are dependent on energy and co-transport of Na^+ along an intact Na^+ gradient for optimal function (Danbolt, 2001) and during energy deprivation serve to release glutamate along its concentration gradient. It should be noted that although the glutamate content is larger in the neuronal compartment than in the glial compartment (Ottersen et al., 1992), a considerable fraction of the glutamate released by reversal of the transporters during energy failure could be of astrocytic origin, because the vast majority of glutamate transporters are localized on astrocytes (Lehre et al., 1995; Lehre and Danbolt, 1998; Levy, 2002). Direct evidence supporting the concept that the metabolic glutamate pool may be quantitatively more important than the neurotransmitter pool in ischemia-induced overflow of glutamate in hippocampus is provided by the demonstration that phenylsuccinate, which selectively inhibits biosynthesis of transmitter glutamate

(Palaiologos et al., 1988), blocks K^+-stimulated overflow of glutamate in the brain in vivo but not the overflow occurring during a 20 min period of global cerebral ischemia (Christensen et al., 1991). It should also be noted that release of glutamate evoked by combined deprivation of glucose and oxygen both in vivo and in vitro can be reduced by threo-β-benzyloxyaspartate (TBOA), a non-transportable inhibitor of the glutamate transporters (Phillis et al., 2000; Anderson et al., 2001; Waagepetersen et al., 2001a; Bonde et al., 2003).

2.2. *Uptake*

As mentioned above, glutamate uptake is an energy dependent process, which is mediated by different plasma membrane carriers, five of which have been cloned and named EAAT1–5 (Gegelashvili and Schousboe, 1998; Danbolt, 2001). Of these, EAAT1 (GLAST) and EAAT2 (GLT), which have a preferential if not exclusive glial expression (Danbolt, 2001; Levy, 2002), are by far the most important for maintenance of low extracellular glutamate levels, and they are able to create an extracellular/intracellular glutamate gradient of $1/10^5$. Since the cycling time of the glutamate transporters is relatively long (Wadiche et al., 1995), binding to the extremely abundant glutamate transporters in the glial plasma membrane is likely to be of fundamental importance for the immediate clearance of glutamate in the synaptic cleft following synaptic release of vesicular glutamate (Lehre and Danbolt, 1998). However, although the binding of glutamate to the carrier is not in itself energy dependent, the capacity for binding is exhausted in case the transport cycle is interrupted by energy failure and collapse of the transmembrane Na^+ gradient (Wadiche et al., 1995). This obviously will have serious consequences for neuronal function due to the excitotoxic action of glutamate (Choi and Rothman, 1990; Schousboe and Frandsen, 1995).

2.3. *Metabolism*

After receptor interaction, astrocytically accumulated neurotransmitter glutamate can be oxidatively degraded in the tricarboxylic acid (TCA) cycle (Fig. 2), amidated to glutamine, or converted into other metabolites, e.g., glutathione. Glutamate dehydrogenase (GDH) and several different aminotransferases, such as aspartate- and alanine-aminotransferases, are capable of catalyzing the reversible conversion of glutamate to its corresponding α-ketoacid, thereby initiating the oxidative degradation of the carbon skeleton (Fig. 2). In astrocytes the entrance of exogenous glutamate into the TCA cycle occurs most likely via the action of GDH (Yu et al., 1982; Schousboe et al., 1993; Westergaard et al., 1996). GDH is a strictly mitochondrial enzyme, located in the inner mitochondrial membrane, whereas the aminotransferases exist in cytosolic as well as mitochondrial isoforms. Both aspartate and lactate are formed during glutamate metabolism via the TCA cycle in astrocytes (Sonnewald et al., 1993b), the latter after exit of glutamate-derived malate from the TCA cycle (Sonnewald et al., 1996; Waagepetersen et al., 2002). This is a quantitatively important pathway (Hertz and Hertz, 2003), necessitating de novo synthesis of glutamate from glucose (see below).

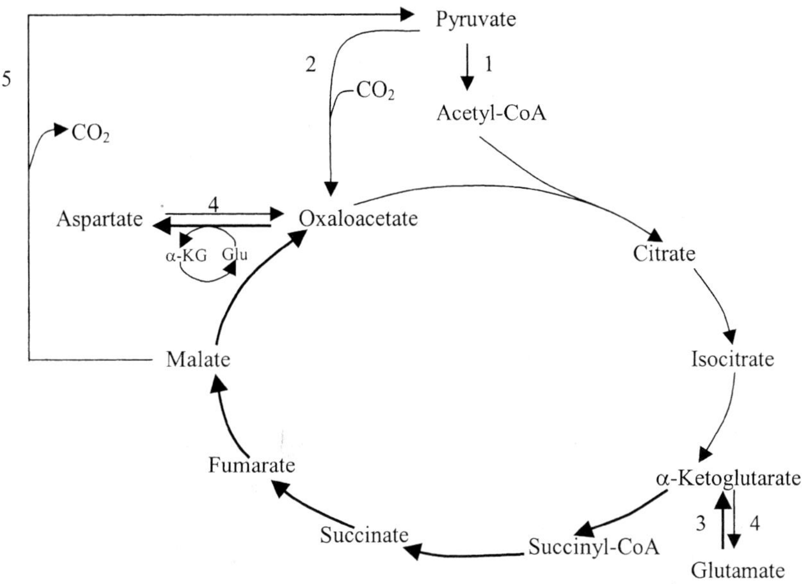

Fig. 2. Schematic illustration of the TCA cycle and closely connected reactions, i.e., those catalyzed by, pyruvate dehydrogenase (1), pyruvate carboxylase (2), glutamate dehydrogenase (3), aspartate aminotransferase (4) and malic enzyme (5). The pathways accounting for the net synthesis of aspartate from glutamate are indicated by bold arrows.

There is a large increase in the synthesis of aspartate from glutamate during hypoglycemia (Bakken et al., 1998), when glutamate becomes an alternative energy substrate. For every molecule of glutamate which is converted to aspartate, 9–12 molecules of ATP are produced in a truncated TCA cycle (3 during GDH-mediated conversion of glutamate to α-ketoglutarate [but none during transamination], 3 during conversion of α-ketoglutarate to succinyl coenzyme A, 1 during conversion of succinyl coenzyme A to succinate, 2 during conversion of succinate to fumarate and 3 during conversion of malate to oxaloacetate as outlined in Fig. 2). In contrast, lactate production from glutamate via the TCA cycle decreases during hypoglycemia as well as hypoxia (Bakken et al., 1998).

The balance between the extent of oxidative consumption of glutamate and synthesis of glutamine is dependent on the extracellular concentration of glutamate, with relatively more glutamate being oxidized at higher glutamate concentrations (McKenna et al., 1996a). Glutamine is synthesized exclusively in astrocytes (and other glial cells) by an ATP dependent amidation of glutamate, catalyzed by the cytosolic enzyme glutamine synthetase (Norenberg and Martinez-Hernandez, 1979; D'Amelio et al., 1990; Tansey et al., 1991). In spite of the energy dependence of the glutamine synthetase reaction it is not sensitive to hypoxia or hypoglycemia in cultured astrocytes (Bakken et al., 1998). Glutamine, a non-neuroactive amino acid, which is the predominant glutamate precursor, is released from astrocytes and taken up in neurons, where it is rapidly and extensively metabolized to glutamate (Bradford et al., 1978; Rothstein and Tabakoff, 1984; Szerb and

O'Reagan, 1985; Hogstad et al., 1988; Shank et al., 1989). Thus, the net flow of glutamate from neurons to astrocytes is counterbalanced by a glutamine flow from astrocytes to neurons, a phenomenon, which is referred to as the glutamate-glutamine cycle, shown in Fig. 3 (Berl and Clarke, 1983). Recently, transporters for glutamine out of astrocytes and into neurons have attracted considerable interest (Boulland et al., 2002). A neuronal high affinity glutamine transporter has been cloned and shown to be preferentially located on glutamatergic neurons (Varoqui et al., 2000), confirming the importance of glutamine as a glutamate precursor.

The supply of a glutamate precursor to the neurons is necessary due to the fact that neurons lack a quantitatively important anaplerotic pathway, since they do not express the enzyme pyruvate carboxylase (Yu et al., 1983; Shank et al., 1985; Kaufman and Driscoll, 1992; Cesar and Hamprecht, 1995), an enzyme which is crucial for net synthesis of TCA cycle intermediates and their derivatives, including glutamate and glutamine. However, on account of the considerable oxidative metabolism of glutamate in astrocytes the transport of glutamine from astrocytes to neurons is not stochiometrically equivalent to the uptake of neuronally released glutamate, and in addition some glutamine is oxidized in neurons (Hertz and Schousboe, 1986; Yudkoff et al., 1988; Hertz et al., 1992a; Westergaard et al., 1995). In order to compensate for the ensuing deficit in neuronal glutamate precursor, additional glutamate must be generated from other sources, probably mainly or exclusively glucose. The de novo synthesis of glutamate from glucose requires pyruvate carboxylation. There is a relatively high pyruvate carboxylase activity in astrocytes (Yu et al., 1983; Kaufman and Driscoll, 1992; Gamberino et al., 1997), and a high rate of pyruvate carboxylation has been established in the brain in vivo, including human brain (see chapter by Gruetter). Since net synthesis of glutamate occurs in astrocytes but not in neurons, either a TCA cycle intermediate, glutamate itself or glutamine must be transported from astrocytes to neurons. Glutamate itself is not suitable on account of its transmitter activity, but it is likely that at least some glutamate is converted in astrocytes to glutamine, which is then carried across to neurons in the glutamate-glutamine cycle.

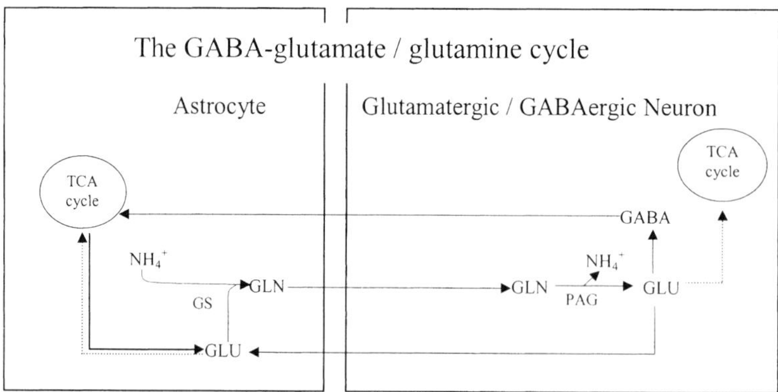

Fig. 3. The GABA-glutamate/glutamine cycle illustrating the flux of the neurotransmitters glutamate and GABA from the neuronal to the astrocytic compartment and the corresponding glutamine flux in the opposite direction. GLN, glutamine; GLU, glutamate; GS, glutamine synthetase; PAG, phosphate activated glutaminase.

The TCA cycle intermediate, α-ketoglutarate has also been shown to serve as precursor for the releasable neurotransmitter pool of glutamate, although not to the same extent as glutamine (Kihara and Kubo, 1989; Shank et al., 1989; Peng et al., 1991). α-Ketoglutarate as well as other TCA cycle intermediates, such as citrate, malate and succinate are released to a larger extent from astrocytes than from neurons. Citrate is the intermediate released at by far the highest rate and the extent of release is sensitive to hypoxia (Sonnewald et al., 1991; Müller et al., 1994; Westergaard et al., 1994a,b). This may be explained by the importance of the energy requiring process, pyruvate carboxylation for the synthesis of releasable citrate (Westergaard et al., 1994b; Waagepetersen et al., 2001c). Evidence for a role of citrate as a neuronal glutamate precursor could, however, not be obtained (Westergaard et al., 1994b). The possible role of malate as a glutamate precursor seems limited, considering a very modest uptake rate of malate into neurons (Shank and Campbell, 1984; Hertz et al., 1992b) compared to the rate of stimulated glutamate release (Drejer et al., 1982). Furthermore, the incorporation of ^{14}C from [^{14}C]malate into glutamate appeared inadequate for malate to be a significant precursor for glutamate (Shank and Campbell, 1984; Hertz et al., 1992b).

Astrocytic TCA cycle and amino acid metabolism is complicated by the existence of mitochondrial heterogeneity presumably within each single cell (Schousboe et al., 1993; Sonnewald et al., 1993a, 1998; McKenna et al., 1996b; Waagepetersen et al., 1999a). This is particularly evident for the metabolic pathways leading to synthesis of citrate and glutamine (Schousboe et al., 1993; Waagepetersen et al., 2001c). Pyruvate carboxylation seems extremely important for synthesis of a releasable pool of citrate. The sequence of reactions leading to synthesis of this pool of citrate is separated from that operating for the synthesis of glutamine (Waagepetersen et al., 2001c). These two energy requiring processes, i.e., pyruvate carboxylation and glutamine synthesis, might a priori both be susceptible to hypoxia but a differential effect of hypoxia on release of citrate and glutamine might be explained by the above mentioned compartmentation (Sonnewald et al., 1994; Waagepetersen et al., 2001c). Moreover, synthesis of releasable glutamine, i.e., the neurotransmitter precursor pool, is compartmentalized from synthesis of a main intracellular pool of glutamine (Waagepetersen et al., 2001c). This compartmentation of the biochemical machinery might serve an important role decreasing the vulnerability of astrocytes during energy failure.

3. GABA

3.1. Release

In analogy with other neurotransmitters, the inhibitory neurotransmitter, GABA, is released by exocytosis from GABAergic nerve endings following depolarization (Sihra and Nicholls, 1987). However, depolarization may additionally lead to release from the cytoplasmic pool via reversal of the plasma membrane carriers (Bernath, 1992; Belhage et al., 1993). This means that the release process, like that for glutamate (see above), is sensitive to energy failure, which favors release from the cytoplasmic pool. In keeping with this, ischemia has been shown to result in an increase in the extracellular concentration of GABA, measured by microdialysis (Hagberg et al., 1985—see also chapter by Håberg and Sonnewald).

3.2. Uptake

The maintenance of a low extracellular GABA concentration is brought about by plasma membrane GABA carriers, four of which have been cloned and named GAT1–4 (Schousboe and Kanner, 2002). It should be noted that the nomenclature for the GABA transporters is confusing, since the numbering differs between transporters cloned from mice and rats (Schousboe and Kanner, 2002). These transporters are mainly expressed in GABAergic neurons, and GAT1 is the most abundant transporter. However, glial cells also express GAT1 in addition to GAT2 and GAT4 (Schousboe and Kanner, 2002). That the GABA transporters are important for the maintenance of low extracellular GABA concentrations is demonstrated by repeated findings that application of inhibitors of GABA transporters, using the microdialysis technique, leads to an increase in the extracellular concentration of GABA (Fink-Jensen et al., 1992; Richards and Bowery, 1996; Juhász et al., 1997). Since the GABA carriers are Na^+ dependent and electrogenic, the uptake process is also highly energy dependent (Schousboe, 1981).

The ability of inhibitors of GABA transport to increase extracellular GABA levels is particularly pronounced for those inhibitors which act primarily on glial GABA uptake (Juhász et al., 1997; White et al., 2002), probably reflecting that GABA which has been taken up into presynaptic nerve endings is primarily re-used as a transmitter, whereas GABA accumulated by astrocytes is either oxidatively degraded or returned to neurons in a rather complex GABA-glutamine-glutamate shuttle (see below). It has therefore been suggested (Schousboe et al., 1983) that inhibitors of glial GABA uptake might be particularly attractive as anticonvulsant agents. It should, however, be noted that other factors may also play important roles in this context, as the antiepileptic drug Tiagabine, which exclusively inhibits GAT1, is only slightly more potent as an inhibitor of glial GABA uptake compared to inhibition of neuronal GABA uptake (White et al., 2002). It may also be of interest to note that seizure sensitive gerbils have a higher expression of GAT1 than their seizure resistant counterparts, which has been interpreted as a compensatory mechanism to provide a higher release of GABA by reversal of the carrier during depolarization (Kang et al., 2001). However, it would be in keeping with the anticonvulsant activity of astrocytic GABA uptake inhibitors, if the excessive expression of GAT1 in these animals mainly were astrocytic (Schousboe et al., 1983; White et al., 2002).

3.3. Metabolism

Although astrocytes lack the enzyme responsible for synthesis of GABA from glutamate they do play an important role in GABA metabolism (Schousboe, 1980; Waagepetersen et al., 1999b). Thus, glutamine also functions as a precursor for GABA via glutamate (Reubi et al., 1978; Battaglioli and Martin, 1990, 1991; Sonnewald et al., 1993c), whereas there is no evidence that the TCA cycle intermediates, which can support synthesis of transmitter glutamate (see above), also can serve as precursors for GABA synthesis (Hertz et al., 1992b). This suggests that the only route for de novo formation of GABA is via formation of glutamine in astrocytes, followed by its transfer to GABAergic neurons, hydrolysis to glutamate, and decarboxylation to GABA by glutamate decarboxylase, GAD (Waagepetersen et al., 1999b). This series of metabolic processes occurs in a glutamate-glutamine cycle, which has been expanded to a GABA-glutamate-glutamine cycle (Van den Berg and Garfinkel, 1971; Berl and Clarke, 1983). This cycle is illustrated in Fig. 3. It involves several steps requiring a high energy level, such as the glutamine synthetase reaction in the astrocytic compartment, and a need for entry of at least part of the newly synthesized glutamate into an operational TCA cycle in the neuronal compartment (Waagepetersen et al., 1999b, 2001b). This means that although GAD is capable of operating under hypoxic conditions, maintenance of a functional GABA pool in GABAergic neurons is hampered by oxygen deprivation.

The finding that TCA cycle intermediates are not precursors for GABA is in agreement with a lower quantitative demand for precursors in GABAergic neurons, since the net flow of glutamate into astrocytes from neurons is much higher than that of GABA (Hertz and Schousboe, 1987) as illustrated in Fig. 1.

Oxidative degradation of GABA in astrocytes, facilitated by a high activity of GABA-transaminase (Schousboe et al., 1977a), is also dependent on oxidative metabolism, since the subsequent catabolism requires oxidation of generated succinic acid semialdehyde by succinic acid semialdehyde dehydrogenase to succinate. As shown in Fig. 2, succinate can be further metabolized in the TCA cycle to malate, which can exit the cycle to form pyruvate, a process which preferentially takes place in astrocytes (Sonnewald et al., 1996; Waagepetersen et al., 1999b, 2002; Lieth et al., 2001). The finding that extracellular GABA levels increase during energy failure (Hagberg et al., 1985) is therefore likely to reflect not only the failure of the GABA carriers to maintain a steep extra/intracellular concentration gradient (see above) but also a decreased glial as well as neuronal catabolic capacity. Alternatively malate may remain in the TCA cycle and eventually be converted to α-ketoglutarate (Fig. 2), followed by formation of glutamate and glutamine in astrocytes and transfer of glutamine to neurons, where it can be hydrolyzed to glutamate, which is then decarboxylated by GAD to form GABA. Thus, a possible return of astrocytically accumulated GABA to neurons follows a much more complex pathway than return of glutamate in the glutamate-glutamine cycle, and it is directly dependent upon oxidative metabolism.

4. Concluding remarks

The metabolic relationships between glutamine, glutamate and GABA have always been a cornerstone in studies of metabolic interactions between different neural cells. The very concept of metabolic compartmentation in brain was developed from the discovery of an anomalous precursor product relationship between glutamate and glutamine in whole brain, and the operation of a GABA-glutamate-glutamine cycle was deduced from the formation of GABA in one metabolic compartment and its metabolism, at least partly, in a different compartment (Berl and Clarke, 1983). Studies of kinetics for uptake of amino acids and of metabolic fluxes in primary cultures of astrocytes and of different types of neurons were instrumental in establishing present-day understanding of these interactions, as outlined in the present chapter. During recent years, modern techniques, such a nuclear magnetic resonance spectroscopy, have enabled quantitative determination of metabolic fluxes in the intact brain during both physiological and pathophysiological conditions (see chapters by Håberg and Sonnewald and by Gruetter), and they have been instrumental in the development of the concept that de novo synthesis, especially of glutamate, is a quantitatively important pathway, even compared with return of astrocytically accumulated glutamate and GABA via the glutamate-glutamine and the GABA-glutamate-glutamine cycle (a concept that had been proposed earlier, based on the high rate of oxidative degradation of glutamate in cultured astrocytes). What we still do not know is the dynamics of these interactions during increased brain activity, e.g., whether brain stimulation might be associated with enhanced de novo synthesis of glutamate from glucose, and cessation of brain activity with oxidative degradation of glutamate and glutamine. Rapid technical developments of in vivo methods for determination of metabolic fluxes in brain, may provide answers to this question within the coming years.

Acknowledgements

The expert secretarial assistance by Ms Hanne Danø is highly appreciated. The work has received financial support from the Danish MRC (grants 22-00-1011 and 1747), as well as the Lundbeck and NOVO Nordisk Foundations.

References

Anderson, C.M., Bridges, R.J., Chamberlin, A.R., Shimamoto, K., Yasuda-Kamatani, Y., Swanson, R.A., 2001. Differing effects of substrate and non-substrate transport inhibitors on glutamate uptake reversal. J. Neurochem. 79, 1207–1216.

Bak, L., Schousboe, A., Waagepetersen, H.S., 2003. Characterization of depolarization-coupled release of glutamate from cultured mouse cerebellar granule cells using DL-threo-β-benzyloxyaspartate (DL-TBOA) do distinguish between the vesicular and cytoplasmic pools. Neurochem. Int. 43, 417–424.

Bakken, I.J., White, L.R., Unsgard, G., Aasly, J., Sonnewald, U., 1998. [U-^{13}C]glutamate metabolism in astrocytes during hypoglycemia and hypoxia. J. Neurosci. Res. 51, 636–645.

Balcar, V.J., Johnston, G.A., 1973. High affinity uptake of transmitters: studies on the uptake of L-aspartate, GABA, L-glutamate and glycine in cat spinal cord. J. Neurochem. 20, 529–539.

Battaglioli, G., Martin, D.L., 1990. Stimulation of synaptosomal γ-aminobutyric acid synthesis by glutamate and glutamine. J. Neurochem. 54, 1179–1187.

Battaglioli, G., Martin, D.L., 1991. GABA synthesis in brain slices is dependent on glutamine produced in astrocytes. Neurochem. Res. 2, 151–156.

Belhage, B., Hansen, G.H., Schousboe, A., 1993. Depolarization by K^+ and glutamate activates different neurotransmitter release mechanisms in GABAergic neurons: vesicular versus non-vesicular release of GABA. Neuroscience 54, 1019–1034.

Belhage, B., Rehder, V., Hansen, G.H., Kater, S.B., Schousboe, A., 1992. ^{3}H-D-Aspartate release from cerebellar granule neurons is differentially regulated by glutamate- and K^+-stimulation. J. Neurosci. Res. 33, 436–444.

Benveniste, H., Drejer, J., Schousboe, A., Diemer, N.H., 1984. Elevation of the extracellular concentrations of glutamate and aspartate in rat hippocampus during transient cerebral ischemia monitored by intracerebral microdialysis. J. Neurochem. 43, 1369–1374.

Berl, S., Clarke, D.D., 1983. The metabolic compartmentation concept. In: Hertz, L., Kvamme, E., McGeer, E.G., Schousboe, A., (Eds.), Glutamine, Glutamate and GABA in the Central Nervous System, Alan R. Liss, Inc, New York, pp. 205–217.

Bernath, S., 1992. Calcium-independent release of amino acid neurotransmitters: fact or artifact? Prog. Neurobiol. 38, 57–91.

Bonde, C., Sarup, A., Schousboe, A., Gegelashvili, G., Zimmer, J., Noraberg, J., 2003. Neurotoxic and neuroprotective effects of the glutamate transporter inhibitor DL-threo-beta-benzylaspartate (DL-TBOA) during physiological and ischemia-like conditions. Neurochem. Int. 43, 371–380.

Boulland, J.L., Osen, K.K., Levy, L.M., Danbolt, N.C., Edwards, R.H., Storm-Mathisen, J., Chaudhry, F.A., 2002. Cell-specific expression of the glutamine transporter SN1 suggests differences in dependence on the glutamine cycle. Eur. J. Neurosci. 15, 1615–1631.

Bradford, H.F., Ward, H.K., Thomas, A.J., 1978. Glutamine—a major substrate for nerve endings. J. Neurochem. 30, 1453–1459.

Cesar, M., Hamprecht, B., 1995. Immunocytochemical examination of neural rat and mouse primary cultures using monoclonal antibodies raised against pyruvate carboxylase. J. Neurochem. 64, 2312–2318.

Choi, D.W., Rothman, S.M., 1990. The role of glutamate neurotoxicity in hypoxic–ischemic neuronal death. Annu. Rev. Neurosci. 13, 171–182.

Christensen, T., Bruhn, T., Diemer, N.H., Schousboe, A., 1991. Effect of phenylsuccinate on potassium- and ischemia-induced release of glutamate in rat hippocampus monitored by microdialysis. Neurosci. Lett. 134, 71–74.

D'Amelio, F., Eng, L.F., Gibbs, M.A., 1990. Glutamine synthetase immunoreactivity is present in oligodendroglia of various regions of the central nervous system. Glia 3, 335–341.

Danbolt, N.C., 2001. Glutamate uptake. Prog. Neurobiol. 65, 1–105.

Drejer, J., Larsson, O.M., Schousboe, A., 1982. Characterization of glutamate uptake into and release from astrocytes and neurons cultured from different brain regions. Exp. Brain Res. 47, 259–269.

Fink-Jensen, A., Suzdak, P.D., Sweberg, M.D., Judge, M.E., Hansen, L., Nielsen, P.G., 1992. The GABA uptake inhibitor, TGB, increases extracellular brain levels of GABA in awake rats. Eur. J. Pharmacol. 20, 197–201.

Gamberino, W.C., Berkich, D.A., Lynch, C.J., Xu, B., LaNoue, K.F., 1997. Role of pyruvate carboxylase in facilitation of synthesis of glutamate and glutamine in cultured astrocytes. J. Neurochem. 69, 2312–2325.

Gegelashvili, G., Schousboe, A., 1998. Cellular distribution and kinetic properties of high-affinity glutamate transporters. Brain Res. Bull. 45, 233–238.

Gram, L., Larsson, O.M., Johnsen, A.H., Schousboe, A., 1988. Effects of valproate, vigabatrin and aminooxyacetic acid on release of endogenous and exogenous GABA from cultured neurons. Epilepsy Res. 2, 87–95.

Hagberg, H., Lehmann, A., Sandberg, M., Nyström, B., Jacobsen, I., Hamberger, A., 1985. Ischemia-induced shift of inhibitory and excitatory amino acids from intra- to extracellular compartments. J. Cereb. Blood Flow Metab. 5, 413–419.

Hertz, L., 1979. Functional interactions between neurons and astrocytes. I. Turnover and metabolism of putative amino acid transmitters. Prog. Neurobiol. 13, 277–323.

Hertz, L., Dringen, R., Schousboe, A., Robinson, S.R., 1999. Astrocytes: glutamate producers for neurons. J. Neurosci. Res. 57, 417–428.

Hertz, L., Hertz, E., 2003. Cataplerotic TCA cycle flux determined as glutamate-sustained oxygen consumption in primary cultures of astrocytes. Neurochem. Int. 43, 355–361.

Hertz, L., Peng, L., Westergaard, N., Yudkoff, M., Schousboe, A., 1992. Neuronal-astrocytic interactions in metabolism of transmitter amino acids of the glutamate family. In: Schousboe, a., Diemer, N.H., Kofod, H., (Eds.), Drug Research Related to Neuroactive Amino Acids. Alfred Benzon Symposium 32. Munksgaard, Copenhagen, pp. 30–48.

Hertz, L., Schousboe, A., 1986. Role of astrocytes in compartmentation of amino acid and energy metabolism. In: Fedoroff, S., Vernadakis, A. (Eds.), Astrocytes, vol. 2. Academic Press, New York, pp. 179–208.

Hertz, L., Schousboe, A., 1987. Primary cultures of GABAergic and glutamatergic neurons as model systems to study neurotransmitter functions. I. Differentiated cells. In: Vernadakis, A., Privat, L.M., Lauder, J.M., Timiras, P.S., Giacobini, E., (Eds.), Model Systems of Development and Aging of the Nervous System. Martinus Nijhoff Publishing, Boston, pp. 19–31.

Hertz, L., Yu, A.C., Schousboe, A., 1992. Uptake and metabolism of malate in neurons and astrocytes in primary cultures. J. Neurosci. Res. 33, 289–296.

Hogstad, S., Svenneby, G., Torgner, I.Aa., Kvamme, E., Hertz, L., Schousboe, A., 1988. Glutaminase in neurons and astrocytes cultured from mouse brain: kinetic properties and effects of phosphate, glutamate and ammonia. Neurochem. Res. 13, 383–388.

Jensen, J.B., Pickering, D.S., Schousboe, A., 2000. Depolarization-induced release of [^{3}H]D-aspartate from GABAergic neurons caused by reversal of glutamate transporters. Int. J. Dev. Neurosci. 18, 309–315.

Juhász, G., Kékesi, K.A., Nyitrai, G., Dobolyi, A., Krogsgaard-Larsen, P., Schousboe, A., 1997. Differential effects of nipecotic acid and 4,5,6,7-tetrahydroisoxazolo[4,5-*c*]pyridin-3-ol on extracellular γ-aminobutyrate levels in rat thalamus. Eur. J. Pharmacol. 331, 139–144.

Kang, T.-C., Kim, H.S., Seo, M.O., Park, S.K., Kwon, O.S., Kang, J.H., Won, M.H., 2001. The changes in the expressions of gamma-aminobutyric acid transporters in the gerbil hippocampal complex following spontaneous seizures. Neurosci. Lett. 310, 29–32.

Kaufman, E.E., Driscoll, B.F., 1992. Carbon dioxide fixation in neuronal and astoglial cells in culture. J. Neurochem. 58, 258–262.

Keyser, D.O., Pellmar, T.C., 1994. Synaptic transmission in the hippocampus: critical role for glial cells. Glia 10, 237–243.

Kihara, M., Kubo, T., 1989. Aspartate aminotransferase for synthesis of transmitter glutamate in the medulla oblongata: effect of aminooxyacetic acid and 2-oxoglutarate. J. Neurochem. 52, 1127–1134.

Larsson, O.M., Thorbek, P., Krogsgaard-Larsen, P., Schousboe, A., 1981. Effect of homo-∃-proline and other heterocyclic GABA analogues on GABA uptake in neurons and astroglial cells and on GABA receptor binding. J. Neurochem. 37, 1509–1516.

Lehre, K.P., Danbolt, N.C., 1998. The number of glutamate transporter subtype molecules at glutamatergic synapses: chemical and stereological quantification in young adult rat brain. J. Neurosci. 18, 8751–8757.

Lehre, K.P., Levy, L.M., Ottersen, O.P., Storm-Mathisen, J., Danbolt, N.C., 1995. Differential expression of two glial glutamate transporters in the rat brain: quantitative and immunocytochemical observations. J. Neurosci. 15, 1835–1853.

Levy, L.M., 2002. Structure, function and regulation of glutamate transporters. In: Egebjerg, J., Schousboe, A., Krogsgaard-Larsen, P., (Eds.), Glutamate and GABA Receptors and Transporters. Structure, Function and Pharmacology. Taylor and Francis, London, pp. 307–336.

Lieth, E., LaNoue, K.F., Berkich, D.A., Xu, B., Ratz, M., Taylor, C., Hutson, S.M., 2001. Nitrogen shuttling between neurons and glial cells during glutamate synthesis. J. Neurochem. 76, 1712–1723.

McMahon, H.T., Nicholls, D.G., 1991. Transmitter glutamate release from isolated nerve terminals: evidence for biphasic release and triggering by localized Ca^{2+}. J. Neurochem. 56, 86–94.

McKenna, M.C., Sonnewald, U., Huang, X., Stevenson, J., Zielke, R.H., 1996. Exogenous glutamate concentration regulates the metabolic fate of glutamate in astrocytes. J. Neurochem. 66, 386–393.

McKenna, M.C., Tildon, J.T., Stevenson, J.H., Huang, X., 1996. New insight into the compartmentation of glutamate and glutamine in cultured rat brain astrocytes. Dev. Neurosci. 18, 380–391.

Nicholls, D., Attwell, D., 1990. The release and uptake of excitatory amino acids. Trends Pharmacol. Sci. 11, 462–468.

Norenberg, M.D., Martinez-Hernandez, A., 1979. Fine structural localization of glutamine synthetase in astrocytes of rat brain. Brain Res. 161, 303–310.

Müller, T.B., Sonnewald, U., Westergaard, N., Schousboe, A., Petersen, S.B., Unsgård, G., 1994. ^{13}C NMR spectroscopy study of cortical nerve cell cultures exposed to hypoxia. J. Neurosci. Res. 38, 319–326.

Ottersen, O.P., Zhang, N., Walberg, F., 1992. Metabolic compartmentation of glutamate and glutamine: morphological evidence obtained by quantitative immunocytochemistry in rat cerebellum. Neuroscience 46, 519–534.

Palaiologos, G., Hertz, L., Schousboe, A., 1988. Evidence that aspartate amino transferase activity and ketodicarboxylate carrier function are essential for biosynthesis of transmitter glutamate. J. Neurochem. 51, 317–320.

Peng, L., Schousboe, A., Hertz, L., 1991. Utilization of alpha-ketoglutarate as a precursor for glutamate in cultured cerebellar granule cells. Neurochem. Res. 16, 29–34.

Phillis, J.W., Ren, J., O'Regan, M.H., 2000. Transporter reversal as a mechanism of glutamate release from the ischemic rat cerebral cortex: studies with DL-threo-beta-benzyloxyaspartate. Brain Res. 868, 105–112.

Reubi, J.-C., Van der Berg, C., Cuénod, M., 1978. Glutamine as precursor for the GABA and glutamate transmitter pools. Neurosci. Lett. 10, 171–174.

Richards, D.A., Bowery, N.G., 1996. Comparative effects of the GABA uptake inhibitors, tiagabine and NNC-711, on extracellular GABA levels in the rat ventrolateral thalamus. Neurochem. Res. 21, 135–140.

Rothstein, J.D., Tabakoff, B., 1984. Alteration of striatal glutamate release after glutamine synthetase inhibition. J. Neurochem. 43, 1438–1446.

Sandberg, M., Butcher, S.P., Hagberg, H., 1986. Extracellular overflow of neuroactive amino acids during severe insulin-induced hypoglycemia: in vivo dialysis of rat hippocampus. J. Neurochem. 47, 178–184.

Schousboe, A., 1980. Primary cultures of astrocytes from mammalian brain as a tool in neurochemical research. Cell. Mol. Biol. 26, 505–513.

Schousboe, A., 1981. Transport and metabolism of glutamate and GABA in neurons and glial cells. Int. Rev. Neurobiol. 22, 1–45.

Schousboe, A., 2000. Pharmacological and functional characterization of astrocytic GABA transport: a short review. Neurochem. Res. 25, 1241–1244.

Schousboe, A., 2003. Role of astrocytes in the maintenance and modulation of glutamatergic and GABAergic neurotransmission. Neurochem. Res. 28, 347–352.

Schousboe, A., Frandsen, Aa., 1995. Glutamate receptors and neurotoxicity. In: Stone, T.W. (Ed.), CNS Neurotransmitters and Neuromodulators: Glutamate. CRC Press, Boca Raton, FL, pp. 239–251.

Schousboe, A., Hertz, L., Svenneby, G., 1977. Uptake and metabolism of GABA in astrocytes cultured from dissociated mouse brain hemispheres. Neurochem. Res. 2, 217–229.

Schousboe, A., Kanner, B., 2002. GABA transporters: functional and pharmacological properties. In: Egebjerg, J., Schousboe, A., Krogsgaard-Larsen, P., (Eds.), Glutamate and GABA Receptors and Transporters. Taylor and Francis, London, pp. 337–349.

Schousboe, A., Larsson, O.M., Wood, J.D., Krogsgaard-Larsen, P., 1983. Transport and metabolism of GABA in neurons and glia: implications for epilepsy. Epilepsia 24, 531–538.

Schousboe, A., Svenneby, G., Hertz, L., 1977. Uptake and metabolism of glutamate in astrocytes cultured from dissociated mouse brain hemispheres. J. Neurochem. 29, 999–1005.

Schousboe, A., Westergaard, N., Sonnewald, U., Petersen, S.B., Huang, R., Peng, L., Hertz, L., 1993. Glutamate and glutamine metabolism and compartmentation in astrocytes. Dev. Neurosci. 15, 359–366.

Shank, R.P., Baldy, W.J., Ash, C.H., 1989. Glutamine and 2-oxoglutarate as metabolic precursors of the transmitter pools of glutamate and GABA: correlation of regional uptake by rat brain synaptosomes. Neurochem. Res. 14, 371–376.

Shank, R.P., Bennett, G.S., Freytag, S.O., Campbell, G.L., 1985. Pyruvate carboxylase: an astrocyte-specific enzyme implicated in the replenishment of amino acid neurotransmitter pools. Brain Res. 329, 364–367.

Shank, R.P., Campbell, G.L., 1984. Alfa-ketoglutarate and malate uptake and metabolism by synaptosomes: further evidence for an astrocytes-to-neuron metabolic shuttle. J. Neurochem. 42, 1153–1161.

Sihra, T.S., Nicholls, D.G., 1987. 4-Aminobutyrate can be released exocytotically from guinea-pig cerebral cortical synaptosomes. J. Neurochem. 49, 261–267.

Sonnewald, U., Hertz, L., Schousboe, A., 1998. Mitochondrial heterogeneity in the brain at the cellular level. J. Cereb. Blood Flow Metab. 18, 231–237.

Sonnewald, U., Müller, T.B., Westergaard, N., Unsgard, G., Schousboe, A., 1994. Neuronal–glial interactions in glutamate and GABA homeostasis: effects of hypoxia. In: Krieglstein, H., Oberpichler-Schwenk (Eds.), Pharmacology of Cerebral Ischemia, Wissenschaftliche Verlagsgesellschaft mbH, Stuttgart, pp. 177–190.

Sonnewald, U., Westergaard, N., Hassel, B., Muller, T.B., Unsgard, G., Fonnum, F., Hertz, L., Schousboe, A., Petersen, S.B., 1993. NMR spectroscopic studies of ^{13}C acetate and ^{13}C glucose metabolism in neocortical astrocytes: evidence for mitochondrial heterogeneity. Dev. Neurosci. 15, 351–358.

Sonnewald, U., Westergaard, N., Jones, P., Taylor, A., Bachelard, H.S., Schousboe, A., 1996. Metabolism of [U-$^{13}C_5$]glutamine in cultured astrocytes studied by NMR spectroscopy: first evidence of astrocytic pyruvate recycling. J. Neurochem. 67, 2566–2572.

Sonnewald, U., Westergaard, N., Krane, J., Unsgård, G., Petersen, S.B., Schousboe, A., 1991. First direct demonstration of preferential release of citrate from astrocytes using [^{13}C]NMR spectroscopy of cultured neurons and astrocytes. Neurosci. Lett. 128, 235–239.

Sonnewald, U., Westergaard, N., Petersen, S.B., Unsgård, G., Schousboe, A., 1993. Metabolism of [U-^{13}C]glutamate in astrocytes studied by ^{13}C NMR spectroscopy: incorporation of more label into lactate than into glutamine demonstrates the importance of the TCA cycle. J. Neurochem. 61, 1179–1182.

Sonnewald, U., Westergaard, N., Schousboe, A., Svendsen, J.S., Unsgård, G., Petersen, S.P., 1993. Direct demonstration by [^{13}C]NMR spectroscopy that glutamine from astrocytes is a precursor for GABA synthesis in neurons. Neurochem. Int. 1, 19–29.

Szerb, J.C., O'Regan, P.A., 1985. Effect of glutamine on glutamate release from hippocampal slices induced by high K^+ or by electrical stimulation: interaction with different Ca^{2+} concentrations. J. Neurochem. 44, 1724–1731.

Tansey, F.A., Farooq, M., Cammer, W., 1991. Glutamine synthetase in oligodendrocytes and astrocytes: new biochemical and immunocytochemical evidence. J. Neurochem. 56, 266–272.

Van den Berg, C.J., Garfinkel, D., 1971. A simulation study of brain compartments: metabolism of glutamate and related substances in mouse brain. Biochem. J. 23, 211–218.

Varoqui, H., Zhu, H., Yao, D., Ming, H., Erickson, J.D., 2000. Cloning and functional identification of a neuronal glutamine transporter. J. Biol. Chem. 275, 4049–4054.

Waagepetersen, H.S., Qu, H., Hertz, L., Sonnewald, U., Schousboe, A., 2002. Demonstration of pyruvate recycling in primary cultures of neocortical astrocytes but not in neurons. Neurochem. Res. 27, 1431–1437.

Waagepetersen, H.S., Shimamoto, K., Schousboe, A., 2001. Comparison of effects of DL-threo-β-benzyloxyaspartate (DL-TBOA) and L-*trans*-pyrrolidine-2,4-dicarboxylate (*t*-2,4-PDC) on uptake and release of [^{3}H]D-aspartate in astrocytes and glutamatergic neurons. Neurochem. Res. 26, 661–666.

Waagepetersen, H.S., Sonnewald, U., Gegelashvili, G., Larsson, O.M., Schousboe, A., 2001. Metabolic distinction between vesicular and cytosolic GABA in cultured GABAergic neurons using ^{13}C MRS. J. Neurosci. Res. 63, 347–355.

Waagepetersen, H.S., Sonnewald, U., Larsson, O.M., Schousboe, A., 2001. Multiple compartments with different metabolic characteristics are involved in biosynthesis of intracellular and released glutamine and citrate in astrocytes. Glia 35, 246–252.

Waagepetersen, H.S.; Sonnewald, U., Qu, H., Schousboe, A., 1999. Mitochondrial compartmentation at the cellular level—astrocytes and neurons. Ann. N. Y. Acad. Sci. 893, 421–425.

Waagepetersen, H.S., Sonnewald, U., Schousboe, A., 1999. The GABA paradox: multiple roles as metabolite, neurotransmitter, and neurodifferentiative agent. J. Neurochem. 73, 1335–1342.

Wadiche, J.I., Arriza, J.L., Amara, S.G., Kavanaugh, M.P., 1995. Kinetics of a human glutamate transporter. Neuron 14, 1019–1027.

Westergaard, N., Drejer, J., Schousboe, A., Sonnewald, U., 1996. Evaluation of the importance of transamination versus deamination in astrocytic metabolism of [U-^{13}C]glutamate. Glia 17, 160–168.

Westergaard, N., Sonnewald, U., Schousboe, A., 1994. Release of alfa-ketoglutarate, malate and succinate from cultured astrocytes: possible role in amino acid neurotransmitter homeostasis. Neurosci. Lett. 176, 105–109.

Westergaard, N., Sonnewald, U., Schousboe, A., 1995. Metabolic trafficking between neurons and astrocytes: the glutamate/glutamine cycle revisited. Dev. Neurosci. 17, 203–211.

Westergaard, N., Sonnewald, U., Unsgård, G., Peng, L., Hertz, L., Schousboe, A., 1994. Uptake, release and metabolism of citrate in neurons and astrocytes in primary cultures. J. Neurochem. 62, 1727–1733.

White, H.S., Sarup, A., Bolvig, T., Kristensen, A.S., Petersen, G., Nelson, N., Pickering, D.S., Larsson, O.M., Frølund, B., Krogsgaard-Larsen, P., Schousboe, A., 2002. Correlation between anticonvulsant activity and inhibitory action on glial GABA uptake of the highly selective mouse GAT1 inhibitor 3-hydroxy-4-amino-4,5,6,7-tetrahydro-1,2-benzisoxazole (*exo*-THPO) and its N-alkylated analogs. J. Pharmacol. Exp. Ther. 302, 636–644.

Yu, A.C.H., Drejer, J., Hertz, L., Schousboe, A., 1983. Pyruvate carboxylase activity in primary cultures of astrocytes and neurons. J. Neurochem. 41, 1484–1487.

Yu, A.C.H., Schousboe, A., Hertz, L., 1982. Metabolic fate of (^{14}C)-labelled glutamate in astrocytes. J. Neurochem. 39, 954–966.

Yudkoff, M., Nissim, I., Pleasure, D., 1988. Astrocyte metabolism of [^{15}N]glutamine: implications for the glutamine-glutamate cycle. J. Neurochem. 51, 843–850.

Advances in
Molecular and
Cell Biology

Astrocytic receptors and second messenger systems

Elisabeth Hansson* and Lars Rönnbäck

Institute of Clinical Neuroscience, Göteborg University, Medicinaregatan 5, SE 405 30 Göteborg, Sweden
**Correspondence address: Tel.: +46-31-773-3363; fax: +46-31-773-3330.*
E-mail: elisabeth.hansson@anatcell.gu.se(E.H.)

Contents

1. Introduction
2. Astrocytic G-protein- and/or ion channel-coupled membrane receptors
   - 2.1. Glutamate receptors
   - 2.2. Gamma-aminobutyric acid receptors
   - 2.3. Glycine receptors
   - 2.4. Noradrenaline receptors
   - 2.5. Dopamine receptors
   - 2.6. Serotonin receptors
   - 2.7. Histamine receptors
   - 2.8. Acetylcholine receptors
   - 2.9. Purine receptors
   - 2.10. Endothelin receptors
   - 2.11. Opioid receptors
   - 2.12. Peptide receptors
   - 2.13. Protease receptors
   - 2.14. Prostanoid receptors
   - 2.15. Chemokine receptors
3. Astrocytic cytokine and growth factor receptors
   - 3.1. Cytokine receptors
   - 3.2. Growth factor receptors
4. Astrocytic intracellular receptors
   - 4.1. Nuclear receptors
   - 4.2. Mitochondrial receptors
5. Functional consequences of astrocytic receptor activation
   - 5.1. Physiological activity
   - 5.2. Neurotrauma
   - 5.3. Development
6. Concluding remarks

Advances in Molecular and Cell Biology, Vol. 31, pages 475–501

ISBN: 0-444-51451-1

Abbreviations

$A_{1,\ 2B,\ 3,}$: adenosine $analog_{1,\ 2B,\ 3}$; ADNF: activity-dependent neurotrophic factor; ADP: adenosine phosphate; AMPA: α-amino-3-hydroxy-5-methyl-4-isoxazoleproprionic acid; ATP: adenosine triphosphate; $B_{1\text{-}2}$: $bradykinin_{1\text{-}2}$: bFGF: basic fibroblast growth factor; Ca^{2+}: calcium; $[Ca^{2+}]i$: intracellular calcium; cAMP: cyclic adenosine monophosphate; CCR1: chemokine receptor 1; CCR3: chemokine receptor 3; CCR5: chemokine receptor 5; CNTF: ciliary neurotrophic factor; CT-1: cardiotrophin-1; CXCR2, chemokine receptor 2; CXCR4: chemokine receptor 4; $D_{1\text{-}5}$: dopamine $subtypes_{1\text{-}5}$: DAG: diacylglycerol; EGF: epidermal growth factor; EP_3: prostaglandin receptor EP_3 subtype; ET: endothelin; FP: prostaglandin receptor FP; G_i: inhibitory G protein; G_o: o for other G protein; G_q: subtype for a G protein; G_s: stimulating G protein; GABA: gamma-aminobutyric acid; GFAP: glial fibrillary acidic protein; GluR: glutamate receptor; GM-CSF: granulocyte/-macrophage colony-stimulating factor; GR: glucocorticoid receptor; H: histamine; 5-HT: 5-hydroxytryptamine; iGluR: ionotropic glutamate receptor; IL: interleukin; IL-1β: interleukin-1beta; INF: interferon; IP: inositolphosphate; IP_3: inositoltrisphosphate; KA: kainate; LIF: leukemia inhibitory factor; $M_{1\text{-}2}$: muscarinic receptor1-2; MAPK: mitogen-activated protein kinase; M-CSF: macrophage colony-stimulating factor; mGluR: metabotropic glutamate receptor; MR: mineralcorticoid receptor; mRNA: messenger ribonucleic acid; $NK_{1\text{-}3}$: neurokinin1-3; NMDA: N-methyl-D-aspartic acid; NMDA R1: N-methyl-D-aspartic acid receptor 1; NMDA R2: N-methyl-D-aspartic acid receptor 2; NPY: neuropeptide Y; $NT_{1\text{-}3}$: neurotrophin1-3; OSM: oncostatin M; P1: purine receptor 1; P2: purine receptor 2; $P2Y_{1\text{-}2}$, subclasses $P2Y_{1\text{-}2}$ of purine receptor 2; $P2X_7$: subclass $P2X_7$ of purine receptor 2; PAR: protease-activated receptor; PAR-1: protease-activated receptor 1; PAR-2: protease-activated receptor 2; PAR-3: protease-activated receptor 3; PAR-4: protease-activated receptor 4; PCR: polymerase chain reaction; PGD_2: phosphogluconate $dehydrogenase_2$; PGDF: platelet-derived growth factor; PGE_2: prostaglandin E_2; $PGF_{2\alpha}$, prostaglandin F_{2alpha}; PIP_2: phosphatidyl inositol 4,5-biphosphate; PKA: protein kinase A; PKC: protein kinase C; PLA_2: phospholipase A_2; PLC: phospholipase C; PLD: phospholipase D; PTBR: peripheral-type benzodiazepine receptor; RNA: ribonucleic acid; Subst P: substance P; TGF-α: transforming growth factor-alpha; TGF-β: transforming growth factor-beta; TNF-α: tumor necrosis factor-alpha; TP: thromboxane receptor TP; TPA: tissue plasminogen activator; TXA_2, thromboxane aprotinin A_2; TXB: thromboxane; VIP: vasoactive intestinal peptide.

Astrocytes, the most numerous glial cells in the CNS, express a large number of receptors for neurotransmitters, peptides, purines, cytokines and other neuroactive substances, many of them previously thought to be present only on neurons. These astrocytic receptors are coupled to G proteins or ion channels, to intracellular protein kinases, or associated with mitochondria or cell nuclei. Even if most evidence for the existence of this extensive repertoire of functional astrocytic receptors comes from in vitro studies, convincing studies have demonstrated the presence of astrocytic receptor groups also in the intact nervous system. The astrocytic receptors and the signaling systems with which they are associated, set the stage for extensive neuronal–glial signaling, which in turn is essential for neuroplasticity, both in the intact nervous system and after trauma and in degenerative disease. Pharmacological manipulation of astrocytic receptor systems might provide a new strategy to reinforce the reparative processes within the CNS after injury or in disease.

1. Introduction

Glial cells support neuronal activity and create opportunities for the dynamics and plasticity of the central nervous system (CNS). After injury, metabolic disturbance, or toxic influence, and in degenerative disease, glial cell reactions protect neurons and facilitate rebuilding. When driven out of physiological control, however, glial reactions can be destructive and cytotoxic. Glial functions of physiological importance include astroglial glutamate uptake, transport of low molecular weight substances in the astroglial gap junction-coupled network, and astroglial volume regulation, with secondary effects on the size and shape of the extracellular space, and consequently, on volume transmission. Extensive inter- and intracellular signaling regulates and integrates these and other processes, and astrocytic receptors, both membrane-bound and intracellular, are of utmost importance. Calcium (Ca^{2+})-mediated signaling is one of the mechanisms by which CNS cells communicate with, and modulate the activity of, adjacent cells after stimulation by specific receptors. Astrocytic Ca^{2+} signals can propagate within the astroglial network as well as to neighboring cells in the CNS. Information on signals, in the form of ions, molecules, proteins, and peptides, which mediate receptor-triggered responses, is crucial for our understanding of how different cell types communicate with each other. After receptor stimulation, transmitters induce transient elevations of internal Ca^{2+} levels and other second messengers and change ion fluxes in astrocytes.

Up to now, most studies of glial cell reactions and signaling involving the receptors in one way or another have been performed and evaluated in primary cultures. However, some studies have also been performed in more 'in vivo-like systems', such as brain slices. Due to the complex morphology of the CNS, with tightly interwoven cellular networks and a large array of substances active in cell signaling, the use of cell cultures enriched with defined cell types have made it easier to evaluate the role of individual substances at the single cell level. Most astrocytic receptors, e.g., glutamate receptors and receptors for monoamines, such as noradrenaline, dopamine, serotonin and histamine, are coupled to heterotrimeric G proteins or ion channels, with some receptor subtypes for a given transmitter being coupled to G proteins and other to ion channels. However, other membrane receptors are coupled to intracellular protein kinases, and a third group of astrocytic receptors are nuclear, associated with either mitochondria or cell nuclei.

2. Astrocytic G-protein- and/or ion channel-coupled membrane receptors

2.1. Glutamate receptors

Glutamate is the major excitatory neurotransmitter in the CNS and glutamate receptors are expressed in many different locations throughout the CNS (Hollmann and Heinemann, 1994; Nakanishi, 1994). They play an important role in neuronal plasticity, neuronal development, and neurodegeneration. Glutamate receptors can be divided into metabotropic and ionotropic receptors.

2.1.1. *Metabotropic glutamate receptors*

Eight different metabotropic glutamate receptors (mGluRs) have been identified and cloned, mGluR1–mGluR8. The mGluR1 and the mGluR5 are coupled via G proteins with phospholipase C (PLC). They are activators of the inositoltrisphosphate (IP_3)-mediated intracellular signaling pathway and release calcium from intracellular calcium ($[Ca^{2+}]_i$) stores (Pin and Duvoisin, 1995—see also chapter by Shuai et al.). Glutamate evokes a non-oscillatory $[Ca^{2+}]_i$ response in single-cell $[Ca^{2+}]_i$ recordings in mGluR1-expressing cells and an oscillatory $[Ca^{2+}]_i$ response in mGluR5-expressing cells (Nakanishi et al., 1998). The difference results from a single amino acid substitution, aspartate in mGluR1 and threonine in mGluR5, in the G protein-interacting carboxy-terminal domains. Protein kinase C (PKC) phosphorylation of the threonine of mGluR5 is responsible for inducing $[Ca^{2+}]_i$ oscillations in mGluR5-expressing cells and cultured glial cells, because phosphorylation of threonine by PKC abolishes the $[Ca^{2+}]_i$ increase by interfering with the signal transduction between the receptor and the intracellular effector, IP_3, and the subsequent dephosphorylation regenerates the $[Ca^{2+}]_i$ increase by restoring the signal transduction. In contrast, aspartate of mGluR1 is not phosphorylated by PKC, resulting in a non-oscillatory, single-peak $[Ca^{2+}]_i$ increase. However, in general, astrocytes fail to express mGluR1, whereas immunoreactivity for mGluR5 as well as for the mGluR5 transcript and protein has been detected on astrocytes from the hippocampus, hypothalamus, and cerebral cortex (Romano et al., 1995; Van den Pol et al., 1995; Nakahara et al., 1997; Muyderman et al., 2001a), as well as from glial fibrillary acidic protein (GFAP)-positive astrocytes in the cerebral cortex, both in brain slices (Muyderman et al., 2001a) and in vivo (Steinhäuser and Gallo, 1996). The mGluR5 expression is upregulated by specific growth factors, such as basic fibroblast growth factor (bFGF), epidermal growth factor (EGF), and transforming growth factor-α (TGF-α) (Miller et al., 1995).

The other six subtypes, mGluR2–4 and mGluR6–8, are all coupled to the inhibition of adenylate cyclase. Using in situ hybridization techniques, mGluR3 has been detected on astrocytes from different brain regions (Ohishi et al., 1993; Fotuhi et al., 1994). It has been speculated that glutamate released at the synaptic clefts may either stimulate or reduce the proliferation rate of surrounding astrocytes through activation of distinct mGluR subtypes. Metabotropic GluR5 enhances proliferation and mGluR3 reduces proliferation in cultured astrocytes (Ciccarelli et al., 1997). Metabotropic GluR3 has been found to be co-localized with the water channel Aquaporin 4 and upon activation by glutamate, can act as an osmoregulation sensor (Shigemoto et al., 1999). Messenger ribonucleic acid (mRNA) for mGluR2, mGluR4 and mGluR6–8 has not been detected on astrocytes (Tanabe et al., 1993; Testa et al., 1994).

2.1.2. *Ionotropic glutamate receptors*

Ionotropic glutamate receptors (iGluRs) include three main groups: *N*-methyl-D-aspartic acid (NMDA), consisting of NMDA R1 and NMDA R2A–D subunits, and non-NMDA ionotropic receptors, which can be subdivided into α-amino-3-hydroxy-5-methyl-4-isoxazolepropionic acid (AMPA) and kainate (KA) receptors. They are all ligand-gated ion channels through which Na^+, K^+, and Ca^{2+} can pass. The iGluR subunits GluR1–4 form AMPA-sensitive receptors, and GluR5–7 form KA-sensitive receptors.

Stimulation of the AMPA/KA receptors leads to depolarization and an influx of Ca^{2+} across the plasma membrane (Jabs et al., 1994; Porter and McCarthy, 1995). Four AMPA receptor subunits (GluR1–4) and five KA receptor subunits (GluR5–7 and KA1–2) have been cloned. Expression of GluR4 has been detected on astrocytes from different brain regions. For the other iGluRs, it has been hypothesized that subpopulations of astrocytes express some of the receptors, preferentially GluR1 and GluR3 (for a review, see Porter and McCarthy, 1997).

In many studies on astrocytes in culture, no expression of NMDA receptors was recorded. However, other data suggest that astrocytes in slices express NMDA receptors (for a discussion and references, see Porter and McCarthy, 1997). Moreover, the radial glial cells, namely, the Bergmann glia in the cerebellum and the Müller cells in the retina, which are not converted into conventional astrocytes after birth, express the NMDA R1 and NMDA R2A or B subunits (López et al., 1997).

2.2. *Gamma-aminobutyric acid receptors*

Gamma-aminobutyric acid (GABA), the main inhibitory neurotransmitter in the CNS, can mediate its effects through three different GABA receptors: $GABA_A$, $GABA_B$, and $GABA_C$. Gamma-aminobutyric $acid_A$ and $GABA_C$ are intrinsic ligand-gated Cl^- channels, whereas the $GABA_B$ receptor is coupled via G proteins to its effectors, K^+- or Ca^{2+}-permeable plasmalemmal channels (Johnston, 1994; Nakayasu et al., 1995). In astrocytes, GABA depolarizes the membrane. By contrast, in neurons, it hyperpolarizes the membrane. The difference is due to different intracellular Cl^- levels in the two cell types. Astrocytes have a more positive Cl^- equilibrium potential than do neurons due to a higher intracellular Cl^- concentration, which may be due to the activity of two inwardly directed Cl^- transporters, the Na^+–K^+–Cl^--cotransporter and the Cl^-/HCO_3^- exchanger (Von Blankenfeld and Kettenmann, 1991; Fraser et al., 1994). Depolarization of the astrocytes takes place mainly through the $GABA_A$ receptors, which have been found on most types of astrocytes (for a review, see Porter and McCarthy, 1997). Activation of membrane-associated benzodiazepine receptors on cultured astrocytes (expressed in additional to the mitochondrial-type benzodiazepine receptors (see Section III.b)) by conventional benzodiazepines or by endogenous agonists, called endozepines, causes an increase in $[Ca^{2+}]_i$ which mainly, if not exclusively is due to an enhancement of Ca^{2+} entry through L-channels (in dibutyryl cyclic AMP-treated cultures), possibly triggered by benzodiazepine mediated enhancement of astrocytic $GABA_A$ receptors and resulting depolarization (Zhao et al., 1996; Gandolfo et al., 2001—see also chapter by Hertz, Peng et al.). Gamma-aminobutyric $acid_C$ receptors on astrocytes have not yet been described (Verkhratsky and Steinhäuser, 2000). Gamma-aminobutyric $acid_B$ receptors induce $[Ca^{2+}]_i$ increases, which may be due to Ca^{2+} entry via voltage-gated channels or to plasmalemmal Ca^{2+} entry (Nilsson et al., 1993).

2.3. Glycine receptors

Functional glycine receptors have been detected on astrocytes. These receptors mediate a Cl^- conductance (Betz et al., 1994) and show similarities with the $GABA_A$ receptor.

2.4. Noradrenaline receptors

The monoamines noradrenaline and dopamine share a common biosynthetic pathway that starts with the amino acid tyrosine. Tyrosine is converted to L-dopa by tyrosine hydroxylase and, in noradrenergic neurons, further transformed to noradrenaline by dopamine β-hydroxylase. The noradrenergic fibers originate from the locus coeruleus and spread through the entire cerebral cortex, hippocampus, and cerebellum, where they terminate in varicosities rather than in genuine synapses (Lindvall and Björklund, 1984). Astrocytes express adrenergic receptors both in vitro and in vivo (Hösli and Hösli, 1982; Salm and McCarthy, 1989; Shao and Sutin, 1992). The proportion of astrocytes surrounding noradrenergic varicosities is fairly high and, consequently, the noradrenergic system is believed to exert a large part of its effect on astrocytes (Stone and Ariano, 1989). There are several basic subtypes of adrenergic receptors, α_1, α_2, β_1, and β_2. They are coupled to different signal transduction systems and have distinct pharmacological profiles, and they have all been localized on astrocytes (for a review, see McCarthy et al., 1995; Verkhratsky et al., 1998).

The expression of α_1 adrenoceptors has been demonstrated with radiolabeled α_1 adrenoceptor agonists on astrocytes (Shao and Sutin, 1992). The α_1 receptor activates a pertussis toxin-insensitive protein, G_q, which in turn triggers PLC to generate IP_3 and diacylglycerol (DAG) from phosphatidyl inositol 4,5-bisphosphate (PIP_2) in the plasma membrane (Berridge and Irvine, 1984). Diacylglycerol elicits a cascade of intracellular events, such as the gating of ion channels, mobilization of Ca^{2+}, and the activation of intracellular kinases such as protein kinase A (PKA) and PKC, respectively, leading to alterations in the degree of protein phosphorylation (Nilsson et al., 1991). In astrocyte cultures from cerebral cortex the mobilization of Ca^{2+} leads to an increase in $[Ca^{2+}]_i$, which frequently is oscillatory (Muyderman et al., 2001a).

The α_2 receptor is negatively linked via an inhibitory G protein (G_i) to the enzyme adenylate cyclase and thus decreases cyclic adenosine monophosphate (cAMP) production (Hansson, 1988). However, stimulation of α_2 receptors on astrocytes is also associated with increases in IP_3 (Enkvist et al., 1996) and in $[Ca^{2+}]_i$ (Zhao et al., 1992) in astrocytes in primary cultures, which in turn can lead to stimulation of glucose metabolism (see chapter by Hertz, Peng et al.). Alternatively, and dependent on the subtype of the α_2 receptor, stimulation of α_2 receptors can lead to opening of Ca^{2+} channels in the cell membrane, causing a non-oscillating increase in $[Ca^{2+}]_i$ (Muyderman et al., 2001b).

The presence of β adrenoceptors on astrocytes has been demonstrated by binding of β adrenoceptor antagonists (Salm and McCarthy, 1989; Shao and Sutin, 1992). The β receptor can be divided into three subgroups, β_1, β_2 and β_3, although the β_3 receptor (Evans et al., 1999) has not been localized to astrocytes so far. The β_1 and β_2 receptors are linked via the stimulating G protein (G_s) to the enzyme adenylate cyclase and increase the cAMP production (Hansson, 1985). Immunoreactivity for β adrenoceptors has been found in both gray and white matter, indicating that both protoplasmic and fibrous astrocytes have β receptors, preferentially the β_2 subtype (Aoki, 1992; Liu et al., 1992). The β_2 adrenoceptors have been detected on astrocytic processes surrounding axons, dendrites, and immunoreactive synapses, which suggests that astrocytes may be able to respond to noradrenaline released from synaptic terminals (Aoki, 1992). On the other hand, neuronally released noradrenaline may diffuse and activate receptors at significant distances from the adrenergic synapses and thus take part in volume transmission (Fuxe and Agnati, 1991). Stimulation of β adrenoceptors by isoproterenol can also elicit $[Ca^{2+}]_i$ elevations in astrocytes, a non-oscillating response that is found in relatively few cells compared with the reaction to receptors coupled to the inositolphosphate (IP) system, and is due to opening of Ca^{2+} channels in the cell membrane (Muyderman et al., 2001a).

2.5. Dopamine receptors

The dopaminergic system in the CNS plays a crucial role in the regulation of physiological actions, such as the control of locomotion, cognition, emotion, and neuroendocrine secretion. These actions are mediated by five distinct G protein-coupled receptor subtypes, which are classified into two main groups (Jaber et al., 1996). The first, D_1-like dopamine receptor subtype (D_1 and D_5) activates adenylate cyclase, enhancing cAMP formation, while D_2-like dopamine receptor subtype (D_2, D_3, and D_4) inhibits adenylate cyclase, decreasing cAMP formation, and also activates K^+ channels. Dopamine receptors have been demonstrated on groups of astrocytes in astroglial primary cultures cultured from striatum, cerebral cortex, and the spinal cord (Hansson et al., 1984; Hansson, 1985; Hösli and Hösli, 1986). With in situ hybridization techniques and polymerase chain reaction (PCR), dopamine D_2 receptor mRNA was found to be expressed by astrocytes from striatum (Bal et al., 1994). It has, however, also been shown that dopamine is able to elicit transient increases in $[Ca^{2+}]_i$ in cultured or bulk-separated astrocytes, which can be blocked by D_1 and D_2 receptor-specific antagonists (Reuss et al., 2000; Khan et al., 2001). Our previous results, that there is regional heterogeneity in the expression of D_1 receptor mRNA in cultured astrocytes from different brain regions, with a maximum response to dopaminergic agonists in striatal astrocytes, a lower effect in cortical astrocytes, and no effect in cerebellar astrocytes, have since been confirmed by PCR and Southern blot hybridization (Zanassi et al., 1999).

At the electromicroscopic level, it was observed that cortical interneurons are surrounded by astrocytic processes, which strongly express D_2 receptors and communicate with dopaminergic interneurons in the same region, as indicated by an increase in astrocytic $[Ca^{2+}]_i$ by exposure to dopaminergic agonists (Khan et al., 2001). This finding confirms earlier observations of 'cross-talk' between astrocytes and neurons, made in a co-cultivation system, where the sensitivity of cultured striatal astrocytes to application of dopamine was enhanced by co-culturing with neurons from substantia nigra, one of the natural projection areas from substantia nigra (Hansson and Rönnbäck, 1988).

2.6. Serotonin receptors

Serotonin [5-hydroxytryptamine (5-HT)] is an important neurotransmitter in the CNS and is synthesized from tryptophan. Disturbances in the 5-HT system have been demonstrated in several mental and neurological disorders, such as depression, anxiety, schizophrenia, and migraine. 5-HT-containing neuronal cell bodies are concentrated in the raphe nuclei, from where fibers project to virtually all parts of the CNS. Not all fibers establish synaptic contacts with target neurons, but instead some mediate a volume transmission, a process, which is slower than synaptic signaling (Jacobs and Azmitia, 1992). The 5-HT receptor family consists of seven major classes, based on structural and pharmacological properties, the 5-HT_{1-7} receptors, which are further divided into several subtypes (Hoyer and Martin, 1997). The 5-HT_3 receptor is a ligand-operated cationic channel, but the others (i.e., 5-HT_{1-2} and 5-HT_{4-7}) are coupled to G proteins. 5-HT_1 and 5-HT_4 are coupled to adenylate cyclase, whereas 5-HT_2 receptors are coupled to PLC, thus regulating IP_3 production (Hoyer et al., 1994). In cultured astrocytes, 5-HT receptors were originally detected by Hertz and Schousboe (Hertz et al., 1979). Several different subtypes of 5-HT receptors have been found on astrocytes: the 5-HT_{1A} (Azmitia et al., 1996), the 5-HT_{2A} (Nilsson et al., 1991; Deecher et al., 1993; Hagberg et al., 1998), the 5-HT_{2B} (Hirst et al., 1998; Sandén et al., 2000; Kong et al., 2002), the 5-HT_{5A} (Carson et al., 1996), and the 5-$HT_{6,7}$ receptors (Hirst et al., 1997). When the 5-HT_{1A} receptor is stimulated, the astrocytes respond by releasing S-100β and attaining a mature morphology, with a shift from a flattened to a process-bearing morphology (Whitaker-Azmitia et al., 1990). Stimulation of the 5-HT_2 receptor on astrocytes results in glycogenolysis (Poblete and Azmitia, 1995) and increases the $[Ca^{2+}]_i$ levels in astrocytes (Nilsson et al., 1991—see also chapter by Hertz, Peng et al.).

2.7. Histamine receptors

Three different types of histamine receptors have been detected: $histamine_1$ (H_1) receptor, which is coupled to PLC and mobilization of $[Ca^{2+}]_i$, $histamine_2$ (H_2) receptor, which increases adenylate cyclase activity, and $histamine_3$ (H_3) receptor, which controls histamine turnover and release (Leurs et al., 1995). Astrocytes express both H_1 and H_2 receptors (Hösli et al., 1984; Inagaki and Wada, 1994; Carman-Krzan and Lipnik-Stangelj, 2000). The H_1 receptor-mediated $[Ca^{2+}]_i$ increases occur mainly in cultures of the so-called type 2 astrocytes, but have also been demonstrated in subpopulations of conventional astrocytes, occasionally called type-1 astrocytes (McCarthy and Salm, 1991; Inagaki and Wada, 1994). In an experimental study, cultured rat cerebellar astrocytes responded with $[Ca^{2+}]_i$ increases after stimulation with histamine, an effect which was antagonized by H_1 receptor blocker (Jung et al., 2000).

2.8. Acetylcholine receptors

Progress in studying the role of acetylcholine in neuronal–glial interactions has not been as fast as it has in the case of glutamate and GABA and of the aminergic receptors.

The reason for this may be that acetylcholine is less widely used in central synaptic transmission than are either glutamate or GABA. Two distinct families of acetylcholine receptors have been described: ionotropic nicotinic cholinoreceptors and metabotropic muscarinic cholinoreceptors. Nicotinic cholinoreceptors contain an integral cationic channel and metabotropic muscarinic cholinoreceptors are coupled to G proteins (Changeux, 1995). There is some evidence that nicotinic receptors exist on astrocytes (Hösli and Hösli, 1988; Sharma and Vijayaraghavan, 2001), and muscarinic receptors are widely found on astrocytes. There are five major subtypes of metabotropic muscarinic cholinoreceptors, M_1–M_5. Type 1, 3, and 5 are coupled to PLC controlling the IP_3 turnover, and type 2 and 4 are coupled to adenylate cyclase and decrease cAMP activity (Caulfield, 1993). In vivo, about 20% of the astrocytes in the rat brain label for muscarinic receptors (Van der Zee et al., 1993). These astrocytes are mainly found in the cerebral cortex and in the corpus callosum. In vitro, most cultured astrocytes possess muscarinic receptors (Hamprecht et al., 1976; Repke and Maderspach, 1982). The two muscarinic receptor subtypes found on astrocytes have been identified as muscarinic $receptor_1$ (M_1) and muscarinic $receptor_2$ (M_2) (Murphy et al., 1986). Activation of M_1 stimulates PLC (Murphy et al., 1986), which leads to an IP_3 increase and a Ca^{2+} mobilization (Enkvist et al., 1989; Araque et al., 2002). Stimulation of muscarinic receptors also leads to phospholipase D (PLD) activation (Gustavsson et al., 1993).

2.9. Purine receptors

Glial cells represent a very important source of purines in the CNS, both under physiological conditions and in pathological states. Astrocytes are the main source of cerebral purines. They release the adenine-based purines adenosine and adenosine triphosphate (ATP) and the guanine-based purines guanosine and guanosine triphosphate. There are several specific purine receptors for adenosine and adenine nucleotides (adenosine phosphate (ADP) and ATP), which are called purine receptor 1 (P1) and purine receptor 2 (P2), respectively (Burnstock, 1978). The P1 purinoceptors are divided into two classes, A_1 and A_2, on the basis of the abilities of their adenosine agonists to either increase (A_2) or decrease (A_1) cAMP production, a distinction originally made on the basis of experiments using cultured astrocytes (Van Calker et al., 1979). The P1 receptors are characterized as G protein-coupled receptors. The A_2 receptor has been divided into A_{2A} and A_{2B}, both of which subclasses stimulate the G_s protein and thus increase cAMP. The more recently demonstrated A_3 receptor is coupled to G_i or G_o proteins and thus inhibits cAMP production. This receptor also stimulates IP_3 formation.

The P2 receptors are divided into the subclasses P2X and P2Y (Burnstock and Kennedy, 1985). The P2X receptor complex comprises a ligand-gated ion channel (i.e., it is an ionotropic receptor), which is opened by receptor activation and promotes influx of Na^+ and Ca^{2+}, and efflux of K^+, followed by membrane depolarization and fast synaptic transmission (Bean, 1992). These receptors comprise a family of seven members that have been cloned and named $P2X_{1-7}$ (Fredholm et al., 1997). The P2Y receptors are G protein-linked receptors (i.e., metabotropic receptors), which are coupled to PLC and activation of IP_3, with $[Ca^{2+}]_i$ mobilization. Seven P2Y receptors have also been cloned and named

$P2Y_{1-7}$ (Fredholm et al., 1997). Astrocytes express several types of purinoceptors, mainly A_1, A_{2B}, A_3, $P2Y_1$, $P2Y_2$, and $P2X_7$ (for a review, see Di Iorio et al., 1998; Ciccarelli et al., 2001).

2.10. Endothelin receptors

Astrocytes represent a major target for endothelins, a family of peptides, released by several cell types in the brain (Kuwaki et al., 1997), that have potent and multiple effects on signal transduction pathways, such as activation of the IP_3 system followed by $[Ca^{2+}]_i$ transients, and activation of mitogen-activated protein kinases (MAPKs), phospholipase A_2 (PLA_2), and PLD. Endothelins constitute a peptide family composed of at least three isoforms termed endothelin-1 (ET-1), endothelin-2 (ET-2), and endothelin-3 (ET-3) (Inoue et al., 1989). They have been found to have multiple biological activities in both vascular and non-vascular tissues (Yanagisawa et al., 1988). Endothelin mRNA expression is widely distributed in the brain (Giaid et al., 1991). Endothelin-induced cellular reactions are mediated via three major receptor subtypes, displaying different sensitivity to the various endothelin isoforms. The most abundant form in the brain is the non-selective $endothelin_B$ (ET_B) receptor, which is equally sensitive to all three endothelins. The $endothelin_A$ (ET_A) receptor, preferentially sensitive to ET-1 and ET-2, but not to ET-3, is much less expressed; the $endothelin_C$ (ET_C) receptor has been described only in non-mammalian tissues (Sokolovsky, 1995). All three subtypes are coupled to heterotrimeric G proteins and they are significantly involved in the regulation of $[Ca^{2+}]_i$ (Goldman et al., 1991; Blomstrand et al., 1999) and release of arachidonic acid (Wu-Wong et al., 1996). $Endothelin_A$ receptors mediate vasoconstriction, whereas ET_B receptors mediate vasodilation. In the brain, ET_A receptors are mostly expressed in vascular cells (Hori et al., 1992), whereas ET_B receptors are mainly expressed on glial cells (Lazarini et al., 1996; see also chapter by Chen and Spatz). However, expression of both ET_A and ET_B receptor mRNA has been detected in astrocytes in culture (Ehrenreich et al., 1993; Schinelli et al., 2001). The endothelin-induced $[Ca^{2+}]_i$ responses in astrocytes have been reported to have heterogeneous kinetic properties, varying from simple peaks to polyphasic peaks (Blomstrand et al., 1999).

2.11. Opioid receptors

The opioid receptor family consists of three pharmacologically distinct subtypes, μ, δ, and κ (Leslie, 1987). They are heterogeneously distributed through the CNS and produce a multitude of behavioral, neuroendocrinological, and autonomic effects. All three receptors are G protein-coupled, and all three have been cloned (Uhl et al., 1994). The classic opioid agonist, morphine acts primarily on μ receptors. The first endogenous opioid peptides discovered were the enkephalins, which act preferentially on the δ receptors. Other endogenous opioids, the β endorphins and the dynorphins, affect μ and κ receptors (Borsodi and Tóth, 1995). Protein and/or mRNA for μ, δ, and κ receptors have been found in astrocyte cultures derived from different rat brain regions. There is heterogeneity of opioid receptor expression in astrocytes from different brain regions, but the sum of the three subtypes is largest in brain cortex. The expression of δ and κ receptors on astrocytes appears to be relatively similar, whereas μ receptors are scarce (Ruzicka et al., 1995). Spinal astrocytes also express δ and μ receptors, although most spinal receptors are on neurons (Cheng et al., 1997). Application of δ- and κ-selective agonists decrease forskolin-stimulated levels of cAMP in astrocytes (Rougon et al., 1983; Eriksson et al., 1990). However, increases in $[Ca^{2+}]_i$ have also been demonstrated after stimulation of κ receptors (Eriksson et al., 1993) and δ receptors (Thorlin et al., 1998).

2.12. Peptide receptors

Some transmitter peptides have already been described, i.e., endothelins, enkephalins, β endorphins and dynorphins, and others will be described below (growth factors, cytokines). However, the receptors for these are generally not regarded as peptide transmitter receptors. The term 'peptide receptors' defines receptors for a relatively well delineated group of neuroactive peptides, some of which are also hormones (e.g., vasopressin and somatostatin), and many of which were first known for their action in the gastrointestinal system (e.g., secretin). The first studies of peptide receptors on astrocytes were performed by Hamprecht and coworkers (Van Calker et al., 1980), who showed that astrocytes in primary cultures have receptors for vasoactive intestinal peptide (VIP), secretin, and somatostatin, which regulate cAMP. Later, Cholewinski et al. (1988) reported that some peptides, namely, oxytocin, vasopressin, bradykinin, and tachykinin, activate IPs and increase $[Ca^{2+}]_i$. Furthermore, angiotensin II receptors and atrial natriuretic peptide receptors have been found on astrocytes (Sumners et al., 1994).

Two different receptors for VIP have been identified: one with a low-affinity binding site coupled to activation of adenylate cyclase (Gozes et al., 1991), and the other with a high-affinity site, which is coupled to IP_3 generation and changes in $[Ca^{2+}]_i$ (Fatatis et al., 1994). Receptors for neurotensin and neuropeptide Y (NPY) have been identified on the basis of ligand binding and electrophysiology (Gimpl et al., 1993; Hösli et al., 1995). Two subtypes of bradykinin receptors, B_1 and B_2, have been described. In cultured astrocytes, mainly the B_2 receptor has been reported to trigger $[Ca^{2+}]_i$ elevations, but a minor subpopulation of astrocytes may also contain B_1 receptors (Stephens et al., 1993). Finally, glucagon receptor stimulation leads to increases in cAMP (Cockram et al., 1995). For more extensive information on peptide receptors can be referred to Deschepper (1998). The role of vasopressin in regulation of fluid homeostasis in the CNS is discussed in the chapter by Chen and Spatz.

Substance P belongs to the family of neurokinins, and three subtypes of receptors have been identified, neurokinin 1 (NK_1), neurokinin 2 (NK_2), and neurokinin 3 (NK_3) (Dam et al., 1988). They are G protein-coupled metabotropic receptors. Binding of the NK_1 receptor to substance P results in phosphoinositide hydrolysis, and substance P increases phosphatidylinositide turnover in cerebellar astrocytes (Marriott and Wilkin, 1992). Both brain astrocytes and spinal astrocytes express NK-1 receptors, but the receptor density is much higher on the spinal cells (Palma et al., 1997). Besides stimulating formation of IP_3, substance P also potentiates the inducing effects of interleukin-1β (IL-1β) and tumor necrosis factor-α (TNF-α) on secretion of interleukin-6 (IL-6) by spinal cord astrocytes (Palma et al., 1997). These effects may play a role for the influence of astrocytes on persistent pain, which is described in the chapter by Raghavendra and DeLeo.

2.13. Protease receptors

Proteases, also called 'proteinases', are believed to be involved in development and in repair processes in the CNS, and they regulate several functions of the CNS by activating protease-activated receptors (PARs). PARs are cell surface receptors that mediate cellular signaling through heterotrimeric G proteins (Wang et al., 2002). Some of the receptors, PAR-1, PAR-2, PAR-3, and PAR-4, are functionally active on astrocytes. Stimulation of these receptors on astrocytes with trombin, trypsin, or peptides induces maintained $[Ca^{2+}]_i$ elevations (Wang et al., 2002).

2.14. Prostanoid receptors

Evidence from in vitro studies suggests that the predominant cell responsible for brain prostanoid formation is the astrocyte (Murphy et al., 1988), although neurons also produce prostanoids (Inagaki and Wada, 1994). Astrocytes in primary cultures have been shown to produce prostaglandins (PGDs) and thromboxanes (TXBs), including phosphogluconate dehydrogenase$_2$ (PGD_2), prostaglandin E_2 (PGE_2), and prostaglandin $F_{2\alpha}$ ($PGF_{2\alpha}$), as well as thromboxane aprotinin A_2 (TXA_2) and thromboxane 2 (TXB_2). The release of prostanoids from astrocytes in primary cultures is stimulated by ATP, substance P, interleukin-1β (IL-1β) and tissue plasminogen activator (TPA) (Inagaki and Wada, 1994). Prostanoid receptors present on astrocytes are the $PGF_{2\alpha}$ receptor (FP receptor), thromboxane A_2 receptor (TP receptor), and PGE receptor EP_3 subtypes (EP_3 receptor) (Kitanaka et al., 1996a). The EP_3 receptor is known to be G protein-coupled (Satoh et al., 1999), and a thromboxane analog increases $[Ca^{2+}]_i$ in primary cultures of astrocytes (Kitanaka et al., 1996b).

2.15. Chemokine receptors

Chemokines are a family of cytokines, which originally were described in the immune system, where they regulate motility of immune cells. Chemokines are synthesized in the CNS by astrocytes, microglia, and neurons, and a chemotactic response of these cells to chemokines may play a role during brain development (Rezaie et al., 2002). This is consistent with a role for chemokines not only in the immune system but also in development and migration of many cell types in the absence of inflammation, e.g., in cell growth, angiogenesis, hematopoiesis, free radical production, apoptosis, neoplasia, wound healing, tissue repair, and interactions with pathogens, including viruses (Bacon and Harrison, 2000). The chemokines have been classified into the four subfamilies α, β, γ, and δ, or CXC, CC, C, and CX3C (Zlotnik and Yoshie, 2000). In contrast to other cytokines, chemokines signal via G protein-coupled receptors, i.e., 7TM receptors of the serpentine superfamily, and elicit biological activities at nanomolar concentrations

(Murphy et al., 2000). Astrocytes express a number of chemokine receptors, including CCR1, CCR3, CCR5, CXCR2, and CXCR4, some of which have been shown to have a functional expression (Dorf et al., 2000—see also chapter by Nakagawa and Schwartz).

3. Astrocytic cytokine and growth factor receptors

3.1. Cytokine receptors

Neurotrophic factors (growth factors acting on neurons and/or glial cells) and cytokines (factors originally described in hematopoietic cells and cells of the immune system, where they regulate development and function) are of fundamental importance in the CNS, and astrocytes express functional receptors for many of them (Otero and Merrill, 1994—see also chapter by Nakagawa and Schwartz). Growth factors and cytokines play a huge role in the CNS, where cytokines can be either pro-inflammatory, neuropoietic, i.e., acting as growth factors, and/or anti-inflammatory. Their receptors are not G protein-coupled, but directly associated with cascades of protein kinases, often receptor protein-tyrosine kinases. Receptor activation is followed by distinct protein kinase pathways (e.g., the Ras, Raf, MAP kinase pathway) and eventual translocation of a phosphorylated kinase to the cell nucleus, leading to transcriptional induction of immediate early genes.

Interleukins (IL) constitute a functionally heterogenous group, with some ILs being pro-inflammatory, other neuropoietic or anti-inflammatory. Interleukin-6 (IL-6) belongs to the family of neuropoietic cytokines, a family that includes ciliary neurotrophic factor (CNTF), leukemia inhibitory factor (LIF), oncostatin M (OSM), IL-6, interleukin-11 (IL-11), and cardiotrophin-1 (CT-1) (Taga and Kishimoto, 1997). All IL-6-type cytokine family members act via receptor complexes that contain the glycoprotein gp130. Some astrocytes express gp130, but respond weakly to IL-6 alone (Van Wagoner and Benveniste, 1999). Interleukin-6 exerts neurotrophic and neuroprotective effects, but it can also function as a mediator of inflammation, demyelination, and astrogliosis (Gadient and Otten, 1997).

Astrocytes play a key role in brain inflammation in response to trauma, ischemia and infection and, along with microglia, belong to the most important cell types for local regulation of inflammatory reactions and tissue repair in the CNS (Ridet et al., 1997). Astrocytes respond to the pro-inflammatory interleukin IL-1 and express IL-I receptor type 1 (Juric and Čarman-Kržan, 2001). The type I receptor is biologically active, while type II receptor functions as a decoy. IL-receptors display an equivalent number of both receptor types (Pousset et al., 2001). Interleukin-1 receptor expression is regulated by multiple control mechanisms, such as IL-1 itself, anti-inflammatory cytokines, such as interleukin-2 (IL-2), IL-4, IL-10, IL-13, and interferon (INF), growth factors, such as platelet-derived growth factor (PDGF), and glucocorticoids (Pousset et al., 2001). Binding of IL-1β to the IL-1 receptor type 1 is followed by recruitment and phosphorylation of receptor associated kinases and eventually by translocation of nuclear factor-κB (NF-κB) to the nucleus (Mercurio and Manning, 1999. IL-1β also activates a distinct MAP kinase cascade that results in DNA binding of activator protein-1 (AP-1) (Minden and Karin, 1997). Effects of peripherally generated IL-1β on CNS IL-1 receptors are discussed in the chapter by Mercier and Hatton and the role of IL-1β in persistent pain in the chapter by Raghavendra and DeLeo.

A second pro-inflammatory cytokine is tumor necrosis factor (TNF). Tumor necrosis factor-α (TNF-α) is a polypeptide, first identified as an inducer of necrosis of some tumor cells, but today considered as an endogenous modulator of many normal physiological parameters. The signaling for biological effects of this cytokine is exerted through cell surface receptors, activating similar cascades as IL-1, although cell-type and species-specific differences exist in the relative potency of these two cytokines. The presence of receptors for TNF-α has been identified both on cultured astrocytes (Aránguez et al., 1995) and on astrocytes in the brain (see chapter by Nagakawa and Schwartz).

Autocrine/paracrine signaling via P2 receptors modulates both IL-1β- and TNF-α-mediated activation of NF-κB and AP-1 in human fetal astrocytes (Liu et al., 2000).

3.2. *Growth factor receptors*

Members of several families of growth factors are produced by astrocytes, which also express corresponding receptors (see chapter by Nakagawa and Schwartz). There is recent evidence that growth factors can be released from astrocytes as a result of stimulation with G protein-coupled receptors, e.g., α_2-adrenergic receptors, in a 'transactivation' process (see chapter by Peng). There are several factors released by astrocytes, including transforming growth factor-β (TGF-β, macrophage colony-stimulating factor (M-CSF), and granulocyte/macrophage colony-stimulating factor (GM-CSF), which can induce ramification and upregulation of K^+DR in microglia (Schilling et al., 2001). Astrocytes are also capable of suppressing phagocytosis (De Witt et al., 1998) and production of the pro-inflammatory cytokine interleukin-12 (IL-12) by microglia (Aloisi et al., 1997).

4. Astrocytic intracellular receptors

4.1. *Nuclear receptors*

Several receptors, including those for many hormones (e.g., glucocorticoid receptors (GR)), translocate from the membrane to the nucleus after activation. In the brain, glucocorticoids bind to two types of receptors, a high-affinity mineralcorticoid receptor (MR) and a lower-affinity GR with affinities for endogenous glucocorticoids that differ by an order of magnitude (De Kloet, 1991). They are both expressed in astrocytes (Bohn et al., 1994). After treatment with glucocorticoids, an increase in $[Ca^{2+}]_i$ concentration has been observed with a Ca^{2+} wave propagation (Simard et al., 1999). This may appear peculiar, because the glucocorticoid receptors are not G-protein coupled, but the explanation appears to be that glucocorticoids enhance the response to such transmitters as ATP and bradykinin.

Astrocytes express receptors for different steroid hormones as described in the chapter by Melcangi et al. Neither these receptors, nor those for polypeptide hormones, such as growth hormone receptors will be discussed here. The same applies to thyroid hormone receptors (see chapter by Gomes and Rehen).

4.2. Mitochondrial receptors

The astrocytic, peripheral-type, benzodiazepine receptor is mainly localized on the mitochondrial membrane (see, however, also Section 2.2). 1,4-Benzodiazepines bind to two distinct classes of receptors. The central benzodiazepine receptor is membrane-associated and coupled to $GABA_A$ receptor-gated Cl^- channels on neurons, and possibly also on astrocytes (Section 2.2). It is generally assumed to be responsible for all anti-anxiety, sedative, and anti-convulsant actions of benzodiazepines (Richards et al., 1987). The peripheral-type benzodiazepine receptor (PTBR), by contrast is localized on astrocytes and absent on neurons (McCarthy and Harden, 1981) and, as in other tissues, it appears to be present chiefly on the outer mitochondrial membrane (Itzhak et al., 1993). A nuclear localization has, however, also been found recently (Kuhlmann and Guilarte, 2000). The PTBR is involved in the intracellular transport and metabolism of cholesterol and the production of neurosteroids (Rao et al., 2001). The PTBR has been demonstrated to increase after brain injury in both activated microglia and astrocytes (Kuhlmann and Guilarte, 2000), and in acute hyperammonemia (Felipo and Butterworth, 2002—see also chapter by Bélanger et al.).

5. Functional consequences of astrocytic receptor activation

5.1. Physiological activity

Since astrocytes are intimately associated with the synapse, enwrapping many pre- and postsynaptic terminals, and since they are in close contact with the capillaries, they are in a position to shuffle nutrients and metabolites between the blood supply and the active neuron. Astrocytes can also transfer information to neighboring neurons and one single astrocyte can have contact with multiple neurons (Ventura and Harris, 1999). The coordination of neuronal and glial activity requires that appropriate signals pass from one cell type to another. Astrocytic functions are probably to a considerable extent regulated by monoaminergic neurons, from which transmitters are released not only in the privacy of synapses, but also from varicosities, from where they reach their targets (e.g., neurons, astrocytes, oligodendrocytes, abluminal surface of endothelial cells) by diffusion. In astrocytes, the released monoamines exert a multitude of functional effects, including regulation of energy metabolism (see chapter by Hertz, Peng et al.) and of channel opening. However, monoamines are not the only transmitters which influence astrocytic activities. As demonstrated in this chapter, glutamatergic and GABAergic transmission can no longer be regarded as an exclusive neuronal phenomenon, since astrocytes express a multitude of different subtypes of receptors for both glutamate and GABA. The same applies to purinergic receptors. A common feature of all of these receptors is that they are membrane-associated and G protein coupled.

Activation of G protein-coupled receptors is primarily linked to stimulation of PLC, with associated increases in IP_3 and DAG, as well as in $[Ca^{2+}]_i$, and/or to increases/decreases in cAMP. The prototypical receptor mediating an increase in cAMP is the β-adrenergic receptor, and a large fraction of β-adrenergic receptors in the CNS are

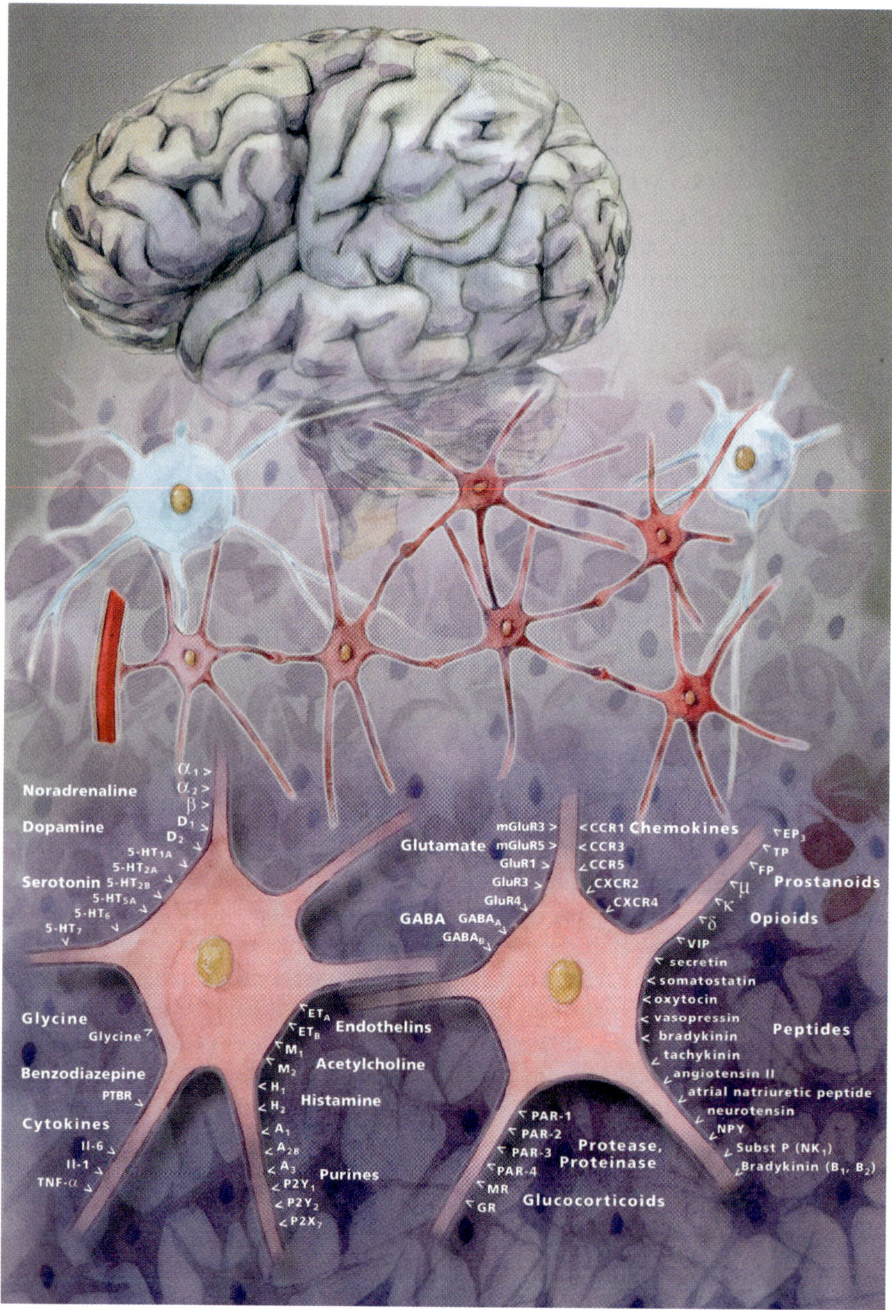
Noradrenaline
α_1
α_2
β
Dopamine
D_1
D_2
Serotonin
5-HT_{1A}
5-HT_{2A}
5-HT_{2B}
5-HT_{5A}
5-HT_6
5-HT_7
Glutamate
mGluR3
mGluR5
GluR1
GluR3
GluR4
GABA
$GABA_A$
$GABA_B$
CCR1
CCR3
CCR5
CXCR2
CXCR4
Chemokines
EP_3
TP
FP
Prostanoids
μ
κ
δ
Opioids
VIP
secretin
somatostatin
oxytocin
vasopressin
bradykinin
tachykinin
angiotensin II
atrial natriuretic peptide
neurotensin
NPY
Subst P (NK_1)
Bradykinin (B_1, B_2)
Peptides
Glycine
Glycine
Benzodiazepine
PTBR
Cytokines
Il-6
Il-1
TNF-α
ET_A
ET_B
Endothelins
M_1
M_2
Acetylcholine
H_1
H_2
Histamine
A_1
A_{2B}
A_3
$P2Y_1$
$P2Y_2$
$P2X_7$
Purines
PAR-1
PAR-2
PAR-3
PAR-4
Protease, Proteinase
MR
GR
Glucocorticoids

found on astrocytes (Stone and Ariano, 1989). Whereas many developmental effects of cAMP have been described in astrocytes, we know relatively little about the consequences of cAMP increases, except for such events as stimulation of glycogenolysis (see chapter by Hertz, Peng et al.) and effects on ion channels. With respect to activation of PLC the situation is different: During the last decade it has become firmly established that astrocytes carry out long-distance signaling, mediated by changes in $[Ca^{2+}]_i$. Increases in $[Ca^{2+}]_i$ not only spread in an astrocytic network, connected by gap junctions, but also across long distances from one network to another, with which it is not anatomically coupled, by aid of ATP, released from activated astrocytes (see below).

Astrocytes are able to synthesize and release many neuroactive compounds, both at rest and perhaps especially during exposure to various stimuli, including receptor activation and astrocytically mediated increases in $[Ca^{2+}]_i$. They release compounds such as taurine, GABA, glutamate, and aspartate (Kimelberg et al., 1990—see also chapter by Cornell-Bell et al.), as well as purine nucleotides and nucleosides (Di Iorio et al., 1998). The released transmitters are able to either enhance or inhibit activity in adjacent neurons. Since on one hand neuronally released transmitters can activate astrocytic $[Ca^{2+}]_i$ signals, and on the other hand these signals can activate or deactivate adjacent neurons, astrocytes mediate a neuronal–astrocytic–neuronal impulse transmission system. This system is not even dependent upon anatomical contiguity, since $[Ca^{2+}]_i$ signals can be transferred by ATP release. Therefore, synaptic sensitivity could be modulated by astrocytes even with no cellular bridges in between the synaptic regions involved.

The synthesis and storage of a wide variety of peptides within the CNS is well established. Specific roles for the peptides are less well known. However, angiotensin II and NPY have been shown to modulate noradrenergic transmission (see Sumners et al., 1994), and interactions also take place between cAMP-activated β receptors on astrocytes and VIP receptors, which results in a decrease in the cAMP level (Hansson and Rönnbäck, 1988). Angiotensin II and atrial natriuretic peptide are involved in the CNS control of water and Na^+ balance and in blood pressure/baroreflex regulation (Sumners et al., 1994). The involvement of receptors on astrocytes, brain endothelial cells and choroid plexus epithelial cells in regulation of fluid balance within the CNS by vasopressin is discussed in the chapter by Chen and Spatz.

Signaling by cytokines and growth factors to and from astrocytes is generally believed to be of importance mainly during development and after injury to the nervous system

---

Fig. 1. The astrocytes (red in the figure) occupy a strategic position in the central nervous system (CNS). The cells are coupled in networks, and come in contact with blood vessels (to the left in the figure) and synaptic regions. The astrocytes are known to support neuronal activity (neurons are depicted in white to pale blue). They create opportunities for the dynamics and plasticity in the CNS, both in physiological and in pathological conditions. The astrocytes express a large number of receptors and signaling systems. These receptors and their second messengers play specific roles in the signaling between individual astrocytes and in the communication between astrocytes and other cells. In this figure, astroglial receptors detected thus far are marked. The figure is a theoretical illustration, since we do not yet know in detail how the receptors are arranged on the cells. μ, mu opioid receptor; δ, delta opioid receptor; κ, kappa opioid receptor. For further information and abbreviations, see text.

(see Hansson and Rönnbäck, 2003). However, it is completely feasible that these compounds also may subserve normal physiological activity. That this is the case in the regulation of the secretion of hypothalamic hormones is discussed in the chapter by Mercier and Hatton.

5.2. Neurotrauma

In neurotrauma, including cerebral ischemia, glutamate is released at the synapse and by reversal of astrocytic glutamate uptake, and the astrocytes are stimulated, which results in increased $[Ca^{2+}]_i$ and Ca^{2+} wave propagations and in an increased release of ATP. Many studies have addressed the question of the relevance for extracellular ATP in glial functions. Besides their involvement in $[Ca^{2+}]_i$ signaling described above, activation of purinergic receptors is functionally coupled to prostanoid formation in astrocytes (Bruner et al., 1994). There is evidence supporting the importance of a glial arachidonic acid cascade in certain pathophysiological states, including an autocrine effect, since pharmacological studies suggest that astrocytes may also be among the target cells for prostanoids in the CNS (Ito et al., 1992). The purines participate in many vital intracellular processes, including astrocytic energy metabolism (formation of ATP and GTP) and the synthesis of cholesterol and of nucleic acids. In CNS disorders, such as seizures, trauma, ischemia, and hypoxia, both neurons and astrocytes greatly increase their release of purine nucleotides and nucleosides. Purinomimetic drugs, such as propentophylline, may become of importance in the treatment of the glial contribution to persistent pain (see chapter by Raghavendra and DeLeo).

Injury, as well as neurodegenerative disease, have profound effects on cytokines. In CNS injury, the pro-inflammatory cytokines TNF-α is upregulated and released by the astrocytes (Ben-Adani et al., 2001) and microglia (Kim et al., 2000; Hansson and Rönnbäck, 2003). The release promotes gliosis (Selmaj et al., 1991), inhibits astrocytic glutamate uptake (Fine et al., 1996) and induces apoptosis, particularly in oligodendrocytes (Hisahara et al., 1997). The cAMP-mediated action of VIP appears to reduce the release of TNF-α, at least by the microglia (Kim et al., 2000). Neurodegenerative diseases such as Alzheimer's disease and parkinsonism, are associated with large increases in both TNF-α and IL-1β (see chapters by Barger and by Przedborski and Goldman).

5.3. Development

It is well established that agents which in the mature organism function as transmitters may play a major developmental role as 'morphogens' during ontogenesis (Lauder, 1988). This applies to 'classical' transmitters like serotonin and noradrenaline acting via stimulation of adenylyl cyclase or PLC (Narumi et al., 1978; Mobley et al., 1986; Lauder, 1988). It has also been speculated whether opioids directly modify brain development at the cellular level or whether they act indirectly by causing alterations in hormone levels, respiration, or nutritional status. There is an upregulation of astrocytic δ receptors during the mitotic phase (Thorlin et al., 1999), and this type of expression during critical periods may support the theories of direct regulatory mechanisms.

VIP is thought to stimulate astrocytes to generate neurotrophic factors, the most potent and neuroprotective protein of which appears to be activity-dependent neurotrophic factor (ADNF) (Brenneman and Gozes, 1996). ADNF then acts directly on neurons to promote glutamate responses and morphological development. ADNF also decreases oxidative stress (Guo and Mattson, 2000) and causes secretion of neurotrophin 3 (NT-3), which regulates the NMDA receptor subunits 2A and 2B (Blondel et al., 2000).

6. Concluding remarks

It is of utmost importance that we learn to understand the cellular signaling underlying the plasticity of the CNS both in physiological regulatory activity and in functional disturbances resulting from damage and disease. It is now well known that neurons signal not only to each other, but also to the glial cells, informing the astroglial networks about neuronal activity and enabling them to support the neurons metabolically and trophically, according to their specific requirements. The astroglial cells also signal back to the neurons and modulate synaptic activity. ATP is an important signal substance, mediating contact between neurons, astroglial cells and microglial cells. During physiological conditions, microglia synthesize and release trophic factors; however, as soon as there is a functional disturbance of any kind in the CNS, microglia react, proliferate, and express protective or, in some situations, cytotoxic properties. These reactions are to a large extent mediated via receptors for cytokines and growth factors on astrocytes. Microglial cells also participate in restorative work, again often in conjunction with astrocytes. By identifying intercellular signaling, which mediates protective and restorative processes and reinforcing such signaling following tissue damage or during degeneration, it may eventually be possible to limit the extent of permanent damage in nervous tissue.

Acknowledgements

The work performed in the authors' lab was supported by grants from the Swedish Research Council (Project no. 33X-06812 and 21X-13015), Edith Jacobson's Foundation, and the Swedish Council for Working Life and Social Research. Fig. 1 was designed by Eva Kraft, Göteborg, Sweden.

References

Aloisi, F., Penna, G., Cerase, J., Iglesias, B.M., Adorini, L., 1997. IL-12 production by central nervous system microglia is inhibited by astrocytes. J. Immunol. 159, 1604–1612.

Aoki, C., 1992. Beta-adrenergic receptors: astrocytic localization in the adult visual cortex and their relation to catecholamine axon terminals as revealed by electron microscopic immunocytochemistry. J. Neurosci. 12, 781–792.

Aránguez, I., Torres, C., Rubio, N., 1995. The receptor for tumor necrosis factor on murine astrocytes: characterization, intracellular degradation, and regulation by cytokines and Theiler's murine encephalomyelitis virus. Glia 13, 185–194.

Araque, A., Martin, E.D., Perea, G., Arellano, J.I., Buno, W., 2002. Synaptically released acetylcholine evokes Ca^{2+} elevations in astrocytes in hippocampal slices. J. Neurosci. 22, 2443–2450.

Azmitia, E.C., Gannon, P.J., Kheck, N.M., Whitaker-Azmitia, P.M., 1996. Cellular localization of the 5-HT1A receptor in primate brain neurons and glial cells. Neuropsychopharmacology 14, 35–46.

Bacon, K.B., Harrison, J.K., 2000. Chemokines and their receptors in neurobiology: perspectives in physiology and homeostasis. J. Neuroimmunol. 104, 92–97.

Bal, A., Bachelot, T., Savasta, M., Manier, M., Verna, J.M., Benabid, A.L., Feuerstein, C., 1994. Evidence for dopamine D_2 receptor mRNA expression by striatal astrocytes in culture: in situ hybridization and polymerase chain reaction studies. Mol. Brain Res. 23, 204–212.

Bean, B.P., 1992. Pharmacology and electrophysiology of ATP-activated ion channels. Trends Pharmacol. Sci. 13, 87–90.

Ben-Adani, L., Gozes, I., Cohen, Y., Assaf, Y., Steingart, R.A., Brenneman, D.E., Eizenberg, O., Trembolver, V., Shohami, E., 2001. A peptide derived from activity-dependent neuroprotective protein (ADNP) ameliorates injury response in closed head injury in mice. J. Pharmacol. Exp. Ther. 296, 57–63.

Berridge, M.J., Irvine, R.F., 1984. Inositol trisphosphate, a novel second messenger in cellular signal transduction. Nature 312, 315–321.

Betz, H., Kuhse, J., Fischer, M., Schmieden, V., Laube, B., Kuryatov, A., Langosch, D., Meyer, G., Bormann, J., Rundstrom, N., 1994. Structure, diversity and synaptic localization of inhibitory glycine receptors. J. Physiol. 88, 243–248.

Blomstrand, F., Giaume, C., Hansson, E., Rönnbäck, L., 1999. Distinct pharmacological properties of ET-1 and ET-3 on astroglial gap junctions and Ca^{2+} signaling. Am. J. Physiol. 277, C616–C627.

Blondel, O., Collin, C., McCarran, W.J., Zhu, S., Zamostiano, R., Gozes, I., Brenneman, D.E., McKay, R.D.G., 2000. A glia-derived signal regulating neuronal differentiation. J. Neurosci. 20, 8012–8020.

Bohn, M.C., O'Banion, M.K., Young, D.A., Giuliano, R., Hussain, S., Dean, D.O., Cunningham, L.A., 1994. In vitro studies of glucocorticoid effects on neurons and astrocytes. Ann. N. Y. Acad. Sci. 746, 243–258.

Borsodi, P., Tóth, G., 1995. Characterization of opioid receptor types and subtypes with new ligands. Ann. N. Y. Acad. Sci. 757, 339–352.

Brenneman, D.E., Gozes, I., 1996. A femtomolar-acting neuroprotective peptide. J. Clin. Invest. 97, 2299–2307.

Bruner, G., Simmons, M.L., Murphy, S., 1994. Astrocytes: targets and sources for purines, eicosanoids and nitrosyl compounds. In: Murphy, S. (Ed.), Astrocytes: Pharmacology and Function. Academic Press, New York, NY, pp. 89–108.

Burnstock, G., 1978. A basis for distinguishing two types of purinergic receptors In: Bolis, L., Staub, W. (Eds.), Cell Membrane Receptors for Drugs and Hormones. Raven Press, New York, NY, pp. 107–118.

Burnstock, G., Kennedy, C., 1985. Is there a basis for distinguishing two types of P2-purinoceptors? Gen. Pharmacol. 16, 433–440.

Carman-Krzan, M., Lipnik-Stangelj, M., 2000. Molecular properties of central and peripheral histamine H_1 and H_2 receptors. Pflügers Arch. 439, R131–R132.

Carson, M.J., Thomas, E.A., Danielson, P.E., Sutcliffe, J.G., 1996. The 5-HT_{5A} serotonin receptor is expressed predominantly by astrocytes in which it inhibits cAMP accumulation: a mechanism for neuronal suppression of reactive astrocytes. Glia 17, 317–326.

Caulfield, M.P., 1993. Muscarinic receptors—characterization, coupling and function. [Review]. Pharmacol. Ther. 58, 319–379.

Changeux, J.P., 1995. Thudichum medal lecture. The acetylcholine receptor: a model for allosteric membrane proteins. Biochem. Soc. Trans. 23, 195–205.

Cheng, P.Y., Liu-Chen, L.Y., Pickel, V.M., 1997. Dual ultrastructural immunocytochemical labeling of mu and delta opioid receptors in the superficial layers of the rat cervical spinal cord. Brain Res. 778, 367–380.

Cholewinski, A.J., Hanley, M.R., Wilkin, G.P., 1988. A phosphoinositide-linked peptide response in astrocytes: evidence for regional heterogeneity. Neurochem. Res. 13, 389–394.

Ciccarelli, R., Ballerini, P., Sabatino, G., Rathbone, M.P., D'Onofrio, M., Caciagli, F., Di Iorio, P., 2001. Involvement of astrocytes in purine-mediated reparative processes in the brain. Review. Int. J. Dev. Neurosci. 19, 395–414.

Ciccarelli, R., Sureda, F., Casabona, G., Di Iorio, P., Caruso, A., Spinella, F., Condorelli, D.F., Nicoletti, F., Caciagli, F., 1997. Opposite influence of the metabotropic glutamate receptor subtypes mGlu3 and -5 on astrocyte proliferation in culture. Glia 21, 390–398.

Cockram, C.S., Kum, W., Ho, S.K., Zhu, S.O., Young, J.D., 1995. Binding and action of glucagon in cultured mouse astrocytes. Glia 13, 141–146.

Dam, T.V., Escher, E., Quirion, R., 1988. Evidence for the existence of three classes of neurokinin receptors in brain. Differential ontogeny of neurokinin-1, neurokinin-2 and neurokinin-3 binding sites in rat cerebral cortex. Brain Res. 453, 372–376.

Deecher, D.C., Wilcox, B.D., Dave, V., Rossman, P.A., Kimelberg, H.K., 1993. Detection of 5-hydroxytryptamine$_2$ receptors by radioligand binding, northern blot analysis, and Ca^{2+} responses in rat primary astrocyte cultures. J. Neurosci. Res. 35, 246–256.

De Kloet, E.R., 1991. Brain corticosteroid receptor balance and homeostatic control. Front. Neuroendocrinol. 12, 95–164.

Deschepper, C.F., 1998. Peptide receptors on astrocytes. Front. Neuroendocrinol. 19, 20–46.

De Witt, D.A., Perry, G., Cohen, M., Doller, C., Silver, J., 1998. Astrocytes regulate microglial phagocytosis of senile plaque cores of Alzheimer's disease. Exp. Neurol. 149, 329–340.

Di Iorio, P., Ballerini, P., Caciagli, F., Ciccarelli, R., 1998. Purinoceptor-mediated modulation of purine and neurotransmitter release from nervous tissue. Pharmacol. Res. 37, 169–178.

Dorf, M.E., Berman, M.A., Tanabe, S., Heesen, M., Luo, Y., 2000. Astrocytes express functional chemokine receptors. J. Immunol. 111, 109–121.

Ehrenreich, H., Costa, T., Clouse, K.A., Pluta, R.M., Ogino, Y., Coligan, J.E., Burd, P.R., 1993. Thrombin is a regulator astrocytic endothelin-1. Brain Res. 600, 201–207.

Enkvist, M.O.K., Hamalainen, H., Jansson, C.C., Kukkonen, J.P., Hautala, R., Courtney, M.J., Åkerman, K.E., 1996. Coupling of astroglial alpha$_2$-adrenoreceptors to second messenger pathways. J. Neurochem. 66, 2394–2401.

Enkvist, M.O.K., Holopainen, I., Åkerman, K.E.O., 1989. α-Receptor and cholinergic receptor-linked changes in cytosolic Ca^{2+} and membrane potential in primary rat astrocytes. Brain Res. 500, 46–54.

Eriksson, P.S., Hansson, E., Rönnbäck, L., 1990. δ and κ opiate receptors in primary cultures. Neuropharmacology 29, 799–804.

Eriksson, P.S., Nilsson, M., Wågberg, M., Hansson, E., Rönnbäck, L., 1993. Kappa-opioid receptors on astrocytes stimulate L-type Ca^{2+} channels. Neuroscience 54, 401–407.

Evans, B.A., Papaioannou, M., Hamilton, S., Summers, R.J., 1999. Alternative splicing generates two isoforms of the beta3-adrenoceptor which are differentially expressed in mouse tissues. Br. J. Pharmacol. 127, 1525–1531.

Fatatis, A., Holtzclaw, L.A., Avidor, R., Brenneman, D.E., Russell, J.T., 1994. Vasoactive intestinal peptide increases intracellular calcium in astroglia: synergism with α-adrenergic receptors. Proc. Natl Acad. Sci. USA 91, 2036–2040.

Felipo, V., Butterworth, R.F., 2002. Mitochondrial dysfunction in acute hyperammonemia. Neurochem. Int. 40, 487–491.

Fine, S.M., Angel, R.A., Perry, S.W., Epstein, L.G., Rothstein, J.D., Dewhurst, S., Gelbard, H.A., 1996. Tumor necrosis factor-α inhibits glutamate uptake by primary human astrocytes. Implications for pathogenesis of HIV-1 dementia. J. Biol. Chem. 271, 15303–15306.

Fotuhi, M., Standaert, D.G., Testa, C.M., Penney, J.B. Jr., Young, A.B., 1994. Differential expression of metabotropic glutamate receptors in the hippocampus and entorhinal cortex of the rat. Brain. Res. Mol. Brain Res. 21, 283–292.

Fraser, D.D., Mudrick Donnon, L.A., MacVicar, B.A., 1994. Astrocytic GABA receptors. Glia 11, 83–93.

Fredholm, B.B., Abbracchio, M.P., Burnstock, G., Dubyak, G.R., Harden, T.K., Jacobson, K.A., Schwabe, U., Williams, M., 1997. Towards a revised nomenclature for P1 and P2 receptors. Trends Pharmacol. Sci. 18, 79–82.

Fuxe, K., Agnati, L.F.(Eds.). 1991. Volume Transmission in the Brain. Novel Mechanisms for Neural Transmission. Raven Press, New York, NY.

Gadient, R.A., Otten, U.H., 1997. Interleukin-6 (IL-6)—a molecule with both beneficial and destructive potentials. Prog. Neurobiol. 52, 379–390.

Gandolfo, P., Louiset, E., Patte, C., Leprince, J., Masmoudi, O., Malagon, M., Gracia-Navarro, F., Vaudry, H., Tonon, M.C., 2001. The triakontatetraneuropeptide TTN increases $[Ca^{2+}]_i$ in rat astrocytes through activation of peripheral-type benzodiazepine receptors. Glia 35, 90–1000.

Giaid, A., Gibson, S.J., Herrero, M.T., Gentleman, S., Legon, S., Yanagisawa, M., Masaki, T., Ibrahim, N.B.N., Roberts, G.W., Rossi, M.L., Polak, J.M., 1991. Topographical localization of endothelin mRNA and peptide immunoreactivity in neurones of the human brain. Histochemistry 95, 303–314.

Gimpl, G., Kirchoff, F., Lang, R.E., Kettenmann, H., 1993. Identification of neuropeptide Y receptors in cultured astrocytes from neonatal rat brain. J. Neurosci. Res. 34, 198–205.

Goldman, R.S., Finkbeiner, S.M., Smith, S., 1991. Endothelin induces a sustained rise in intracellular calcium in hippocampal astrocytes. Neurosci. Lett. 123, 4–8.

Gozes, I., McCune, S.K., Jacobson, L., Warren, D., Moody, T.W., Fridkin, M., Brenneman, D.E., 1991. An antagonist to vasoactive intestinal peptide affects cellular functions in the central nervous system. J. Pharmacol. Exp. Ther. 257, 959–966.

Guo, Z.H., Mattson, M.P., 2000. Neurotrophic factors protect cortical synaptic terminals against amyloid and oxidative stress-induced impairment of glucose transport, glutamate transport and mitochondrial function. Cereb. Cortex 10, 50–57.

Gustavsson, L., Lundquist, C., Hansson, E., 1993. Receptor-mediated phospholipase D activity in primary astroglial cultures. Glia 8, 249–255.

Hagberg, G.-B., Blomstrand, F., Nilsson, M., Tamir, H., Hansson, E., 1998. Stimulation of 5-HT_{2A} receptors on astrocytes in primary culture opens voltage-independent Ca^{2+} channels. Neurochem. Int. 32, 153–162.

Hamprecht, B., Kemper, W., Amano, T., 1976. Electrical response of glioma cells to acetylcholine. Brain Res. 101, 129–135.

Hansson, E., 1985. Primary cultures from defined brain areas: effects of seeding time on the development of β-adrenergic and dopamine-stimulated cAMP-activity during cultivation. Dev. Brain Res. 21, 187–192.

Hansson, E., 1988. Astroglia from defined brain regions as studied with primary cultures. Prog. Neurobiol. 30, 369–397.

Hansson, E., Rönnbäck, L., 1988. Neurons from substantia nigra increase the efficacy and potency of second messenger arising from striatal astroglia dopamine receptor. Glia 1, 393–397.

Hansson, E., Rönnbäck, L., 2003. Glial neuronal signaling in the central nervous system. FASEB J. 17, 341–348.

Hansson, E., Rönnbäck, L., Sellström, Å., 1984. Is there a dopaminergic glial cell? Neurochem. Res. 9, 679–689.

Hertz, L., Baldwin, F., Schousboe, A., 1979. Serotonin receptors on astrocytes in primary cultures: effects of methylsergide and fluoxetine. Can. J. Physiol. Pharmacol. 57, 223–226.

Hirst, W., Cheung, N., Rattray, M., Price, G., Wilkin, G., 1998. Cultured astrocytes express messenger RNA for multiple serotonin receptor subtypes, without functional coupling of 5-HT_1 receptor subtypes to adenyl cyclase. Mol. Brain Res. 61, 90–99.

Hirst, W., Price, G.W., Rattray, M., Wilkin, G.P., 1997. Identification of 5-hydroxytryptamine receptors positively coupled to adenyl cyclase in rat cultured astrocytes. Br. J. Pharmacol. 120, 509–515.

Hisahara, S., Shoji, S., Okano, H., Miura, M., 1997. ICE/CED-3 family executes oligodendrocyte apoptosis by tumor necrosis factor. J. Neurochem. 69, 10–20.

Hollmann, M., Heinemann, S., 1994. Cloned glutamate receptors. Annu. Rev. Neurosci. 17, 31–108.

Hori, S., Komatsu, Y., Shigemoto, R., Mizuno, N., Nakanishi, S., 1992. Distinct tissue distribution and cellular localization of two messenger ribonucleic acids encoding different subtypes of rat endothelin receptors. Endocrinology 130, 1885–1895.

Hösli, E., Hösli, L., 1982. Evidence for the existence of alpha- and beta-adrenoceptors on neurones and glial cells of cultured rat central nervous system—an autoradiographic study. Neuroscience 7, 2873–2881.

Hösli, E., Hösli, L., 1986. Binding sites for $[^3H]$dopamine and dopamine antagonists on cultured astrocytes of rat striatum and spinal cord: an autoradiographic study. Neurosci. Lett. 65, 177–182.

Hösli, E., Hösli, L., 1988. Autoradiographic localization of binding sites for muscarinic and nicotinic agonists and antagonists on cultured astrocytes. Exp. Brain Res. 71, 450–454.

Hösli, L., Hösli, E., Schneider, U., Wiget, W., 1984. Evidence for the existence of histamine H_1- and H_2-receptors on astrocytes of cultured rat central nervous system. Neurosci. Lett. 48, 287–291.

Hösli, E., Stauffer, S., Hösli, L., 1995. Autoradiographic and electrophysiological evidence for the existence of neurotensin receptors on cultured astrocytes. Neuroscience 66, 627–633.

Hoyer, D., Clarke, D.E., Fozard, J.R., Hartig, P.R., Martin, G.R., Mylecharane, E.J., Saxena, P.R., Humphrey, P.P., 1994. International Union of Pharmacology classification of receptors for 5-hydroxytryptamine (serotonin). Pharmacol. Rev. 46, 157–203.

Hoyer, D., Martin, G., 1997. 5-HT receptor classification and nomenclature: towards a harmonization with the human genome. Neuropharmacology 36, 419–428.

Inagaki, N., Wada, H., 1994. Histamine and prostanoid receptors on glial cells. Glia 11, 102–109.

Inoue, A., Yanagisawa, M., Kimura, S., Kasuya, Y., Miyauchi, T., Goto, K., Masaki, T., 1989. The human endothelin family: three structurally and pharmacologically distinct isopeptides predicted by three separate genes. Proc. Natl Acad. Sci. USA 86, 2863–2867.

Ito, S., Sugama, K., Inagaki, N., Fukui, H., Giles, H., Wada, H., Hayaishi, O., 1992. Type-1 and type-2 astrocytes are distinct targets for prostaglandins D_2, E_2 and $F_{2\alpha}$. Glia 6, 67–74.

Itzhak, Y., Baker, L., Norenberg, M.D., 1993. Characterization of the peripheral-type benzodiazepine receptor in cultured astrocyte: evidence for multiplicity. Glia 9, 211–218.

Jaber, M., Robinson, S.W., Missale, C., Caron, M.G., 1996. Dopamine receptors and brain functions. Neuropharmacology 35, 1503–1519.

Jabs, R., Kirchhoff, F., Steinhauser, C., 1994. Kainate activates Ca^{2+}-permeable receptors and blocks voltage-gated K^+ currents in glial cells of mouse hippocampal slices. Pflugers Arch. Eur. J. Physiol. 426, 310–319.

Jacobs, B.L., Azmitia, E.C., 1992. Structure and function of the brain serotonin system. Physiol. Rev. 72, 165–229.

Johnston, G.A., 1994. $GABA_C$ receptors. Prog. Brain Res. 100, 61–65.

Jung, S., Pfeiffer, F., Deitmer, J.W., 2000. Histamine-induced calcium entry in rat cerebellar astrocytes: evidence for capacitative and non-capacitative mechanisms. J. Physiol. 527, 549–561.

Jurǐc, D.M., Čarman-Kržan, M., 2001. Interleukin-1β, but not IL-1α, mediates nerve growth factor secretion from rat astrocytes via type 1 IL-1 receptor. Int. J. Dev. Neurosci. 19, 675–683.

Khan, Z.U., Koulen, P., Rubinstein, M., Grandy, D.K., Goldman-Rakic, P.S., 2001. An astroglia-linked dopamine D2-receptor action in prefrontal cortex. Proc. Natl Acad. Sci. 98, 1964–1969.

Kim, W.-K., Kan, Y., Ganea, D., Hart, R.P., Gozes, I., Jonakit, G.M., 2000. Vasoactive intestinal peptide and pituitary adenylyl cyclase-activating polypeptide inhibit tumor necrosis factor-α production in injured spinal cord and in activated microglia via a cAMP-dependent pathway. J. Neurosci. 20, 3622–3630.

Kimelberg, H.K., Goderie, S.K., Higman, S., Pang, S., Waniewski, R.A., 1990. Swelling-induced release of glutamate, aspartate, and taurine from astrocyte cultures. J. Neurosci. 10, 1583–1591.

Kitanaka, J., Hashimoto, H., Gotoh, M., Kondo, K., Sakata, K., Hirasawa, Y., Sawada, M., Suzumura, A., Marunouchi, T., Matsuda, T., Baba, A., 1996a. Expression pattern of messenger RNAs for prostanoid receptors in glial cell cultures. Brain Res. 707, 282–287.

Kitanaka, J., Takuma, K., Kondo, K., Baba, A., 1996b. Prostanoid receptor-mediated calcium signaling in cultured rat astrocytes. Jpn. J. Pharmacol. 71, 85–87.

Kong, E.K., Peng, L., Chen, Y., Yu, A.C.H., Hertz, L., 2002. Up-regulation of $5\text{-}HT_{2B}$ receptor density and receptor-mediated glycogenolysis in mouse astrocytes by long-term fluoxetine administration. Neurochem. Res. 27, 113–120.

Kuhlmann, A.C., Guilarte, T.R., 2000. Cellular and subcellular localization of peripheral benzodiazepine receptors after trimethyltin neurotoxicity. J. Neurochem. 74, 1694–1704.

Kuwaki, T., Kurihara, H., Cao, W.H., Kurihara, Y., Unekawa, M., Yazaki, Y., Kumada, M., 1997. Physiological role of brain endothelin in the central autonomic control: from neuron to knockout mouse. Prog. Neurobiol. 51, 545–579.

Lauder, J.M., 1988. Neurotransmitters as morphogens. Prog. Brain Res. 73, 365–387.

Lazarini, F., Strosberg, A.D., Couraud, P.O., Cazaubon, S.M., 1996. Coupling of ET_B endothelin receptor to mitogen-activated protein kinase stimulation and DNA synthesis in primary cultures of rat astrocytes. J. Neurochem. 66, 459–465.

Leslie, F.M., 1987. Methods used for the study of opioid receptors. Pharmacol. Rev. 39, 197–249.

Leurs, R., Smit, M.J., Timmerman, H., 1995. Molecular pharmacological aspects of histamine receptors. Pharmacol. Ther. 66, 413–463.

Lindvall, O., Björklund, A., 1984. General organization of the cortical monoamine system. In: Descarries, L., Reader, T., Jasper, H., Alan, R., (Eds.), Monoamine Innervation of Cerebral Cortex. Liss, New York, NY, pp. 9–40.

Liu, Y., Jia, W.G., Strosberg, A.D., Cynader, M., 1992. Morphology and distribution of neurons and glial cells expressing beta-adrenergic receptors in developing kitten visual cortex. Dev. Brain Res. 65, 269–273.

Liu, J.S.H., John, G.R., Sikora, A., Lee, S.C., Brosnan, C.F., 2000. Modulation of interleukin-1β and tumor necrosis α signaling by P2 purinergic receptors in human fetal astrocytes. J. Neurosci. 20, 5292–5299.

López, T., López-Colomé, A.M., Ortega, A., 1997. NMDA receptors in cultured radial glia. FEBS Lett. 405, 245–248.

Marriott, D.R., Wilkin, G.P., 1992. Preincubation with substance P induces substance P-stimulated phosphatidylinositol turnover in cultured cerebellar astrocytes. J. Neurochem. 59, 443–448.

McCarthy, K.D., Enkvist, K., Shao, Y., 1995. Astroglial adrenergic receptors: expression and function. In: Kettenmann, H., Ransom, B.R. (Eds.), Neuroglia. Oxford University Press, Oxford, pp. 354–366.

McCarthy, K.D., Harden, T.K., 1981. Identification of two benzodiazepine binding sites on cells cultured from rat cerebral cortex. J. Pharmacol. Exp. Ther. 216, 183–191.

McCarthy, K.D., Salm, A.K., 1991. Pharmacologically-distinct subsets of astroglia can be identified by their calcium response to neuroligands. Neuroscience 41, 325–333.

Mercurio, F., Manning, A.M., 1999. Multiple signals converging on NF-kappaB. Curr. Opin. Cell. Biol. 11, 226–232.

Miller, S., Romano, C., Cotman, C.W., 1995. Growth factor upregulation of a phosphoinositide-coupled metabotropic glutamate receptor in cortical astrocytes. J. Neurosci. 15, 6103–6109.

Minden, A., Karin, M., 1997. Regulation and function of the JNK subgroup of MAP kinases. Biochim. Biophys. Acta 1333, F85–104.

Mobley, P.L., Scott, S.L., Cruz, E.G., 1986. Protein kinase C in astrocytes: a determinant of cell morphology. Brain Res. 398, 366–369.

Murphy, P.M., Baggiolini, M., Charo, I.F., Herbert, C.A., Horuk, R., Matsushima, K., Miller, L.H., Oppenheim, J.J., Power, C.A., 2000. International Union of Pharmacology. XXII. Nomenclature for chemokine receptors. Pharmacol. Rev. 52, 145–176.

Murphy, S., Pearce, B., Jeremy, J., Dandona, P., 1988. Astrocytes as eicosanoid-producing cells. Glia 1, 214–245.

Murphy, S., Pearce, B., Morrow, C., 1986. Astrocytes have both M_1 and M_2 muscarinic receptor subtypes. Brain Res. 364, 177–180.

Muyderman, H., Ängehagen, M., Sandberg, M., Björklund, U., Olsson, T., Hansson, E., Nilsson, M., 2001a. α_1-Adrenergic modulation of metabotropic glutamate receptor-induced calcium oscillations and glutamate release in astrocytes. J. Biol. Chem. 276, 46504–46514.

Muyderman, H., Sinclair, J., Jardemark, K., Hansson, E., Nilsson, M., 2001b. Activation of β-adrenoceptors opens calcium-activated potassium channels in astroglial cells. Neurochem. Int. 38, 269–276.

Nakahara, K., Okada, M., Nakanishi, S., 1997. The metabotropic glutamate receptor mGluR5 induces calcium oscillations in cultured astrocytes via protein kinase C phosphorylation. J. Neurochem. 69, 1467–1475.

Nakanishi, S., 1994. Metabotropic glutamate receptors: synaptic transmission, modulation and plasticity. Neuron 13, 1031–1037.

Nakanishi, S., Nakajima, Y., Masu, M., Ueda, Y., Nakahara, K., Watanabe, D., Yamaguchi, S., Kawabata, S., Okada, M., 1998. Glutamate receptors: brain function and signal transduction. Brain Res. Rev. 26, 230–235.

Nakayasu, H., Kimura, H., Kuriyama, K., 1995. Cerebral $GABA_A$ and $GABA_B$ receptors. Structure and function. Ann. N. Y. Acad. Sci. 757, 516–527.

Narumi, S., Kimelberg, H.K., Bourke, R.S., 1978. Effects of norepinephrine on the morphology and some enzyme activities of primary monolayer cultures from rat brain. J. Neurochem. 31, 1479–1490.

Nilsson, M., Eriksson, P.S., Rönnbäck, L., Hansson, E., 1993. GABA induces Ca^{2+} transients in astrocytes. Neuroscience 54, 605–614.

Nilsson, M., Hansson, E., Rönnbäck, L., 1991. Adrenergic and 5-HT_2 receptors on the same astroglial cell. A microspectrofluorimetric study on cytosolic Ca^{2+} responses in single cells in primary culture. Dev. Brain Res. 63, 33–41.

Ohishi, H., Shigemoto, R., Nakanishi, S., Mizuno, N., 1993. Distribution of the mRNA for a metabotropic glutamate receptor (mGluR3) in the rat brain: an *in situ* hybridization study. J. Comp. Neurol. 335, 252–266.

Otero, G.C., Merrill, J.E., 1994. Cytokine receptors on glial cells. Glia 11, 117–128.

Palma, C., Minghetti, L., Astolfi, M., Ambrosini, E., Silberstein, F.C., Manzini, S., Levi, G., Aloisi, F., 1997. Functional characterization of substance P receptors on cultured human spinal cord astrocytes: synergism of substance P with cytokines in inducing interleukin-6 and prostaglandin E2 production. Glia 21, 183–193.

Pin, J.P., Duvoisin, R., 1995. The metabotropic glutamate receptors: structure and functions. Neuropharmacology 34, 1–26.

Poblete, J.C., Azmitia, E.C., 1995. Activation of glycogen phosphorylase by serotonin and 3,4-methylenedioxy-methamphetamine in astroglial-rich primary cultures: involvement of the 5-HT2A receptor. Brain Res. 680, 9–15.

Porter, J.T., McCarthy, K.D., 1995. GFAP-positive hippocampal astrocytes in situ respond to glutamatergic neuroligands with increases in $[Ca^{2+}]_i$. Glia 13, 101–112.

Porter, J.T., McCarthy, K.D., 1997. Astrocytic neurotransmitter receptors in situ and in vivo. Prog. Neurobiol. 51, 439–455.

Pousset, F., Cremona, S., Dantzer, R., Kelley, K.W., Parnet, P., 2001. IL-10 and IL-4 regulate type-I and Type-II IL-1 receptors expression on IL-1β-activated mouse primary astrocytes. J. Neurochem. 79, 726–736.

Rao, V.L., Bowen, K.K., Rao, A.M., Dempsey, R.J., 2001. Up-regulation of the peripheral-type benzodiazepine receptor expression and $[^{(3)}H]$PK11195 binding in gerbil hippocampus after transient forebrain ischemia. J. Neurosci. Res. 64, 493–500.

Repke, H., Maderspach, K., 1982. Muscarinic acetylcholine receptors on cultured glial cells. Brain Res. 232, 206–211.

Reuss, B., Leung, D.S., Ohlemeyer, C., Kettenmann, H., Unisicker, K., 2000. Regionally distinct regulation of astroglial neurotransmitter receptors by fibroblast growth factor-2. Mol. Cell. Neurosci. 16, 42–58.

Rezaie, P., Pazos-Trillo, G., Everall, I.P., Male, D.K., 2002. Expression of β-chemokines and chemokine receptors in human fetal astrocyte and microglial co-cultures: potential role of chemokines in the developing CNS. Glia 37, 64–75.

Richards, J.G., Schoch, P., Häring, P., Takacs, B., Möhler, H., 1987. Resolving $GABA_A$/benzodiazepine receptors: cellular and subcellular localization in the CNS with monoclonal antibodies. J. Neurosci. 7, 1866–1886.

Ridet, J.L., Malhotra, S., Privat, A., Gage, F., 1997. Reactive astrocytes: cellular and molecular cues to biological function. Trends Neurosci. 20, 570–577.

Romano, C., Sesma, M.A., McDonald, C.T., O'Malley, K., Van den Pol, A.N., Olney, J.W., 1995. Distribution of metabotropic glutamate receptor mGluR5 immunoreactivity in rat brain. J. Comp. Neurol. 355, 455–469.

Rougon, G., Noble, M., Mudge, A.W., 1983. Neuropeptides modulate the β-adrenergic response of purified astrocytes in vitro. Nature 305, 715–717.

Ruzicka, B.B., Fox, C.A., Thompson, R.C., Meng, F., Watson, S.J., Akil, H., 1995. Primary astroglial cultures derived from several rat brain regions differentially express μ, δ and κ opioid receptor mRNA. Mol. Brain Res. 34, 209–220.

Salm, A.K., McCarthy, K.D., 1989. Expression of beta-adrenergic receptors by astrocytes isolated from adult rat cortex. Glia 2, 346–352.

Sandén, N., Thorlin, T., Blomstrand, F., Persson, P.A.I., Hansson, E., 2000. $5\text{-Hydroxytryptamine}_{2B}$ receptors stimulate Ca^{2+} increases in cultured astrocytes from three different brain regions. Neurochem. Int. 36, 427–434.

Satoh, S., Chang, C., Katoh, H., Hasegawa, H., Nakamura, K., Aoki, J., Fujita, H., Ichikawa, A., Negishi, M., 1999. The key amino acid residue of prostaglandin EP_3 receptor for governing G protein association and activation steps. Biochem. Biophys. Res. Commun. 255, 164–168.

Schilling, T., Nitsch, R., Heinemann, U., Haas, D., Eder, C., 2001. Astrocyte-released cytokines induce ramification and outward K^+ channel expression in microglia via distinct signalling pathways. Eur. J. Neurosci. 14, 463–473.

Schinelli, S., Zanassi, P., Paolillo, M., Wang, H., Feliciello, A., Gallo, V., 2001. Stimulation of endothelin B receptors in astrocytes induces cAMP response element-binding protein phosphorylation and c-fos expression via multiple mitogen-activated protein kinase signaling pathways. J. Neurosci. 21, 8842–8853.

Selmaj, K., Shaft-Zagardo, B., Aquino, D.A., Farooq, M., Raine, C.S., Norton, W.T., Brosnan, C.F., 1991. Tumor necrosis factor-induced proliferation of astrocytes from mature brain is associated with down-regulation of glial fibrillary acidic protein mRNA. J. Neurochem. 57, 823–830.

Shao, Y., Sutin, J., 1992. Expression of adrenergic receptors in individual astrocytes and motor neurons isolated from the adult rat brain. Glia 6, 108–117.

Sharma, G., Vijayaraghavan, S., 2001. Nicotinic cholinergic signaling in hippocampal astrocytes involves calcium-induced calcium release from intracellular stores. Proc. Natl Acad. Sci. USA 98, 4148–4153.

Shigemoto, R., Masugi, M., Fujimoto, K., 1999. Assembly-disassembly of metabotropic glutamate receptor 3 and water channel aquaporin 4 in astrocyte cell membrane. Neuropharmacology 38(A42), 135.

Simard, M., Couldwell, W.T., Zhang, W., Song, H., Liu, S., Cotrina, M.L., Goldman, S., Nedergaard, M., 1999. Glucocorticoids—potent modulators of astrocytic calcium signaling. Glia 28, 1–12.

Sokolovsky, M., 1995. Endothelin receptor subtypes and their role in transmembrane signaling mechanisms. Pharmacol. Ther. 68, 435–471.

Steinhäuser, C., Gallo, V., 1996. News on glutamate receptors in glial cells. Trends Neurosci. 19, 339–345.

Stephens, G.J., Cholewinski, A.J., Wilkin, G.P., Djamgoz, M.B., 1993. Calcium-mobilizing and electrophysiological effects of bradykinin on cortical astrocyte subtypes in culture. Glia 9, 269–279.

Stone, E.A., Ariano, M.A., 1989. Are glial cells targets of the central noradrenergic system? A review of the evidence. Brain Res. Rev. 14, 297–309.

Sumners, C., Tang, W., Paulding, W., Raizada, M.K., 1994. Peptide receptors in astroglia: focus on angiotensin II and atrial natriuretic peptide. Glia 11, 110–116.

Taga, T., Kishimoto, T., 1997. gp130 and the interleukin-6 family of cytokines. Annu. Rev. Immunol. 15, 797–819.

Tanabe, Y., Nomura, A., Masu, M., Shigemoto, R., Mizuno, N., Nakanishi, S., 1993. Signal transduction, pharmacological properties, and expression patterns of two rat metabotropic glutamate receptors, mGluR3 and mGluR4. J. Neurosci. 13, 1372–1378.

Testa, C.M., Standaert, D.G., Young, A.B., Penney, J.B. Jr., 1994. Metabotropic glutamate receptor mRNA expression in the basal ganglia of the rat. J. Neurosci. 14, 3005–3018.

Thorlin, T., Eriksson, P.S., Perssson, P.A.I., Åberg, N.D., Hansson, E., Rönnbäck, L., 1998. Delta-opioid receptors on astroglial cells in primary culture: mobilization of intracellular free calcium via pertussis sensitive G protein. Neuropharmacology 37, 299–311.

Thorlin, T., Perssson, P.A.I., Eriksson, P.S., Hansson, E., Rönnbäck, L., 1999. Delta-opioid receptor immunoreactivity on astrocytes is upregulated during mitosis. Glia 25, 370–378.

Uhl, G.R., Childers, S., Pak, G., 1994. An opiate-receptor gene family reunion. Trends Neurosci. 17, 89–93.

Van Calker, D., Müller, M., Hamprecht, B., 1979. Adenosine regulates via two different types of receptors, the accumulation of cyclic AMP in cultured brain cells. J. Neurochem. 33, 999–1005.

Van Calker, D., Müller, M., Hamprecht, B., 1980. Regulation by secretin, vasoactive intestinal peptide, and somatostatin of cyclic AMP accumulation in cultured brain cells. Proc. Natl Acad. Sci. USA 77, 6907–6911.

Van den Pol, A.N., Romano, C., Ghosh, P., 1995. Metabotropic glutamate receptor mGluR5 subcellular distribution and developmental expression in hypothalamus. J. Comp. Neurol. 362, 134–150.

Van der Zee, E.A., De Jong, G.I., Strosberg, A.D., Lutten, P.G.M., 1993. Muscarinic acetylcholine receptor-expression in astrocytes in the cortex of young and aged rats. Glia 8, 42–50.

Van Wagoner, N.J., Benveniste, E.N., 1999. Interleukin-6 expression and regulation in astrocytes. J. Neuroimmunol. 100, 124–139.

Ventura, R., Harris, K.M., 1999. Three-dimensional relationships between hippocampal synapses and astrocytes. J. Neurosci. 19, 6897–6906.

Verkhratsky, A., Orkand, R.K., Kettenmann, H., 1998. Glial calcium: homeostasis and signaling function. Physiol. Rev. 78, 99–140.

Verkhratsky, A., Steinhäuser, C., 2000. Ion channels in glial cells. Brain Res. Rev. 32, 380–412.

Von Blankenfeld, G., Kettenmann, H., 1991. Glutamate and GABA receptors in vertebrate glial cells. Mol. Neurobiol. 5, 31–43.

Wang, H., Ubl, J.J., Reiser, G., 2002. Four subtypes of protease-activated receptors, co-expressed in rat astrocytes, evoke different physiological signaling. Glia 37, 53–63.

Whitaker-Azmitia, P.M., Murphy, R.B., Azmitia, E.C., 1990. S-100 protein release from astrocytic glial cells by stimulation of 5-HT_{1A} receptors and regulates the development of serotonergic neurons. Brain Res. 528, 155–158.

Wu-Wong, J.R., Dayton, B.D., Opgenorth, T.J., 1996. Endothelin-1-evoked arachidonic acid release: a Ca^{2+}-dependent pathway. Am. J. Physiol. 271, C869–C877.

Yanagisawa, M., Kurihara, H., Kimura, S., Tomobe, Y., Kobayashi, M., Mitsui, Y., Yazaki, Y., Goto, K., Masaki, T., 1988. Endothelin: a novel potent vasoconstrictor peptide by vascular endothelial cells. Nature 332, 411–415.

Zanassi, P., Paolillo, M., Montecucco, A., Avvedimento, E.V., Schinelli, S., 1999. Pharmacological and molecular evidence for dopamine D_1 receptor expression by striatal astrocytes in culture. J. Neurosci. Res. 58, 544–552.

Zhao, Z., Code, W.E., Hertz, L., 1992. Dexmedetomidine, a potent and highly specific $alpha_2$ agonist, evokes cytosolic calcium surge in astrocytes but not in neurons. Neuropharmacology 31, 1077–1079.

Zhao, Z., Hertz, L., Code, W.E., 1996. Effects of benzodiazepines on potassium induced increase in free cytosolic calcium concentration in astrocytes and neurons in primary cultures. Can. J. Physiol. Pharmacol. 74, 273–277.

Zlotnik, A., Yoshie, O., 2000. Chemokines: a new classification system and their role in immunity. Immunity 12, 121–127.

Advances in
Molecular and
Cell Biology

Transactivation in astrocytes as a novel mechanism of neuroprotection

Liang Peng

Professor, College of Basic Medical Sciences, China Medical University, No. 92 Beier Boad, Heping Distric, Shenyang, P.R. China
E-mail: sharkfin039@yahoo.com

Contents

1. Introduction
2. EGFR transactivation pathways
3. HB-EGF and EGFR expression in brain
4. Is transactivation in astrocytes receptor selective?
    4.1. EGF transactivation in astrocytes
    4.2. α_2-Adrenergic receptors
    4.3. Metabotropic glutamate receptors
    4.4. Thrombin
    4.5. ATP
    4.6. Other transmitters
5. Cytoprotective effect of HB-EGF
6. Concluding remarks

Epithelial growth factor receptor (EGFR) transactivation in response to stimulation of a G protein-coupled receptor (GPCR) is a novel signaling pathway in cross-talk from cell to cell and from receptor to receptor. The presently reviewed material suggests that transactivation may be specific for the EGFR and not a phenomenon common to many growth factor receptors. However, different GPCRs can lead to EGFR transactivation, albeit probably utilizing different transduction pathways in the donor cells, as exemplified by similar effects of the group I metabotropic glutamate receptor agonist, [*RS*]-3,5-dihydroxyphenylglycine, and the α_{2A}-adrenergic agonist dexmedetomidine. Both drugs were found to act on astrocytes, and these cells are likely to be a major source of soluble heparin-binding EGF-like growth factor (HB-EGF), since GPCRs related to EGF activation and HB-EGF are expressed at high level in astrocytes in primary cultures and in the brain in vivo, and since expression of the HB-EGF gene may mainly occur in astrocytes in neurodegenerative diseases.

Advances in Molecular and Cell Biology, Vol. 31, pages 503–518

ISBN: 0-444-51451-1

1. Introduction

Transactivation is a process in which signaling to a G-protein-coupled receptor (GPCR) leads to release ('shedding') of a growth factor, which in turn stimulates receptor tyrosine-kinases (RTK) on the same or adjacent cells and evokes a response that is indistinguishable from that which can be directly obtained by direct RTK stimulation, e.g., phosphorylation of mitogen-activated protein (MAP) kinases, such as extracellular-signal regulated kinase (ERK) 1 and 2, and subsequent down-stream signaling. Transactivation may play a major role in not only astrocytic functions but also in interactions between astrocytes and neurons, including astrocyte-mediated neuroprotection. Both astrocytes and neurons respond to neurotransmitters and growth factors by specific receptors (see chapters by Hansson and Rönnbäck and by Nakagawa and Schwartz), and both cell types also express and release a number of neurotrophic factors, such as heparin-binding EGF-like growth factor (HB-EGF), ciliary neurotrophic factor, nerve growth factor, brain-derived neurotrophic factor and neurotrophin 3 (Rudge et al., 1992), and transforming growth factor-beta (TGF-β) (Bruno et al., 1998). This arrangement sets the stage for extremely complicated communications from cell to cell, and from receptor to receptor, which may be crucial for the precise regulation of astrocytic effects on proliferation, differentiation, function, and survival, not only of astrocytes but also of their neuronal neighbors.

Cross-talk at the molecular level between different signaling pathways, especially between GPCRs and RTKs, was first uncovered in transfected cell lines (Daub et al., 1997), and has later been demonstrated in primary cultures of different cell types (Peavy et al., 2001; Kalmes et al., 2000; Grewal et al., 2001; Saito et al., 2002; Uchiyama-Tanaka et al., 2001). Recently, we have established that dexmedetomidine, a specific agonist of the α_2-adrenergic receptor, stimulates the transactivation pathway in primary cultures of mouse astrocytes by shedding of HB-EGF. This conclusion was based on inhibition of dexmedetomidine-induced ERK phosphorylation in astrocytes by tyrphostin AG 1478, a specific inhibitor of the EGFR tyrosine kinase (Peng et al., 2002). Since HB-EGF has neuroprotective properties (see below), and since released HB-EGF in the brain in situ is likely to act not only on astrocytes but also on adjacent neurons, such a transactivation may be one of the underlying mechanisms of the well-established neuroprotective effect by dexmedetomidine (Jolkkonen et al., 1999). There is also evidence that dexmedetomidine in the brain may decrease neuronal glutamate release and/or enhance glutamate removal (Jolkkonen et al., 1999; Talke and Bickler, 1996; Huang and Hertz, 2000; Li and Eisenach, 2001). However, its neuroprotective effects can be established even in the absence of reduced extracellular glutamate concentration (Engelhard et al., 2002).

Many neuroscientists have been interested for a long time in the potential use of growth factors as neuroprotective agents (Gozes, 2001). However, clinical application of growth factors with large molecular size is, at best, difficult, because these polypeptides are unable to cross the blood–brain-barrier (Hefti, 1994). Instead, using a small molecule, such as the α_2-adrenergic agonist dexmedetomidine, to induce HB-EGF would provide a novel therapeutic strategy for treatment of neurodegenerative and ischemic brain diseases. However, other transmitters, e.g., agonists of certain metabotropic receptors, have also been found to have neuroprotective properties (Bruno et al., 1998, 2000), and it cannot be excluded that neuroprotection by transactivation and resulting release of growth factor(s) may be a phenomenon of more general importance than presently realized. It is the purpose of the present chapter to discuss this possibility.

2. EGFR transactivation pathways

Transactivation following GPCR stimulation has been observed with many different kinds of GPCRs and in many different cell types. The best studied RTK transactivation by GPCRs is activation of the EGFR. Activation of this receptor represents the paradigm for cross-talk between GPCRs and RTKs (Gschwind et al., 2001), but a priori it cannot be excluded that other growth factor receptors may be activated in a similar manner. The involvement of EGFR in transactivation was demonstrated by Ullrich's group in 1997, who showed that MAPK activity induced by the GPCR agonists endothelin-1, lysophosphatic acid and thrombin was reduced by tyrphostin AG1478 or in a dominant-negative EGFR receptor mutant (Daub et al., 1997). For a period of time, it was not clear whether a RTK ligand was involved, since the onset of transactivation is rapid, and no free ligand was detected in the extracellular medium. However, by now, HB-EGF shedding has been successfully demonstrated with several different GPCR agonists (or activators of second messengers) and in many different cell types.

Phorbol ester-induced HB-EGF shedding from monkey kidney Vero-H cells (stable transfectants of Vero cells, overexpressing human pro-HB-EGF) was first demonstrated immunocytochemically by Goishi et al. (1995), and it has been confirmed by Western blotting, using antibodies specific to the tail fragment of pro-HB-EGF in the cell lysate or to HB-EGF in the medium and cell lysate (Umata et al., 2001). The shedding is dependent on metalloproteinases and therefore blocked in mutants lacking the metalloproteinase domain (Izumi et al., 1998). Phorbol esters activate protein kinase C (PKC), which accordingly can trigger HB-EGF shedding. However, in NbMC-2 prostate epithelial cells it has been shown that this can also be achieved independently of PKC activation by an increase in free cytosolic Ca^{2+} concentration evoked by administration of a Ca^{2+} ionophore and dependent on the presence of extracellular calcium (Dethlefsen et al., 1998). Secondarily, released HB-EGF stimulates phosphorylation of EGFR, a response which in primary cultures of vascular smooth muscle (VSMCs) stimulated by thrombin was shown to be blocked by heparin (which inactivates HB-EGF) and neutralizing HB-EGF antibody, as well as by a metalloproteinase inhibitor (Kalmes et al., 2000).

Transactivation of EGF can be induced by GPCR-mediated activation of either $G_{i/o}$- and G_q-coupled receptors (Table 1). In the conventional signaling pathway of GCPRs, receptor stimulation leads to exchange of GDP, bound to the alpha subunit of the G-protein, with GTP. This exchange triggers dissociation of the 'activated' alpha subunit and initiation of the typical receptor response, e.g., hydrolysis of PIP_2 to generate the two second messengers inositol trisphosphate (IP_3), which stimulates release of Ca^{2+} from intracellular stores, and diacylglycerol (DAG), which stimulates PKC activity. The importance of these mechanisms in the signaling mechanisms of HB-EGF shedding and EGFR transactivation are not completely clear, and could be different in different cell

Table 1
Cell types in which transactivation has been demonstrated, the G protein-coupled receptors involved, and the types of G protein activated by receptor stimulation

Receptor	G protein	Cell type
α_{2A}-Adrenergic receptor	$G_{i/o}$ (Antaa et al., 1995)	Astrocytes (Peng et al., 2003)
		Retinal Müller cells (Peng et al., 1998)[a]
		Transfected cell line (Pierce et al., 2001)
α_{2B}-adrenergic receptor	$G_{i/o}$ (Antaa et al., 1995)	Renal tubular cell line; proximal tubule cells (Cussac et al., 2002a,b)
mGluR5	G_q (Miura et al., 2002)	Astrocytes (Peavy et al., 2001)
Thrombin receptor	G_q and $G_{i/o}$ (Gardner et al., 2002; Debeiret et al., 1996)	Astrocytes (Daub et al., 1997)
		VSMC (Kalmes et al., 2000)
		Transfected cell line (Daub et al., 1997)
P2Y	G_q and $G_{i/o}$ (Lin and Chuang, 1994)	Astrocytes (Daub et al, 1997)
5-HT_{2A} receptor	G_q and G_i (Miyoshi et al., 2001; Maayani et al., 2001)	Renal mesangial cells (Grewal et al., 2001)
Angiotensin II receptor	G_q (Keys et al., 2002)	VSMC (Saito et al., 2002; Voisin et al., 2002)
		Glomerular mesangial cells (Uchiyama-Tanaka et al., 2001)
		Transfected cell lines (Goishi et al., 1995; Umata et al., 2001)
$Beta_2$-adrenergic receptor	G_i (Zamah et al., 2002)	Cardiac fibroblasts (Kim et al., 2002)
Estrogen receptor	GPR 30 (Filardo et al., 2002)	Breast cell lines (Filardo, 2002; Filardo et al., 2002)
Endothelin-1 receptor	G_q (Takigawa et al., 1995)	Tranfected cell line (Daub et al., 1997)
Bradykinin receptor	G_q (Graness et al., 1998)	PC12 cells (Zwick et al., 1997)

VSMC: vascular smooth muscle cells.
[a]The involvement of transactivation has not been directly established in these cells.

types and after stimulation of different receptors. Consistent with the dependence of HB-EGF shedding on either PKC activity or Ca^{2+} entry in some cell types, it has been demonstrated (i) that phosphorylation of EGFR by bradykinin in PC12 cells, a neuronal cell line developed from a pheochrocytoma, is dependent upon the presence of extracellular Ca^{2+} (Zwick et al., 1997); and (ii) that serotonin (5-HT) in renal mesangial cells induces a concentration-dependent EGFR phosphorylation (by stimulation of 5-HT_{2A} receptors), which can be attenuated by a PKC inhibitor, GF109203X, but not by PD98059, an inhibitor of MEK (MAP kinase/ERK kinase), and thus of ERK phosphorylation (Grewal et al., 2001). Whether GPCR-induced and metalloprotease-mediated proteolytic cleavage of HB-EGF plays a key role in these cases needs, however, to be further studied. It has also been reported that calmodulin is involved in phorbol ester-induced EGF transactivation in a monkey kidney cell line, COS-1 (Tebar et al., 2002). In a renal tubular cell line and in primary cultures of proximal tubule cells, α_{2B}-adrenergic receptor stimulation by dexmedetomidine causes a release of arachidonic acid, which is PKC-independent, but abolished by prior treatment with pertussis toxin (indicating the involvement of a GPCR) and leads to ERK phosphorylation, which is reduced by tyrphostin AG 1478, showing the dependence on EGF shedding (Cussac et al., 2002a,b). However, this pathway is specific to the α_{2B}-adrenergic receptor, since agonist stimulation of $\alpha_{2A/D}$ and α_{2C} receptors has no effect on archidonic acid release (Audubert et al., 1999).

Using α_2-adrenergic receptor stimulation of two types of transfected COS-7 cells (another monkey kidney cell line), one serving as a HB-EGF donor and expressing α_{2A}-adrenergic receptors, and the other expressing no α_{2A}-adrenergic receptors and serving as a HB-EGF acceptor, many details of α_{2A}-adrenergic receptor-mediated transactivation in these cells have been elegantly elucidated in co-culture experiments (Pierce et al., 2001). α_{2A}-Adrenergic stimulation was found to cause ERK phosphorylation by two different, major pathways (Pierce et al., 2001). One of these leads directly to ERK phosphorylation in the stimulated cell, the second is the two-stage transactivation pathway (Fig. 1). In its first stage the beta,gamma subunits of the activated, heterotrimeric G_i protein lead, via activation of cytosolic Src tyrosine kinases, to dynamin-dependent, metalloprotease-catalyzed shedding of HB-EGF from its extracellular transmembrane-spanning HB-EGF precursor; in the second stage the shedded HB-EGF transactivates EGFR in the same and adjacent cells in conventional manners, i.e., the EGFR is phosphorylated by a RTK and internalized. This leads to the recruitment of Grb2–Sos1 complexes to the activated RTK and the exchange of GDP for GTP on the low molecular weight G protein, Ras, catalyzed by Ras guanine nucleotide exchange factor, Sos1. Ras activation, in turn, initiates the phosphorylation cascade consisting of Raf, MEK, and ERK. Recent work has demonstrated that beta-arrestins, i.e., cytosolic proteins that mediate desensitization and internalization of GPCRs, can increase the phosphorylation of ERK (in angiotensin II-stimulated COS-7 cells), presumably by functioning as adaptors and scaffolds for MAP kinase activation, bringing sequentially acting kinases into proximity with each other and with the receptor (Tohgo et al., 2002). In addition, phosphatidylinositol-3-kinase (PI3K) can be activated by EGFR stimulation of astrocytes (Zelenaia et al., 2000), but the pathways connecting PI3K activation with MAP phosphorylation are uncertain and may vary between cell types and from transmitter to transmitter. The PI3K pathway may be important for the neuroprotective effect of

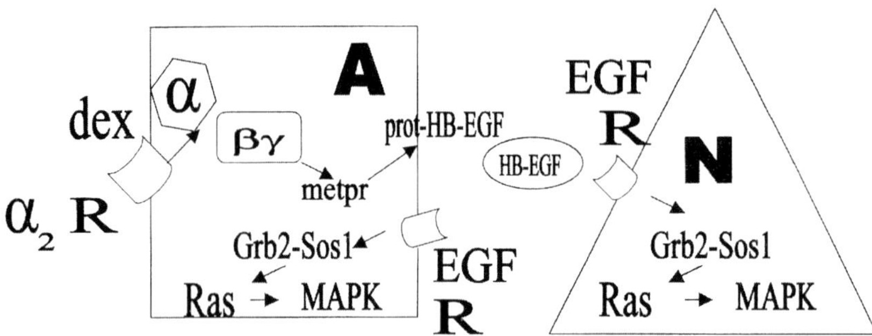

Fig. 1. Schematic representation of key signaling mechanisms presumably involved in transactivation by exposure of astrocytes (A) to the α_2-adrenergic agonist dexmedetomidine. The beta, gamma subunits of the activated, heterotrimeric G_i protein lead to shedding of HB-EGF from its extracellular transmembrane-spanning HB-EGF precursor, catalyzed by a metalloprotease (metpr). The shedded HB-EGF transactivates EGFR in the same (A) and adjacent cells, presumably including neurons (N), in conventional manners, i.e., the EGFR is phosphorylated by a RTK and internalized. This leads to the recruitment of Grb2–Sos1 complexes to the activated RTK and the exchange of GDP for GTP on the low molecular weight G protein, Ras. Ras activation of Ras, in turn, initiates the phosphorylation cascade consisting of Raf, MEK, and the MAP kinases (MAPK) ERK. The proposed signaling sequences are based upon determinations by Pierce et al. (2001) of α_2-activated signaling pathways in a transfected cell line and the observation by Peng et al. (2003) that inhibition of HB-EGF stimulation in astrocytes reduces ERK phosphorylation in these cells.

HB-EGF (Brunet et al., 2001), because of its ability to phosphorylate the membrane phospholipid phosphatidylinositide bisphosphate (PIP_2) to phosphatidylinositide trisphosphate (PIP_3). PIP_3 activates the protein-serine/threonine kinase Akt, which then phosphorylates and de-activates a number of pro-apoptotic proteins, including Bad, a Bcl-2 family member, which induces cell death by stimulating release of cytochrome c from mitochondria (Yamaguchi and Wang, 2001; Xue et al., 2000). However, ERK1/2 phosphorylation has also been linked to cytoprotection (Kim et al., 2000).

3. HB-EGF and EGFR expression in brain

HB-EGF is a 22 kDa, *O*-glycosylated protein, which was isolated from conditioned medium of a human macrophage-like cell line in 1991 (Higashiyama et al., 1991). It is a member of the EGF family, which consists of EGF, transforming growth factor-alpha (TGF-α) and heparin-binding EGF (HB-EGF). A number of membrane-anchored growth factors, including members of the EGF family, have dual functions. As membrane glycosylated proteins, they may be important for cell-to-cell adhesion. After cleavage by proteases, they are released into the extracellular fluid as soluble growth factors. In addition to HB-EGF, TGF-α may act in this fashion. In a Chinese hamster ovary cell line transfected with rat pro-TGF-α gene, TGF-α shedding can be activated with serum factors and tumor-promoting phorbol esters via both PKC-dependent and PKC-independent pathways (Pandiella and Massague, 1991). TGF-α specifically binds to HER1. In brain, HB-EGF promotes proliferation of astrocytes and stem cells, neurogenesis (Jin et al., 2002; Michalsky et al., 2002) and survival and differentiation of postmitotic neurons (Kornblum et al., 1999).

HB-EGF is expressed in both neurons and astrocytes (Nakagawa et al., 1998). In agreement with its function during early developmental stages, the level of HB-EGF expression is much higher in brains of young animals than in adult brain (Opanashuk et al., 1999). Gene expression of HB-EGF is upregulated in cerebral ischemia (Grewal et al., 2001; Tanaka et al., 1999) and by excitotoxicity and brain injury (Opanashuk et al., 1999). In cultured cerebral cortical neurons, expression of HB-EGF is increased to 150% of control level after hypoxia (Jin et al., 2002). Although gene expression of TGF-α also is quite high in brain, it is not upregulated by ischemia (Grewal et al., 2001). The presence of EGF has only rarely been reported in brain (Scalabrino et al., 1999), and in contrast to HB-EGF, there seems to be no reports that EGF gene expression is increased in ischemic brain.

The mitogenic, and non-mitogenic (including neuroprotective) effects of HB-EGF could be stage- and cell type-specific, and dependent upon the type of EGF receptor expressed by the particular cells at the specific period of time. The EGF receptor is widespread in brain and plays a major role both under physiological and pathological conditions, including carcinogenesis (Prenzel et al., 2001). There are four EGF receptors, HER1, HER2, HER3 and HER4 (Prenzel et al., 2001). According to studies in transfected cell lines that express either an individual receptor or a combination of receptors, HB-EGF binds only to HER1 and HER4 (Raab and Klagsbrun, 1997), although binding of HB-EGF to those two receptors may indirectly active other HER family member (Raab and Klagsbrun, 1997). Expression of EGF receptors varies among different types of neurons (Gerecke et al., 2001). Astrocytes express HER1 and maybe small amounts of HER4 (Pinkas-Kramarski et al., 1997; Kornblum et al., 1999). HER4 has been detected at high level in the postsynaptic density, indicating the possibility of its involvement in synaptic plasticity (Huang et al., 2001). Head injury induces an increase of HER4 expression in neurons, but not in astrocytes (Erlich et al., 2000), suggesting a role of neuronal HER4 under pathological conditions. Nevertheless, HB-EGF-promoted survival of cultured dopaminergic neurons in neuronal-astrocytic co-cultures appears to be exerted by stimulation of astrocytic EGFR (Farkas and Krieglstein, 2002).

4. Is transactivation in astrocytes receptor selective?

4.1. EGF transactivation in astrocytes

Several different G-protein coupled transmitters appear to be capable of inducing EGF transactivation in astrocytes (Table 1), and the question arises whether HB-EGF shedding is a common underlying mechanism. Conceivably, transactivation might also stimulate release of other neurotrophic factors, such as basic fibroblast growth factor (basic FGF) or transforming growth factor-beta (TGF-β), which belongs to a different growth factor family than TGF-α. Several growth factors other than those acting on the EGFR provide neuroprotection. Below, it will be discussed whether they may act by transactivation.

4.2. α_2-Adrenergic receptors

The α_2-adrenergic receptor comprises three subtypes, $\alpha_{2A/D}$, α_{2B} and α_{2C}, which all are coupled to pertussis-sensitive $G_{i/o}$-coupled receptors (Bylund and Chacko, 1999). In brain, only a minor fraction of α_2-adrenergic receptors may be presynaptic (Hein et al., 1999). Since the major target cells in the CNS expressing α_2-adrenergic receptors include astrocytes, we tested effects of dexmedetomidine on ERK phosphorylation in these cells. Indeed, dexmedetomidine induced EGF transactivation in astrocytes, demonstrated by inhibition of ERK phosphorylation with tyrphostin AG1478 (Peng et al., 2002). It confirms the involvement of HB-EGF shedding that phosphorylation of ERK is also inhibited by heparin, which inactivates HB-EGF.

In retina, α_2-adrenergic agonists stimulate basic FGF expression in photoreceptors and protects photoreceptors against light damage (Wen et al., 1996); ERK phosphorylation induced by α_2-adrenergic receptor activity has been observed in Müller cells, an astrocyte-like cell of the retina (Peng et al., 1998). However, no experiments are available that unequivocally indicate that transactivation is involved. Unlike EGFR agonists there is also no extensive literature suggesting transactivation by shedding of basic FGF in other cell types. However, it cannot be excluded that the α_2-adrenergic receptor stimulation may have caused EGFR transactivation, and that released HB-EGF secondarily has stimulated release of basic FGF. Moreover, as in the brain, reduction of glutamate excitotoxicity may also play a role in the neuroprotective properties of α_2-adrenergic agonists (Baptiste et al., 2002).

4.3. Metabotropic glutamate receptors

There are three groups of metabotropic glutamate receptors (mGluRs). Group I mGluRs, i.e., mGluR1 and 5, are linked to G_q protein, and activation of these receptors stimulate hydrolysis of PIP_2, leading to release of DAG and IP_3 and an increase in free cytosolic Ca^{2+} concentration; Group II mGluRs, i.e., mGluR2 and 3, and Group III, i.e., mGluR4, 6, 7 and 8, are coupled to $G_{i/o}$ protein, and are linked to opening/closing of ion channels and inhibition of adenylyl cyclase (Bruno et al., 1997). Several of these receptors are expressed in astrocytes (see chapter by Hansson and Rönnbäck). The mechanisms of the neuroprotective effects differ among the three groups of mGluRs. Agonists of the presynaptically located mGluR III are thought to protect neurons against excitotoxicity by inhibition of glutamate release from presynaptic terminals (Gasparini et al., 1999; Bruno et al., 2000). The neuroprotective effect of mGluR II agonists requires synthesis of new protein (Bruno et al., 1997, 1998). Stimulation of mGluR II receptors induces a several-fold increase of TGF-β release from glial cultures, but it is unknown whether EGF transactivation is involved in this release (Bruno et al., 1998). It has, however, been clearly demonstrated in cultured astrocytes that the activity of mGluR5 is associated with EGF transactivation (Peavy et al., 2001). Like the α_2-adrenoceptor-induced EGF transactivation in transfected COS-7 cells, phosphorylation of EGFR by mGluR5 activity in astrocytes is dependent on Src family tyrosine kinase, but unlike α_2-adrenergic transactivation in COS-7 cells, it is not dependent on G-protein beta,gamma subunits. Thus the transactivation pathway in the donor cells is either cell type-specific or

GCPR-specific. The EGF transactivation in astrocytes by mGluR5 agonists may be responsible for activity-induced upregulation of EGF, EGFR and mGluR in these cells. It will be very interesting to establish definitively whether activation of mGluR5 actually stimulates HB-EGF shedding from astrocytes and whether the transactivation pathways in astrocytes are GCPR-specific.

4.4. Thrombin

Thrombin is a serine proteinase that at low doses has been reported to have protective effect in brain injury. It binds to a GPCR called protease-activated receptor (PAR). Astrocytes express three subtypes of thrombin receptor, PAR-1, PAR-2, and PAR-4 (Wang et al., 2002). Stimulation with thrombin induces an increase of intracellular calcium concentration in astrocytes (Ubl and Reiser, 1997), and enhances PKC activity in rat glioma C6 cells (Kaufmann et al., 1996). In 1321 astrocytoma cells, thrombin increases Ras and MAPK activation (Post et al., 1996). Not surprisingly, it was shown that thrombin stimulates EGF transactivation by EGFR phosphorylation in astrocytes (Daub et al., 1997). There is no information whether thrombin induces HB-EGF shedding in astrocytes, but it is known to be the case in smooth muscle cells (SMCs) (Kalmes et al., 2000).

4.5. ATP

ATP functions both as a trophic factor and a transmitter (see chapter by Hansson and Rönnbäck), and ATP-induced EGF transactivation has been demonstrated in primary cultures of astrocytes (Daub et al., 1997). Astrocytes express both ionotropic (P2X) and metabotropic (P2Y) receptors for purine nucleotides. There are seven subtypes of P2X receptors. With specific antibodies, astrocytes were found to stain for P2X1–P2X4, P2X6, and P2X7 receptors, but not for P2X5 (Kukley et al., 2001). ATP induces a significant increase of intracellular Ca^{2+} concentration in astrocytes by both influx of extracellular Ca^{2+} and G-protein-mediated Ca^{2+} release from intracellular stores (Shiga et al., 2001). P2Y receptors are GPCR. Stimulation of P2Y receptor in astrocytes leads to an IP_3-induced Ca^{2+} release from intracellular Ca^{2+} stores, PKC activation, ERK phosphorylation and cell proliferation (Neary et al., 1999; Lenz et al., 2000). Interestingly, P2X7 seems also to be involved in MAPK activation in astrocytes (Panenka et al., 2001). Also, extracellular ATP stimulates the release of heparin-binding epidermal and platelet-derived growth factors from cultured Müller cells (see chapter by Bringmann et al.).

4.6. Other transmitters

Some GPCRs, that have been demonstrated on astrocytes in the brain in vivo, have been shown to stimulate EGF transactivation in other tissues and cells, but have not been tested in astrocytes. The β_2-adrenergic receptor, the angiotensin II receptor and the estrogen receptor may be of particular interest. DNA synthesis induced by β_2-adrenergic receptor activity in cardiac fibroblasts is dependent on EGF transactivation (Kim et al., 2002). Since the β_2-adrenergic receptor traditionally is believed to be exclusively linked to G_s

protein, its activation should not be expected to lead to EGF transactivation. However, β_2-adrenergic linkage to G_i protein has recently been reported in cardiac myocytes, where PKA-mediated phosphorylation of the β_2-adrenergic receptor regulates its coupling to G_s and G_i, respectively (Zamah et al., 2002).

The angiotensin II receptor is coupled to G_q protein, and mediates IP_3-induced Ca^{2+} release from intracellular Ca^{2+} stores. Stimulation of protein synthesis by angiotensin II in vascular SMCs is mediated by EGF transactivation (Voisin et al., 2002). Furthermore, angiotensin II-induced EGF transactivation in vascular SMCs was inhibited by a metalloprotease inhibitor, indicating the involvement of EGF shedding, whereas the concomitant increase of intracellular Ca^{2+} concentration was unaffected (Saito et al., 2002). Even more relevant in the present context, in glomerular mesangial cells, EGF transactivation might be necessary for angiotensin II-stimulated TGF-β release (Uchiyama-Tanaka et al., 2001), a phenomenon which might be similar to that observed in astrocytes stimulated with a group II mGluR agonist.

Estrogen increases cAMP and IP_3, elevates free intracellular Ca^{2+} concentration, causes MAPK phosphorylation and stimulates EGF transactivation in normal and cancer breast cell lines via a GPCR (Filardo, 2002; Filardo et al., 2002). In astrocytes, estrogen induces extension of processes and expression of glial fibrillary acidic protein (Garcia-Segura et al., 1989). It has also been suggested that estrogen is involved in astrocyte-directed neuronal plasticity (Garcia-Segura et al., 1999) and that it may be neuroprotective (Dhandapani and Brann, 2002). An estrogen-producing enzyme, aromatase is induced in astrocytes after brain injury (see also chapter by Melcangi et al.). It is an important question whether or not these effects of estrogen are mediated by a typical steroid receptor or whether they are unrelated to the conventional effects of estrogen.

5. Cytoprotective effect of HB-EGF

HB-EGF shedded from the cell surface and released into the extracellular space may exert both autocrine and paracrine cytoprotective effects. It is protective against apoptosis and necrosis in intestinal epithelial cells (Michalsky et al., 2001) and against intestinal ischemia/reperfusion injury, probably by down-regulation of cytokine-induced nitric oxide synthase (Lara-Marquez et al., 2002) and enhancement of hexokinase expression (Bryson et al., 2002). In spite of the up-regulation of HB-EGF gene expression after brain injury, relatively little is presently known about cytoprotective effects of HB-EGF in the CNS. In cultured hippocampal neurons, pretreatment of cells with HB-EGF prevents kainate-induced cell damage without affecting intracellular Ca^{2+} concentration (Opanashuk et al., 1999). As previously mentioned, its cytoprotective effect on cultured dopaminergic neurons seems to depend on astrocytes (Farkas and Krieglstein, 2002). It is also well established that EGFR activity induced by TGF-α and EGF protects neurons from injury (Staecker et al., 1997; Maiese and Boccone, 1995).

6. Concluding remarks

Our understanding of the astrocytic signaling pathways involved in dexmedetomidine-induced neuronal rescue by transactivation in astrocytes is, at best, sketchy, and to a large extent based on information about pathways stimulated by α_2-adrenergic agonists in other cell types or about the effects of other GPCR agonists in astrocytes. It is known that stimulation of the $G_{i/o}$-coupled α_{2A}-adrenergic receptors in astrocytes leads to hydrolysis of PIP_2, triggered by GTP binding to alpha units of the G-protein, and a resulting increase in $[Ca^{2+}]_i$ and IP_3, and presumably also to stimulation of PKC. In other cells expressing α_{2A}-adrenergic receptors naturally or after transfection, both an increase in $[Ca^{2+}]_i$ and stimulation of PKC activity have been correlated with phosphorylation of EGFR and/or release of HB-EGF. However, at least in transfected COS-7 cell, shedding of HB-EGF may be more directly mediated by the cytosolic tyrosine kinase Src following its activation mediated by G-protein beta,gamma subunits. Since the signaling pathway(s) may be GPCR and/or cell type-specific, they must be established in astrocytes. What is known, is that dexmedetomidine-induced stimulation in astrocytes causes release of growth factor(s), capable of causing tyrphostin AG1478-sensitive ERK1/2 activation and that the effect is at least partly neutralized by heparin. It is likely that HB-EGF is the only or major growth factor involved, but no attempt has been made to identify the released growth factor(s), which constitutes a high-priority goal.

It has been assumed in this review that neuroprotection by released HB-EGF is exerted by paracrine stimulation of neurons, but it cannot be excluded that another autocrine loop exists, in which an action of HB-EGF on astrocytes leads to release of additional growth factor(s), which then may act on neurons. This might explain the apparent involvement of additional growth factors during exposure of retina to α_2-adrenergic agonists. Also, in spite of an almost unanimous conclusion that dexmedetomidine has neuroprotective capability in the brain in vivo, no attempts have been made to clarify the participating signaling pathways in culture preparations highly enriched in neurons. Such cultures are readily available and would allow verification of the involvement of Ras–Raf and MEK in the phosphorylation of ERK1/2 and also examination of the possible importance of PI3K and IP_3 in neuroprotection and the transduction processes linking PI3K activation to HB-EGFR stimulation.

The neuroprotective role of α_2-adrenergic agonists in the retina seems on the verge of becoming of major importance in the treatment of glaucoma. Being a chronic, slowly developing condition, this is a different situation than ischemic brain damage, with a much broader window of opportunity for drug intervention. It is possible, but yet unproven, that the α_2-adrenergic agonists may act in part by exerting a Müller cell-mediated neuroprotection of neuronal elements in the retina. Confirmation whether transactivation is, indeed, involved in retinal neuroprotection by α_2-adrenergic agonists is urgently needed.

References

Antaa, R., Marjamaki, A., Scheinin, M., 1995. Molecular pharmacology of alpha$_2$-adrenoceptor subtypes. Ann. Med. 27, 439–449.

Audubert, F., Klapisz, E., Berguerand, M., Gouache, P., Jouniaux, A.M., Bereziat, G., Masliah, J., 1999. Differential potentiation of arachidonic acid release by rat alpha$_2$-adrenergic receptor subtypes. Biochim. Biophys. Acta 1437, 265–276.

Baptiste, D.C., Hartwick, A.T., Jollimore, C.A., Baldridge, W.H., Chauhan, B.C., Tremblay, F., Kelly, M.E., 2002. Comparison of the neuroprotective effects of adrenoceptor drugs in retinal cell culture and intact retina. Invest. Ophthalmol. Vis. Sci. 43, 2666–2676.

Brunet, A., Datta, S.R., Greenberg, M.E., 2001. Transcription-dependent and -independent control of neuronal survival by the PI13K-Akt signal pathway. Curr. Opin. Neurobiol. 11, 297–305.

Bruno, V., Battaglia, G., Casabona, G., Copani, A., Caciagli, F., Nicoletti, F., 1998. Neuroprotection by glial metabotropic glutamate receptors is mediated by transforming growth factor-beta. J. Neurosci. 18, 9594–9600.

Bruno, V., Battaglia, G., Ksiazek, I., Putten, van der, Catania, M.V., Giuffrida, R., Lukic, S., Leonhardt, T., Inderbitzin, W., Gasparini, F., Kuhn, R., Hampson, D.R., Nicoletti, F., Flor, P.J., 2000. Selective activation of mGlu4 metabotropic glutamate receptors is protective against excitotoxic neuronal death. J. Neurosci. 20, 6413–6420.

Bruno, V., Sureda, F.X., Storto, M., Casabona, G., Caruso, A., Knopfel, T., Kuhn, R., Nicoletti, F., 1997. The neuroprotective activity of group-II metabotropic glutamate receptors requires new protein synthesis and involved a glial–neuronal signaling. J. Neurosci. 17, 1891–1897.

Bryson, J.M., Coy, P.E., Gottlob, K., Hay, N., Robey, R.B., 2002. Increased hexokinase activity, or either ectopic or endogenous origin, protects renal epithelial cells against acute oxidant-induced cell death. J. Biol. Chem. 277, 11392–11400.

Bylund, D.B., Chacko, D.M., 1999. Characterization of alpha$_2$-adrenergic receptor subtypes in human ocular tissue homogenates. Invest. Ophthalmol. Vis. Sci. 40, 2299–2306.

Cussac, D., Schaak, S., Denis, C., Paris, H., 2002a. Alpha$_{2B}$-Adrenergic receptor activates MAPK via a pathway involving arachidonic acid metabolism, matrix metalloproteinases, and epidermal growth factor receptor transactivation. J. Biol. Chem. 277, 19882–19888.

Cussac, D., Schaak, S., Gales, C., Flordellis, C., Denis, C., Paris, H., 2002b. Alpha$_{2B}$-Adrenergic receptors activate MAPK and modulate proliferation of primary cultured proximal tubule cells. Am. J. Physiol. Renal Physiol. 282, F943–F952.

Daub, H., Wallasch, C., Lankenau, A., Herrlich, A., Ullrich, A., 1997. Signal characteristics of G protein-transactivated EGF receptor. EMBO J. 16, 7032–7044.

Debeir, T., Gueugnon, J., Vige, X., Benavides, J., 1996. Transduction mechanisms involved in thrombin receptor-induced nerve growth factor secretion and cell division in primary cultures of astrocytes. J. Neurochem. 66, 2320–2328.

Dethlefsen, S.M., Raab, G., Moses, M.A., Adam, R.M., Klagsbrun, M., Freeman, M.R., 1998. Extracellular calcium influx stimulates metalloproteinase cleavage and secretion of heparin-binding EGF-like growth factor independently of protein kinase C. J. Cell. Biochem. 69, 143–153.

Dhandapani, K.M., Brann, D.W., 2002. Estrogen–Astrocyte interactions: implications for neuroprotection. BMC Neurosci. 3, 6.

Engelhard, K., Werner, C., Kaspar, S., Mollenberg, O., Blobner, M., Bachl, M., Kochs, E., 2002. Effect of the alpha$_2$-agonist dexmedetomidine on cerebral neurotransmitter concentrations during cerebral ischemia in rats. Anesthesiology 96, 450–457.

Erlich, S., Shohami, E., Pinkas-Kramarski, R., 2000. Closed head injury induces up-regulation of ErbB-4 receptor at the site of injury. Mol. Cell. Neurosci. 16, 597–608.

Farkas, L.M., Krieglstein, K., 2002. Heparin-binding epidermal growth factor-like growth factor (HB-EGF) regulates survival of midbrain dopaminergic neurons. J. Neural. Transm. 109, 267–277.

Gardner, A., Phillips-Mason, P.J., Raben, D.M., Baldassare, J.J., 2002. A novel role for G_q alpha in alpha-thrombin-mediated mitogenic signalling pathways. Cell. Signal 14, 499–507.

Hefti, F., 1994. Development of effective therapy for Alzheimer's disease based on neurotrophic factors. Neurobiol. Aging 15(Suppl. 2), S193–S194.

Filardo, E.J., 2002. Epidermal growth factor receptor (EGFR) transactivation by estrogen via the G-protein-coupled receptor, GPR30: a novel signaling pathway with potential significance for breast cancer. J. Steroid Biochem. Mol. Biol. 80, 231–238.

Filardo, E.J., Quinn, J.A., Bland, K.I., Frackelton, A.R. Jr., 2002. Estrogen-induced activation of ERK-1 and ERK-2 required the G protein-coupled receptor homolog, GPR30, and occurs via transactivation of epidermal growth factor receptor through release of HB-EGF. Mol. Endocrinol. 14, 1649–1660.

Garcia-Segura, L.M., Naftolin, F., Hutchison, J.B., Azcoitia, I., Chowen, J.A., 1999. Role of astroglia in estrogen regulation of synaptic plasticity and brain repair. J. Neurobiol. 40, 574–584.

Garcia-Segura, L.M., Torres-Aleman, I., Naftolin, F., 1989. Astrocytic shape and glial fibrillary acidic protein immunoreactivity are modified by estradiol in primary rat hypothalamic cultures. Brain Res. Dev. Brain Res. 47, 298–302.

Gasparini, F., Bruno, V., Battaglia, G., Lukic, S., Leonhardt, T., Inderbitzin, W., Laurie, D., Sommer, B., Varney, M.A., Hess, S.D., Johnson, E.C., Kuhn, R., Urwyler, S., Sauer, D., Portet, C., Schmutz, M., Nicoletti, F., Flor, P.J., 1999. (*R*,*S*)-4-Phosphonophenylglycine, a potent and selective group III metabotropic glutamate receptor agonist, is anticonvulsive and neuroprotective in vivo. J. Pharmacol. Exp. Therap. 290, 1678–1687.

Gerecke, K.M., Wyss, J.M., Karavanova, I., Buonanno, A., Carroll, S.L., 2001. ErbB transmembrane tyrosine kinase receptors are differentially expressed throughout the adult rat central nervous system. J. Comp. Neurol. 433, 86–100.

Goishi, K., Higashiyama, S., Klagsbrun, M., Nakano, N., Umata, T., Ishikawa, M., Mekada, E., Taniguchi, N., 1995. Phorbol ester induces the rapid processing of cell surface heparin-binding EGF-like growth factor: conversion from juxtacrine to paracrine growth factor activity. Mol. Biol. Cell 6, 967–980.

Gozes, I., 2001. Neuroprotective peptide drug delivery and development: potential new therapeutics. Trends Neurosci. 24, 700–705.

Graness, A., Adomeit, A., Heinze, R., Wetzker, R., Liebmann, C., 1998. A novel mitogenic signaling pathway of bradykinin in the human colon carcinoma cell line SW-480 involves sequential activation of a G_q/11 protein, phosphatidylinositol 3-kinase beta, and protein kinase Cepsilon. J. Biol. Chem. 273, 32016–32022.

Grewal, J.S., Luttrell, L.M., Raymond, J.R., 2001. G protein-coupled receptors desensitize and down-regulate epidermal growth factor receptors in renal mesangial cells. J. Biol. Chem. 276, 27335–27344.

Gschwind, A., Zwick, E., Prenzel, N., Leserer, M., Ullrich, A., 2001. Cell communication networks: epidermal growth factor receptor transactivation as the paradigm for interreceptor signal transmission. Oncogene 20, 1594–1600.

Hein, L., Altman, J.D., Kobilka, B.K., 1999. Two functionally distinct alpha$_2$-adrenergic receptors regulate sympathetic neurotransmission. Nature 402, 181–184.

Higashiyama, S., Abraham, J.A., Miller, J., Fiddes, J.C., Klagsbrun, M., 1991. A heparin-binding growth factor secreted by macrophage-like cells that is related to EGF. Science 251, 936–939.

Huang, R., Hertz, L., 2000. Receptor subtype and dose dependence of dexmedetomidine-induced accumulation of [^{14}C]glutamine in astrocytes suggests glial involvement in its hypnotic-sedative and anesthetic-sparing effects. Brain Res. 873, 297–301.

Huang, Y.Z., Wang, Q., Xiong, W.C., Mei, L., 2001. Erbin is a protein concentrated at postsynaptic membranes that interacts with PSD-95. J. Biol. Chem. 276, 19318–19326.

Izumi, Y., Hirata, M., Hasuwa, H., Iwamoto, R., Umata, T., Miyado, K., Tamai, Y., Kurisaki, T., Sehara-Fujisawa, A., Ohno, S., Mekada, E.A., 1998. Metalloprotease-disintegrin, MDC9/meltrin-gamma/ADAM9 and PKCdelta are involved in TPA-induced ectodomain shedding of membrane-anchored heparin-binding EGF-like growth factor. EMBO J. 17, 7260–7272.

Jin, K., Mao, X.O., Sun, Y., Xie, L., Jin, L., Nishi, E., Klagsbrun, M., Greenberg, D.A., 2002. Heparine-binding epidermal growth factor-like growth factor: hypoxia-inducible expression in vitro and stimulation of neurogenesis in vitro and in vivo. J. Neurosci. 22, 5365–5373.

Jolkkonen, J., Puurunen, K., Koistinaho, J., Kauppinen, R., Haapalinna, A., Nieminen, L., Sivenius, J., 1999. Neuroprotection by the alpha$_2$-adrenoceptor agonist, dexmedetomidine, in rat focal cerebral ischemia. Eur. J. Pharmacol. 372, 31–36.

Kalmes, A., Vesti, B.R., Daum, G., Abraham, J.A., Clowes, A.W., 2000. Heparin blockage of thrombin-induced smooth muscle cell migration involves inhibition of epidermal growth factor (EGF) receptor transactivation by heparin-binding EGF-like growth factors. Circ. Res. 87, 92–98.

Kaufmann, R., Lindschau, C., Hoer, A., Henklein, P., Haller, H., Liebmann, C., Oberdisse, E., Nowak, G., 1996. Signaling effects of alpha-thrombin and SFLLRN in rat glioma C6 cells. J. Neurosci. Res. 46, 641–651.

Keys, J.R., Greene, E.A., Koch, W.J., Eckhart, A.D., 2002. G_q-coupled receptor agonists mediate cardiac hypertrophy via the vasculature. Hypertension 40(5), 660–666.

Kim, J., Eckhart, A.D., Eguchi, S., Koch, W.J., 2002. Beta-adrenergic receptor mediated DNA synthesis in cardiac fibroblasts is dependent on transactivation of the epidermal growth factor receptor and subsequent activation of extracellular signal-regulated kinases. J. Biol. Chem. 277, 32116–32123.

Kim, M.S., So, H.S., Park, J.S., Lee, K.M., Moon, B.S., Lee, H.S., Kim, T.Y., Moon, S.K., Park, R., 2000. Hwansodan protects PC12 cells against serum-deprivation-induced apoptosis via a mechanism involving Ras and mitogen-activated protein (MAP) kinase pathway. Gen. Pharmacol. 34, 227–235.

Kornblum, H.I., Zurcher, S.D., Werb, Z., Derynck, R., Seroogy, K.B., 1999. Multiple trophic actions of heparin-binding epidermal growth factor (HB-EGF) in the central nervous system. Eur. J. Neurosci. 11, 3236–3246.

Kukley, M., Barden, J.A., Steinhauser, C., Jabs, R., 2001. Distribution of P2X receptors on astrocytes in juvenile rat hippocampus. Glia 36, 11–21.

Lara-Marquez, M.L., Mehta, V., Michalsky, M.P., Fleming, J.B., Besner, G.E., 2002. Heparin-binding EGF-like growth factor down regulates proinflammatory cytokin-induced nitric oxide and inducible nitric oxide synthase production in intestinal epithelial cells. Nitric Oxide 6, 142–152.

Lenz, G., Gottfried, C., Luo, Z., Avruch, J., Rodnight, R., Nie, W.J., Kong, Y., Neary, J.T., 2000. P(2Y) purinoceptor subtypes recruit different mek activators in astrocytes. Br. J. Pharmacol. 129, 927–936.

Li, X., Eisenach, J.C., 2001. $Alpha_{2A}$-adrenoceptor stimulation reduces capsaicin-induced glutamate release from spinal cord synaptosomes. J. Pharmacol. Exp. Ther. 299, 939–944.

Lin, W.W., Chuang, D.M., 1994. Different signal transduction pathways are coupled to the nucleotide receptor and the P2Y receptor in C6 glioma cells. J. Pharmacol. Exp. Ther. 269, 926–931.

Maayani, S., Schwarz, T., Martinez, R., Tagliente, T.M., 2001. Activation of G_i-coupled receptors releases a tonic state of inhibited platelet aggregation. Platelets 12, 94–98.

Maiese, K., Boccone, L., 1995. Neuroprotection by peptide growth factors against anoxia and nitric oxide toxicity requires modulation of protein kinase C. J. Cereb. Blood Flow Metab. 15, 440–449.

Michalsky, M.P., Kuhn, A., Mehta, V., Besner, G.E., 2001. Heparin-binding EGF-like growth factor decreases apotosis in intestinal epithelial cells in vitro. J. Pediatr. Surg. 36, 1130–1135.

Michalsky, M.P., Lara-Marquez, M., Chun, L., Besner, G.E., 2002. Heparin-binding EGF-like growth factor is present in human amniotic fluid and breast milk. J. Pediatr. Surg. 37, 1–6.

Miura, M., Watanabe, M., Offermanns, S., Simon, M.I., Kano, M., 2002. Group I metabotropic glutamate receptor signaling via Galpha q/Galpha 11 secures the induction of long-term potentiation in the hippocampal area CA1. J. Neurosci. 22, 8379–8390.

Miyoshi, I., Kagaya, A., Kohchi, C., Morinobu, S., Yamawaki, S., 2001. Characterization of 5-HT2A receptor desensitization and the effect of cycloheximide on it in C6 cells. J. Neural. Transm. 108, 249–260.

Nakagawa, T., Sasahara, M., Hayase, Y., Haneda, M., Yasuda, H., Kikkawa, R., Higashiyama, S., Hazama, F., 1998. Neuronal and glial expression of heparin-binding EGF-like growth factor in central nervous system of prenatal and early-postnatal rat. Dev. Brain Res. 108, 263–272.

Neary, J.T., Kang, Y., Bu, Y., Yu, E., Akong, K., Peters, C.M., 1999. Mitogenic signaling by ATP/P2Y purinergic receptors in astrocytes: involvement of a calcium-independent protein kinase C, extracellular signal-regulated protein kinase pathway distinct from the phosphatidylinositol-specific phospholipase C/calcium pathway. J. Neurosci. 19, 4211–4220.

Opanashuk, L.K., Mark, R.J., Porter, J., Damm, D., Mattson, M.P., Seroogy, K.B., 1999. Heparin-binding epidermal growth factor-like growth factor in hippocampus: modulation of expression by seizures and anti-excitotoxic action. J. Neurosci. 19, 133–146.

Pandiella, A., Massague, J., 1991. Cleavage of the membrane precursor for transforming growth factor alpha is a regulated process. Proc. Natl Acad. Sci. USA 88, 1726–1730.

Panenka, W., Jijon, H., Herx, L.M., Armstrong, J.N., Feighan, D., Wei, T., Young, V.W., Ransohoff, R.M., MacViicar, B.A., 2001. P2X7-like receptor activation in astrocytes increases chemokine monocyte chemoattractant protein-1 expression via mitogen-activated protein kinase. J. Neurosci. 21, 7135–7142.

Peavy, R.D., Chang, M.S.S., Sanders-Bush, E., Conn, P.J., 2001. Metabotropic glutamate receptor 5-induced phosphorylation of extracellular signal-regulated kinase in astrocytes depends on transactivation of the epidermal growth factor receptor. J. Neurosci. 21, 9619–9628.

Peng, M., Li, Y., Luo, Z., Liu, C., Laties, A.M., Wen, R., 1998. $Alpha_2$-adrenergic agonists selectively activate extracellular signal-regulated kinases in Muller cells in vivo. Invest. Ophthalmol. Vis. Sci. 39, 1721–1726.

Peng, L., Yu, A.C.H., Fung, K.Y., Hertz, L., 2002. Dexmedetomidine-induced ERK phosphorylation in astrocytes is mediated by 'transactivation'. Submitted to Brain Res.

Peng, L., Yu, A.C.H., Fung, K.Y., Prevot, V., Hertz, L., 2003. Adrenergic stimulation of ERK phosphorylation in astrocytes in alpha$_2$-specific and may be mediated by transactivation. Brain Res. 978, 65–71.

Pierce, K.L., Tohgo, A., Ahn, S., Field, M.E., Luttrell, L.M., Lefkowitz, R.J., 2001. Epidermal growth factor (EGF) receptor-dependent ERK activation by G protein-coupled receptors. J. Biol. Chem. 276, 23155–23160.

Pinkas-Kramarski, R., Eilam, R., Alroy, I., Levkowitz, G., Lonai, P., Yarden, Y., 1997. Differential expression of NDF/neuregulin receptors ErbB-3 and ErbB-4 and involvement in inhibition of neuronal differentiation. Oncogene 15, 2803–2815.

Post, G.R., Collins, L.R., Kennedy, E.D., Moskowitz, S.A., Aragay, A.M., Goldstein, D., Brown, J.H., 1996. Coupling of the thrombin receptor to G12 may account for selective effects of thrombin on gene expression and DNA synthesis in 1321N1 astrocytoma cells. Mol. Biol. Cell 7, 1679–1690.

Prenzel, N., Fischer, O.M., Streit, S., Hart, S., Ullrich, A., 2001. The epidermal growth factor receptor family as a central element for cellular signal transduction and diversification. Endocr.-Relat. Cancer 8, 11–31.

Raab, G., Klagsbrun, M., 1997. Heparin-binding EGF-like growth factor. Biochim. Biophys. Acta 1333, F179–F199.

Rudge, J.S., Alderson, R.F., Pasnikowski, E., McClain, J., Ip, N.Y., Lindsay, R.M., 1992. Expression of ciliary neurotrophic factor and the neurotrophins-nerve growth factor, brain-derived neurotrophic factor and neurotrophin 3 in cultured rat hippocampal astrocytes. Eur. J. Neurosci. 4, 459–471.

Saito, S., Frank, G.D., Motley, E.D., Dempsey, P.J., Utsunomiya, H., Inag, T., Eguchi, S., 2002. Metalloprotease inhibitor blocks angiotensin II-induced migration through inhibition of epidermal growth factor receptor transactivation. Biochem. Biophys. Res. Commun. 294, 1023–1029.

Scalabrino, G., Nicolini, G., Buccellato, F.R., Peracchi, M., Tredici, G., Manfridi, A., Pravettoni, G., 1999. Epidermal growth factor as a local mediator of the neurotrophic action of vitamin B(12) (cobalamin) in the rat central nervous system. FASEB J. 13, 2083–2090.

Shiga, H., Tojima, T., Ito, E., 2001. Ca^{2+} signaling regulation by an ATP-dependent autocrine mechanism in astrocytes. Neuroreport 12, 2619–2622.

Staecker, H., Dazert, S., Malgrange, B., Lefebvre, P.P., Ryan, A.F., Van de Water, T.R., 1997. Transforming growth factor alpha treatment alters intracellular calcium levels in hair cells and protects them from ototoxic damage in vitro. Int. J. Dev. Neurosci. 15, 553–562.

Takigawa, M., Sakurai, T., Kasuya, Y., Abe, Y., Masaki, T., Goto, K., 1995. Molecular identification of guanine-nucleotide-binding regulatory proteins which couple to endothelin receptors. Eur. J. Biochem. 228, 102–108.

Talke, P., Bickler, P.E., 1996. Effects of dexmedetomidine on hypoxia-evoked glutamate release and glutamate receptor activity in hippocampal slices. Anesthesiology 85, 551–557.

Tanaka, N., Sasahara, M., Ohno, M., Higashiyama, S., Hayase, Y., Shima, M., 1999. Heparin-binding epidermal growth factor-like growth factor mRNA expression in neonatal rat brain with hypoxic/ischemic injury. Brain Res. 827, 130–138.

Tebar, F., Llado, A., Enrich, C., 2002. Role of calmodulin in the modulation of the MAPK signaling pathway and the transactivation of epidermal growth factor mediated by PKC. FEBS Lett. 517, 206–210.

Tohgo, A., Pierce, K.L., Choy, E.W., Lefkowitz, R.J., Luttrell, L.M., 2002. beta-Arrestin scaffolding of the ERK cascade enhances cytosolic ERK activity but inhibits ERK-mediated transcription following angiotensin AT1a receptor stimulation. J. Biol. Chem. 277, 9429–9436.

Ubl, J.J., Reiser, G., 1997. Characteristics of thrombin-induced calcium signals in rat astrocytes. Glia 21, 361–369.

Uchiyama-Tanaka, Y., Matsubara, H., Nozawa, Y., Murasawa, S., Mori, Y., Kosaki, A., Maruyama, K., Masaki, H., Shibasaki, Y., Fujiyama, S., Nose, A., Iba, O., Hasagawa, T., Tateishi, E., Higashiyama, S., Iwasaka, T., 2001. Angiotensin II signaling HB-EGF shedding via metalloproteinase in glomerular messangial cells. Kidney Int. 60, 2153–2163.

Umata, T., Hirata, M., Takahashi, T., Ryu, F., Shida, S., Takahashi, Y., Tsuneoka, M., Miura, M., Masuda, M., Horiguchi, Y., Mekada, E., 2001. A dual signaling cascade that regulates the ectodomain shedding of heparin-binding epidermal growth factor-like growth factor. J. Biol. Chem. 276, 30475–30482.

Voisin, L., Foisy, S., Giasson, E., Lambert, C., Moreau, P., Meloche, S., 2002. EGF receptor transactivation is obligatory for protein synthesis stimulation by G protein-coupled receptor. Am. J. Physiol. Cell Physiol. 283, C446–C455.

Wang, H., Ubl, J.J., Reiser, G., 2002. Four subtypes of protease-activated receptors, co-expressed in rat astrocytes, evoked different physiogical signaling. Glia 37, 53–63.

Wen, R., Cheng, T., Li, Y., Cao, W., Steinberg, R.H., 1996. Alpha$_2$-adrenergic agonists induce basic fibroblast growth factor expression in photoreceptors in vivo and ameliorate light damage. J. Neurosci. 16, 5986–5992.

Xue, L., Murray, J.H., Tolkovsky, A.M., 2000. The Ras/phosphatidylinositol 3-kinase and Ras/ERK pathways function as independent survival modules each of which inhibits a distinct apoptotic signaling pathway in sympathetic neurons. J. Biol. Chem. 275, 8817–8824.

Yamaguchi, H., Wang, H.G., 2001. The protein kinase PKB/Akt regulates cell survival and apoptosis by inhibiting Bax conformational change. Oncogene 20, 7779–7786.

Zamah, A.M., Delahunty, M., Luttrell, L.M., Lefkowitz, R.J., 2002. PKA-mediated phosphorylation of the beta$_2$-adrenergic receptor regulates its coupling to G_s and G_i: demonstration in a reconstituted system. J. Biol. Chem. 277, 31249–31256.

Zelenaia, O., Schlag, B.D., Gochenauer, G.E., Ganel, R., Song, W., Beesley, J.S., Grinspan, J.B., Rothstein, J.D., Robinson, M.B., 2000. Epidermal growth factor receptor agonists increase expression of glutamate transporter GLT-1 in astrocytes through pathways dependent on phosphatidylinositol 3-kinase and transcription factor NF-kappaB. Mol. Pharmacol. 57, 667–678.

Zwick, E., Daub, H., Aoki, N., Yamaguchi-Aoki, Y., Tinhofer, I., Maly, K., Ullrich, A., 1997. Critical role of calcium-dependent epidermal growth factor receptor transactivation in PC12 cell membrane depolarization and bradykinin signaling. J. Biol. Chem. 272, 24767–24779.

Advances in
Molecular and
Cell Biology

Roles of glia cells in cholesterol homeostasis in the brain

Jin-ichi Ito and Shinji Yokoyama*

Biochemistry, Cell Biology and Metabolism, Graduate School of Medical Sciences, Nagoya City University, Kawasumi 1, Mizuho-cho, Mizuho-ku, Nagoya 467-8601, Japan
Correspondence address: E-mail: syokoyam@med.nagoya-cu.ac.jp

Contents

1. Introduction
2. Cholesterol in brain
3. Lipoproteins and their receptors in brain
4. ApoE synthesis and secretion by astrocytes
5. Production of HDL by astrocytes
6. Brain injury and apoE production
7. Concluding remarks: glia and cholesterol homeostasis in the brain

Abbreviations

apo: apolipoprotein; HDL: high density lipoprotein; LDL: low density lipoprotein; BBB: blood–brain barrier; CSF: cerebrospinal fluid; CNS: central nervous system; SR-B1, scavenger receptor B1; ABCA1: ATP binding cassette transporter protein A1; VLDL: very low density lipoprotein; LRP: LDL receptor-related protein; FGF: fibroblast growth factor

Astrocytes generate high density lipoprotein (HDL) with apolipoprotein (apo) E and apoJ synthesized by astrocytes themselves and with other, exogenous, apolipoproteins, such as apoA-I. These HDLs are thought to function in the transport of cholesterol between brain cells, whether to supply cholesterol when it is needed, such as during recovery from damage, or to remove it from the cells for its homeostasis. The physiological importance and many pathophysiological roles of brain HDL are discussed in this chapter.

1. Introduction

The brain is a cholesterol-rich organ, and increasing evidence suggests that cholesterol plays a number of key roles in the central nervous system (CNS) through regulation of membrane functions and by other mechanisms. What makes cholesterol metabolism in the

Advances in Molecular and Cell Biology, Vol. 31, pages 519–534

ISBN: 0-444-51451-1

brain special is that brain is an autonomous organ with respect to cholesterol homeostasis, since the blood–brain barrier prevents lipoproteins in circulating plasma from entering the CNS. Therefore, a unique system for cholesterol metabolism is found in the brain, operating through high density lipoprotein (HDL), generated from endogenously synthesized apolipoproteins E and J (apoE and apoJ) by astrocytes and microglia. Brain cells can also react with extracellular apoA-I from unknown sources. ApoE–HDL reactivity increases in the brain in response to acute and chronic injury of the nervous system, which is believed to play a role in brain recovery by supporting neurite outgrowth and synapse formation. Since glia cells are the main source of brain HDL, they play a key role in the maintenance of brain structure and function by cholesterol-dependent mechanisms.

2. Cholesterol in brain

Cholesterol is an important lipid component of the plasma membrane. While its hydroxyl group is believed to be localized close to the hydrophilic head groups of phospholipid molecules, its steroid backbone with a flat and rigid structure fills a potential space among the hydrocarbon chains of the phospholipids, regulating their mobility and accordingly a physicochemical property of the membrane. In addition, cholesterol preferably interacts with sphingomyelin molecules to form cholesterol/sphingomyelin-rich domains in the plasma membrane (rafts or caveolae) (Radhakrishnan et al., 2000), which offer a special environment for accumulation of specific proteins involved in many specific membrane functions, including signal transduction and intercellular interaction (Schnitzer et al., 1995; Shin and Abraham, 2001). Because cholesterol plays such key roles in regulating membrane functions, it is essential for cells to maintain its homeostasis.

Approximately 25% of the total cholesterol in the human body is found in the brain, where it accounts for 15% of the dry weight (Anonymous, 1981). Cholesterol is abundant in the plasma membranes of neurons and astrocytes, and myelin membranes also contain large amounts of cholesterol. Suppression of cholesterol synthesis in cultured neurons significantly lowers their viability and reduces axonal growth, indicating a critical dependence on cholesterol supply (de Chaves et al., 1997; Michikawa and Yanagisawa, 1999). A raft-like domain, rich in cholesterol–sphingomyelin, appears to be present in neurons (Gorodinsky et al., 1994; Gorodinsky and Harris, 1995; Maekawa et al., 1999; Masserini et al., 1999), and to be a preferred localization for tyrosine kinase receptors (Wu et al., 1997). Also, AMPA-type glutamate receptor subunits can be recovered in a low density, Triton-insoluble fraction of rat synaptic membranes, which is considered to represent dendritic rafts (Suzuki et al., 2001). These findings indicate that neurotransmitter-mediated signal transduction at least partly takes place in raft-like domains, and that cholesterol accordingly is important for neuronal responsiveness to neurotransmitters (Scanlon et al., 2001). It has also recently been reported that formation of synapses, development of synaptic vesicles and synthesis of appropriate amounts of synapsin I and synaptophysin are dependent on supply of cholesterol complexed to apoE-containing HDL (Mauch et al., 2001). Therefore, cholesterol and its transport system are essential for the very fundamental functions of the CNS (Barres and Smith, 2001; Lang et al., 2001; Mauch et al., 2001; Nagler et al., 2001; Ullian et al., 2001).

All somatic cells synthesize cholesterol, and they are also capable of incorporating it by uptake of cholesterol-containing low density lipoprotein (LDL) through the LDL receptor. In contrast, cholesterol is not metabolized in most somatic cells, except for limited partial hydroxylation at a few locations, and its major catabolic sites are the liver and steroidogenic cells. Therefore, peripheral cells require a cholesterol export system for their cholesterol homeostasis. Such a system is also required for whole-body cholesterol homeostasis, as cholesterol has to be transported to its catabolic sites. HDL is believed to play a central role in this transport system (Yokoyama, 2000).

The brain is segregated from the systemic circulation by the blood–brain barrier (BBB), and brain cells can accordingly not utilize the extracellular cholesterol transport system by plasma lipoprotein. Therefore, cholesterol homeostasis in the brain is thought to depend largely on endogenous biosynthesis within the brain (Dietschy and Turley, 2001). Intercellular transport of cholesterol in the brain also depends on its own lipoprotein system. HDL is the only lipoprotein identified in cerebrospinal fluid (CSF) for this function, and it contains apolipoproteins E, A-I, D and J (Roheim et al., 1979; Pitas et al., 1987a,b; Borghini et al., 1995; Koch et al., 2001). ApoE and J are synthesized in astrocytes and microglia and are secreted as HDL (Boyles et al., 1985; Pitas et al., 1987a,b; Poirier et al., 1991; Krul and Tang, 1992; LaDu et al., 1998; DeMattos et al., 2001). The source of apoA-I is not certain, but it is also found in HDL in the brain (Weiler-Guttler et al., 1990; Mockel et al., 1994).

Clearance of cholesterol from the CNS is not well understood. A part of cholesterol in the brain appears to be catabolized to 24S-hydroxycholesterol by cholesterol 24-hydroxylase localized in microsomes, at least in the neurons (Bjorkhem et al., 1997, 1998; Lund et al., 1999), but it is unclear how much brain cholesterol is catabolized by this pathway. It is also not known how this cholesterol metabolite is transported out from the CNS. A very small portion of cholesterol can be used for the synthesis of steroid hormones, but this amount is negligible compared to the turnover rate of cholesterol in brain. Thus, the major pathway for cholesterol clearance from CNS may be transport by HDL that carry cholesterol from CSF to the systemic circulation. The turnover of cholesterol in human brain is estimated to be 0.02% per day, much slower than that of 0.4% per day in the rat (Dietschy and Turley, 2001). It has been estimated that 2/3 of the brain cholesterol are catabolized to 24S-hydroxycholesterol in the rat brain (Bjorkhem et al., 1998).

3. Lipoproteins and their receptors in brain

The delivery and removal of cholesterol to and from cells in the periphery are mainly mediated by lipoproteins in the systemic circulation. However, it is very unlikely that CNS cells could directly exchange cholesterol with plasma lipoproteins, since brain cells are isolated by the blood–brain barrier from the macromolecular components of plasma. Therefore, there must be an autonomous, CNS-specific mechanism for intercellular cholesterol transport and regulation of cholesterol homeostasis among neural cells. The main apolipoproteins in CSF are apoE and apoA-I, which are present in CSF as HDL-like particles, suggesting that these two apolipoproteins play central roles in mediating intercellular cholesterol transport in the brain. There have been reports that plasma LDL undergoes transcytosis in capillary endothelial cells and thereby crosses the BBB to enter the brain, and that astrocytes regulate the transcytosis of LDL by secreting some trophic factor (Dehouck et al., 1994, 1997). However, neither LDL nor apoB have been detected in CSF (Koch et al., 2001). Therefore, even though LDL may be able to cross the BBB, it may immediately be incorporated and processed by astrocytes that surround endothelial cells at the BBB. Nevertheless, patients suffering from abetalipoproteinemia (apoB-lipoprotein assembly deficiency due to a genetic defect of microsomal triglyceride transfer protein) show severe congenital CNS symptoms (Kane and Havel, 2001), suggesting that apoB-lipoproteins are required for CNS development before the BBB is established, perhaps for cholesterol supply.

Although there is one report that apoA-I in CNS is produced by the brain capillary endothelial cells (Mockel et al., 1994), it is unclear whether this production is significant and whether apoA-I is secreted by the endothelial cells directly into brain parenchyma. On the other hand, some authors maintain that the apoA-I synthesized and secreted by the liver or the intestine is transported into the brain through the BBB (Ritas, 1997). In contrast to the ambiguity of the source of apoA-I in CSF, apoE has clearly been shown to be produced in the brain, mainly by astrocytes (Boyles et al., 1985; Pitas et al., 1987a,b; Poirier et al., 1991; Krul and Tang, 1992; Patel et al., 1995; LaDu et al., 1998; Fagan et al., 1999; Fujita et al., 1999; Ito et al., 1999; Xu et al., 2000; DeMattos et al., 2001, Ueno et al., 2002) and also by microglia (Nakai et al., 1996; Stone et al., 1997; Xu et al., 2000). Accordingly, the phenotype of human brain apoE does not change after liver transplantation, proving that brain apoE is indeed produced in the CNS and functions as an intercellular cholesterol transporter in the CNS (Linton et al., 1991). Astrocytes synthesize and secrete apoE along with the cellular cholesterol and phospholipid as cholesterol-rich HDLs. On the other hand, the cells also depend upon extracellular apoA-I to generate HDLs (Ito et al., 1999), but these HDLs have a much lower content of cholesterol (Fig. 1, Table 1). Thus, apoE–HDL and apoA-I–HDL in the brain may be produced by slightly different mechanisms and carry out different functions in intercellular lipid transport in CNS.

Interactions of apoA-I/HDL with cells are mediated by cellular proteins. Scavenger receptor type-B1 (SR-B1) binds apoA-I/HDL particles and mediates selective uptake of its cholesteryl ester (Acton et al., 1996). Cubilin has been shown to be the HDL binding protein necessary for the uptake of HDL particles or of free apoA-I (Hammad et al., 1999; Kozyraki et al., 1999). The gene encoding ATP-binding cassette transporter protein A-1 (ABCA1) has been identified as a causative gene for HDL deficiency (Bodzioch et al., 1999; Brooks-Wilson et al., 1999; Rust et al., 1999). Its interaction with apolipoprotein for generation of HDL from cellular lipid is under intense investigation by many laboratories (Oram et al., 2000; Chambenoit et al., 2001; Fitzgerald et al., 2002) including our own (Abe-Dohmae et al., 2000; Arakawa and Yokoyama, 2002). SR-B1 (Posse De Chaves et al., 2000) and ABCA1 (Fukumoto et al., 2003) seem to be present in the brain, but there are no data for cubilin, and no

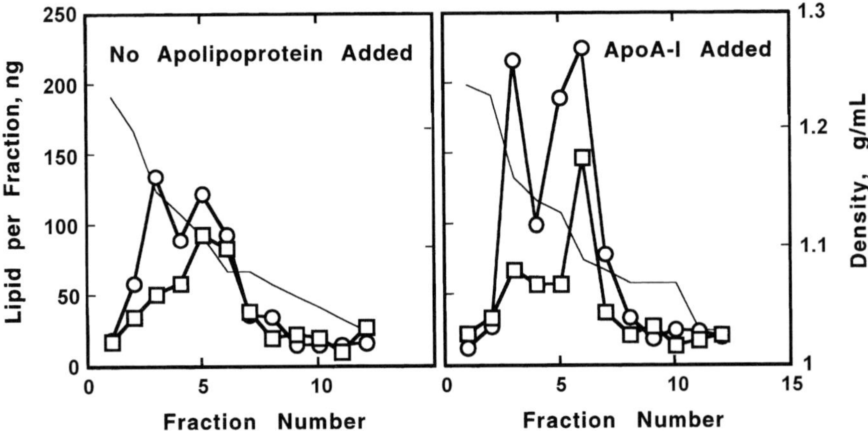

Fig. 1. Density gradient ultracentrifugation analysis of the culture medium of astrocytes after incubation with apoA-I. The cells in two 3 cm plates (150 ± 17 mg protein per plate) were incubated with and without apoA-I, 5 mg/1 mL medium, and the medium was analyzed by sucrose density gradient ultracentrifugation. Preloaded radioactivity in cholesterol and phosphatidylcholine was determined for each of twelve fractions from the bottom to the top, and mass of each lipid was calculated from the specific radioactivity of the respective lipid in the cells compared to that in the precursor. Open circles represent phosphatidylcholine and open squares represent cholesterol in each fraction. The thin solid lines indicate density of each fraction. Taken from Ito et al. (1999), with permission.

brain-specific HDL-binding protein has so far been identified. On the other hand, four types of apoE-binding proteins have been identified in the brain: (i) LDL receptor; (ii) very low density lipoprotein (VLDL) receptor; (iii) LDL receptor-related protein (LRP); and (iv) apoE receptor-2 (Pitas et al., 1987a,b; Bu et al., 1994; Oka et al., 1994; Kim et al., 1996). All of these apoE-binding proteins are found in neurons. LDL receptor, VLDL receptor and LRP are found also in astrocytes, but expression of apoE receptor-2 has not been confirmed in astrocytes (Nimpf and Schneider, 2000; Herz, 2001; Herz and

Table 1
Lipid composition of prebeta-HDL-like particles generated by astrocytes. From Ito et al. (1999), with permission

Exogenous apolipoprotein	Cholesterol (pmol/3 mL)	Phosphatidylcholine (pmol/3 mL)	Chol/PC[a] (mol/mol)
None	1226	823	1.49
ApoA-I	1339	1235	(1.08)
Increment	113	412	0.27

The cells in two 3 cm plates (150 mg protein per plate) were incubated with and without apoA-I, 5 mg/mL medium, and the medium was analyzed for its lipid composition. Preloaded radioactivity in cholesterol and phosphatidylcholine was determined and mass of each lipid was calculated from the specific radioactivity of the respective lipid in the cells compared to that in the precursor. 'Increment' is the difference between the control (without apoA-I) and the sample with apoA-I.
[a]Molar ratio of cholesterol/phosphatidylcholine.

Strickland, 2001; Riddell et al., 2001). Recently, Trommsdorff et al. (1999) reported that double deficiency of VLDL receptor and apoE receptor-2 results in reeler/disabled-like disruption of neuronal migration, suggesting that these receptors may also play a role in intracellular signaling.

4. ApoE synthesis and secretion by astrocytes

Human apoE is a glycoprotein composed of 299 amino acids with a molecular weight of 37 kDa. While the main source of plasma apoE seems to be the liver, it is also produced by various types of cells such as macrophages and steroidogenic cells. In CNS, apoE is mainly produced and secreted by astrocytes (Boyles et al., 1985; Pitas et al., 1987a,b; Poirier et al., 1991; Krul and Tang, 1992; Patel et al., 1995; LaDu et al., 1998; Fagan et al., 1999; Fujita et al., 1999; Ito et al., 1999; Xu et al., 2000; DeMattos et al., 2001; Ueno et al., 2002), but microglial cells are also capable of doing so (Nakai et al., 1996; Stone et al., 1997; Xu et al., 2000). When apoE is secreted from astrocytes, most of it forms HDL-like particles with diameters of 10–17 nm with cholesterol and phospholipid (Pitas et al., 1987a,b; Borghini et al., 1995; Shanmugaratnam et al., 1997; LaDu et al., 1998; Ito et al., 1999; Ueno et al., 2002). Astrocytes also secrete apoJ as HDL but in a slightly different form with smaller diameters of 7.5–12 nm (Fagan et al., 1999). ApoE synthesis and secretion, presumably as HDL, increase as the cells differentiate (Zhang et al., 2000). They also increase during brain development and in the brain recovering from injury, whether acute or chronic (Dawson et al., 1986; Ignatius et al., 1986; Snipes et al., 1986; Mahley, 1988; Boyles et al., 1989; Harel et al., 1989; Popko et al., 1993; Goodrum et al., 1995). It is therefore believed that apoE plays an important role either in the recovery of cholesterol from damaged cell debris or in the supply of cholesterol to regenerating cells, especially neurons. However, the mechanism for upregulation of apoE synthesis and secretion by astrocytes has not been established.

We recently reported that apoE secretion by astrocytes is regulated by one or more trophic factor(s) produced in the brain (Ueno et al., 2002). When rat fetal brain astrocytes were grown in primary cultures for 4 weeks followed by 1 week of culturing as secondary cultures, the cells had a lower cholesterol content than corresponding cells prepared by our conventional method of culturing for 1 week as secondary cultures after only 1 week in primary cultures. The older cells synthesized and secreted apoE and cholesterol as HDL more actively, and their HMG-CoA reductase activity was upregulated. Conditioned medium from the cells which had been grown for 4 weeks in primary cultures stimulated astrocytes prepared by our conventional method to increase their synthesis and secretion of apoE and cholesterol. This effect could be reproduced by fibroblast growth factor-1 (FGF-1) (acidic FGF), but not by any other cytokine examined, including FGF-2, and anti-FGF-1 antibody canceled the stimulatory activity of the conditioned medium (Fig. 2). The fetal rat brain astrocytes expressed large amounts of FGF-1 mRNA after 3 weeks in primary culture. These experimental results showed that the FGF-1-like factor is secreted by rat fetal brain cells during long-term primary culture and stimulates production of HDL in astrocytes.

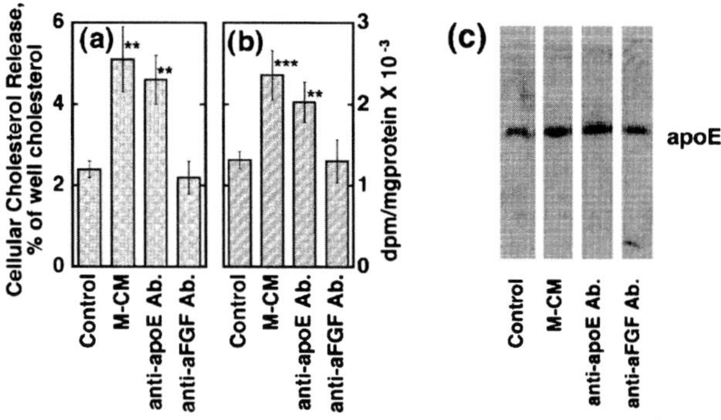

Fig. 2. Enhancement of cholesterol release (a, b) and apoE secretion (c) from conventionally prepared rat astrocytes by conditioned medium from one-month-old primary culture (M-CM) and inhibition of the enhancement by an anti-FGF-1 antibody, but not by an Anti-apoE antibody. For the antibody experiments the M-CM was treated with a goat anti-human aFGF antibody (Santa Cruz Biotech. Inc.) or a rabbit anti-rat apoE antibody conjugated on protein G-Sepharose (Amersham Pharmacia Biotech.) for 4 h at room temperature, and the gels were removed by centrifugation. W/W cells were incubated for 5 days with 0.5 mL/mL of M-CM, which had been pretreated with either antibody or 0.02% BSA/F-10 as a control. Before the experiment the cells had been labeled with 30 mCi/mL of (^{3}H)-acetate for 12 h, washed 3 times with DPBS and incubated in 0.02% BSA/F-10 containing 1 mM acetate for a further 12 h period. The release of newly synthesized cholesterol (from (^{3}H)-acetate) into the medium was determined by counting of the radioactivity in cholesterol. The results were expressed as percentage of labeled cholesterol in total cellular cholesterol in the well (a) and as the radioactivity per mg cell protein (b). Each value represents the average and standard error of triplicate experiments, and ** and *** indicate $p < 0.05$ and 0.01 from control in the same panel. (c) Stimulation of apoE secretion from rat astrocytes by M-CM, and its inhibition by an anti-aFGF antibody. Rat astrocytes were incubated for 5 days in the fresh 0.02% BSA/F-10 medium containing the indicated conditioned medium. The cells were washed and incubated for further 24 h, and the conditioned media were analyzed by immunoblotting. The gels represent one of a total of three independent experiments. Taken from Ueno et al. (2002), with permission.

The primary cultures that highly express FGF-1 mRNA after 3 weeks in culture, consisted almost exclusively of astrocytes and neurons could hardly be identified. This finding indicates that FGF-1 is produced by astrocytes and stimulates astrocytes in an autocrine manner.

Epidermal growth factor is also known to stimulate apoE secretion (Baskin et al., 1997). It reportedly increases apoA-I expression through the Ras-MAP kinase cascade and Sp1 (Zheng et al., 2001). As FGF-1 is known to induce signaling through P21ras/Erk cascade in astrocytes (Asada et al., 1999), it should be investigated if the mechanism for its upregulation of apoE–HDL production involves stimulation of this intracellular signaling pathway. FGF-1 has no signal peptide, so the pathway for its secretion is unknown at present. FGF-1-transfected cells release FGF-1 into the culture medium only during heat-shock conditions (Jackson et al., 1992). FGF-9 produced in the brain and kidney is also known to be secreted, although it lacks a signal peptide, and it is assumed to be secreted via the Golgi apparatus, as it is N-glycosylated (Miyamoto et al., 1993). Thus, specific mechanism(s) may exist for such cytokines to be secreted without any signal peptides.

5. Production of HDL by astrocytes

Astrocytes produce cholesterol-rich HDL with endogenous apoE and generate cholesterol-poor HDL with exogenous apolipoproteins such as apoA-I (Ito et al., 1999) (Fig. 1, Table 1). It is unknown whether these two types of HDL have different functions. Cholesterol-poor HDL or small HDL with apoJ may have a higher capacity to accumulate additional cholesterol and may therefore rather accept cholesterol from other cells, while cholesterol-rich HDL can deliver cholesterol to target cells via apoE-recognizing receptors. It would be of importance to investigate the cholesterol transport network by various HDLs among the CNS cells in normal cholesterol homeostasis and during pathophysiological conditions.

Exogenous apoA-I removes less cholesterol and produces cholesterol-poor HDL. However, the cholesterol content in apoA-I–HDL in cultured astrocytes can be increased by digestion of cellular sphingomyelin by extracellular sphingomyelinase (SMase) (Fig. 3), indicating that the mobility of the cholesterol molecule in the cells is restricted by sphingomyelin. This observation suggests that sphingomyelin/cholesterol-rich domains of the cell surface are likely to be used as a source of cholesterol for HDL assembly

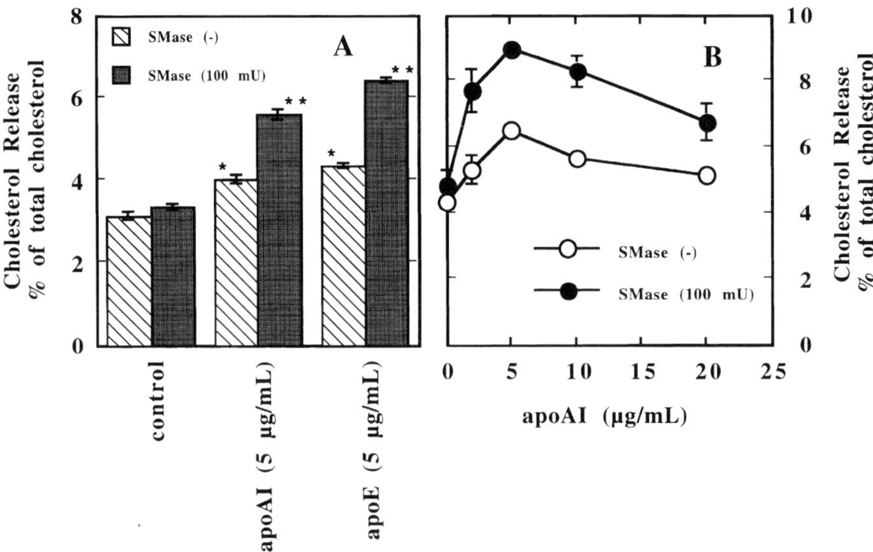

Fig. 3. Effect of sphingomyelinase (SMase) treatment on the cholesterol release from the rat astrocytes mediated by apolipoprotein. The astrocytes were cultured in the 3 cm culture plates. LDL labeled with [^{3}H]cholesteryl oleate was added to the cells in fresh 0.1% BSA/F-10 tissue culture medium, followed by incubation at 37 °C for 24 h. After washing with cell buffer and replacement with 0.1% BSA/F-10 medium, the cells were treated with 100 mU SMase (SMase (100 mU)) or left untreated (SMase (−)) for 1 h, and incubated in the presence or absence of apoA-I or apoE at the indicated concentration for another 8 h. Lipid was extracted with organic solvent from the cells and the conditioned medium and analyzed after separation by TLC. Each value represents the average and standard error (if larger than the symbol) of triplicate measurements. In panel A, the single asterisk indicates a significant difference from the respective control in the absence of any apolipoprotein ($p < 0.01$), and the double asterisk indicates a difference from the respective SMase (−) ($p < 0.01$). Modified from Ito et al. (2000).

(Ito et al., 2000, 2002, 2003). Although apoA-I does not remove much cholesterol (Table 1), it induces translocation of newly synthesized cholesterol, phospholipid and caveolin-1 to the cytosol in rat astrocytes prior to the appearance of these lipids in HDL in the medium (Ito et al., 2002, 2003). The lipids and caveolin-1 translocated to the cytosol can be recovered as lipid–protein complex particles along with cyclophilin A and are found within the typical density range of HDL. Cyclosporin A, a cyclophilin A-specific binding agent, inhibited both the translocation and the apoA-I-mediated cholesterol release. Thus, at least in astrocytes, cyclophilin A may contribute to the intracellular response to apoA-I that occurs in association with the apoA-I-mediated cholesterol release. This finding is consistent with our previous observation that down-regulation of caveolin-1 resulted in the decrease in cholesterol content in the HDL generated by apoA-I in THP-1 cells (Arakawa et al., 2000). Caveolin-1 is generally thought to play an important role for the intracellular cholesterol trafficking (Fielding and Fielding, 1997). Our findings indicated that caveolin-1 participates in the intracellular cholesterol trafficking linked to the apolipoprotein-mediated generation of HDL in astrocytes.

The mechanism(s) triggering such a cholesterol trafficking system is unknown, but the involvement of protein kinase C has been implicated (Li and Yokoyama, 1995; Li et al., 1997). After removal of sphingomyelin from the cells by SMase it is rapidly replenished by transfer of phosphorylcholine from phorphatidylcholine to ceramide, a process, which generates diacylglyceride as a potential signal initiator (Ito et al., 2002, 2003).

ABCA1 has been shown to play a key role in the cell–apolipoprotein interaction and subsequent generation of HDL. Mutations of ABCA1 have been shown to result in extremely low plasma HDL levels and lack of apolipoprotein-mediated HDL production in familial HDL deficiency patients, such as Tangier disease (Bodzioch et al., 1999; Brooks-Wilson et al., 1999; Rust et al., 1999) and in ABCA1 knock-out mice (Christiansen-Weber et al., 2000), indicating that the apolipoprotein–cell interaction is the major source of the production of plasma HDL. It is unclear whether ABCA1 mediates production of HDL in CNS cells or whether there is any back-up system for ABCA1 in the CNS (Dean et al., 2001). Despite the extremely low plasma HDL level in the homozygotes with Tangier disease, their risk for atherosclerotic disease is not increased as much as the homozygotes of familial hypercholesterolemia, who have three to five times increased plasma LDL level and almost 100% risk for coronary heart disease by their second decade of life. Also, there is no clear indication that patients with Tangier disease exhibit any CNS symptoms though some patients exhibit peripheral neuropathy (Assmann et al., 2001). It is necessary to analyze the function of ABCA1 and other proteins that may function in generation of HDL in the brain.

6. Brain injury and apoE production

There are many reports that the production and secretion of apoE are stimulated in CNS and peripheral nerves after injury. Twenty years ago Skene and Shooter (1983) observed an increase of the production of a 37 kDa protein after injury of the sciatic nerve. The protein was identified as apoE in 1986 (Ignatius et al., 1986; Snipes et al., 1986). Increase of the LDL receptor was also observed in the cells around the injury of the sciatic nerve.

Since then, numerous papers have been published about the relation between apoE production and nerve injury, including acute and chronic brain damage, such as cerebrovascular infarction and degenerative brain diseases. They indicate a role of apoE in intercellular cholesterol transport in nerve reconstruction and remyelination after the damage (Dawson et al., 1986; Ignatius et al., 1986; Snipes et al., 1986; Mahley, 1988; Boyles et al., 1989; Harel et al., 1989; Popko et al., 1993; Goodrum et al., 1995).

As discussed above, the mechanism for upregulation of apoE synthesis in astrocytes is not clear. There are several reports indicating an increase in transcription and translation of FGF superfamily members after injury in the nervous system (Barotte et al., 1989; Tooyama et al., 1991; Eckenstein et al., 1994; Bugra and Hicks, 1997; Yoshimura et al., 2001). Thus hippocampal administration of kainic acid to adult mice induces production of FGF-2 (basic FGF) and increases the density of BrdUrd/NeuN-positive cells (dividing progenitor cells differentiating into neurons) (Yoshimura et al., 2001). This response is attenuated in the FGF-2-deficient mouse, but it can be restored by vector-mediated delivery of FGF-2 to the hippocampus. The choline acetyltransferase activity after brain injury is also enhanced by the microinjection of FGF-2 (Barotte et al., 1989). Our in vitro observation that FGF-1 stimulates apoE–HDL production by a presumably autocrine mechanism (Ueno et al., 2002) suggests the possibility that this growth factor also plays an important role in the recovery of brain injury and therefore suggests a function of FGF-1 as a post-injury survival factor.

There are three major isoproteins of human apoE: apoE2, E3 and E4, corresponding to gene alleles of *e*2, *e*3 and *e*4. It is a well established observation that *e*4 is a strong risk factor for Alzheimer's disease (Corder et al., 1993; Poirier et al., 1993; Strittmatter et al., 1993a,b; Nathan et al., 1994; Roses et al., 1994), and for the deposition of amyloid protein after head injury (Nicoll et al., 1995; Jordan et al., 1997) or intracranial hemorrhage (Alberts et al., 1995). Since this relationship was discovered, the relationship between apoE and the pathogenesis of Alzheimer's disease has been extensively studied from various viewpoints. Many approaches were made to characterize isoform specific apoE functions, such as different affinity for amyloid peptides (Strittmatter et al., 1993a,b; LaDu et al., 1994), anti-oxidative activity (Miyata and Smith, 1996), and neurite outgrowth-stimulating activity (Fagan et al., 1996; DeMattos et al., 1998). In addition, changes of cellular cholesterol level appear to influence the production of amyloid 1–40 and amyloid 1–42 (Chochina et al., 2001; Fassbender et al., 2001).

7. Concluding remarks: glia and cholesterol homeostasis in the brain

There is no doubt that astrocytes play a key role in cholesterol metabolism in the brain. This function can be attributed to their ability to synthesize apoE and perhaps apoJ, and thereby generate HDL with their own lipid. HDL produced in this manner is an intercellular carrier of cholesterol in the brain, and seems to function as the major cholesterol supplier for neurons. Efficient neurite growth and synapse formation require external supply of cholesterol. Before the BBB is established, plasma lipoproteins, perhaps mainly LDL, may supply neural cells with cholesterol (Kane and Havel, 2001). However, after the formation of the BBB, glial synthesis of HDL takes over. It is rational to postulate that HDL production by astrocytes is stimulated by increased needs of cholesterol, such as after acute and chronic neuronal damage. Our results indicate that apoE and cholesterol synthesis are upregulated by FGF-1, which is released from brain cells (Ueno et al., 2002). Astrocytes or related cells may sense one or more microenvironmental signal(s) to release FGF-1 in order to stimulate apoE–HDL production.

It is also of interest that HDL produced by extracellular, rather than endogenous, apolipoprotein is relatively cholesterol-poor (Ito et al., 1999, 2000). Since such HDL particles are capable of accepting more cholesterol released from cells by diffusion, this HDL may function as an acceptor of cholesterol from neurons or cell debris after the damage.

HDL is the only carrier of cholesterol to export it from the CNS via the flow of CSF to the systemic circulation. A remaining question is to what extent cholesterol catabolism contributes to 24S-hydroxycholesterol in human brain. Two issues must be addressed: (i) what portion of brain cholesterol is metabolized by this pathway; and (ii) how is 24S-hydroxycholesterol transported out of CNS. Regarding the second question, 24S-hydroxycholesterol has been reported to appear in HDL, when it is released from the cells (Babiker and Diczfalusy, 1998), but a recent report also indicated that it may be directly exported to the systemic circulation through endothelial cells (Panzenboeck et al., 2002).

It is important to investigate regulation of HDL production in astrocytes in order to understand cholesterol homeostasis in the brain. This will lead to further understanding of the regulation of the function of the brain and its individual cell types, and of repair mechanisms following acute and chronic brain damage, and eventually assist in the development of technology to enhance regeneration.

References

Abe-Dohmae, S., Suzuki, S., Wada, Y., Aburatani, H., Vance, D.E., Yokoyama, S., 2000. Characterization of apolipoprotein-mediated HDL generation induced by cAMP in a murine macrophage cell line. Biochemistry 39, 11092–11099.

Acton, S., Rigotti, A., Landschulz, K.T., Xu, S., Hobbs, H.H., Krieger, M., 1996. Identification of scavenger receptor SR-BI as a high density lipoprotein receptor. Science 271, 518–520.

Alberts, M.J., Graffagnino, C., McClenny, C., DeLong, D., Strittmatter, W., Saunders, A.M., Roses, A.D., 1995. ApoE genotype and survival from intracerebral hemorrhage. Lancet 346, 575.

Anonymous, 1981. Composition of the body. In: Lentner, C. (Ed.), Geigy Scientific Tables. Ciba-Geigy, Basle, pp. 217–225.

Arakawa, R., Abe-Dohmae, S., Asai, M., Ito, J.-i., Yokoyama, S., 2000. Involvement of caveolin-1 in cholesterol enrichment of high density lipoprotein during its assembly by apolipoprotein and THP-1 cells. J. Lipid Res. 41, 1952–1962.

Arakawa, R., Yokoyama, S., 2002. Helical apolipoproteins stabilize ATP-binding cassette transporter A1 by protecting it from thiol protease-mediated degradation. J. Biol. Chem. 277, 22426–22429.

Asada, S., Kasuya, Y., Hama, H., Masaki, T., Goto, K., 1999. Cytodifferentiation potentiates aFGF-induced p21ras/Erk signaling pathway in rat cultured astrocytes. Biochem. Biophys. Res. Commun. 260, 441–445.

Assmann, G., von Eckardstein, A., Brewer, B.H. Jr., 2001. Familial analphalipoproteinemia: Tangier disease. In: Sciriber, C.R., Beaudet, A.L., Sly, W.S., Valle, D. (Eds.), The Metabolic & Molecular Basis of Inherited Disease. 8th ed. McGraw-Hill, New York, pp. 2937–2960.

Babiker, A., Diczfalusy, U., 1998. Transport of side-chain oxidized oxysterols in the human circulation. Biochim. Biophys. Acta 1392, 333–339.

Barotte, C., Eclancher, F., Ebel, A., Labourdette, G., Sensenbrenner, M., Will, B., 1989. Effects of basic fibroblast growth factor (bFGF) on choline acetyltransferase activity and astroglial reaction in adult rats after partial fimbria transection. Neurosci. Lett. 101, 197–202.

Barres, B.A., Smith, S.J., 2001. Cholesterol-making or breaking the synapse. Science 294, 1296–1354.

Baskin, F., Smith, G.M., Fosmire, J.A., Rosenberg, R.N., 1997. Altered apolipoprotein E secretion in cytokine treated human astrocyte cultures. J. Neurol. Sci. 148, 15–18.

Bjorkhem, I., Lutjohann, D., Breuer, O., Sakinis, A., Wennmalm, A., 1997. Importance of a novel oxidative mechanism for elimination of brain cholesterol. J. Biol. Chem. 272, 30178–30184.

Bjorkhem, I., Lutjohann, D., Diczfalusy, U., Stahle, L., Ahlborg, G., Wahren, J., 1998. Cholesterol homeostasis in human brain: turnover of 24S-hydroxycholesterol and evidence for a cerebral origin of most of this oxysterol in the circulation. J. Lipid Res. 39, 1594–1600.

Bodzioch, M., Orso, E., Klucken, J., Langmann, T., Böttcher, A., Diederich, W., Drobnik, W., Barlage, S., Büchler, C., Porsch-Özcürümez, M., Kaminski, W.E., Hahmann, H.W., Oette, K., Rothe, G., Aslanidis, C., Lackner, K.J., Schmitz, G., 1999. The gene encoding ATP-binding cassette transporter 1 is mutated in Tangier disease. Nat. Genet. 22, 347–351.

Borghini, I., Barja, F., Pometta, D., James, R.W., 1995. Characterization of subpopulations of lipoprotein particles isolated from human cerebrospinal fluid. Biochim. Biophys. Acta 1255, 192–200.

Boyles, J.K., Pitas, R.E., Wilson, E., Mahley, R.W., Taylor, J.M., 1985. Apolipoprotein E associated with astrocytic glia of the central nervous system and with nonmyelinating glia of the peripheral nervous system. J. Clin. Invest. 76, 1501–1513.

Boyles, J.K., Zoellner, C.D., Anderson, L.J., Kosik, L.M., Pitas, R.E., Weisgraber, K.H., Hui, D.Y., Mahley, R.W., Gebicke-Haerter, P.J., Ignatius, M.J., Shooter, E.M., 1989. A role for apolipoprotein E, apolipoprotein A-I and low density lipoprotein receptors in cholesterol transport during regeneration and remyelination of the rat sciatic nerve. J. Clin. Invest. 83, 1015–1031.

Brooks-Wilson, A., Marcil, M., Clee, S.M., Zhang, L.-H., Roomp, K., Dam, M.v., Brewer, C., Collins, J.A., Molhuizen, H.O.F., Loubser, D., Ouelette, B.F.F., Fchter, K., Asbourne-Excoffon, K.J.D., Sensen, C., Scherer, S., Mott, S., Denis, M., Martindale, D., Frohlich, J., Morgan, K., Koop, B., Pimstone, S., Kastelein, J.J.P., Genest, J.J., Hayden, M.R., 1999. Mutations in ABC1 in tangier disease and familial high-density lipoprotein deficiency. Nat. Genet. 22, 336–345.

Bu, G., Maksymovitch, E.A., Nerbonne, J.M., Schwartz, A.L., 1994. Expression and function of the low density lipoprotein receptor-related protein (LRP) in mammalian central neurons. J. Biol. Chem. 269, 18521–18528.

Bugra, K., Hicks, D., 1997. Acidic and basic fibroblast growth factor messenger RNA and protein show increased expression in adult compared to developing normal and dystrophic rat retina. J. Mol. Neurosci. 9, 13–25.

Chambenoit, O., Hamon, Y., Marguet, D., Rigneault, H., Rosseneu, M., Chimini, G., 2001. Specific docking of apolipoprotein A-I at the cell surface requires a functional ABCA1 transporter. J. Biol. Chem. 276, 9955–9960.

de Chaves, E.I., Rusinol, A.E., Vance, D.E., Campenot, R.B., Vance, J.E., 1997. Role of lipoproteins in the delivery of lipids to axons during axonal regeneration. J. Biol. Chem. 272, 30766–30773.

Chochina, S.V., Avdulov, N.A., Igbavboa, U., Cleary, J.P., O'Hare, E.O., Wood, W.G., 2001. Amyloid-peptide1–40 increases neuronal membrane fluidity: role of cholesterol and brain fergion. J. Lipid Res. 42, 1292–1297.

Christiansen-Weber, T.A., Voland, J.R., Wu, Y., Ngo, K., Roland, B.L., Nguyen, S., Peterson, P.A., Fung-Leung, W.P., 2000. Functional loss of ABCA1 in mice causes severe placental malformation, aberrant lipid distribution, and kidney glomerulonephritis as well as high-density lipoprotein cholesterol deficiency. Am. J. Pathol. 157, 1017–1029.

Corder, E.H., Saunders, A.M., Strittmatter, W.J., Schmechel, D.E., Gaskell, P.C., Small, G.W., Roses, A.D., Haines, J.L., Pericak-Vance, M.A., 1993. Gene dose of apolipoprotein E type 4 allele and the risk of Alzheimer's disease in late onset families. Science 261, 921–923.

Dawson, P.A., Schechter, N., Williams, D.L., 1986. Induction of rat E and chicken A-I apolipoproteins and mRNAs during optic nerve degeneration. J. Biol. Chem. 261, 5681–5684.

Dean, M., Hamon, Y., Chimini, G., 2001. The human ATP-binding cassette (ABC) transporter superfamily. J. Lipid Res. 42, 1007–1017.

Dehouck, B., Dehouck, M.-P., Fruchart, J.-C., Cecchelli, R., 1994. Upregulation of the low density lipoprotein receptor at the blood–brain barrier: intercommunications between brain capillary endothelial cells and astrocytes. J. Cell Biol. 126, 465–473.

Dehouck, B., Fenart, L., Dehouck, M.-P., Pierce, A., Torpier, G., Cecchelli, R., 1997. A new function for the LDL receptor: transcytosis of LDL across the blood–brain barrier. J. Cell Biol. 138, 877–889.

DeMattos, R.B., Brendza, R.P., Heuser, J.E., Kierson, M., Cirrito, J.R., Fryer, J., Sullivan, P.M., Fagan, A.M., Han, X., Holtzman, D.M., 2001. Purification and characterization of astrocyte-secreted apolipoprotein E and J-containing lipoproteins from wild-type and human apoE transgenic mice. Neurochem. Int. 39, 415–425.

DeMattos, R.B., Curtiss, L.K., Williams, D.L., 1998. A minimally lipidated form of cell-derived apolipoprotein E exhibits isoform-specific stimulation of neurite outgrowth in the absence of exogenous lipids or lipoproteins. J. Biol. Chem. 273, 4206–4212.

Dietschy, J.M., Turley, S.D., 2001. Cholesterol metabolism in the brain. Curr. Opin. Lipid. 12, 105–112.

Eckenstein, F.P., Andersson, C., Kuzis, K., Woodward, W.R., 1994. Distribution of acidic and basic fibroblast growth factors in the mature, injured and developing rat nervous system. Prog. Brain Res. 103, 55–63.

Fagan, A.M., Bu, G., Sun, Y., Daugherty, A., Holzman, D.M., 1996. Apolipoprotein E-containing high density lipoprotein promotes neurite outgrowth and is a ligand for the low density lipoprotein receptor-related protein. J. Biol. Chem. 271, 30121–30125.

Fagan, A.M., Holtzman, D.M., Munson, G., Mathur, T., Schneider, D., Chang, L.K., Getz, G.S., Reardon, C.A., Lukens, J., Shah, J.A., LaDu, M.J., 1999. Unique lipoproteins secreted by primary astrocytes from wild type, apoE (−/−), and human apoE transgenic mice. J. Biol. Chem. 274, 30001–30007.

Fassbender, K., Simons, M., Bergmann, C., Stroick, M., Lutjohann, D., Keller, P., Runz, H., Kuhl, S., Bertsch, T., Bergmann, K.v., Hennerici, M., Beyreuther, K., Hartmann, T., 2001. Simvastatin strongly reduces levels of Alzheimer's disease-amyloid peptides A_42 and A_40 in vitro and in vivo. Proc. Natl Acad. Sci. USA, 1–6 (Early Edition).

Fielding, C.J., Fielding, P.E., 1997. Intracellular cholesterol transport. J. Lipid Res. 38, 1503–1521.

Fitzgerald, M.L., Morris, A.L., Rhee, J.S., Anderson, L.P., Mendez, A.J., Freeman, M.W., 2002. Naturally occurring mutations in the largest extracellular loops of ABCA1 can disrupt its direct ınteractıon with apolipoprotein A-I. J. Biol. Chem. 277, 33178–33187.

Fujita, S.C., Sakuta, K., Tsuchiya, R., Hamamaka, H., 1999. Apolipoprotein E is found in astrocytes but not in microglia in the normal mouse brain. Neurosci. Res. 35, 123–133.

Fukumoto, H., Deng, A., Irizarry, M.C., Fitzgerald, M.L., Rebeck, G.W., 2003. Induction of the cholesterol transporter ABCA1 in CNS cells by LXR agonists increases secreted A{β} levels. J. Biol. Chem. 277, 48508–48513.

Goodrum, J.F., Bouldin, T.W., Zhang, S.H., Maeda, N., Popko, B., 1995. Nerve regeneration and cholesterol reutilization occur in the absence of apolipoproteins E and A-I in mice. J. Neurochem. 64, 408–416.

Gorodinsky, A., Harris, D.A., 1995. Glycolipid-anchored proteins in neuroblastoma cells from detergent-resistant complexes without caveolin. J. Cell Biol. 129, 619–627.

Gorodinsky, A., Wu, G.F., Harris, D.A., 1994. GPI-anchored proteins in neuronal cells form detergent-resistant complexes without caveolin. Mol. Biol. Cell 5(Suppl.), 320a.

Hammad, S.M., Stefansson, S., Twal, W.O., Drake, C.J., Fleming, P., Remaley, A., Brewer, H.B. Jr., Argraves, W.S., 1999. Cubilin, the endocytic receptor for intrinsic factor-vitamin B(12) complex, mediates high-density lipoprotein holoparticle endocytosis. Proc. Natl Acad. Sci. USA 96, 10158–10163.

Harel, A., Fainaru, M., Shafer, Z., Hernandez, M., Cohen, A., Schwartz, M., 1989. Optic nerve regeneration in adult fish and apolipoprotein A-I. J. Neurochem. 52, 1218–1228.

Herz, J., 2001. The LDL receptor gene family: (un)expected signal transducers in the brain. Neuron 29, 571–581.

Herz, J., Strickland, D.K., 2001. LRP: a multifunctional scavenger and signaling receptor. J. Clin. Invest. 108, 779–784.

Ignatius, M.J., Gebicke-Härter, P.J., Skene, J.H.P., Schilling, J.W., Weisgraber, K.H., Mahley, R.W., Shooter, E.M., 1986. Expression of apolipoprotein E during nerve degeneration and regeneration. Proc. Natl Acad. Sci. USA 83, 1125–1129.

Ito, J., Nagayasu, Y., Kato, K., Sato, R., Yokoyama, S., 2002. Apolipoprotein A-I induces translocation of cholesterol, phospholipid, and caveolin-1 to cytosol in rat astrocytes. J. Biol. Chem. 277, 7929–7935.

Ito, J., Nagayasu, Y., Ueno, S., Yokoyama, S., 2003. Apolipoprotein-mediated cellular lipid release requires replenishment of sphingomyelin in a phosphatidylcholine-specific phospholipase C-dependent manner. J. Biol. Chem. 277, 44709–44714.

Ito, J., Nagayasu, Y., Yokoyama, S., 2000. Cholesterol–sphingomyelin interaction in membrane and apolipoprotein-mediated cellular cholesterol efflux. J. Lipid Res. 41, 894–904.

Ito, J., Zhang, L.-Y., Asai, M., Yokoyama, S., 1999. Differential generation of high-density lipoprotein by endogenous and exogenous apolipoproteins in cultured fetal rat astrocytes. J. Neurochem. 72, 2362–2369.

Jackson, A., Friedman, S., Zhan, X., Engleka, K.A., Forough, R., 1992. Heat shock induces the release of fibroblast growth factor 1 from NIH 3T3 cells. Proc. Natl Acad. Sci. USA 89, 10691–10695.

Jordan, B.D., Relkin, N.R., Ravdin, L.D., Jacobs, A.R., Bennett, A., Gandy, S., 1997. Apolipoprotein E epsilon 4 associated with chronic traumatic injury in boxing. J. Am. Med. Assoc. 278, 136–140.

Kane, J.P., Havel, R.J., 2001. Disorders of the biogenesis and secretion of lipoproteins containing the B apolipoproteins. In: Sciriber, C.R., Beaudet, A.L., Sly, W.S., Valle, D., (Eds.), The Metabolic & Molecular Basis of Inherited Disease. 8th ed. McGraw-Hill, New York, pp. 2717–2752.

Kim, D.H., Iijima, H., Goto, K., Sakai, J., Ishii, H., Kim, H.J., Suzuki, H., Kondo, H., Saeki, S., Yamamoto, T., 1996. Human apolipoprotein E receptor 2. A novel lipoprotein receptor of the low density lipoprotein receptor family predominantly expressed in brain. J. Biol. Chem. 271, 8373–8380.

Koch, S., Donarski, N., Goetze, K., Kreckel, M., Stuerenburg, H.-J., Buhmann, C., Beisiegel, U., 2001. Characterization of four lipoprotein classes in human cerebrospinal fluid. J. Lipid Res. 42, 1143–1151.

Kozyraki, R., Fyfe, J., Kristiansen, M., Gerdes, C., Jacobsen, C., Cui, S., Christensen, E.I., Aminoff, M., de la Chapelle, A., Krahe, R., Verroust, P.J., Moestrup, S.K., 1999. The intrinsic factor-vitamin B12 receptor, cubilin, is a high-affinity apolipoprotein A-I receptor facilitating endocytosis of high-density lipoprotein. Nat. Med. 5, 656–661.

Krul, E.S., Tang, J., 1992. Secretion of apolipoprotein E by an astrocytoma cell line. J. Neurosci. Res. 32, 227–238.

LaDu, M.J., Falduto, M.T., Manelli, A.M., Reardon, C.A., Getz, G.S., Frail, D.E., 1994. Isoform-specific binding of apolipoprotein E to beta-amyloid. J. Biol. Chem. 269, 23403–23406.

LaDu, M.J., Gilligan, S.M., Lukens, J.R., Cabana, V.G., Reardon, C.A., Van Eldik, L.J., Holtzman, D.M., 1998. Nascent astrocyte particles differ from lipoproteins in CSF. J. Neurochem. 70, 2070–2081.

Lang, T., Bruns, D., Wenzel, D., Riedel, D., Holroyd, P., Thiele, C., Jahn, R., 2001. SNAREs are concentrated in cholesterol-dependent clusters that define docking and fusion sites for exocytosis. EMBO J. 20, 2202–2213.

Li, Q., Tsujita, M., Yokoyama, S., 1997. Selective down-regulation by protein kinase C inhibitors of apolipoprotein-mediated cellular cholesterol efflux in macrophages. Biochemistry 36, 12045–12052.

Li, Q., Yokoyama, S., 1995. Independent regulation of cholesterol incorporation into free apolipoprotein-mediated cellular lipid efflux in rat vascular smooth muscle cells. J. Biol. Chem. 270, 26216–26223.

Linton, M.F., Gish, R., Hubl, S.T., Butler, E., Esquivel, C., Bry, W.L., Boyles, J.K., Wardell, M.R., Young, S.G., 1991. Phenotypes of apolipoprotein B and E after liver transplantation. J. Clin. Invest. 88, 270–281.

Lund, E.G., Guileyardo, J.M., Russell, D.W., 1999. cDNA cloning of cholesterol 24-hydroxylase, a mediator of cholesterol homeostasis in the brain. Proc. Natl Acad. Sci. USA 96, 7238–7243.

Maekawa, S., Sato, C., Kitajima, K., Funatsu, N., Kumanogoh, H., Sokawa, Y., 1999. Cholesterol-dependent localization of NAP-22 on a neuronal membrane microdomain (raft). J. Biol. Chem. 274, 21369–21374.

Mahley, R.W., 1988. Apolipoprotein E: cholesterol transport protein with expanding role in cell biology. Science 240, 622–630.

Masserini, M., Palestini, P., Pitto, M., 1999. Glycolipid-enriched caveolae and caveolae-like domains in the nervous system. J. Neurochem. 73, 1–11.

Mauch, D.H., Nagler, K., Schumacher, S., Goritz, C., Muller, E.-C., Otto, A., Pfrieger, F.W., 2001. CNS synaptogenesis promoted by glia-derived cholesterol. Science 294, 1354–1357.

Michikawa, M., Yanagisawa, K., 1999. Inhibition of cholesterol production but not of nonsterol isoprenoid products induces neuronal cell death. J. Neurochem. 72, 2278–2285.

Miyamoto, M., Naruo, K.-i., Seko, C., Matsumoto, S., Kondo, T., Kurokawa, T., 1993. Molecular cloning of a novel cytokine cDNA encoding the ninth member of the fibroblast growth factor family, which has a unique secretion property. Mol. Cell. Biol. 13, 4251–4259.

Miyata, M., Smith, J.D., 1996. Apolipoprotein E allele-specific antioxidant activity and effects on cytotoxicity by oxidative insults and beta-amyloid peptides. Nat. Genet. 14, 55–61.

Mockel, B., Zinke, H., Flach, R., Weis, B., Weiler-Guttler, H., Gassen, H.G., 1994. Expression of apolipoprotein A-I in porcine brain endothelium in vitro. J. Neurochem. 62, 788–798.

Nagler, K., Mauch, D.H., Pfrieger, F.W., 2001. Glia-derived signals induces synapse formation in neurones of the rat central nervous system. J. Physiol. 533, 665–679.

Nakai, M., Kawamata, T., Taniguchi, T., Maeda, K., Tanaka, C., 1996. Expression of apolipoprotein E mRNA in rat microglia. Neurosci. Lett. 211, 41–44.

Nathan, B.P., Bellosta, S., Sanan, D.A., Weisgraber, K.H., Mahley, R.W., Pitas, R.E., 1994. Differential effects of apolipoproteins E3 and E4 on neuronal growth in vitro. Science 264, 850–852.

Nicoll, J.A.R., Roberts, G.W., Graham, D., 1995. Apolipoprotein E_4 allele is associated with deposition of amyloid-protein following head injury. Nat. Med. 1, 135–137.

Nimpf, J., Schneider, W.J., 2000. From cholesterol transport to signal transduction: low density lipoprotein receptor, very low density lipoprotein receptor, and apolipoprotein E receptor-2. Biochim. Biophys. Acta 1529, 287–298.

Oka, K., Ishimura-Oka, K., Chu, M.J., Sullivan, M., Krushkal, J., Li, W.H., Chan, L., 1994. Mouse very-low-density-lipoprotein receptor (VLDLR) cDNA cloning, tissue-specific expression and evolutionary relationship with the low-density-lipoprotein receptor. Eur. J. Biochem. 224, 975–982.

Oram, J.F., Lawn, R.M., Garvin, M.R., Wade, D.P., 2000. ABCA1 is the cAMP-inducible apolipoprotein receptor that mediates cholesterol secretion from macrophages. J. Biol. Chem. 275, 34508–34511.

Panzenboeck, U., Balazs, Z., Sovic, A., Hrzenjak, A., Levak-Frank, S., Wintersperger, A., Malle, E., Sattler, W., 2002. ABCA1 and SR-B1 are modulators of reverse sterol transport at an in vitro blood–brain barrier constituted of porcine brain capillary endotherial cells. J. Biol. Chem. 277, 42781–42789.

Patel, S.C., Asotra, K., Patel, Y.C., McConathy, W.J., Patel, R.C., Suresh, S., 1995. Astrocytes synthesize and secrete the lipophilic ligand carrier apolipoprotein D. Neuroreport Mar 6. 4, 653–657.

Pitas, R.E., Boyles, J.K., Lee, S.H., Foss, D., Mahley, R.W., 1987. Astrocytes synthesize apolipoprotein E and metabolize apolipoprotein E-containing lipoproteins. Biochim. Biophys. Acta 917, 148–161.

Pitas, R.E., Boyles, J.K., Lee, S.H., Hui, D., Weisgraber, K.H., 1987. Lipoproteins and their receptors in the central nervous system. Characterization of the lipoproteins in cerebrospinal fluid and identification of apolipoprotein B,E(LDL) receptors in the brain. J. Biol. Chem. 262, 14352–14360.

Poirier, J., Davignon, J., Bouthillier, D., Kogan, S., Bertrand, P., Gauthier, S., 1993. Apolipoprotein E polymorphism and Alzheimer's disease. Lancet 342, 697–699.

Poirier, J., Hess, M., May, P.C., Finch, C.E., 1991. Astrocytic apolipoprotein E mRNA and GFAP mRNA in hipocampus after entorhinal cortex lesioning. Mol. Brain Res. 11, 97–106.

Popko, B., Goodrum, J.F., Bouldin, T.W., Zhang, S.H., Maeda, N., 1993. Nerve regeneration occurs in the absence of apolipoprotein E in mice. J. Neurochem. 60, 1155–1158.

Posse De Chaves, E.I., Vance, D.E., Campenot, R.B., Kiss, R.S., Vance, J.E., 2000. Uptake of lipoproteins for axonal growth of sympathetic neurons. J. Biol. Chem. 275, 19883–19890.

Radhakrishnan, A., Anderson, T.G., McConnell, H.M., 2000. Condensed complexes, raft, and the chemical activity of cholesterol in membrane. Proc. Natl Acad. Sci. USA 97, 12422–12427.

Riddell, D.R., Sun, X.-M., Stannard, A.K., Soutar, A.K., Owen, J.S., 2001. Localization of apolipoprotein E receptor 2 to caveolae in the plasma membrane. J. Lipid Res. 42, 998–1002.

Ritas, R.E., 1997. Cerebrospinal fluid lipoproteins, lipoprotein receptors, and neurite outgrowth. Nutr. Metab. Cardiovasc. Dis. 7, 202–209.

Roheim, P.S., Carey, M., Forte, T., Vega, G.L., 1979. Apolipoproteins in human cerebrospinal fluid. Proc. Natl Acad. Sci. USA 76, 4646–4649.

Roses, A.D., Strittmatter, W.J., Pericak-Vance, M.A., Corder, E.H., Saunders, A.M., Schmechel, D.E., 1994. Clinical application of apolipoprotein E genotyping to Alzheimer's disease. Lancet 343, 1564–1565.

Rust, S., Rosier, M., Funke, H., Real, J., Amoura, Z., Piette, J.-C., Deleuze, J.-F., Brewer, H.B., Duverger, N., Denefle, P., Assmann, G., 1999. Tangier disease is caused by mutations in the gene encoding ATP-binding cassette transporter 1. Nat. Genet. 22, 352–355.

Scanlon, S.M., Williams, D.C., Schloss, P., 2001. Membrane cholesterol modulates serotonin transporter activity. Biochemistry 40, 10507–10513.

Schnitzer, J.E., McIntosh, D.P., Dvorak, A.M., Liu, J., Oh, P., 1995. Separation of caveolae from associated microdomains of GPI-anchored proteins. Science 269, 1435–1439.

Shanmugaratnam, J., Berg, E., Kimerer, L., Johnson, R.J., Amaratunga, A., Schreiber, B.M., Fine, R.E., 1997. Retinal Muller glia secrete apolipoproteins E and J which are efficiently assembled into lipoprotein particles. Brain Res. Mol. Brain Res. 50, 113–120.

Shin, J.-S., Abraham, S.N., 2001. Caveolae—not just craters in the cellular landscape. Science 293, 1447–1448.

Skene, J.H.P., Shooter, E.M., 1983. Denervated sheath cells secrete a new protein after nerve injury. Proc. Natl Acad. Sci. USA 80, 4169–4173.

Snipes, G.J., McGuire, C.B., Norden, J.J., Freeman, J.A., 1986. Nerve injury stimulates the secretion of apolipoprotein E by nonneuronal cells. Proc. Natl Acad. Sci. USA 83, 1130–1134.

Stone, D.J., Rozovsky, I., Morgan, T.E., Anderson, C.P., Hajian, H., Finch, C.E., 1997. Astrocytes and microglia respond to estrogen with increased apoE mRNA in vivo and in vitro. Exp. Neurol. 143, 313–318.

Strittmatter, W.J., Saunders, A.M., Donald, S., Pericak-Vance, M., Enghild, J., Salvesen, G.S., Roses, A.D., 1993. Apolipoprotein E: high-avidity binding to amyloid and increased frequency of type 4 allele in late-onset familial Alzheimer disease. Proc. Natl Acad. Sci. USA 90, 1977–1981.

Strittmatter, W.J., Weisgraber, K.H., Huang, D.Y., Dong, L.-M., Salvesen, G.S., Pericak-Vance, M., Schmechel, D., Saunders, A.M., Goldgaber, D., Roses, A.D., 1993. Binding of human apolipoprotein E to synthetic amyloid peptide: isoform-specific effects and implications for late-onset Alzheimer disease. Proc. Natl Acad. Sci. USA 90, 8098–8102.

Suzuki, T., Ito, J., Takagi, H., Saitoh, F., Nawa, H., Shimizu, H., 2001. Biochemical evidence for localization of AMPA-type glutamate receptor subunits in the dendritic raft. Mol. Brain Res. 89, 20–28.

Tooyama, I., Akiyama, H., McGeer, P.L., Hara, Y., Yasuhara, O., Kimura, H., 1991. Acidic fibroblast growth factor-like immunoreactivity in brain of Alzheimer patients. Neurosci. Lett. 121, 155–158.

Trommsdorff, M., Gotthardt, M., Hiesberger, T., Shelton, J., Stockinger, W., Nimpf, J., Hammer, R.E., Richardson, J.A., Herz, J., 1999. Reeler/disabled-like disruption of neural migration in knockout mice lacking the VLDL receptor and apoE receptor 2. Cell 97, 689–701.

Ueno, S., Ito, J., Nagayasu, Y., Fueukawa, T., Yokoyama, S., 2002. An acidic fibroblast growth factor-like factor secreted into the brain cell culture medium upregulates apoE synthesis, HDL secretion and cholesterol metabolism in rat astrocytes. Biochim. Biophys. Acta 1589, 261–272.

Ullian, E.M., Sapperstein, S.K., Christopherson, K.S., Barres, B.A., 2001. Control of synapse number by glia. Science 291, 657–661.

Weiler-Guttler, H., Sommerfeldt, M., Papandrikopoulou, A., Mischek, U., Bonitz, D., Frey, A., Grupe, M., Scheerer, J., Gassen, H.G., 1990. Synthesis of apolipoprotein A-I in pig brain microvascular endothelial cells. J. Neurochem. 54, 444–450.

Wu, C., Butz, S., Ying, Y.-S., Anderson, R.G.W., 1997. Tyrosine kinase receptors concentrated in caveolae-like domains from neuronal plasma membrane. J. Biol. Chem. 272, 3554–3559.

Xu, Q., Li, Y., Cyras, C., Sanan, D.A., Cordell, B., 2000. Isolation and characterization of apolipoproteins from murine microglia. J. Biol. Chem. 275, 31770–31777.

Yokoyama, S., 2000. Release of cellular cholesterol: molecular mechanism for cholesterol homeostasis in cells and in the body. Biochim. Biophys. Acta 1529, 231–244.

Yoshimura, S., Takagi, Y., Harada, J., Teramoto, T., Thomas, S.S., Waeber, C., Bakowska, J.C., Breakefield, X.O., Moskowitz, M.A., 2001. FGF-2 regulation of neurogenesis in adult hippocampus after brain injury. Proc. Natl Acad. Sci. USA 98, 5874–5879.

Zhang, L.Y., Ito, J.I., Kato, T., Yokoyama, S., 2000. Cholesterol homeostasis in rat astrocytoma cells GA-1. J. Biochem. (Tokyo) 128, 837–845.

Zheng, X.-L., Matsubara, S., Diao, C., Hollenberg, M.D., Wong, N.C.W., 2001. Epidermal growth factor induction of apolipoprotein A-I is mediated by the Ras-MAP kinase cascade and Sp1. J. Biol. Chem. 276, 13822–13829.

Advances in
Molecular and
Cell Biology

Non-neuronal cells in the nervous system: sources and targets of neuroactive steroids

Roberto C. Melcangi,[a] Iñigo Azcoitia,[b] Mariarita Galbiati,[a] Valerio Magnaghi,[a] Daniel Garcia-Ovejero[c] and Luis M. Garcia-Segura[c,*]

[a]*Department of Endocrinology and Center of Excellence on Neurodegenerative Diseases, University of Milan, 20133 Milan, Italy*
[b]*Departamento de Biología Celular, Facultad de Biología, Universidad Complutense, E-28040 Madrid, Spain*
[c,*]*Instituto Cajal, CSIC, E-28002 Madrid, Spain*
E-mail: lmgs@cajal.csic.es

Contents

1. Introduction
2. Steroid hormone synthesis and metabolism
    2.1. Steroidogenesis by non-neuronal cells
    2.2. Role of 5α-reductase and the 3α-hydroxysteroid dehydrogenase in glia steroidogenesis
    2.3. Aromatase and the production of estradiol by astrocytes
    2.4. Functional implications of glial steroidogenesis
3. Non-neuronal cells as targets for steroids
    3.1. Steroid hormone receptors in non-neuronal cells
    3.2. Effects of steroids on brain endothelial cells
    3.3. Effects of steroids on astroglia
    3.4. Effects of steroids on microglia
    3.5. Effects of steroids on oligodendrocytes and Schwann cells
4. Concluding remarks

Non-neuronal cells synthethize and metabolize steroid hormones and produce local neuroactive steroids that exert neuromodulatory and neurotrophic actions under physiological and pathological conditions. In the central nervous system, the steroids produced by non-neuronal cells, such as pregnenolone, dehydroepiandrosterone (DHEA), testosterone, estradiol, progesterone and other steroid metabolites, regulate synaptic function, affect anxiety, cognition, sleep and behavior and exert neuroprotective and reparative roles. In the peripheral nervous system, progesterone and progesterone

Advances in Molecular and Cell Biology, Vol. 31, pages 535–559

ISBN: 0-444-51451-1

derivatives produced by Schwann cells, promote myelin formation and the remyelination and regeneration of injured nerves. Non-neuronal cells are also targets for steroids and mediate or participate in many of the actions of these substances in the nervous system. These include: (i) the regulation of blood–brain barrier and cerebrovascular permeability by actions of steroids on endothelial cells; (ii) the regulation of synaptogenesis, synaptic plasticity, neuritic growth and neuroendocrine secretion by actions of steroids on astroglia and (iii) the regulation of neuronal survival and regeneration by actions of steroids on microglia, astroglia and Schwann cells.

1. Introduction

The concept that the brain is a target for steroid hormones has abundant experimental support. Steroid hormones regulate the development and function of the nervous system and affect mood, behavior and cognition. For many years, it was considered that actions of steroid hormones in the brain were restricted to specific areas involved in neuroendocrine regulation and the control of behavior. Today it has become evident that steroid hormones exert a broad spectrum of actions both in the central (CNS) and the peripheral nervous systems (PNS). Furthermore, it has been discovered that the nervous system is able to synthethize and metabolize steroid hormones and produce local steroids that affect neural function. Thus, two new concepts have emerged. One is defined by the term neurosteroids, introduced by the laboratory of Baulieu (1998). This term describes those steroids synthesized in the brain directly from cholesterol (which itself is synthesized from glucose within the brain (Morell and Jurevics, 1996)), and gives a name to the concept that nervous tissue is steroidogenic. The other useful concept is defined by the term neuroactive steroids, to comprise those steroids that are able to regulate neural function. It is obvious that these terms are partially overlapping, since several steroids produced by the brain are neuroactive. Neurosteroids and neuroactive steroids include sex steroids and sex steroid metabolites. Some of these steroids act as neuromodulators, regulating the function of ion channels and neurotransmitter receptors and affecting mood, behavior and cognition (Baulieu, 1998; Compagnone and Mellon, 2000; Melcangi and Panzica, 2001). Non-neuronal cells play a central role in the synthesis, metabolism and action of steroids in the nervous system. Steroids, acting on non-neuronal cells and/or produced by them, regulate neural development, neural function and the response of nervous tissue to injury. Therefore, steroids may be considered a new class of molecular signals, with neuromodulatory and neurotrophic actions, involved in the communication of non-neuronal and neuronal cells in the nervous system. An overview of the most recent advances in this recent and fascinating emerging field is provided in this chapter.

2. Steroid hormone synthesis and metabolism

2.1. *Steroidogenesis by non-neuronal cells*

The capability to synthesize steroid hormones is not only a peculiarity of the classical steroidogenic tissues, like for instance the gonads and adrenal gland, but may be also

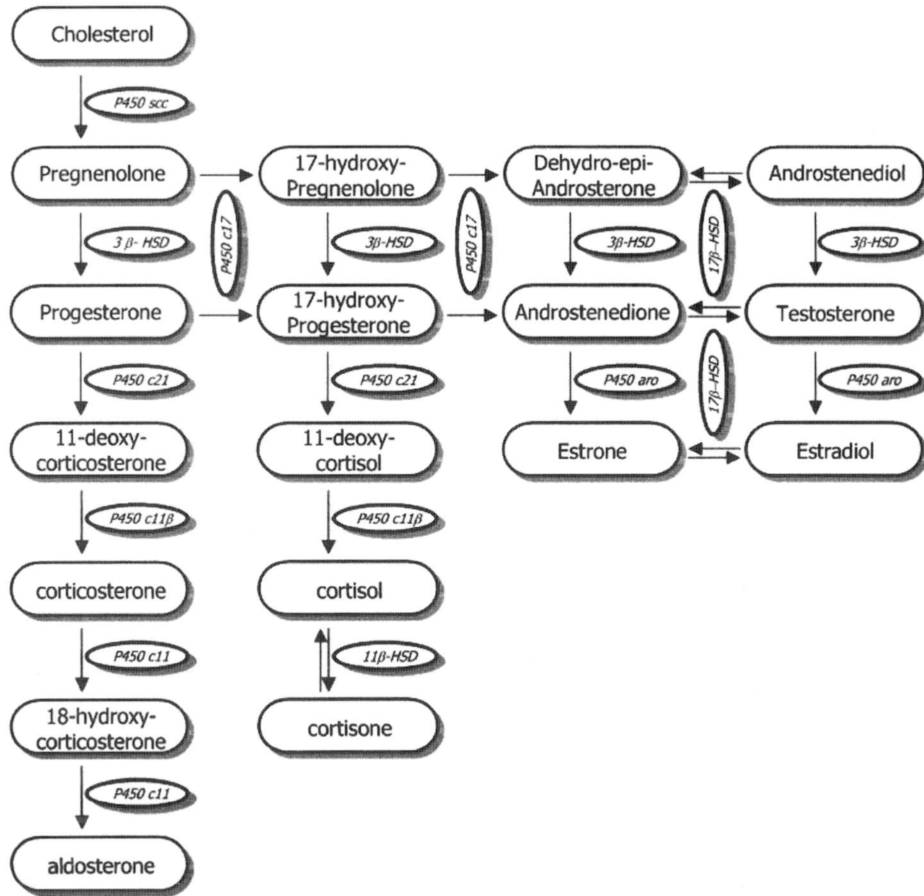

Fig. 1. Biosynthesis of steroids in the nervous system. Cholesterol is converted to pregnenolone, catalyzed by the cytochrome P450 side-chain cleavage [P450 scc], and pregnenolone is converted to progesterone, catalyzed by the 3β-hydroxy-steroid dehydrogenase [3β-HSD]. Pregnenolone and progesterone can be further converted, via 17-hydroxy-pregnenolone and 17-hydroxy-progesterone, respectively, to dehydroepiandrosterone and androstenedione, catalyzed by the 17α-hydroxylase/C17-20-lyase [P450 c17], and from there to androstenediol and testosterone, respectively, catalyzed by17β-hydroxylase/C17-20-lyase [17β-HSD]). Androstenedione and testosterone, can be converted to estrone and estradiol, respectively, catalyzed by aromatase (P450 aro). In addition to these pathways, producing androgens and estrogens, progesterone and its derivative 17-hydroxy-progesterone can be converted to the glucocorticoids corticosterone and cortisol, respectively. Cortisol can be further converted to the glucocorticoid cortisone, catalyzed by 11β-hydroxysteroid dehydrogenase (11β-HSD), and corticosterone via 18-hydroxy-corticosterone to the mineralocorticoid aldosterone.

ascribed to the nervous system. Steroidogenesis (Fig. 1) seems to take place prevalently in the non-neuronal compartment. In fact, it has been demonstrated that in the CNS, astrocytes and oligodendrocytes express several enzymes involved in the steroidogenic process (e.g., the cytochrome P450 side-chain cleavage [P450 scc], the 17α-hydroxylase/C17-20-lyase [P450 c17], the 3β-hydroxy-steroid dehydrogenase [3β-HSD], the 17β-hydroxylase/C17-20-lyase [17β-HSD]), and consequently are able to produce different kinds of steroids (Hu et al., 1987; Akwa et al., 1993; Kimoto et al., 1997; Mensah-Nyagan et al., 1999, 2001; Zwain and Yen, 1999a,b; Mellon et al., 2001). In particular, as shown by Zwain and Yen (1999a,b), astrocytes appear to be the most active steroidogenic cells in the brain, since cultures of these cells produce pregnenolone, progesterone, DHEA, androstenedione, testosterone and estradiol. On the contrary, oligodendrocytes seem to be able to form only pregnenolone and androstenedione. The steroidogenic activity of microglia has not been assessed. However, systemic macrophages that share many properties and a common origin with microglia, express aromatase and produce androgenic and estrogenic derivatives (Schmidt et al., 2000).

The capability to synthesize steroids seems not to be a peculiarity of the CNS, since also the PNS is able to convert pregnenolone into progesterone; also in this case, this transformation occurs in the glial component, namely in the Schwann cells (Koenig et al., 1995; Guennoun et al., 1997; Schumacher et al., 2001). It is interesting to note that in Schwann cells the formation of progesterone, by 3β HSD activity, is neuronal dependent. As recently shown by Robert and co-workers (2001), the expression and activity of the 3β HSD present in Schwann cells cultured alone is very low; however, when these cells are cultured in contact with sensory neurons, both expression and activity of this steroidogenic enzyme is induced.

2.2. Role of 5α-reductase and the 3α-hydroxysteroid dehydrogenase in glia steroidogenesis

As mentioned above, glial cells of the CNS and of the PNS possess several enzymes able to convert steroids into neuroactive steroids. In particular, glial cells are able to metabolize native steroid hormones into their 5α- and 3α-hydroxy-5α reduced derivatives via the enzymatic complex formed by the 5α-reductase (5α-R) and the 3α-hydroxysteroid dehydrogenase (3α-HSD) (Fig. 2) (for review, see Melcangi et al., 1999a, 2001b,c). This enzymatic complex is very versatile, since every steroid possessing the delta 4-3keto configuration may be first 5α-reduced and subsequently 3α-hydroxylated. In particular, testosterone can be converted into dihydrotestosterone and then into 5α-androstane-3α, 17β-diol (3α-diol), progesterone into dihydroprogesterone and subsequently into tetrahydroprogesterone, corticosterone into dihydrocorticosterone, and 11-deoxycorticosterone into dihydrodeoxycorticosterone and then into tetrahydrodeoxycorticosterone (Melcangi et al., 1999a, 2001c). The distribution of the 5α-R activity in the different cell types of the rat brain has been analyzed utilizing either testosterone or progesterone as substrates. In particular, the ability to metabolize testosterone was first studied in freshly isolated cell preparations (Melcangi et al., 1990); subsequent studies were performed in cultures of neurons, of the so-called type-1 astrocytes (the conventional astrocyte culture), of the so-called type-2 astrocytes, and of oligodendrocytes (Melcangi et al., 1993, 1994a). The two different groups of experiments have provided similar data. In particular, utilizing testosterone as substrate, it has been observed that fetal rat neurons possess significantly higher 5α-R activity than neonatal oligodendrocytes and astrocytes (Melcangi et al., 1993). Among glial cells, type 2 astrocytes possess a considerable 5α-R activity, while

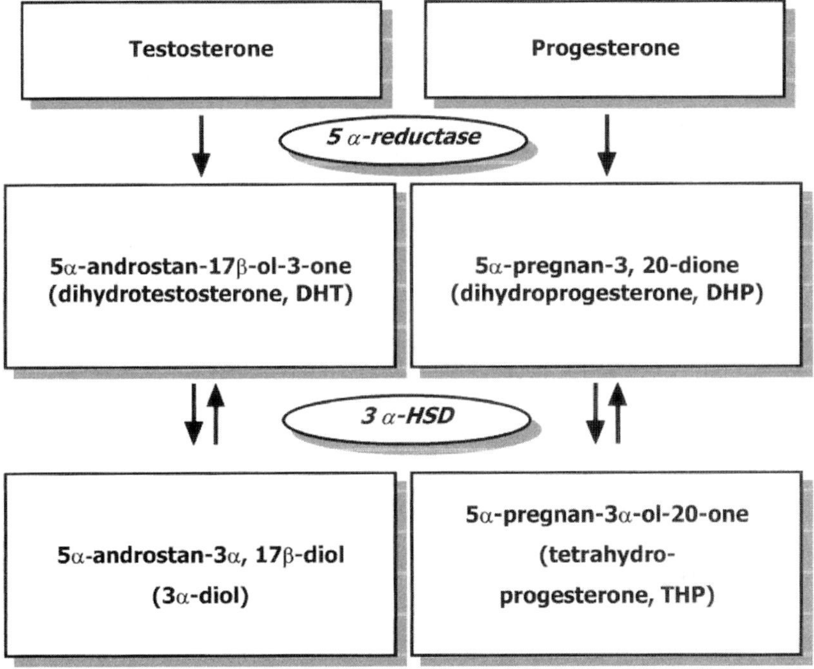

Fig. 2. Metabolism of testosterone and progesterone by the action of the enzymes 5α-reductase and 3α-hydroxysteroid-dehydrogenase (3α-HSD). In a similar manner the 5α-reductase converts corticosterone to dihydrocorticosterone (DHC) and 11-deoxycorticosterone to dihydrodeoxycorticosterone (DHDOC), and the 3α-HSD further converts DHDOC to tetrahydrodeoxycorticosterone (THDOC).

conventional astrocytes are almost devoid of such an activity. On the contrary, the enzyme 3α-HSD appears to be mainly localized in conventional astrocytes (Melcangi et al., 1993). Analogously, also when progesterone was used as substrate, neurons were shown to possess a significantly higher 5α-R activity than any glial cell analyzed. However, it is important to point out that, due to the higher affinity of the 5α-R for progesterone than for testosterone (Melcangi et al., 1999a, 2001c), the formation rate of dihydroprogesterone was about 2 times higher than that of dihydrotestosterone. Also in this case the 3α-HSD activity appeared predominantly concentrated in conventional astrocytes (Melcangi et al., 1994a). However, a consistent formation of tetrahydroprogesterone was also present in oligodendrocytes; the amounts of tetrahydroprogesterone measured in the cultures of these cells were significantly higher than those formed in cultures of type-2 astrocytes and neurons, but significantly lower than those found in conventional astrocytes (Melcangi et al., 1994a). In this context it is important to underline that Gago et al. (2001) have demonstrated that the formation of dihydroprogesterone in fully differentiated oligodendrocytes is 5-fold higher than in oligodendrocyte pre-progenitors and in oligodendrocyte progenitors. On the contrary, the formation of tetrahydroprogesterone is higher in oligodendrocyte pre-progenitors and decreases with oligodendrocyte differentiation (Gago et al., 2001). These findings underline that not only differentiated

CNS cells possess the 5α-R/3α-HSD system, but that considerable enzymatic activities for conversion of steroid hormones are also present in undifferentiated cells, as shown for the first time on stem cells originating from the mouse striatum (Melcangi et al., 1996a). Interestingly, both the 5α-R activity and the 3α-HSD activity present in conventional astrocytes are stimulated by the simultaneous presence of neurons (Melcangi et al., 1994b), indicating a possible interaction of the two populations of cells in the metabolism of steroid hormones.

The ability to convert steroid hormones into 5α- and 3α-hydroxy-5α-reduced derivatives is also present in glial cells of the PNS. Schwann cells possess both 5α-R and 3α-HSD activity (Melcangi et al., 1998, 1999b; Yokoi et al., 1998). In particular, the 5α-R activity present in these cells is at least four times higher than in oligodendrocytes, while the 3α-HSD activity is lower than in oligodendrocytes (Melcangi et al., 1998).

2.3. Aromatase and the production of estradiol by astrocytes

The enzyme aromatase (P450 aro), that is able to convert androgens into estrogens, is not present in glial cells of the CNS of mammals under normal circumstances, since only neurons possess such an activity (Negri-Cesi et al., 1992). However, it has been recently demonstrated that rodent astrocytes isolated from the cerebral cortex of neonatal rats are able to produce estradiol and estrone (Zwain et al., 1997; Zwain and Yen, 1999a). It is possible that specific culture conditions may induce aromatase expression in astrocytes. Indeed, recent results (Azcoitia, unpublished) indicate that stressful conditions, such as serum deprivation, induce aromatase expression in cultured astrocytes. Furthermore, aromatase is expressed by astrocytes in the brain of birds and mammals after brain injury (Garcia-Segura et al., 1999b; Peterson et al., 2001), indicating that the enzyme may be induced de novo in these cells under specific circumstances. In contrast to mammals, aromatase is expressed in radial glia in the brain of teleost fish under normal conditions (Forlano et al., 2001).

2.4. Functional implications of glial steroidogenesis

The steroids produced by glia may serve a paracrine role regulating synaptic function. These steroids modulate anxiety, cognition, sleep, ingestion, aggression, and reinforcement. Some of them are positive modulators of *N*-methyl-D-aspartate receptors and enhance cognitive performance. Other steroids produced by glia, such as tetrahydroprogesterone and terahydrocorticosterone, are highly selective and extremely potent modulators of the $GABA_A$ receptor, and they elicit marked anxiolytic and anti-stress effects and increase feeding and sleeping (Barbaccia et al., 2001; Engel and Grant, 2001; Lambert et al., 2001; Vallee et al., 2001).

Steroids produced by glia may also exert a neuroprotective or reparative role. For instance, steroids produced by Schwann cells (progesterone and its derivatives) regulate the expression of myelin proteins and promote axonal regeneration (Koenig et al., 1995; Magnaghi et al., 2001) (see Section 3.5). Steroids produced by central glia, such as DHEA (Kimonides et al., 1998; Bastianetto et al., 1999; Cardounel et al., 1999) and estradiol (Green and Simpkins, 2000; Wise et al., 2000; Garcia-Segura et al., 2001) are neuroprotective. In this regard, it is particularly interesting that P450 aro is expressed by reactive astroglia. Aromatase-expressing astrocytes are observed in all injured brain areas, including the cortex, corpus callosum, striatum, hippocampus, thalamus and hypothalamus (Garcia-Segura et al., 1999b). This indicates that astrocytes from most brain areas have the potential for expressing P450 aro, and therefore to produce estradiol, in response to injury. Estrogen formed by astrocytes may be released as a trophic factor for damaged neurons and may be involved in the compensatory restructuring of injured brain tissue. Estrogen released by astroglia may affect synaptic function, selective regeneration of neuronal processes and local cerebral blood flow, contributing to facilitation of neuronal recovery and reduction of neuronal death.

Recent studies indicate that aromatase knock out mice are more sensitive to excitotoxic neurodegenerative injury than wild type mice (Azcoitia et al., 2001b). Furthermore, the susceptibility to kainic acid excitotoxic injury in male rats is greatly enhanced after the intracerebral infusion of the P450 aro inhibitor fadrozole (Azcoitia et al., 2001b). This finding indicates that local cerebral P450 aro activity is involved in neuroprotection. Therefore, the induction of P450 aro in astroglia after brain injury and the local formation of estradiol by these cells may represent a response of the injured neural tissue to limit neurodegenerative damage.

3. Non-neuronal cells as targets for steroids

3.1. Steroid hormone receptors in non-neuronal cells

Steroid hormone receptors are expressed in the CNS, both in neurons and in non-neuronal cells. Glucocorticoid and mineralocorticoid receptors are expressed by microglia, astroglia, oligodendroglia and Schwann cells (Neuberger et al., 1994; Garcia-Segura et al., 1996; Tanaka et al., 1997). Estrogen receptors (ERs), androgen receptor (AR) and progesterone receptor (PR) expression has also been described in non-neuronal cells of the CNS, in vitro and in vivo, under normal and/or neurodegenerative conditions, as summarized in Table 1. In addition, Schwann cells (Melcangi et al., 1999b) and conventional astrocytes (see chapter by Hansson and Rönnbäck) express the $GABA_A$ receptor and consequently may respond to steroids that interact with this neurotransmitter receptor, such as tetrahydroprogesterone and tetrahydrodeoxycorticosterone (for review, see Melcangi et al., 1999a, 2001b).

3.2. Effects of steroids on brain endothelial cells

As in the periphery, endothelial cells in the brain are affected by steroids. Several studies have examined the effects of glucocorticoids on brain endothelium. Glucocorticoids exert different effects on brain vascular system, affecting vascular morphogenesis, blood–brain barrier properties, cerebrovascular permeability and brain edema. These effects are in part mediated by the regulation of the expression of endothelin receptors,

Table 1
Sex steroid receptors in non-neuronal cells

Receptor	Cell type	Condition	References
ERα	Astroglia	In vitro	Jung-Testas, 1992; Santagati et al., 1994; Buchanan et al., 2000
		In vivo	Langub and Watson, 1992; Milner et al., 2001
		In neurodegeneration	Azcoitia et al., 2001a; Blurton-Jones and Tuszynski, 2001; Garcia-Ovejero et al., 2002
	Ependymoglia	In vivo	Langub and Watson, 1992
	Growth promoting glia[a]	In vivo	Gudiño-Cabrera and Nieto-Sampedro, 1999
	Oligodendroglia	In vitro	Santagati et al., 1994
	Schwann glia	In vitro	Thi et al., 1998
	Microglia	In vitro	Vegeto et al., 2001; Bruce-Keller et al., 2000
	Endothelial cells	In vivo	Langub and Watson, 1992
ERβ	Astroglia	In vitro	Buchanan et al., 2000; Hösli et al., 2001
		In vivo	Azcoitia et al., 1999; Cardona-Gomez et al., 2000b
	Microglia	In vitro	Mor et al., 1999; Li et al., 2000; Vegeto et al., 2001
AR	Astroglia	In vitro	Jung-Testas et al., 1992; Hösli et al., 2001
		In vivo	Finley and Kritzer, 1999
		In neurodegeneration	Puy et al., 1995
	Oligodendroglia	In vivo	Finley and Kritzer, 1999
		In neurodegeneration	Puy et al., 1995
	Microglia	In neurodegeneration	Garcia-Ovejero et al., 2002; Puy et al., 1995
PR	Astroglia	In vitro	Jung-Testas et al., 1992
	Schwann glia	In vitro	Jung-Testas et al., 1996; Thi et al., 1998
		In vivo	Magnaghi et al., 1999, 2001

[a]Olfactory ensheathing cells, tanycytes, pituicytes, pineal glia, retinal Müller cells, and cerebellar Bergmann glia.

histamine receptors or matrix metalloproteinases in vascular endothelium (Stanimirovic et al., 1994; Karlstedt et al., 1999; Harkness et al., 2000).

The effect of estradiol on brain endothelium has been studied as well. As shown in Table 1, Langub and Watson (1992) reported ER alpha immunoreactivity in guinea pig brain endothelial cells by electron microscopy. Furthermore, estrogen increases rat brain endothelial nitric oxide synthase via ERs (McNeill et al., 2002). Part of the neuroprotective effects of estrogen may be due to hormonal actions on brain endothelium (Shi et al., 1997; Watanabe et al., 2001; Galea et al., 2002). For instance, estrogen treatment increases the endothelial cell glucose transporter GLUT1 and protects against brain capillary endothelial cell loss, which may in turn reduce focal ischemic brain damage (Shi et al., 1997). Estradiol may also exert protective effects in cerebral ischemia by blockade of leukocyte adhesion in cerebral endothelial cells (Santizo and Pelligrino, 1999), an effect that is a consequence of the hormonal down-regulation of the expression of adhesion molecules in cerebral endothelium (Galea et al., 2002).

3.3. Effects of steroids on astroglia

There is an abundant literature showing that steroids affect astroglia cell shape and gene expression (Garcia-Segura et al., 1996, 1999a, Jones et al., 1999; Jordan, 1999; Mong and McCarthy, 1999; Nichols, 1999; Vardimon et al., 1999). Steroid hormone precursors and neurosteroids, such as pregnenolone and DHEA, affect astroglia cell shape and glial fibrillary acidic protein (GFAP) expression in different brain areas (Del Cerro et al., 1996; Legrand and Alonso, 1998; Garcia-Estrada et al., 1999). Sex steroids affect GFAP expression as well. For instance, castration decreases GFAP immunoreactivity in the hypothalamus of male rats (Day et al., 1993), and this phenomenon is counteracted by testosterone or dihydrotestosterone administration, but not by estradiol (Day et al., 1993). The situation is quite different in the hippocampus, since it has been demonstrated that in male rats castration increases the levels of GFAP mRNA and protein in this cerebral structure (Day et al., 1990, 1993). Among the gonadal steroids tested (estradiol, testosterone and dihydrotestosterone), only estradiol proved able to counteract the effect of castration (Day et al., 1993). After a penetrating brain injury, sex steroids (estradiol and progesterone in females, or testosterone in males) are able to decrease the process of gliosis and astrocyte proliferation, resulting in a decrease in the number of reactive astrocytes in the cerebral cortex and in the hippocampus (Garcia-Estrada et al., 1993).

It has also been observed that the levels of GFAP mRNA and immunoreactivity show sex differences in the arcuate nucleus of the rat, lower levels being found in females than in males (Chowen et al., 1995). Androgenization of neonatal females increases GFAP mRNA to male levels, while castration of newborn males decreases GFAP mRNA to the levels found in females (Chowen et al., 1995). In the arcuate nucleus and in the hilus of the dentate gyrus of adult female rats, the surface density of GFAP-immunoreactive cells fluctuates throughout the estrous cycle, decreases after ovariectomy and increases after the administration of estradiol (Luquin et al., 1993; Garcia-Segura et al., 1994a,c). More recently, Stone et al. (1998) have analyzed GFAP transcription and the GFAP mRNA levels in the hypothalamus and hippocampus of rats during the estrous cycle. They observed that on the afternoon of proestrous, both parameters increase in the arcuate nucleus of the hypothalamus and in the outer molecular layer of the dentate gyrus. On the contrary, neither GFAP transcription nor GFAP mRNA exhibits variations in the hilus of the hippocampus during the estrous cycle. This is particularly interesting since, as mentioned above, the extension of GFAP-immunoreactive astrocytic processes shows striking changes during the estrous cycle in the hilus (Luquin et al., 1993), suggesting that ovarian hormones may affect GFAP redistribution and growth or retraction of astrocytic processes without affecting GFAP synthesis in this brain area. Furthermore, it should be mentioned that although estrogens may act directly on astrocytes, the direction of the transcriptional response is influenced by interactions of astrocytes with neurons. Thus, estradiol, when added to cortical astrocytes in culture, increases GFAP transcription, while the same treatment, performed in astrocytes co-cultured with neurons, induces a decrease of this parameter (Stone et al., 1998). Cell adhesion molecules, such as PSA-NCAM (Garcia-Segura et al., 1995a), and neurotransmitters, such as GABA (Mong et al., 2002), have been shown to be involved in estrogen-induced neuron to astrocyte signaling.

Corticosteroids are also able to influence GFAP expression. Treatments with corticosterone and other glucorticoids (e.g., dexamethasone) inhibit GFAP expression in the neonatal and adult rat brain (Nichols et al., 1990a,b; Laping et al., 1991; Tsuneishi et al., 1991). In line with these observations, adrenalectomy, performed in adult rats, has been shown to increase GFAP mRNA and protein, an effect, which is reversed by corticosterone administration (Laping et al., 1994). However, a different pattern of activity is shown by corticosterone in vitro, since exposure of cultured astrocytes to this steroid induces an increase in GFAP mRNA and protein (Rozovsky et al., 1995). Surprisingly, this effect of corticosterone on GFAP is reversed if astrocytes are co-cultured with neurons, a finding which suggests that there is a cross-talk between the two types of cells, and provides a possible explanation for the discrepancies between the results obtained in vivo and in vitro (Rozovsky et al., 1995).

Glucocorticoids influence other important functional parameters of astrocytes as well. For instance, methylprednisolone and dexamethasone enhance astrocytic calcium signaling, increasing both resting cytosolic calcium ($[Ca^{2+}]_i$) levels and the extent and the amplitude of Ca^{2+} wave propagation (Simard et al., 1999). The glucocorticoid-associated potentiation of Ca^{2+} signaling may result from up-regulation of the cellular ability to mobilize Ca^{2+} and to release the purinergic transmitter ATP (see chapter by Hansson and Rönnbck), because both ATP-induced $[Ca^{2+}]_i$ increments and ATP release were proportionally enhanced by glucocorticoids. Moreover, glucocorticoids are also able to influence the metabolism of glutamate in astrocytes, increasing the expression of both glutamine synthetase and glutamate dehydrogenase in these cells (Hardin-Pouzet et al., 1996; Vardimon et al., 1999). Finally, corticosteroids can modulate the expression of growth factors, such as basic fibroblast growth factor (bFGF) and acidic fibroblast growth factor, in astrocytes (Magnaghi et al., 2000).

In many cases the effect exerted by steroid hormones on GFAP expression are due to their conversion into active metabolites (i.e., neuroactive steroids). For instance, metabolites of testosterone and progesterone (respectively, dihydrotestosterone and dihydroprogesterone or tetrahydroprogesterone) are able to modulate gene expression of GFAP in cultures of conventional astrocytes (Melcangi et al., 1996b). This seems to be the case also for corticosteroid metabolites, since while the mineralocorticoid deoxycorticosterone is ineffective, its 5α-reduced derivative, dihydrodeoxycorticosterone, strongly inhibits GFAP gene expression (Melcangi et al., 1997).

3.3.1. Implications of steroid effects on astroglia for synaptogenesis and synaptic plasticity

The effects of sex steroids on astrocytes are linked to the regulation of synaptic connectivity. In the hypothalamic arcuate nucleus there are sex differences in the number of axo-somatic and axo-dendritic synapses that are accompanied by sex differences in the levels of GFAP, the differentiation of astrocytes and the amount of neuronal membrane surface covered by astroglial cell processes (Chowen et al., 1995; Garcia-Segura et al., 1995b; Mong et al., 1996, 1999). These sex differences are induced by the perinatal secretion of testosterone in male rats. This hormone induces stellation of astrocytes, an increased expression of GFAP and an increased coverage of neuronal membrane by astrocytic processes. Coincident with these changes in astrocytic morphology there is a strong reduction in the density of dendritic spines and axo-somatic synapses on arcuate neurons (Garcia-Segura et al., 1995b; Mong et al., 1996, 1999). The effect of testosterone on astrocytes and synapses is probably mediated by its conversion to estradiol. In the arcuate nucleus of adult female rats estradiol has similar effects as those of testosterone during development, with the important difference that estradiol effects in adults are transient and those of testosterone during development are permanent.

In the afternoon of proestrus, the surge of estradiol induces a transient growth of astrocyte processes on the surface of arcuate neurons (Garcia-Segura et al., 1994c). As a consequence, there is a transient disconnection of inhibitory GABAergic synapses from arcuate neuronal *somata* by the interposed glial processes (Garcia-Segura et al., 1994b). These changes are also elicited by the administration of estradiol to adult ovariectomized rats (Garcia-Segura et al., 1994a,b). Estradiol also increases the number of synapses on dendritic spines in the arcuate nucleus during the estrous cycle. A similar effect in the hippocampus is accompanied by an increased expression of tyrosine kinase A (TrkA) receptors in astrocytes (McCarthy et al., 2002). Since nerve growth factor, the ligand for TrkA, stimulates astrocytes to function as substrates for axon growth (Kawaja and Gage, 1991), McCarthy et al. (2002) have proposed that estradiol may regulate axonal growth and synaptic plasticity in the hippocampus by the induction of TrkA receptors in astrocytes. In addition, soluble factors released by astrocytes from target brain areas have been shown to influence the neuritogenic effect of estradiol on cultured hypothalamic neurons (Cambiasso et al., 2000). Finally, astrocytes secrete laminin in response to estrogen, facilitating neurite extension when co-cultured with neurons (Rozovsky et al., 2002).

3.3.2. Implications of steroid effects on astroglia for neuroendocrine regulation

It has recently been observed that astrocytes modulate the function of the hypothalamic neurons synthesizing and secreting the luteinizing hormone releasing hormone (LHRH). The most important principles, released from glial cells and apparently responsible for these effects, are transforming growth factor α (TGFα), transforming growth factors β1

(TGFβ1), transforming growth factor β2 (TGFβ2), bFGF and insulin-like growth factor-I (IGF-I). The astrocytic involvement provides an additional and new mode of control of LHRH secretion, which is also regulated by neuronal inputs, as well as by steroid hormones acting via positive or negative feedback signals (see also chapter by Prévot et al.). The possibility of a functional co-operation between growth factors and steroid hormones at the level of astrocytes and tanycytes in the control of the LHRH-secreting neurons has recently been taken in consideration (for review, see Melcangi et al., 2002). For instance, it has been demonstrated that estradiol is able to induce an increase in TGFα and bFGF mRNA levels in cultures of hypothalamic astrocytes (Ma et al., 1994, Galbiati et al., 2002). It is interesting to note that in hypothalamic astrocytes, estradiol is also able to facilitate the effect of prostaglandin PGE2, one of the humoral factors intervening in the control of the secretion of LHRH (Rage et al., 1997). In particular, it has been demonstrated that hypothalamic astrocytes treated with estradiol produce a conditioned medium that after application to a cell line of LHRH-secreting neurons (i.e., GT1 cells), induces a selective up-regulation of two of the four known members of the PGE2 receptor family (i.e., EP1-R and EP3γ-R) (Ojeda and Ma, 1999).

Estrogens and DHEA are able to influence the release of TGFβ1 from cultures of hypothalamic astrocytes (Buchanan et al., 2000; Zwain et al., 2002). Moreover, in this case, an effect of progesterone is also evident. In particular, it has been observed that treatment of cultured hypothalamic astrocytes for 6 h with either progesterone or one of its derivatives (dihydroprogesterone and tetrahydroprogesterone) results in a stimulation of TGFβ1 gene expression (Melcangi et al., 2001a). However, with longer treatments (24 h), only dihydroprogesterone and tetrahydroprogesterone are able to increase the mRNA levels of TGFβ1. On the basis of this time-dependent effect, it has been hypothesized that the effect of progesterone might be due to its conversion into dihydroprogesterone and tetrahydroprogesterone, since, as mentioned above, 5α-reductase and 3α-hydroxysteroid dehydrogenase activities are present in conventional astrocytes (Melcangi et al., 1993, 1994a).

In contrast, dihydrotestosterone, the 5α-reduced metabolite of testosterone, reduces the expression of bFGF in astrocytes (Melcangi et al., 2001a). This effect may be due to an interaction with the AR, which is expressed in the astrocytes in vitro (Melcangi et al., 2001b). However, testosterone, which is also able to bind to this receptor, albeit with a lower affinity than dihydrotestosterone, does not affect bFGF expression (Melcangi et al., 2001a).

Another factor involved in the regulation of LHRH neurons is IGF-I. Tanycytes in the hypothalamus and the median eminence accumulate IGF-I from cerebrospinal fluid and plasma and may therefore contribute to LHRH regulation by modulating local IGF-I levels (Garcia-Segura et al., 1999a). Gonadal steroids regulate the expression and distribution of IGF-I receptors and IGF binding proteins in tanycytes (Cardona-Gomez et al., 2000a) and this, in turn, affects the uptake of IGF-I originating from blood or cerebrospinal fluid (Garcia-Segura et al., 1999a). Therefore, tanycytes integrate hormonal signaling from the reproductive axis (gonadal steroids) and the growth hormone axis (IGF-I) to regulate LHRH release.

In summary, sex steroids affect the synthesis, accumulation and release by hypothalamic astrocytes and tanycytes of different growth factors that regulate LHRH neurons. Therefore, astrocytes and tanycytes may play an important role as mediators of the action of peripheral steroid hormones in the control of neuroendocrine regulation.

3.3.3. Implications of steroid effects on astroglia for neuroprotection

The effects of steroids on astroglia have implications for neuroprotection and brain repair as well. Sex steroids and neuroactive steroids may affect brain responses to pathological conditions by regulating reactive gliosis (Del Cerro et al., 1996; Garcia-Estrada et al. 1993, 1999) and the expression of molecules in reactive astroglia that are part of the response of astrocytes to injury. For instance, DHEA inhibits production of tumor necrosis factor alpha and interleukin-6 in astrocytes (Kipper-Galperin et al., 1999). Estrogen down-regulates the expression of bFGF in astrocytes (Flores et al., 1999) and increases the expression of apolipoprotein E (ApoE), a molecule involved in neuroregulation after injury, in astrocytes and microglia (Stone et al., 1997). A very interesting finding with implications for reactive astroglia is that estradiol in cultured astrocytes reduces activation of NF-kappaB induced by amyloid A beta (1-40) and lipopolysacccahride (LPS) (Dodel et al., 1999). Since NF-kappaB is a potent immediate-early transcriptional regulator of numerous pro-inflammatory genes, the hormonal regulation of this molecule in astrocytes may play a crucial role in the neuroprotective effects of estrogen.

Effects of steroids on astrogliosis may be a contributing factor to neural regeneration. This is suggested by studies of the effects of the estradiol precursor testosterone on the regulation of the central astrocytic response to peripheral nerve injury. In adult male hamsters, testosterone propionate administration reduces the increase in GFAP mRNA in the facial nucleus after facial nerve axotomy (Jones et al., 1997b, 1999), attenuates glial-mediated synaptic stripping of axotomized motoneurons (Jones et al., 1997a, 1999) and increases facial nerve regeneration (Kujawa et al., 1991). The relationship between reduced astrogliosis in the facial motor nucleus and increased axonal regeneration is still unknown. However, the results strongly suggest that hormonal regulation of astrogliosis may contribute to the regenerative mechanisms of testosterone and estradiol on facial motoneurons. These effects may also explain why estradiol decreases GFAP expression and promotes neurite outgrowth in the hippocampus after deafferenting lesion of the entorhinal cortex (Rozovsky et al., 2002), rather than causing the increase in GFAP seen in unlesioned animals (see above).

Other effects of steroids on astrocytes may also be relevant under neuropathological conditions. For instance, glucocorticoids may be detrimental to reactive astrocytes, by mechanisms involving depletion of intracellular ATP levels and deterioration of mitochondrial transmembrane potentials (Shin et al., 2001). In contrast, estradiol increases the expression of heat shock proteins in astrocytes (Mydlarski et al., 1995), an effect that has also been observed in striatal astrocytes after global ischemia in gerbils (Lu et al., 2002), and may be related to the protective effects of the hormone in animal models of brain ischemia. Furthermore, estradiol increases glutamate uptake in astrocytes derived from Alzheimer's patients (Liang et al., 2002), which may contribute to the potential protective hormonal effect against this neurodegenerative disease, where the extracellular glutamate concentration appears to be increased (see chapter by Barger).

3.4. Effects of steroids on microglia

The well known anti-inflammatory effects of steroids in peripheral tissues have been also observed in the brain. Steroids exert their anti-inflammatory actions in the brain acting on non-neuronal cells, mainly on microglia. Microglia are, as macrophages, target of steroids and both cell types show similar anti-inflammatory responses to glucocorticoids. For instance, cortisol represses LPS induction of nitric oxide production in primary cultured microglia and transformed N9 microglial cells (Drew and Chavis, 2000a). Therefore, although cortisol may be directly toxic to neurons, it may indirectly protect neurons by blocking the production of cytotoxic molecules by microglia (Drew and Chavis, 2000a). Other steroids, such as DHEA and progesterone, also inhibit nitric oxide production in primary cultures of microglia in response to LPS (Barger et al., 2000; Drew and Chavis, 2000b). The effect of DHEA may be related to the neuroprotective effects of this steroid. Furthermore, Drew and Chavis (2000b) have proposed that the progesterone-mediated inhibition of microglial cell activation may contribute to the decreased severity of multiple sclerosis symptoms commonly observed during with pregnancy.

Several studies have analyzed the effect of estradiol on microglia, in search for a basis for the neuroprotective effects of the hormone. Stone et al. (1997) showed that estradiol enhances ApoE secretion by microglia in vivo. Vegeto et al. (1999) reported that estradiol inhibits apoptosis in microglia cultures by a receptor-mediated enhancement of Nip2 protein production. Subsequent studies in microglia cultures have shown that estradiol inhibits the induction of inducible nitric oxide synthase, and the consequent production of nitric oxide, in response to LPS and to the pro-inflammatory cytokines interferon-γ and TNF-α (Bruce-Keller et al., 2000; Drew and Chavis, 2000b; Vegeto et al., 2001). Furthermore, Vegeto et al. (2001) have found that estradiol also reduces LPS-induced production in microglia of other inflammatory mediators, such as PGE2, and metalloproteinase-9. Estradiol is also able to enhance uptake of amyloid beta-protein (Aβ) by microglia derived from the human cortex (Li et al., 2000), an effect that may be relevant for the protective effect of this hormone against Alzheimer's disease. Bruce-Keller et al. (2000, 2001) have studied the mechanism involved in the anti-inflammatory actions of estradiol on microglia. These authors have found that estrogen receptor-dependent activation of MAP kinase is involved in the hormonal action. This opens the possibility for a co-ordinated regulation of microglia by estradiol and growth factors via activation of the MAP kinase signaling pathway.

3.5. Effects of steroids on oligodendrocytes and Schwann cells

The effects of steroid hormones on central myelin and oligodendroglia have not been sufficiently explored. Very little is known about effects of gonadal hormones on oligodendroglia. Glucocorticoids potentiate oligodendrocyte differentiation and myelinogenesis during neonatal development (see Garcia-Segura et al., 1996, for review).

However, these hormones inhibit the proliferation of oligodendrocyte precursors located throughout the white and gray matter regions of the adult rat brain. Since the proliferation of oligodendrocyte precursors plays a major role during remyelination, these data raise the question of possible detrimental effects of therapeutic treatments of CNS trauma in adult individuals by administration of glucocorticoids (Alonso, 2000). Furthermore, dihydrocorticosterone decreases the expression of myelin basic protein in oligodendrocytes (Melcangi et al., 1997), and it has been reported that prenatal corticosteroid administration delays myelination of the corpus callosum in fetal sheep (Huang et al., 2001).

During the last few years, several studies have been performed to evaluate possible effects of steroid hormones on physiological parameters of Schwann cells. For instance, it has been shown that dexamethasone, a synthetic glucocorticoid, is able to strongly enhance the mitogenic activity exerted by an axolemma-enriched fraction in Schwann cell cultures (Neuberger et al., 1994). In the same experimental model, no co-mitogenic action was exerted by progesterone, testosterone, estradiol, or aldosterone (Neuberger et al., 1994). In partial disagreement with these observations, it has been shown that estrogens are able to promote Schwann cell proliferation in culture in the presence of agents elevating intracellular cAMP (Jung-Testas et al., 1993). The effects of estrogens and progesterone on the proliferation of Schwann cells have also been studied in cultures of segments of the rat sciatic nerve obtained from adult or newborn male and female rats (Svenningsen and Kanje, 1999). In this experimental model, these two sex steroids are able to enhance [^{3}H] thymidine incorporation into Schwann cells, an effect that was blocked by their respective receptor antagonists. However, it is important to underline that this effect depends on the sex of the animals. Thus, estrogens are effective on Schwann cell proliferation only in segments from male rats, while progesterone increases Schwann cell proliferation only in segments obtained from females. An effect of sex steroids is also evident on Schwann cells that cover motoneuron terminals. As shown by Lubischer and Bebinger (1999), the number of this particular type of non-myelinating Schwann cells present in an androgen-sensitive muscle of the rat (i.e., the levator ani) decrease after castration. The decline is reversed by testosterone treatment. This steroid effect seems to be indirect, since Schwann cells do not express AR (Magnaghi et al., 1999), and the involvement of motoneurons and/or muscle fibers expressing AR might be hypothesized (Jordan et al., 1997; Jordan and Williams, 2001).

One of the major products of Schwann cells is myelin. Consequently, several studies have been performed both in vivo (i.e., in the whole sciatic nerve of rat) and in vitro (i.e., in cultures of rat Schwann cells) to evaluate whether steroid hormones and/or neuroactive steroids might influence this cellular component. In particular, the attention of different laboratories has been directed towards the proteins proper of the peripheral myelin, like the glycoprotein Po (Po) and peripheral myelin protein 22 (PMP22) (for review, see Melcangi et al., 2000b; Magnaghi et al., 2001). Data obtained so far strongly suggest an important role of steroid hormones in modulating the expression of these two crucial myelin proteins. In particular, progesterone and its physiological 5α- and 3α-5α-reduced derivatives, dihydroprogesterone and tetrahydroprogesterone, stimulate the expression of Po in intact or transected sciatic nerve of male rats. These effects are also evident in rat Schwann cell cultures, indicating a direct effects of these steroids on the cells producing Po (Désarnaud et al., 1998; Notterpek et al., 1999; Melcangi et al., 2000a,b;

Magnaghi et al., 2001). In the same models, the gene expression of PMP22 is influenced only by tetrahydroprogesterone (Melcangi et al., 1999b). However, at variance with these findings, other observations indicate that the expression of PMP22 may also be stimulated by progesterone itself. In fact, utilizing Schwann cells transiently transfected with a reporter construct in which the expression of the luciferase is controlled by the promoter region of the PMP22 gene, Désarnaud and co-workers (1998) showed that progesterone stimulates the gene expression of this myelin protein, acting on promoter 1, but not on promoter 2 of the PMP22 gene.

It has been proposed that the effects of progesterone, dihydroprogesterone and tetrahydroprogesterone (after retro-conversion into dihydroprogesterone) on gene expression of protein Po are linked to an interaction with the PR, which is present in Schwann cells (Magnaghi et al., 1999), while effects of tetrahydroprogesterone on protein PMP22 appear to be due to an interaction of this steroid with the $GABA_A$ receptor, whose subunits have been found in Schwann cells (Melcangi et al., 1999b). This hypothesis is based on the findings that, in primary cultures of Schwann cells isolated from neonatal rat sciatic nerves: (i) mifepristone (RU38486), an antagonist at the PR, is able to block the effects of progesterone, dihydroprogesterone and tetrahydroprogesterone, on the expression of protein Po; and (ii) the effect of tetrahydroprogesterone on PMP22 is blocked by bicuculline, a classical antagonist of the $GABA_A$ receptor (Magnaghi et al., 2001). The conclusion that the expression of Po seems to be under the control of the PR, while that of PMP22 is under $GABA_A$ receptor influence is further supported by the findings that the $GABA_A$ receptor agonist muscimol does not increase the mRNA levels of protein Po, while it increases the expression of protein PMP22 (for review, see Magnaghi et al., 2001).

An effect of progesterone and its derivatives on Po gene expression through the PR might suggest a genomic mechanism, in which the complex ligand–receptor interacts with steroid responsive elements located in the promoter region of Po. In agreement with this hypothesis, a computer analysis has demonstrated that putative progesterone responsive elements are present on the Po promoter (Magnaghi et al., 1999).

It is also interesting to recall that progesterone stimulates gene expression of the transcription factor Krox-20, which plays an important role in myelination of peripheral nerves, as well as of other transcription factors present in Schwann cells (for review, see Schumacher et al., 2001). In particular, it has been observed that progesterone induces a rapid increase in the gene expression of Krox-20, Krox-24, Egr-3 and Fos B (Guennoun et al., 2001; Mercier et al., 2001), suggesting that this steroid hormone might also co-ordinate the signaling pathways involved in the initiation of myelination.

The possible involvement of the $GABA_A$ receptor in the control of the expression of PMP22 is also evident in experiments, in which rat Schwann cells were exposed to testosterone, dihydrotestosterone and 3α-diol (see Fig. 2). In fact, in this case only 3α-diol was able to influence PMP22 expression (Melcangi et al., 2000a). It is interesting to note that this steroid, which does not interact with the AR, is able to interact with the $GABA_A$ receptor (Frye et al., 1996a,b). On the other hand, it is important to mention that androgen molecules themselves are not completely ineffective on Po expression. Removal of circulating androgens by castration decreases mRNA levels of Po in the sciatic nerve, a phenomenon, which is counteracted by subsequent treatment with dihydrotestosterone

(Magnaghi et al., 1999). However, since Schwann cells as previously mentioned do not express the AR (Magnaghi et al., 1999), it has been hypothesized that the gene expression of Po might be stimulated by androgen-dependent mechanisms acting on Schwann cells through the adjacent neuronal component, which seems to express AR (Magnaghi et al., 1999).

Finally, also glucocorticoids seem to be able to stimulate the transcription from Po and PMP22 promoters in Schwann cells. As shown by Désarnaud et al. (2000), dexamethasone and corticosterone are able to stimulate Po expression and both promoters of the PMP22 gene. Since Po and PMP22 play an important physiological role in the maintenance of the multilamellar structure of peripheral myelin (for review, see Melcangi et al., 2000a; Magnaghi et al., 2001), these observations might suggest the possible utilization of neuroactive steroids after peripheral injury, during aging or in particular in demyelinating diseases (e.g., Charcot-Marie-Tooth type 1a and 1b, Déjérine-Sottas syndrome), in which rebuilding of myelin is needed.

4. Concluding remarks

This chapter has reviewed recent information indicating that non-neuronal cells of the nervous system synthesize and metabolize steroids and are targets for hormonal and neuroactive steroids. We have noted that glia affect neuronal function by producing steroids and metabolizing hormonal steroids into neuroactive steroids. On the other hand, endothelial cells and glia are targets for hormonal and neuroactive steroids and mediate some of the effects of these molecules in the brain. However, there is still an important gap in our knowledge of all the implications and mechanisms involved in the communication between non-neuronal and neuronal cells via steroids. In some cases, there is evidence that steroids produced by glia exert specific functions on adjacent neuronal constituents. The best example is the production of progesterone and its derivatives by Schwann cells and their involvement in peripheral nerve remyelination and regeneration. This is also the case of the neuroprotective actions of estradiol produced by reactive astroglia. There is also considerable evidence for specific consequences of steroid actions on glia. For instance, steroids modulate synaptic plasticity and neuroendocrine secretion by an action on astroglia, and they regulate the response of nerve tissue to injury by an action on microglia. However, there are still many unsolved questions and unexplored areas in relation to steroids and non-neuronal cells. For instance, further studies are necessary to determine to what extent steroids produced by glia affect brain development and CNS myelination, whether physiological or pharmacological modifications of glial steroidogenesis affect mood, cognitive function and behavior, and whether the production of neuroactive steroids by glia is a general mechanism for regulation of synaptic function. The resolution of these unsolved questions may broaden our perception of the manners in which integration of neuronal and non-neuronal mechanisms regulate nervous system function. In addition, understanding of steroid-mediated communication between neurons and non-neuronal cells in the nervous system has obvious implications for the treatment of neurodegenerative diseases and psychiatric disorders.

Acknowledgements

We want to acknowledge financial support from the Commission of the European Communities, specific RTD programme "Quality of Life and Management of Living Resources", QLK6-CT-2000-00179.

References

Akwa, Y., Sananès, N., Gouézou, M., Robel, P., Baulieu, E.E., Le Goascogne, C., 1993. Astrocytes and neurosteroids: metabolism of pregnenolone and dehydroepiandrosterone. Regulation by cell density. J. Cell Biol. 121, 135–143.

Alonso, G., 2000. Prolonged corticosterone treatment of adult rats inhibits the proliferation of oligodendrocyte progenitors present throughout white and gray matter regions of the brain. Glia 31, 219–231.

Azcoitia, I., Garcia-Ovejero, D., Chowen, J.A., Garcia-Segura, L.M., 2001a. Astroglia play a key role in the neuroprotective actions of estrogen. Prog. Brain Res. 132, 469–478.

Azcoitia, I., Sierra, A., Garcia-Segura, L.M., 1999. Localization of estrogen receptor beta-immunoreactivity in astrocytes of the adult rat brain. Glia 26, 260–267.

Azcoitia, I., Sierra, A., Veiga, S., Honda, S., Harada, N., Garcia-Segura, L.M., 2001b. Brain aromatase is neuroprotective. J. Neurobiol. 47, 318–329.

Barbaccia, M.L., Serra, M., Purdy, R.H., Biggio, G., 2001. Stress and neuroactive steroids. Int. Rev. Neurobiol. 46, 243–272.

Barger, S.W., Chavis, J.A., Drew, P.D., 2000. Dehydroepiandrosterone inhibits microglial nitric oxide production in a stimulus-specific manner. J. Neurosci. Res. 62, 503–509.

Bastianetto, S., Ramassamy, C., Poirier, J., Quirion, R., 1999. Dehydroepiandrosterone (DHEA) protects hippocampal cells from oxidative stress-induced damage. Mol. Brain Res. 66, 35–41.

Baulieu, E.E., 1998. Neurosteroids: a novel function of the brain. Psychoneuroendocrinology 23, 963–987.

Blurton-Jones, M., Tuszynski, M.H., 2001. Reactive astrocytes express estrogen receptors in the injured primate brain. J. Comp. Neurol. 433, 115–123.

Bruce-Keller, A.J., Barger, S.W., Moss, N.I., Pham, J.T., Keller, J.N., Nath, A., 2001. Pro-inflammatory and pro-oxidant properties of the HIV protein Tat in a microglial cell line: attenuation by 17 beta-estradiol. J. Neurochem. 78, 1315–1324.

Bruce-Keller, A.J., Keeling, J.L., Keller, J.N., Huang, F.F., Camondola, S., Mattson, M.P., 2000. Anti-inflammatory effects of estrogen on microglial activation. Endocrinology 141, 3646–3656.

Buchanan, C.D., Mahesh, V.B., Brann, D.W., 2000. Estrogen–astrocyte-luteinizing hormone–releasing hormone signaling: a role for transforming growth factor β1. Biol. Reprod. 62, 1710–1721.

Cambiasso, M.J., Colombo, J.A., Carrer, H.F., 2000. Differential effect of oestradiol and astroglia-conditioned media on the growth of hypothalamic neurons from male and female rat brains. Eur. J. Neurosci. 12, 2291–2298.

Cardona-Gomez, G.P., Chowen, J.A., Garcia-Segura, L.M., 2000a. Estradiol and progesterone regulate the expression of insulin-like growth factor-I and insulin-like growth factor binding protein-2 in the hypothalamus of adult female rats. J. Neurobiol. 43, 269–281.

Cardona-Gomez, G.P., DonCarlos, L., Garcia-Segura, L.M., 2000b. Insulin-like growth factor I receptors and estrogen receptors colocalize in female rat brain. Neuroscience 99, 751–760.

Cardounel, A., Regelson, W., Kalimi, M., 1999. Dehydroepiandrosterone protects hippocampal neurons against neurotoxin-induced cell death: mechanism of action. Proc. Soc. Exp. Biol. Med. 222, 145–149.

Chowen, J.A., Busiguina, S., Garcia-Segura, L.M., 1995. Sexual dimorphism and sex steroid modulation of glial fibrillary acidic protein (GFAP) mRNA and immunoreactive levels in the rat hypothalamus. Neuroscience 69, 519–532.

Compagnone, N.A., Mellon, S.H., 2000. Neurosteroids: biosynthesis and function of these novel neuromodulators. Front. Neuroendocrinol. 21, 1–56.

Day, J.R., Laping, N.J., McNeill, T.H., Schreiber, S.S., Pasinetti, G., Finch, C.E., 1990. Castration enhances expression of glial fibrillary acidic protein and sulfated glycoprotein-2 in the intact and lesion-altered hippocampus of the adult male rat. Mol. Endocrinol. 4, 1995–2002.

Day, J.R., Laping, N.J., Lampert-Etchells, M., Brown, S.A., O'Callaghan, J.P., McNeill, T.H., Finch, C.E., 1993. Gonadal steroids regulate the expression of glial fibrillary acidic protein in the adult male rat hippocampus. Neuroscience 55, 435–443.

Del Cerro, S., Garcia-Estrada, J., Garcia-Segura, L.M., 1996. Neurosteroids modulate the reaction of astroglia to high extracellular potassium levels. Glia 18, 293–305.

Désarnaud, F., Bidichandani, S., Patel, P.I., Baulieu, E.E., Schumacher, M., 2000. Glucocorticosteroids stimulate the activity of the promoters of peripheral myelin protein-22 and protein zero genes in Schwann cells. Brain Res. 865, 12–16.

Désarnaud, F., Do Thi, A.N., Brown, A.M., Lemke, G., Schumacher, M., Baulieu, E.E., Schumacher, M., 1998. Progesterone stimulates the activity of the promoters of peripheral myelin protein-22 and protein zero genes in Schwann cells. J. Neurochem. 71, 1765–1768.

Dodel, R.C., Du, Y., Bales, K.R., Gao, F., Paul, S.M., 1999. Sodium salicylate and 17beta-estradiol attenuate nuclear transcription factor NF-kappaB translocation in cultured rat astroglial cultures following exposure to amyloid A beta(1-40) and lipopolysaccharides. J. Neurochem. 73, 1453–1460.

Drew, P.D., Chavis, J.A., 2000a. Inhibition of microglial cell activation by cortisol. Brain Res. Bull. 52, 391–396.

Drew, P.D., Chavis, J.A., 2000b. Female sex steroids: effects upon microglial cell activation. J. Neuroimmunol. 111, 77–85.

Engel, S.R., Grant, K.A., 2001. Neurosteroids and behavior. Int. Rev. Neurobiol. 46, 321–348.

Finley, S.K., Kritzer, M.F., 1999. Immunoreactivity for intracellular androgen receptors in identified subpopulations of neurons, astrocytes and oligodendrocytes in primate prefrontal cortex. J. Neurobiol. 40, 446–457.

Flores, C., Salmaso, N., Cain, S., Rodaros, D., Stewart, J., 1999. Ovariectomy of adult rats leads to increased expression of astrocytic basic fibroblast growth factor in the ventral tegmental area and in dopaminergic projection regions of the entorhinal and prefrontal cortex. J. Neurosci. 19, 8665–8673.

Forlano, P.M., Deitcher, D.L., Myers, D.A., Bass, A.H., 2001. Anatomical distribution and cellular basis for high levels of aromatase activity in the brain of teleost fish: aromatase enzyme and mRNA expression identify glia as source. J. Neurosci. 21, 8943–8955.

Frye, C.A., Duncan, J.E., Basham, M., Erkine, M.S., 1996a. Behavioral effects of 3alpha-androstanediol. II: hypothalamic and preoptic area actions via a GABAergic mechanism. Behav. Brain Res. 79, 119–130.

Frye, C.A., Van Keuren, K.R., Erkine, M.S., 1996b. Behavioral effects of 3alpha-androstanediol. I: modulation of sexual receptivity and promotion of GABA-stimulated chloride flux. Behav. Brain Res. 79, 109–118.

Gago, N., Akwa, Y., Sananes, N., Guennoun, R., Baulieu, E.E., El-Etr, M., Schumacher, M., 2001. Progesterone and the oligodendroglial lineage: stage-dependent biosynthesis and metabolism. Glia 36, 295–308.

Galbiati, M., Martini, L., Melcangi, R.C., 2002. Oestrogens, via transforming growth factor a, modulate basic fibroblast growth factor synthesis in hypothalamic astrocytes: in vitro observations. J. Neuroendocrinol. 14, 829–835.

Galea, E., Santizo, R., Feinstein, D.L., Adamsom, P., Greenwood, J., Koenig, H.M., Pelligrino, D.A., 2002. Estrogen inhibits NFkappaB-dependent inflammationin brain endothelium without interfering withIkappaB degradation. Neuroreport 13, 1469–1472.

Garcia-Estrada, J., Del Rio, J.A., Luquin, S., Soriano, E., Garcia-Segura, L.M., 1993. Gonadal hormones down-regulate reactive gliosis and astrocyte proliferation after a penetrating brain injury. Brain Res. 628, 271–278.

Garcia-Estrada, J., Luquin, S., Fernandez, A., Garcia-Segura, L.M., 1999. Dehydroepiandrosterone, pregnenolone and sex steroids down-regulate reactive astroglia in the male rat brain after a penetrating brain injury. Int. J. Dev. Neurosci. 17, 145–151.

Garcia-Ovejero, D., Veiga, S., Garcia-Segura, L.M., DonCarlos, L.L., 2002. Glial expression of estrogen and androgen receptors after rat brain injury. J. Comp. Neurol. 450, 256–271.

Garcia-Segura, L.M., Azcoitia, I., DonCarlos, L.L., 2001. Neuroprotection by estradiol. Prog. Neurobiol. 63, 29–60.

Garcia-Segura, L.M., Cañas, B., Parducz, A., Rougon, G., Theodosis, D., Naftolin, F., Torres-Aleman, I., 1995a. Estradiol promotion of astroglial cell shape changes depends on the expression of polysisalic acid on neuronal membranes. Glia 13, 209–216.

Garcia-Segura, L.M., Chowen, J.A., Duenas, M., Torres-Aleman, I., Naftolin, F., 1994a. Gonadal steroids as promoters of neuro-glial plasticity. Psychoneuroendocrinology 19, 445–453.

Garcia-Segura, L.M., Chowen, J.A., Naftolin, F., 1996. Endocrine glia: roles of glial cells in the brain actions of steroid and thyroid hormones and in the regulation of hormone secretion. Front. Neuroendocrinol. 17, 180–211.

Garcia-Segura, L.M., Chowen, J.A., Parducz, A., Naftolin, F., 1994b. Gonadal hormones as promoters of structural synaptic plasticity: cellular mechanisms. Prog. Neurobiol. 44, 279–307.

Garcia-Segura, L.M., Dueñas, M., Busiguina, S., Naftolin, F., Chowen, J.A., 1995b. Gonadal hormone regulation of neuronal–glial interactions in the developing neuroendocrine hypothalamus. J. Steroid Biochem. Mol. Biol. 53, 293–298.

Garcia-Segura, L.M., Luquin, S., Parducz, A., Naftolin, F., 1994c. Gonadal hormone regulation of glial fibrillary acidic protein immunoreactivity and glial ultrastructure in the rat neuroendocrine hypothalamus. Glia 10, 59–69.

Garcia-Segura, L.M., Naftolin, F., Hutchison, J.B., Azcoitia, I., Chowen, J.A., 1999a. Role of astroglia in estrogen regulation of synaptic plasticity and brain repair. J. Neurobiol. 40, 574–584.

Garcia-Segura, L.M., Wozniak, A., Azcoitia, I., Rodriguez, J.R., Hutchison, R.E., Hutchison, J.B., 1999b. Aromatase expression by astrocytes after brain injury: implications for local estrogen formation in brain repair. Neuroscience 89, 567–578.

Green, P.S., Simpkins, J.W., 2000. Neuroprotective effects of estrogens: potential mechanisms of action. Int. J. Dev. Neurosci. 18, 347–358.

Gudino-Cabrera, G., Nieto-Sampedro, M., 1999. Estrogen receptor immunoreactivity in Schwann-like brain macroglia. J. Neurobiol. 40, 458–470.

Guennoun, R., Benmessahel, Y., Delespierre, B., Gouezou, M., Rajkowski, K.M., Baulieu, E.E., Schumacher, M., 2001. Progesterone stimulates Krox-20 gene expression in Schwann cells. Mol. Brain Res. 90, 75–82.

Guennoun, R., Schumacher, M., Robert, F., Delespierre, B., Gouezou, M., Eychenne, B., Akwa, Y., Robel, P., Baulieu, E.E., 1997. Neurosteroids: expression of funtional 3β-hydroxysteroid dehydrogenase by rat sensory neurons and Schwann cells. Eur. J. Neurosci. 9, 2236–2247.

Hardin-Pouzet, H., Giraudon, P., Belin, M.F., Didier-Bazes, M., 1996. Glucocorticoid upregulation of glutamate dehydrogenase gene expression in vitro in astrocytes. Mol. Brain Res. 37, 324–328.

Harkness, K.A., Adamson, P., Sussman, J.D., Davies-Jones, G.A., Greenwood, J., Woodroofe, M.N., 2000. Dexamethasone regulation of matrix metalloproteinase expression in CNS vascular endothelium. Brain 123, 698–709.

Hösli, E., Jurasin, K., Ruhl, W., Luthy, R., Hösli, L., 2001. Colocalization of androgen, estrogen and cholinergic receptors on cultured astrocytes of rat central nervous system. Int. J. Dev. Neurosci. 19, 11–19.

Hu, Z.Y., Bourreau, E., Jung-Testas, E., Robel, P., Baulieu, E.E., 1987. Oligodendrocytes mitochondria convert cholesterol to pregnenolone. Proc. Natl Acad. Sci. USA 84, 8215–8229.

Huang, W.L., Harper, C.G., Evans, S.F., Newnham, J.P., Dunlop, S.A., 2001. Repeated prenatal corticosteroid administration delays myelination of the corpus callosum in fetal sheep. Int. J. Dev. Neurosci. 19, 415–425.

Jones, K.J., Coers, S., Storer, P.D., Tanzer, L., Kinderman, N.B., 1999. Androgenic regulation of the central glia response following nerve damage. J. Neurobiol. 40, 560–573.

Jones, K.J., Durica, T.E., Jacob, S.K., 1997a. Gonadal steroid preservation of central synaptic input to hamster facial motoneurons following peripheral axotomy. J. Neurocytol. 26, 257–266.

Jones, K.J., Kinderman, N.B., Oblinger, M.M., 1997b. Alterations in glial fibrillary acidic protein (GFAP) mRNA levels in the hamster facial motor nucleus: effects of axotomy and testosterone. Neurochem. Res. 22, 1359–1366.

Jordan, C.L., 1999. Glia as mediators of steroid hormone action on the nervous system: an overview. J. Neurobiol. 40, 434–445.

Jordan, C.L., Padgett, B., Hershey, J., Prins, G., Arnold, A., 1997. Ontogeny of androgen receptor immunoreactivity in lumber motoneurons and in the sexually dimorphic levator ani muscle of male rats. J. Comp. Neurol. 379, 88–98.

Jordan, C.L., Williams, T.J., 2001. Testosterone regulates terminal Schwann cell number and junctional size during developmental synapse elimination. Dev. Neurosci. 23, 441–451.

Jung-Testas, I., Renoir, M., Bugnard, H., Greene, G.L., Baulieu, E.E., 1992. Demonstration of steroid hormone receptors and steroid action in primary cultures of rat glial cells. J. Steroid Biochem. Mol. Biol. 41, 621–631.

Jung-Testas, I., Schumacher, M., Bugnard, H., Baulieu, E.E., 1993. Stimulation of rat Schwann cell proliferation by estradiol: synergism between estrogen and cAMP. Dev. Brain Res. 72, 282–290.

Jung-Testas, I., Schumacher, M., Robel, P., Baulieu, E.E., 1996. Demonstration of progesterone receptors in rat Schwann cells. J. Steroid Biochem. Mol. Biol. 58, 77–82.

Karlstedt, K., Sallmen, T., Eriksson, K.S., Lintunen, M., Couraud, P.O., Joo, F., Panula, P., 1999. Lack of histamine synthesis and down-regulation of H1 and H2 receptor mRNA levels by dexamethasone in cerebral endothelial cells. J Cereb. Blood Flow Metab. 19, 321–330.

Kawaja, M.D., Gage, F.H., 1991. Reactive astrocytes are substrates for the growth of adult CNS axons in the presence of elevated levels of nerve growth factor. Neuron 7, 1019–1030.

Kimonides, V.G., Khatibi, N.H., Svendsen, C.N., Sofroniew, M.V., Herbert, J., 1998. Dehydroepiandrosterone (DHEA) and DHEA-sulfate (DHEAS) protect hippocampal neurons against excitatory amino acid-induced neurotoxicity. Proc. Natl Acad. Sci. USA 95, 1852–1857.

Kimoto, T., Asou, H., Ohta, Y., Mukai, H., Chernogolov, A.A., Kawato, S., 1997. Digital fluorescence imaging of elementary steps of neurosteroid synthesis in rat brain glial cells. J. Pharm. Biomed. Anal. 15, 1231–1240.

Kipper-Galperin, M., Galilly, R., Danenberg, H.D., Brenner, T., 1999. Dehydroepiandrosterone selectively inhibits production of tumor necrosis factor alpha and interleukin-6 in astrocytes. Int. J. Dev. Neurosci. 17, 765–775.

Koenig, H.L., Schumacher, M., Ferzas, B., Do Thi, A.N., Ressouches, A., Guennoun, R., Jung-Testas, I., Robel, P., Akwa, Y., Baulieu, E.E., 1995. Progesterone synthesis and myelin formation by Schwann cells. Science 268, 1500–1503.

Kujawa, K.A., Emeric, E., Jones, K.J., 1991. Testosterone differentially regulates the regenerative properties of injured hamster facial motoneurons. J. Neurosci. 11, 3898–3906.

Lambert, J.J., Harney, S.C., Belelli, D., Peters, J.A., 2001. Neurosteroid modulation of recombinant and synaptic GABA A receptors. Int. Rev. Neurobiol. 46, 177–205.

Langub, M.C., Watson, R.E., 1992. Estrogen receptor-immunoreactive glia, endothelia, and ependyma in guinea pig preoptic area and median eminence: electron microscopy. Endocrinology 130, 364–372.

Laping, N.J., Nichols, N.R., Day, J.R., Finch, C.E., 1991. Corticosterone differentially regulates the bilateral response of astrocyte mRNA in the hippocampus to entorhinal cortex lesions in male rats. Mol. Brain Res. 10, 291–297.

Laping, N.J., Teter, B., Nichols, N.R., Rozovsky, I., Finch, C.E., 1994. Glial fibrillary acidic protein: regulation by hormones, cytokines, and growth factors. Brain Pathol. 1, 259–275.

Legrand, A., Alonso, G., 1998. Pregnenolone reverses the age-dependent accumulation of glial fibrillary acidic protein within astrocytes of specific regions of the rat brain. Brain Res. 802, 125–133.

Li, R., Shen, Y., Yang, L.B., Lue, L.F., Finch, C., Rogers, J., 2000. Estrogen enhances uptake of amyloid beta-protein by microglia derived from the human cortex. J. Neurochem. 75, 1447–1454.

Liang, Z., Valla, J., Sefidvash-Hockley, S., Rogers, J., Li, R., 2002. Effects of estrogen treatment on glutamate uptake in cultured human astrocytes derived from cortex of Alzheimer's disease patients. J. Neurochem. 80, 807–814.

Lu, A., Ran, R.Q., Clark, J., Reilly, M., Nee, A., Sharp, F.R., 2002. 17-beta-estradiol induces heat shock proteins in brain arteries and potentiates ischemic heat shock protein induction in glia and neurons. J. Cereb. Blood Flow Metab. 22, 183–195.

Lubischer, J.L., Bebinger, D.M., 1999. Regulation of terminal Schwann cell number at the adult neuromuscular junction. J. Neurosci. 19(RC46), 1–5.

Luquin, S., Naftolin, F., Garcia-Segura, L.M., 1993. Natural fluctuation and gonadal hormone regulation of astrocyte immunoreactivity in dentate gyrus. J. Neurobiol. 24, 913–924.

Ma, Y.J., Berg-von der Emde, K., Moholt-Siebert, M., Hill, D.F., Ojeda, S.R., 1994. Region specific regulation of transforming growth factor alpha (TGFα) gene expression in astrocytes of the neuroendocrine brain. J. Neurosci. 14, 5644–5651.

Magnaghi, V., Cavarretta, I., Galbiati, M., Martini, L., Melcangi, R.C., 2001. Neuroactive steroids and peripheral myelin proteins. Brain Res. Rev. 37, 360–371.

Magnaghi, V., Cavarretta, I., Zucchi, I., Susani, L., Rupprecht, R., Hermann, B., Martini, L., Melcangi, R.C., 1999. Po gene expression is modulated by androgens in the sciatic nerve of adult male rats. Mol. Brain Res. 70, 36–44.

Magnaghi, V., Riva, M.A., Cavarretta, I., Martini, L., Melcangi, R.C., 2000. Corticosteroids regulate the gene expression of FGF-1 and FGF-2 in cultured rat astrocytes. J. Mol. Neurosci. 15, 11–18.

McCarthy, J.B., Barker-Gibb, A.L., Alves, S.E., Milner, T.A., 2002. TrkA immunoreactive astrocytes in dendritic fields of the hippocampal formation across estrous. Glia 38, 36–44.

McNeill, A.M., Zhang, C., Stanczyk, F.Z., Duckles, S.P., Krause, D.N., 2002. Estrogen increases endothelial nitric oxide synthase via estrogen receptors in rat cerebral blood vessels: effect preserved after concurrent treatment with medroxyprogesterone acetate or progesterone. Stroke 33, 1685–1691.

Melcangi, R.C., Cavarretta, I., Magnaghi, V., Martini, L., Galbiati, M., 2001a. Interactions between growth factors and steroids in the control of LHRH neurons. Brain Res. Rev. 37, 223–234.

Melcangi, R.C., Celotti, F., Ballabio, M., Castano, P., Massarelli, R., Poletti, A., Martini, L., 1990. 5α-reductase activity in isolated and cultured neuronal and glial cells of the rat. Brain Res. 516, 229–236.

Melcangi, R.C., Celotti, F., Castano, P., Martini, L., 1993. Differential localization of the 5α-reductase and the 3α-hydroxysteroid dehydrogenase in neuronal and glial cultures. Endocrinology 132, 1252–1259.

Melcangi, R.C., Celotti, F., Martini, L., 1994a. Progesterone 5α-reduction in neurons, astrocytes and oligodendrocytes. Brain Res. 639, 202–206.

Melcangi, R.C., Celotti, F., Martini, L., 1994b. Neurons influence the metabolism of testosterone in cultured astrocytes via humoral signals. Endocrine 2, 709–713.

Melcangi, R.C., Froelichsthal, P., Martini, L., Vescovi, A.L., 1996a. Steroid metabolizing enzymes in pluripotential progenitor CNS cells: effect of differentiation and maturation. Neuroscience 72, 467–475.

Melcangi, R.C., Magnaghi, V., Cavarretta, I., Martini, L., Piva, F., 1998. Age-induced decrease of glycoprotein Po and Myelin Basic Protein gene expression in the rat sciatic nerve. Repair by steroid derivatives. Neuroscience 85, 569–578.

Melcangi, R.C., Magnaghi, V., Cavarretta, I., Riva, M.A., Martini, L., 1997. Corticosteroid effects on gene expression of myelin basic protein in oligodendrocytes and of glial fibrillary acidic protein in type 1 astrocytes. J. Neuroendocrinol. 9, 729–733.

Melcangi, R.C., Magnaghi, V., Cavarretta, I., Zucchi, I., Bovolin, P., D'Urso, D., Martini, L., 1999b. Progesterone derivatives are able to influence peripheral myelin protein 22 and Po gene expression: possible mechanisms of action. J. Neurosci. Res. 56, 349–357.

Melcangi, R.C., Magnaghi, V., Galbiati, M., Ghelarducci, B., Sebastiani, L., Martini, L., 2000b. The action of steroid hormones on peripheral myelin proteins: a possible new tool for the rebuilding of myelin? J. Neurocytol. 29, 327–339.

Melcangi, R.C., Magnaghi, V., Galbiati, M., Martini, L., 2001b. Glial cells: a target for steroid hormones. Prog. Brain Res. 132, 31–40.

Melcangi, R.C., Magnaghi, V., Galbiati, M., Martini, L., 2001c. Formation and effects of neuroactive steroids in the central and peripheral nervous system. Int. Rev. Neurobiol. 46, 145–176.

Melcangi, R.C., Magnaghi, V., Martini, L., 1999a. Steroid metabolism and effects in central and peripheral glial cells. J. Neurobiol. 40, 471–483.

Melcangi, R.C., Magnaghi, V., Martini, L., 2000a. Aging in peripheral nerves: regulation of myelin protein genes by steroid hormones. Prog. Neurobiol. 60, 291–308.

Melcangi, R.C., Martini, L., Galbiati, M., 2002. Growth factors and steroid hormones: a complex interplay in the hypothalamic control of reproductive functions. Prog. Neurobiol. 67, 421–449.

Melcangi, R.C., Panzica, G.C., 2001. Steroids in the nervous system: a Pandora box? Trends Neurosci. 24, 311–312.

Melcangi, R.C., Riva, M.A., Fumagalli, F., Magnaghi, V., Racagni, G., Martini, L., 1996b. Effect of progesterone, testosterone and their 5α-reduced metabolites on GFAP gene expression in type 1 astrocytes. Brain Res. 711, 10–15.

Mellon, S.H., Griffin, L.D., 2001. Compagnone N.A. Byosynthesis and action of neurosteroids. Brain Res. Rev. 37, 3–12.

Mensah-Nyagan, A.G., Beaujean, D., Luu-The, V., Pelletier, G., Vaudry, H., 2001. Anatomical and biochemical evidence for the synthesis of unconjugated and sulfated neurosteroids in amphibians. Brain Res. Rev. 37, 13–24.

Mensah-Nyagan, A.G., Do-Rego, J., Beaujean, D., Luu-The, V., Pelletier, G., Vaudry, H., 1999. Neurosteroids: expression of steroidogenic enzymes and regulation of steroid biosynthesis in the central nervous system. Pharmacol. Rev. 51, 63–81.

Mercier, G., Turque, N., Schumacher, M., 2001. Early activation of transcription factor expression in Schwann cells by progesterone. Mol. Brain Res. 97, 137–148.

Milner, T.A., McEwen, B.S., Hayashi, S., Li, C.J., Reagan, L.P., Alves, S.E., 2001. Ultrastructural evidence that hippocampal alpha estrogen receptors are located at extranuclear sites. J. Comp. Neurol. 429, 355–371.

Mong, J.A., Glaser, E., McCarthy, M.M., 1999. Gonadal steroids promote glial differentiation and alter neuronal morphology in the developing hypothalamus in a regionally specific manner. J. Neurosci. 19, 1464–1472.

Mong, J.A., Kurzweil, R.L., Davis, A.M., Rocca, M.S., McCarthy, M.M., 1996. Evidence for sexual differentiation of glia in rat brain. Horm. Behav. 30, 553–562.

Mong, J.A., McCarthy, M.M., 1999. Steroid-induced developmental plasticity in hypothalamic astrocytes: implications for synaptic pattering. J. Neurobiol. 40, 602–619.

Mong, J.A., Nunez, J.L., McCarthy, M.M., 2002. GABA mediates steroid-induced astrocyte differentiation in the neonatal rat hypothalamus. J. Neuroendocrinol. 14, 45–55.

Mor, G., Nilsen, J., Horvath, T., Bechmann, I., Brown, S., Garcia-Segura, L.M., Naftolin, F., 1999. Estrogen and microglia: a regulatory system that affects the brain. J. Neurobiol. 40, 484–496.

Morell, P., Jurevics, H., 1996. Origin of cholesterol in myelin. Neurochem. Res. 21, 463–470.

Mydlarski, M.B., Liberman, A., Schipper, H.M., 1995. Estrogen induction of glial heat shock proteins: implications for hypothalamic aging. Neurobiol. Aging 16, 977–981.

Negri-Cesi, P., Melcangi, R.C., Celotti, F., Martini, L., 1992. Aromatase activity in cultured brain cells: difference between neurons and glia. Brain Res. 589, 327–332.

Neuberger, T.J., Kalimi, O., Regelson, W., Kalimi, M., De Vries, G.H., 1994. Glucocorticoids enhance the potency of Schwann cell mitogens. J. Neurosci. Res. 38, 300–313.

Nichols, N.R., 1999. Glial responses to steroids as markers of brain aging. J. Neurobiol. 40, 585–601.

Nichols, N.R., Masters, J.N., Finch, C.E., 1990. Changes in gene expression in hippocampus in response to glucocorticoids and stress. Brain Res. Bull. 24, 659–662.

Nichols, N.R., Osterburg, H.H., Masters, J.N., Millar, S.L., Finch, C.E., 1990. Messenger RNA for glial fibrillary acidic protein is decreased in rat brain following acute and chronic corticosterone treatment. Mol. Brain Res. 7, 1–7.

Notterpek, L., Snipes, G.J., Shooter, E.M., 1999. Temporal expression pattern of peripheral myelin protein 22 during in vivo and in vitro myelinization. Glia 25, 358–369.

Ojeda, S.R., Ma, Y.J., 1999. Glial-neuronal interactions in the neuroendocrine control of mammalian puberty: facilitatory effects of gonadal steroids. J. Neurobiol. 40, 528–540.

Peterson, R.S., Saldanha, C.J., Schlinger, B.A., 2001. Rapid upregulation of aromatase mRNA and protein following neural injury in the zebra finch (Taeniopygia guttata). J. Neuroendocrinol. 13, 317–323.

Puy, L., MacLusky, N.J., Becker, L., Karsan, N., Trachtenberg, J., Brown, T.J., 1995. Immunocytochemical detection of androgen receptor in human temporal cortex characterization and application of polyclonal androgen receptor antibodies in frozen and paraffin-embedded tissues. J. Steroid Biochem. Mol. Biol. 55, 197–209.

Rage, F., Lee, B.J., Ma, Y.J., Ojeda, S.R., 1997. Estradiol enhances prostaglandin E2 receptor gene expression in luteinizing hormone-releasing hormone (LHRH) neurons and facilitates the LHRH response to PGE2 by activating a glia-to-neuron signaling pathway. J. Neurosci. 17, 9145–9156.

Robert, F., Geuennoun, R., Desarnaud, F., Do-Thi, A., Benmessahel, Y., Baulieu, E.E., Schumacher, M., 2001. Synthesis of progesterone in Schwann cells: regulation by sensory neurons. Eur. J. Neurosci. 13, 916–924.

Rozovsky, I., Laping, N.J., Krohn, K., Teter, B., O'Callaghan, J.P., Finch, C.E., 1995. Transcriptional regulation of glial fibrillary acidic protein by corticosterone in rat astrocytes in vitro is influenced by the duration of time in culture and by astrocyte–neuron interactions. Endocrinology 136, 2066–2073.

Rozovsky, I., Wei, M., Stone, D.J., Zanjani, H., Anderson, C.P., Morgan, T.E., Finch, C.E., 2002. Estradiol (E2) enhances neurite outgrowth by repressing glial fibrillary acidic protein expression and reorganizing laminin. Endocrinology 143, 636–646.

Santagati, S., Melcangi, R.C., Celotti, F., Martini, L., Maggi, A., 1994. Estrogen receptor is expressed in different types of glial cells in culture. J. Neurochem. 63, 2058–2064.

Santizo, R., Pelligrino, D.A., 1999. Estrogen reduces leukocyte adhesion in the cerebral circulation of female rats. J. Cereb. Blood Flow Metab. 19, 1061–1065.

Schmidt, M., Kreutz, M., Loffler, G., Scholmerich, J., Straub, R.H., 2000. Conversion of dehydroepiandrosterone to downstream steroid hormones in macrophages. J Endocrinol. 164, 161–169.

Schumacher, M., Guennoun, R., Mercier, G., Désarnaud, F., Lacor, P., Bénavides, J., Ferzas, B., Robert, F., Baulieu, E.E., 2001. Progesterone synthesis and myelin formation in peripheral nerves. Brain Res. Rev. 37, 343–359.

Shi, J., Zhang, Y.Q., Simpkins, J.W., 1997. Effects of 17beta-estradiol on glucose transporter 1 expression and endothelial cell survival following focal ischemia in the rats. Exp. Brain Res. 117, 200–206.

Shin, C.Y., Choi, J.W., Jang, E.S., Ryu, J.H., Kim, W.K., Kim, H.C., Ko, K.H., 2001. Glucocorticoids exacerbate peroxynitrite mediated potentiation of glucose deprivation-induced death of rat primary astrocytes. Brain Res. 923, 163–171.

Simard, M., Couldwell, W.T., Zhang, W., Song, H., Liu, S., Cotrina, M.L., Goldman, S., Nedergaard, M., 1999. Glucocorticoids—potent modulators of astrocytic calcium signaling. Glia 28, 1–12.

Stanimirovic, D.B., McCarron, R.M., Spatz, M., 1994. Dexamethasone down-regulates endothelin receptors in human cerebromicrovascular endothelial cells. Neuropeptides 26, 145–152.

Stone, D.J., Rozovsky, I., Morgan, T.E., Anderson, C.P., Hajain, H., Finch, C.E., 1997. Astrocytes and microglia respond to estrogen with increased apoE mRNA in vivo and in vitro. Exp. Neurol. 143, 313–318.

Stone, D.J., Song, Y., Anderson, C.P., Krohn, K.K., Finch, C.E., Rozovsky, I., 1998. Bidirectional transcription regulation of glial fibrillary acidic protein by estradiol in vivo and in vitro. Endocrinology 139, 3202–3209.

Svenningsen, A.F., Kanje, M., 1999. Estrogen and progesterone stimulate Schwann cell proliferation in a sex- and age-dependent manner. J. Neurosci. Res. 57, 124–130.

Tanaka, J., Fujita, H., Matsuda, S., Toku, K., Sakanaka, M., Maeda, N., 1997. Glucocorticoid and mineralocorticoid receptors in microglial cells: the two receptors mediate differential effects of corticosteroids. Glia 20, 23–37.

Thi, A.D., Jung-Testas, I., Baulieu, E.E., 1998. Neuronal signals are required for estrogen-mediated induction of progesterone receptor in cultured rat Schwann cells. J. Steroid Biochem. Mol. Biol. 67, 201–211.

Tsuneishi, S., Takada, S., Motoike, T., Ohashi, T., Sano, K., Nakamura, H., 1991. Effects of dexamethasone on the expression of myelin basic protein, proteolipid protein, and glial fibrillary acidic protein genes in developing rat brain. Dev. Brain Res. 61, 117–123.

Vallee, M., Mayo, W., Le Moal, M., 2001. Role of pregnenolone, dehydroepiandrosterone and their sulfate esters on learning and memory in cognitive aging. Brain Res. Rev. 37, 301–312.

Vardimon, L., Ben-Dror, I., Avisar, N., Oren, A., Shiftan, L., 1999. Glucocorticoid control of glial gene expression. J. Neurobiol. 40, 513–527.

Vegeto, E., Bonincontro, C., Pollio, G., Sala, A., Viappiani, S., Nardi, F., Brusadelli, A., Viviani, B., Ciana, P., Maggi, A., 2001. Estrogen prevents the lipopolysaccharide-induced inflammatory response in microglia. J. Neurosci. 21, 1809–1818.

Vegeto, E., Pollio, G., Pellicciari, C., Maggi, A., 1999. Estrogen and progesterone induction of survival of monoblastoid cells undergoing TNF-alpha-induced apoptosis. FASEB J. 13, 793–803.

Watanabe, Y., Littleton-Kearney, M.T., Traystman, R.J., Hurn, P.D., 2001. Estrogen restores postischemic pial microvascular dilation. Am. J. Physiol. 281, H155–H160.

Wise, P.M., Dubal, D.B., Wilson, M.E., Rau, S.W., 2000. Estradiol is a neuroprotective factor in in vivo and in vitro models of brain injury. J. Neurocytol. 29, 401–410.

Yokoi, H., Tsuruo, Y., Ishimura, K., 1998. Steroid 5α-reductase type 1 immunolocalized in the rat peripheral nervous system and paraganglia. Histochem. J. 30, 731–739.

Zwain, I.H., Arroyo, A., Amato, P., Yen, S.S., 2002. A role for hypothalamic astrocytes in dehydroepiandrosterone and estradiol regulation of gonadotropin-releasing hormone (GnRH) release by GnRH neurons. Neuroendocrinology 75, 375–383.

Zwain, I., Yen, S.S.C., 1999. Dehydroepiandrosterone (DHEA): biosynthesis and metabolism in the brain. Endocrinology 140, 880–887.

Zwain, I., Yen, S.S.C., 1999. Neurosteroidogenesis in astrocytes, oligodendrocytes and neurons of cerebral cortex of rat brain. Endocrinology 140, 3843–3852.

Zwain, I.H., Yen, S.S.C., Cheng, C.Y., 1997. Astrocytes cultured in vitro produce estradiol-17β and express aromatase cytochrome P-450 (P-450 AROM) mRNA. Biochim. Biophys. Acta 1334, 338–348.

Advances in
Molecular and
Cell Biology

Expression of neurotrophic factors and cytokines and their receptors on astrocytes in vivo

Takao Nakagawa[1] and Joan P. Schwartz*

Neurotrophic Factors Section, NINDS, National Institutes of Health, Building 1, Room 135, Bethesda, MD 20892-0151, USA

**Correspondence address: Tel.: +1-301-496-1248; fax: +1-301-402-0027.*
E-mail: jps@helix.nih.gov(J.P.S.)

Contents

1. Introduction
2. Neurotrophins
3. TGF-β superfamily
4. Insulin-like growth factor family
5. Fibroblast growth factors
6. Vascular endothelial growth factor
7. Neuropoietic cytokines
8. Pro-inflammatory cytokines
9. Anti-inflammatory cytokines
10. Chemokines
11. Concluding remarks

The functions of astrocytes following injury are mediated in part by synthesis of neurotrophic factors as well as cytokines. For this chapter we have chosen to review data on the in vivo synthesis and expression of various neurotrophic factors and cytokines/chemokines, as well as their receptors, in astrocytes from control and injured brain. We present data only for the central nervous system and discuss a variety of different models of injury, as well as results from human neurological diseases. These data from the literature are compared with our cDNA microarray results for astrocytes isolated in vivo from normal or 6-hydroxy-dopamine-lesioned rat striatum.

1. Introduction

The central nervous system (CNS) is composed of several cell populations, primarily neurons, macroglia and microglia, with astrocytes and oligodendrocytes the principal

[1]Present address: Department of Neurosurgery, Fukui Medical University, Fukui, Japan.

Advances in Molecular and Cell Biology, Vol. 31, pages 561–573

ISBN: 0-444-51451-1

macroglial cell types. Astrocytes constitute nearly 40% of the total CNS cell population (Rutka et al., 1997), yet our knowledge of the function of astrocytes is still incomplete. Astrocytes are stellate cells with multiple fine processes, some of which contact cells of mesodermal origin (most frequently capillary endothelial cells) and some of which are intertwined within the neuropil and ensheath synaptic contacts (Rohlmann and Wolff, 1996). These quantitative and morphological findings support the idea that astrocytes have important functions in maintaining or modulating CNS function. Among these are metabolic support for neurons and uptake of neurotransmitters (see chapter by Schousboe and Waagepetersen), ion homeostasis (see chapter by Walz), guidance for neuronal cell migration and axon outgrowth during development (see chapter by Gomes and Rehen), and preservation of host tissue integrity following injury, a property of activated, or reactive, astrocytes.

Reactive gliosis, which occurs in the CNS in response to virtually all forms of brain injury, has been defined as an increase in the size of the astrocyte cell body and its processes, with an increase of glial fibrillary acidic protein (GFAP) being the prototypical biochemical change (see chapter by Kalman). Prominent reactive astrocytosis is seen in acute traumatic brain injury (Faden, 1993), neurodegenerative diseases such as Alzheimer's and Parkinson's (Delacourte, 1990; Pike et al., 1995; Knott et al., 2002), and following exposure to toxins such as 1-methyl-4-phenyl-1, 2, 3, 6-tetrahydropyridine, 6-hydroxydopamine (6-OHDA) and kainic acid (Forno et al., 1992; O'Callaghan and Miller, 1993). A distinguishing characteristic among these different types of injuries is whether inflammation occurs, thus involving immune cells and/or cytokines. Cytokines and chemokines, important mediators of the host defense system and the inflammatory response, are elevated in a variety of neurological diseases (Eng and Ghirnikar, 1994; Zhao and Schwartz, 1998). Both astrocytes and microglia can produce and respond to cytokines (Giulian et al., 1994a,b). Thus, cytokines or chemokines elevated after brain injury may act as signals among microglia, neurons, and astrocytes but may also be functioning in trophic roles.

The functions of astrocytes during development and following injury are mediated in part by astrocyte synthesis of neurotrophic factors as well as cytokines (Eddleston and Mucke, 1993; McMillian et al., 1994; Ridet et al., 1997). Furthermore, the presence of growth factor and cytokine receptors on astrocytes suggests that these factors modulate astrocyte functions in autocrine or paracrine ways (Ridet et al., 1997). Some of the neurotrophic factors and cytokines have been detected in embryonic and neonatal brain in vivo. Few are present in the normal adult brain, but synthesis of both neurotrophic factors and cytokines/chemokines is turned on following brain injury. Of great interest has been the specific expression of these factors in astrocytes, both developmentally and following injury, since some studies have suggested that reactive astrocytes recapitulate the properties of neonatal astrocytes. Furthermore, both neurotrophic factors and cytokines have been shown to rescue damaged neurons in various injury models.

What factors do astrocytes synthesize in the developing or injured brain? For this chapter we have chosen to review data on the in vivo synthesis and expression of various neurotrophic factors and cytokines/chemokines in astrocytes. We will present data only for the CNS and will compare a variety of different models of injury, as well as results from human neurological diseases. Summarized in Table 1 are the various families of

Table 1
Neurotrophic factors, cytokines, chemokines and their receptors expressed in astrocytes

Family	Factor	Receptor(s)
Neurotrophic factors		
Neurotrophins	NGF[a]	trkA, p75
	BDNF[b]	trkB, p75
	Neurotrophin-3 (NT-3)	trkC, p75
TGF-β superfamily	GDNF[c]	Ret + GFRα1
	Neurturin	Ret + GFRα2
	Artemin	Ret + GFRα3
	Persephin	Ret + GFRα4
Insulin-like growth factors	IGF-1	IGF1R
	IGF-2	IGF2R
		IGFBP-1 to IGFBP-6
Fibroblast growth factors	FGF-1 (acidic)	FGFR-1 to FGFR-4
(23 family members, 10 in CNS)	FGF-2 (basic)	FGFR-1 to FGFR-4
Vascular endothelial growth factor	VEGF	Flt-1 (VEGFR-1)
	VEGF-B to VEGF-E	Flk-1 (VEGFR-2)
Cytokines		
Neuropoietic cytokines	CNTF[d]	gp130 + CNTFRα
	LIF[e]	gp130 + LIFR
	IL-6	gp130 + IL-6Rα
	Oncostatin-M	gp130 + LIFR + OsMR
	Cardiotropin	gp130 + LIFR + CT1Rα
Pro-inflammatory cytokines	IL-1β	IL-1R1, IL-1R2
	TNFα	TNF-R1, TNF-R2
Anti-inflammatory cytokines	TGF-β	TGF-βR1
	IL-1ra	IL-1R1
	IL-4	IL-4R
	IL-10	IL-10R
Chemokines		
CC family	MIP-1α[f]	CCR1, CCR5
	MIP-1β	CCR1, CCR5
	MCP-1[g]	CCR1, CCR2
	MIP-3α	CCR6
	MIP-3β	CCR7
	RANTES	CCR1, CCR3, CCR5
	Eotaxin	CCR3, CCR5
CXC family	IP-10[h]	CXCR3
	Mig[i]	CXCR3
	MIP-2	CXCR2
	IL-8	CXCR1, CXCR2
CX3C family	Fractalkine	CX3CR

[a]Nerve growth factor.
[b]Brain-derived neurotrophic factor.
[c]Glial cell derived neurotrophic factor.
[d]Ciliary neurotrophic factor.
[e]Leukemia inhibitory factor.
[f]Macrophage inflammatory protein.
[g]Monocyte chemoattractant protein.
[h]Interferon-γ-inducible protein.
[i]Macrophage interferon-γ-inducible protein.

neurotrophic factors (growth factors acting on neurons and/or glial cells), cytokines (factors regulating development and function of hematopoietic cells and cells of the immune system), and chemokines (factors regulating motility of immune cells) to be discussed, with their associated receptors. Corresponding factors and receptors that have been found on astrocytes in culture but presently have not been demonstrated on astrocytes in vivo (e.g., the epithelial growth factor family) are not included.

These data from the literature will be compared with our cDNA microarray results (manuscript in preparation). The microarray methodology has become a popular way to examine expression of a variety of genes simultaneously (reviewed in Luo and Geschwind, 2001). We have specifically taken advantage of microarrays to determine striatal astrocyte genes that are affected following substantia nigra lesion with 6-OHDA. Reactive astrocytes are induced by the 6-OHDA lesion, with the peak number of cells and intensity of GFAP increase occurring seven days after the lesion (Sheng et al., 1993). Combining this animal model of Parkinson's disease (PD) with the new methodologies for isolating acutely dissociated astrocytes allowed us to pick up individual cells from normal adult rat brain striatum and from the striatum on the lesioned side of the 6-OHDA-lesioned brain. RNA was prepared and the identity of astrocytes confirmed by single cell RT-PCR for GFAP mRNA (Zhou et al., 2000). Amplification of the cDNA allowed preparation of sufficient material to carry out microarray analysis using Clontech nylon arrays (Eberwine et al., 1992, 2001; Spirin et al., 1999; Wang et al., 2000). The results from the literature will be compared with these results, and we will refer to the astrocytes isolated in vivo as 'in vivo reactive astrocytes'.

2. Neurotrophins

The family of neurotrophins, of which nerve growth factor (NGF) was the first to be isolated and studied, was originally localized to target neurons. More recent data, using immunohistochemistry and in situ hybridization, have shown expression of NGF, brain derived neurotrophic factor (BDNF) and neurotrophin-3 (NT-3) in astrocytes in normal adult brain (Arendt et al., 1995; Dreyfus et al., 1999; Fiedorowicz et al., 2001; Knott et al., 2002) and spinal cord (Dreyfus et al., 1999) in vivo. To date, no studies have examined astrocyte expression of neurotrophins developmentally. Expression of both NGF and BDNF is enhanced in reactive astrocytes induced by a variety of lesions in animals, including spinal cord transection (Krenz and Weaver, 2000), kainic acid (Bakhit et al., 1991; Strauss et al., 1994), electrolytic lesion (Oderfeld-Nowak et al., 1992), stab wound (Goss et al., 1998), ischemia (Shozuhara et al., 1992) and insertion of a nitrocellulose filter (McKeon et al., 1997). Elevated levels of NGF in reactive astrocytes are reported following ethanol or trimethyltin toxicity in rats (Arendt et al., 1995; Koczyk and Oderfeld-Nowak, 2000; Fiedorowicz et al., 2001), while the content of BDNF and NT-3 in reactive astrocytes is elevated in PD and multiple sclerosis (MS) (Knott et al., 2002; Stadelmann et al., 2002). Along with this increased synthesis of the neurotrophins comes increased expression of their receptors. The neurotrophin receptors trkA, trkB and trkC (Table 1) have all been identified on astrocytes in normal adult brain (Knott et al., 2002; Stadelmann et al., 2002), and are present on reactive astrocytes in PD (Knott et al., 2002; Stadelmann et al., 2002). Experimental autoimmune encephalitis (EAE) upregulates trkA expression on reactive astrocytes in the spinal cord (Oderfeld-Nowak et al., 2001), while transection or mechanical trauma increases both trkA and trkB (Frisén et al., 1993; Foschini et al., 1994; Dougherty et al., 2000), and implantation of a nitrocellulose filter elevates expression of trkB (McKeon et al., 1997). In the brain, trkB expression is induced on reactive astrocytes by trimethyltin toxicity (Koczyk and Oderfeld-Nowak, 2000), and by nitrocellulose filter implants (McKeon et al., 1997). These findings suggest autocrine or paracrine effects of NGF and the other neurotrophins on astrocytes. Our analyses of in vivo reactive astrocytes found increased content of trkB and NGF.

3. TGF-β superfamily

Glial cell derived neurotrophic factor (GDNF) was first cloned from a glial cell line (and named 'glial cell-line derived neurotrophic factor'), but like the neurotrophins, it and the other family members, neurturin, artemin and persephin, have been localized primarily to neurons in vivo (Strömberg et al., 1993; Poulsen et al., 1994). However, there is one report of GDNF being expressed in adult astrocytes, measured by in situ hybridization (Ho et al., 1995), and both kainic acid lesions (Bresjanac and Antauer, 2000; Marco et al., 2002) and ischemia (Miyazaki et al., 2001) induce its expression in reactive astrocytes. Kainate also induces expression of the GDNF receptor α1 (GFRα1) on reactive astrocytes (Bresjanac and Antauer, 2000; Marco et al., 2002). Our results for in vivo reactive astrocytes also show increased expression of GDNF mRNA.

4. Insulin-like growth factor family

Insulin-like growth factor-1 (IGF-1) has not been detected in astrocytes under control conditions although the IGF-Binding Protein-2 (IGFBP-2) has (Lee et al., 1993). Cuprizone exposure turns on IGF-1 expression in reactive astrocytes (Mason et al., 2001), while either ischemia or high potassium enhances expression of IGFBP-2 (Lee et al., 1997; Holmin et al., 2001). In contrast, we find that IGF-1 is unchanged in in vivo reactive astrocytes, IGFBP-2 expression is decreased, and IGFBP-1 and IGFBP-3 are increased.

5. Fibroblast growth factors

Astrocyte expression of fibroblast growth factor 2 (basic FGF) and of its receptor FGFR1 has been shown for both neonatal astrocytes (Smith et al., 2001; Ganat et al., 2002) and for adult astrocytes (Clarke et al., 2001) in vivo. Furthermore, it has been well established that FGF-1 and FGF-2 are increased in reactive astrocytes following a variety of insults, including electrolytic lesions (Buytaert et al., 2001), ischemia (Ganat et al., 2002), stab wounds (Smith et al., 2001; Ganat et al., 2002), and spinal cord transection

(Leme and Chadi, 2001). FGF-2 stimulates astrocytic proliferation and differentiation into a reactive phenotype in vivo (Eclancher et al., 1996) and also induces astrocyte migration in vivo (Holland and Varmus, 1998). Thus, FGF-2 may also have autocrine or paracrine effects on astrocytes.

6. Vascular endothelial growth factor

Vascular endothelial growth factor (VEGF) has neurotrophic effects in addition to its better-known angiogenic effects (Sondell et al., 2000; Wick et al., 2002). Moreover, VEGF stimulates proliferation of astrocytes in vivo (Krum et al., 2002). It is expressed only in neurons in control adult brain (Papavassiliou et al., 1997; Krum and Rosenstein, 1998) but expression is turned on in reactive astrocytes following ischemia (Beck et al., 1993; Salhia et al., 2000), stab wound (Papavassiliou et al., 1997; Krum and Rosenstein, 1998; Salhia et al., 2000; Sköld et al., 2000), freeze injury (Papavassiliou et al., 1997) or implantation of grafts (Krum and Rosenstein, 1998). In addition, VEGF expression is elevated in the senile plaques of Alzheimer's disease (Salhia et al., 2000). Two receptors have been identified, Flt-1 or VEGFR-1, and Flk-1 or VEGFR-2. Stab wounds and graft implantation (Krum and Rosenstein, 1998), as well as infusion trauma (Krum et al., 2002), induce expression of Flt-1 on reactive astrocytes, while Sköld et al. (2000) showed the presence of Flk-1 on reactive astrocytes in the spinal cord following transection.

7. Neuropoietic cytokines

The family of neuropoietic cytokines (Allan and Rothwell, 2001), that includes ciliary neurotrophic factor (CNTF), leukemia inhibitory factor (LIF), IL-6, oncostatin-M and cardiotropin, has a variety of effects in the CNS. The family shares one common receptor component, gp130, with different additional components required by each of the various family members (Table 1). CNTF itself has been implicated in induction of reactive gliosis (Winter et al., 1995; Clatterbuck et al., 1996; Levison et al., 1996; Albrecht et al., 2002), although other work has not supported that conclusion (Emerich et al., 1996; Lisovoski et al., 1997). Both CNTF and IL-6 have been demonstrated in neonatal astrocytes in vivo (Acarin et al., 2000; Widenfalk et al., 2001). CNTF was reported to be present in normal adult astrocytes (Widenfalk et al., 2001; Dallner et al., 2002), although Lee et al. (1997) did not find it. Friedman (2001) could not detect IL-6 in normal adult astrocytes. A variety of injuries are capable of inducing either CNTF, e.g., a stab wound (Dallner et al., 2002) or spinal cord transection (Widenfalk et al., 2001), or IL-6, e.g., NMDA administration (Acarin et al., 2000) or ischemia (Friedman, 2001) in reactive astrocytes. In addition, induction of long term potentiation turns on IL-6 expression in astrocytes, although it remains to be established whether they are reactive (Jankowsky et al., 2000). The CNTF-specific receptor subunit CNTFRα was shown to be expressed on normal adult astrocytes (Dallner et al., 2002) and to be increased following a stab wound (Lee et al., 1997). Other receptor subunits have not been examined. Our results show induction of the IL-6Rα on in vivo reactive astrocytes. LIF was also expressed in in vivo reactive astrocytes, in support of results from Banner et al. (1997) using a stab wound model, whereas Lemke et al. (1999) have not detected it in astrocytes after 192IgG-saporin cholinergic lesion.

8. Pro-inflammatory cytokines

IL-1β, TNF-α, and IL-6 have often been referred to as the pro-inflammatory cytokines, although it is clear that IL-6 may be protective under certain conditions in the CNS (Merrill and Benveniste, 1996; Allan and Rothwell, 2001). As discussed in Section 7, IL-6 is induced in reactive astrocytes following various forms of injury. Neither IL-1β nor TNF-α has been detected in either neonatal (Acarin et al., 2000) or adult astrocytes (Friedman, 2001). However, NMDA treatment induces not only IL-6 (see above), but also IL-1β and TNF-α in reactive astrocytes (Acarin et al., 2000). Ischemia (Orzylowska et al., 1999; Friedman, 2001) and cuprizone exposure (Mason et al., 2001) both induce expression of IL-1β and IL-6 in reactive astrocytes. The IL-1β receptor subunit IL-1R1, expressed on neurons developmentally, is induced on reactive astrocytes following a stab wound (Friedman, 2001). We find that both TNF-α and its receptor are expressed on in vivo reactive astrocytes.

9. Anti-inflammatory cytokines

None of the anti-inflammatory cytokines, TGF-β, IL-1ra (the IL-1R antagonist), IL-4, and IL-10 (Allan and Rothwell, 2001), is expressed in neonatal astrocytes (Acarin et al., 2000) although IL-4 and IL-10 have been identified in adult astrocytes (Hulshof et al., 2002). Acarin et al. (2000) have shown that NMDA treatment induces expression of TGF-β in reactive astrocytes, while both IL-4 and IL-10, as well as their receptors, are turned on in reactive astrocytes in MS plaques (Hulshof et al., 2002). The α subunit of IL-4R has also been reported to be present on reactive astrocytes in the vicinity of epileptic foci and CNS tumors (Liu et al., 2000). No changes were detected in our in vivo reactive astrocytes.

10. Chemokines

Chemokines, a large family of proteins that stimulate motility of immune cells, have been classified into four groups, of which three have been found in astrocytes in the brain in vivo, whereas no data are available suggesting any function of the fourth chemokine family member (C) in astrocytes. We will discuss some of the members of the CC family, the CXC family, and the CX3C member fractalkine (Table 1). Corresponding to these families are families of receptors, designated the CC, CXC, and CX3C receptors. The interest in this group of compounds has grown very considerably following the discovery that they play major roles in the neurological complications of HIV (see chapters by Zsembery et al. and by Ghorpade and Gendelman).

Only one of the CC family chemokines, MIP-1β, has been detected in control adult astrocytes (Sanders et al., 1998). Che et al. (2001) showed that MCP-1 is induced by ischemia in rats. MIP-1β has been identified in reactive astrocytes in brains from patients with HIV-encephalitis (Sanders et al., 1998) and Alzheimer's disease (Xia et al., 2000). MCP-1 expression occurs in reactive astrocytes in EAE, the animal model of MS (DeGroot and Woodroofe, 2001), as well as in MS (Simpson et al., 2000; DeGroot and Woodroofe, 2001). The other CC chemokines, including MCP-2, MCP-3, MIP-1α, MIP-1β, and RANTES (regulated on activation, normal T cell expressed and secreted), are also expressed in reactive astrocytes in MS (DeGroot and Woodroofe, 2001). *Toxoplasma* encephalitis results in MCP-1, but not RANTES, expression in reactive astrocytes (Strack et al., 2002). Results on the expression of the CC receptors in control adult brain astrocytes is highly variable, with lack of expression of CCR1, CCR3 and CCR5 reported (Ghorpade et al., 1998; Sanders et al., 1998) but positive expression for CCR3 and CCR5 also reported (Rottman et al., 1997; Ghorpade et al., 1998; Westmoreland et al., 1998; Simpson et al., 2000; Otto et al., 2001). CCR2, CCR3 and CCR5 are all present on reactive astrocytes in MS plaques (Simpson et al., 2000) and in both macaque SIV-encephalitis (Westmoreland et al., 1998) and human HIV-encephalitis brain (Ghorpade et al., 1998). MIP-1α was induced in in vivo reactive astrocytes.

Three members of the CXC family have been localized to astrocytes in the CNS. Of these, Mig is not present under normal conditions (Asensio et al., 2001), IL-8 has been detected (Sanders et al., 1998), and reports about IP-10 are conflicting (Xia et al., 2000; Asensio et al., 2001). All are elevated in human diseases. IP-10 is present in reactive astrocytes in brains from patients with HIV-encephalitis (Sanders et al., 1998) and in an animal model of HIV encephalopathy (Asensio et al., 2001), in MS (and EAE) (DeGroot and Woodroofe, 2001) and Alzheimer's disease (Xia et al., 2000), as well as in *Toxoplasma* encephalitis (Strack et al., 2002). Mig is also found in reactive astrocytes in MS (DeGroot and Woodroofe, 2001) but not following *Toxoplasma* infection (Strack et al., 2002). IL-8 is induced in reactive astrocytes in HIV-encephalitic brain (Sanders et al., 1998). There are reports of expression of the receptors CXCR2 and CXCR4 on astrocytes in adult brain (Westmoreland et al., 1998; Otto et al., 2001), whereas CXCR3 has been localized to neurons (Xia et al., 2000) and is not found on astrocytes (Asensio et al., 2001). Expression of CXCR2 is enhanced on reactive astrocytes after a stab wound (Otto et al., 2001). CXCR3 is expressed on reactive astrocytes in MS plaques (DeGroot and Woodroofe, 2001) while CXCR4 is present on reactive astrocytes in the HIV-encephalitic brain (Sanders et al., 1998). Fractalkine, the only CX3C ligand, is present in adult astrocytes and increased in reactive astrocytes in a prion model of chronic neurodegeneration and inflammation (Hughes et al., 2002). None were detected in our studies.

11. Concluding remarks

This review supports the view that whereas astrocytes produce neurotrophic factors and cytokines in the developing and the injured brain, expression in the normal adult brain is very low. Factors fall into one of three patterns: many are not expressed at all in the normal adult brain, but can be induced in reactive astrocytes (IGF-1, IL-1β, TNF-α, most of the chemokines); some are expressed at low levels in normal astrocytes (NTFs, FGFs, CNTF, MIP-1β, IL-4, IL-8 and IL-10) and upregulated in reactive astrocytes; and some are expressed exclusively in neurons, but can be induced in reactive astrocytes in response to injury (GDNF family, VEGF, IL-6). In addition to their functions as neurotrophic and immunological factors, these factors can have other functions such as stimulating the proliferation, differentiation and migration of astrocytes, and enhancing both angiogenesis and gliogenesis. Induction of many of the receptors on reactive astrocytes suggests a set of autocrine/paracrine functions. By comparing these factors in the normal versus the injured brain, versus the developing brain, we will further understand which functions are important in which circumstances.

Our microarray analysis of gene expression in acutely isolated astrocytes from the rat 6-OHDA model of PD revealed upregulation of a number of the same neurotrophic factor and cytokine genes as are seen in other models of injury. Too few studies have examined brains from PD to provide the data with which to compare ours, but in general the microarray findings support the idea that a specific brain lesion induces a specific increase in a specific set of genes. Understanding which genes those are, and what the functions of their respective products are, may allow us to address treatments for these neurological diseases.

References

Acarin, L., González, B., Castellano, B., 2000. Neuronal, astroglial and microglial cytokine expression after an excitotoxic lesion in the immature rat brain. Eur. J. Neurosci. 12, 3505–3520.

Albrecht, P.J., Dahl, J.P., Stoltzfus, O.K., Levenson, R., Levison, S.W., 2002. Ciliary neurotrophic factor activates spinal cord astrocytes, stimulating their production and release of fibroblast growth factor-2, to increase motor neuron survival. Exp. Neurol. 173, 46–62.

Allan, S.M., Rothwell, N.J., 2001. Cytokines and acute degeneration. Nat. Rev.—Neurosci. 2, 734–744.

Arendt, T., Brückner, M.K., Krell, T., Pagliusi, S., Kruska, L., Heumann, R., 1995. Degeneration of rat cholinergic basal forebrain neurons and reactive changes in nerve growth factor expression after chronic neurotoxic injury—II. Reactive expression of the nerve growth factor gene in astrocytes. Neuroscience 65, 647–659.

Asensio, V., Maier, J., Milner, R., Boztug, K., Kincaid, C., Moulard, M., Phillipson, C., Lindsley, K., Krucker, T., Fox, H.S., Campbell, I.L., 2001. Interferon-independent, human immunodeficiency virus type 1 gp120-mediated induction of CXCL10/IP-10 gene expression by astrocytes *in vivo* and *in vitro*. J. Virol. 75, 7067–7077.

Bakhit, C., Armanini, M., Bennett, G.L., Wong, W.L.T., Hansen, S.E., Taylor, R., 1991. Increase in glia-derived nerve growth factor following destruction of hippocampal neurons. Brain Res. 560, 76–83.

Banner, L.R., Moayeri, N.N., Patterson, P.H., 1997. Leukemia inhibitory factor is expressed in astrocytes following cortical brain injury. Exp. Neurol. 147, 1–9.

Beck, K.D., Lamballe, F., Klein, R., Barbacid, M., Schauwecker, P.E., McNeill, T.H., Finch, C.E., Hefti, F., Day, J.R., 1993. Induction of noncatalytic TrkB neurotrophin receptors during axonal sprouting in the adult hippocampus. J. Neurosci. 13, 4001–4014.

Bresjanac, M., Antauer, G., 2000. Reactive astrocytes of the quinolinic acid-lesioned rat striatum express GFRα1 as well as GDNF *in vivo*. Exp. Neurol. 164, 53–59.

Buytaert, K.A., Kline, A.E., Montañez, S., Likler, E., Millar, C.J., Hernandez, T.D., 2001. The temporal patterns of c-Fos and basic fibroblast growth factor expression following a unilateral anteromedial cortex lesion. Brain Res. 894, 121–130.

Che, X., Ye, W., Panga, L., Wu, D.-C., Yang, G.-Y., 2001. Monocyte chemoattractant protein-1 expressed in neurons and astrocytes during focal ischemia in mice. Brain Res. 902, 171–177.

Clarke, W.E., Berry, M., Smith, C., Kent, A., Logan, A., 2001. Coordination of fibroblast growth factor receptor 1 (FGFR1) and fibroblast growth factor-2 (FGF-2) trafficking to nuclei of reactive astrocytes around cerebral lesions in adult rats. Mol. Cell. Neurosci. 17, 17–30.

Clatterbuck, R.E., Price, D.L., Koliatsos, V.E., 1996. Ciliary neurotrophic factor stimulates the expression of glial fibrillary acidic protein by brain astrocytes *in vivo*. J. Comp. Neurol. 369, 543–551.

Dallner, C., Woods, A.G., Deller, T., Kirsch, M., Hofmann, H.-D., 2002. CNTF and CNTF receptor alpha are constitutively expressed by astrocytes in the mouse brain. Glia 37, 374–378.

DeGroot, C.J.A., Woodroofe, M.N., 2001. The role of chemokines and chemokine receptors in CNS inflammation. Prog. Brain Res. 132, 533–544.

Delacourte, A., 1990. General and dramatic glial reaction in Alzheimer brains. Neurology 40, 33–37.

Dougherty, K.D., Dreyfus, C.F., Black, I.B., 2000. Brain-derived neurotrophic factor in astrocytes, oligodendrocytes, and microglia/macrophages after spinal cord injury. Neurobiol. Dis. 7, 574–585.

Dreyfus, C.E., Dai, X., Lercher, L.D., Racey, B.R., Friedman, W.J., Black, I.B., 1999. Expression of neurotrophins in the adult spinal cord *in vivo*. J. Neurosci. Res. 56, 1–7.

Eberwine, J., Kacharmina, J.E., Andrews, C., Miyashiro, K., McIntosh, T., Becker, K., Barrett, T., Hinkle, D., Dent, G., Marciano, P., 2001. mRNA expression analysis of tissue sections and single cells. J. Neurosci. 21, 8310–8314.

Eberwine, J., Yeh, H., Miyashiro, K., Cao, Y., Nair, S., Finnell, R., Zettel, M., Coleman, P., 1992. Analysis of gene expression in single live neurons. Proc. Natl Acad. Sci. USA 89, 3010–3014.

Eclancher, F., Kehrli, P., Labourdette, G., Sensenbrenner, M., 1996. Basic fibroblast growth factor (bFGF) injection activates the glial reaction in the injured adult rat brain. Brain Res. 737, 201–214.

Eddleston, M., Mucke, L., 1993. Molecular profile of reactive astrocytes-implications for their role in neurologic disease. Neuroscience 54, 15–36.

Emerich, D.F., Lindner, M.D., Winn, S.R., Chen, E.Y., Frydel, B.R., Kordower, J.H., 1996. Implants of encapsulated human CNTF-producing fibroblasts prevent behavioral deficits and striatal degeneration in a rodent model of Huntingdon's disease. J. Neurosci. 16, 5168–5181.

Eng, L.F., Ghirnikar, R.S., 1994. GFAP and astrogliosis. Brain. Pathol. 4, 229–237.

Faden, A.L., 1993. Experimental neurobiology of central nervous system trauma. Crit. Rev. Neurobiol. 7, 175–186.

Fiedorowicz, A., Figiel, I., Kaminska, B., Zaremba, M., Wilk, S., Oderfeld-Nowak, B., 2001. Dentate granule neuron apoptosis and glia activation in murine hippocampus induced by trimethyltin exposure. Brain Res. 912, 116–127.

Forno, L.S., DeLanney, L.E., Irwin, I., DiMonte, D., Langston, J.W., 1992. Astrocytes and Parkinson's disease. Prog. Brain Res. 94, 429–436.

Foschini, D.R., Kestler, A.M., Egger, M.D., Crockett, D.P., 1994. The up-regulation of trkA and trkB in dorsal column astrocytes following dorsal rhizotomy. Neurosci. Lett. 169, 21–24.

Friedman, W.J., 2001. Cytokines regulate expression of the type 1 Interleukin-1 receptor in rat hippocampal neurons and glia. Exp. Neurol. 168, 23–31.

Frisén, J., Verge, V.M.K., Fried, K., Risling, M., Persson, H., Trotter, J., Hökfelt, T., Lindholm, D., 1993. Characterization of glial trkB receptors: differential response to injury in the central and peripheral nervous systems. Proc. Natl Acad. Sci. USA 90, 4971–4975.

Ganat, Y., Soni, S., Chacon, M., Schwartz, M.L., Vaccarino, F.M., 2002. Chronic hypoxia up-regulates fibroblast growth factor ligands in the perinatal brain and induces fibroblast growth factor-responsive radial glial cells in the sub-ependymal zone. Neuroscience 112, 977–991.

Ghorpade, A., Xia, M.Q., Hyman, B.T., Persidsky, Y., Nukuna, A., Bock, P., Che, M., Limoges, J., Gendelman, H.E., Mackay, C.R., 1998. Role of the β-chemokine receptors CCR3 and CCR5 in human immunodeficiency virus type 1 infection of monocytes and microglia. J. Virol. 72, 3351–3361.

Giulian, D., Li, J., Leara, B., Keenen, C., 1994. Phagocytic microglia release cytokines and cytotoxins that regulate the survival of astrocytes and neurons in culture. Neurochem. Int. 25, 227–233.

Giulian, D., Li, J., Li, X., George, J., Rutecki, P.A., 1994. The impact of microglia-derived cytokines upon gliosis in the CNS. Dev. Neurosci. 16, 128–136.

Goss, J.R., O'Malley, M.E., Zou, L., Styren, S.D., Kochanek, P.M., DeKoskey, S.T., 1998. Astrocytes are the major source of nerve growth factor upregulation following traumatic brain injury in the rat. Exp. Neurol. 149, 301–309.

Ho, A., Gore, A.C., Weickert, C.S., Blum, M., 1995. Glutamate regulation of GDNF gene expression in the striatum and primary striatal astrocytes. Neuroreport 6, 1326–1330.

Holland, E.C., Varmus, H., 1998. Basic fibroblast growth factor induces cell migration and proliferation after glia-specific gene transfer in mice. Proc. Natl Acad. Sci. USA 95, 1218–1223.

Holmin, S., Mathiesen, T., Langmoen, I.A., Sandberg Nordqvist, A.-C., 2001. Depolarization induces insulin-like growth factor binding protein-2 expression *in vivo* via NMDA receptor stimulation. Growth Horm. IGF Res. 11, 399–406.

Hughes, P.M., Botham, M.S., Frentzel, S., Mir, A., Perry, V.H., 2002. Expression of fractalkine (CX3CL1) and its receptor, CX3CR1, during acute and chronic inflammation in the rodent CNS. Glia 37, 314–327.

Hulshof, S., Montagne, L., DeGroot, C.J.A., Van der Valk, P., 2002. Cellular localization and expression patterns of Interleukin-10, Interleukin-4, and their receptors in multiple sclerosis lesions. Glia 38, 24–35.

Jankowsky, J.L., Derrick, B.E., Patterson, P.H., 2000. Cytokine responses to LTP induction in the rat hippocampus: a comparison of *in vitro* and *in vivo* techniques. Learn. Mem. 7, 400–412.

Knott, C., Stern, G., Kingsbury, A., Welcher, A.A., Wilkin, G.P., 2002. Elevated glial brain-derived neurotrophic factor in Parkinson's diseased nigra. Parkinsonism Relat. Disord. 8, 329–341.

Koczyk, D., Oderfeld-Nowak, B., 2000. Long-term microglial and astroglial activation in the hippocampus of trimethyltin-intoxicated rat: stimulation of NGF and trkA immunoreactivities in astroglia but not in microglia. Int. J. Dev. Neurosci. 18, 591–606.

Krenz, N.R., Weaver, L.C., 2000. Nerve growth factor in glia and inflammatory cells of the injured rat spinal cord. J. Neurochem. 74, 730–739.

Krum, J.M., Mani, N., Rosenstein, J.M., 2002. Angiogenic and astroglial responses to vascular endothelial growth factor administration in adult rat brain. Neuroscience 110, 589–604.

Krum, J.M., Rosenstein, J.M., 1998. VEGF mRNA and its receptor flt-1 are expressed in reactive astrocytes following neural grafting and tumor cell implantation in the adult CNS. Exp. Neurol. 154, 57–65.

Lee, M.-Y., Dellner, T., Kirsch, M., Frotscher, M., Hofmann, H.-D., 1997. Differential regulation of ciliary neurotrophic factor (CNTF) and CNTF receptor α expression in astrocytes and neurons of the fascia dentata after entorhinal cortex lesion. J. Neurosci. 17, 1137–1146.

Lee, W.H., Michels, K.M., Bondy, C.A., 1993. Localization of insulin-like growth factor binding protein-2 messenger RNA during postnatal brain development: correlation with insulin-like growth factors I and II. Neuroscience 53, 251–265.

Leme, R.J., Chadi, G., 2001. Distant microglial and astroglial activation secondary to experimental spinal cord lesion. Arq. Neuropsiquiatr. 59, 483–492.

Lemke, R., Rossner, S., Schliebs, R., 1999. Leukemia inhibitory factor expression is not induced in activated microglia and reactive astrocytes in response to rat basal forebrain cholinergic lesion. Neurosci. Lett. 267, 53–56.

Levison, S.W., Ducceschi, M.H., Young, G.M., Wood, T.L., 1996. Acute exposure to CNTF *in vivo* induces multiple components of reactive gliosis. Exp. Neurol. 141, 256–268.

Lisovoski, F., Akli, S., Peltekian, E., Vigne, E., Haase, G., Perricaudet, M., Dreyfus, P.A., Kahn, A., Peschanski, M., 1997. Phenotypic alteration of astrocytes induced by ciliary neurotrophic factor in the intact adult brain, as revealed by adenovirus-mediated gene transfer. J. Neurosci. 17, 7228–7236.

Liu, H., Prayson, R.A., Estes, M.L., Drazba, J.A., Barnett, G.H., Bingaman, W., Liu, J., Jacobs, B.S., Barna, B.P., 2000. *In vivo* expression of the Interleukin 4 receptor alpha by astrocytes in epilepsy cerebral cortex. Cytokine 12, 1656–1661.

Luo, Z., Geschwind, D.H., 2001. Microarray applications in neuroscience. Neurobiol. Dis. 8, 183–193.

Marco, S., Canudas, A.M., Canals, J.M., Gavaldà, N., Pérez-Navarro, E., Alberch, J., 2002. Excitatory amino acids differentially regulate the expression of GDNF, neurturin, and their receptors in the adult rat striatum. Exp. Neurol. 174, 243–252.

Mason, J.L., Suzuki, K., Chaplin, D.D., Matsushima, G.K., 2001. Interleukin-1β promotes repair of the CNS. J. Neurosci. 21, 7046–7052.

McKeon, R.J., Silver, J., Large, T.H., 1997. Expression of full-length trkB receptors by reactive astrocytes after chronic CNS injury. Exp. Neurol. 148, 558–567.

McMillian, M.K., Thai, L., Hong, J.-S., O'Callaghan, J.P., Pennypacker, K.R., 1994. Brain injury in a dish: a model for reactive gliosis. Trends Neurosci. 17, 138–142.

Merrill, J.E., Benveniste, E.N., 1996. Cytokines in inflammatory brain lesions: helpful and harmful. Trends Neurosci. 19, 331–338.

Miyazaki, H., Nagashima, K., Okuma, Y., Nomura, Y., 2001. Expression of glial cell line-derived neurotrophic factor induced by transient forebrain ischemia in rats. Brain Res. 922, 165–172.

O'Callaghan, J.P., Miller, D.B., 1993. Quantification of reactive gliosis as an approach to neurotoxicity assessment. NIDA Res. Monogr. 136, 188–212.

Oderfeld-Nowak, B., Bacia, A., Gradkowska, M., Fusco, M., Vantini, G., Leon, A., Aloe, L., 1992. *In vivo* activated brain astrocytes may produce and secrete nerve growth factor-like molecules. Neurochem. Int. 21, 455–461.

Oderfeld-Nowak, B., Zaremba, M., Micera, A., Aloe, L., 2001. The upregulation of nerve growth factor receptors in reactive astrocytes of rat spinal cord during experimental autoimmune encephalomyelitis. Neurosci. Lett. 308, 165–168.

Orzylowska, O., Oderfeld-Nowak, B., Zaremba, M., Januszewski, S., Mossakowski, M., 1999. Prolonged and concomitant induction of astroglial immunoreactivity of interleukin-1beta and interleukin-6 in the rat hippocampus after transient global ischemia. Neurosci. Lett. 263, 72–76.

Otto, V.I., Stahel, P.F., Rancan, M., Kariya, K., Shohami, E., Yatsiv, I., Eugster, H.-P., Kossmann, T., Trentz, O., Morganti-Kossmann, M.C., 2001. Regulation of chemokines and chemokine receptors after experimental closed head injury. Neuroreport 12, 2059–2064.

Papavassiliou, E., Gogate, N., Proescholdt, M., Heiss, J.D., Walbridge, S., Edwards, N.A., Oldfield, E.H., Merrill, M.J., 1997. Vascular endothelial growth factor (vascular permeability factor) expression in injured rat brain. J. Neurosci. Res. 49, 451–460.

Pike, C.J., Cummings, B.J., Cotman, C.W., 1995. Early association of reactive astrocytes with senile plaques in Alzheimer's disease. Exp. Neurol. 132, 172–179.

Poulsen, K.T., Armanini, M.P., Klein, R.D., Hynes, M.A., Phillips, H.S., Rosenthal, A., 1994. TGFβ2 and TGFβ3 are potent survival factors for midbrain dopaminergic neurons. Neuron 13, 1245–1252.

Ridet, J.L., Malhotra, S.K., Privat, A., Gage, F.H., 1997. Reactive astrocytes: cellular and molecular cues to biological function. Trends Neurosci. 20, 570–577.

Rohlmann, A., Wolff, J.R., 1996. Subcellular topography and plasticity of gap junction distribution on astrocytes. In: Spray, D.C., Dermietzel, R. (Eds.), Gap Junctions in the Nervous System. R.G. Landes Co, Austin, TX, pp. 175–192.

Rottman, J.B., Ganley, K.P., Williams, K., Wu, L., Mackay, C.R., Ringler, D.J., 1997. Cellular localization of the chemokine receptor CCR5. Am. J. Pathol. 151, 1341–1351.

Rutka, J.T., Murakami, M., Dirks, P.B., Hubbard, S.L., Becker, L.E., Fukuyama, K., Jung, S., Tsuga, A., Matsuzawa, K., 1997. Role of glial filamants in cells and tumors of glial origin: a review. J. Neurosurg. 87, 420–430.

Salhia, B., Angelov, L., Roncari, L., Wu, X., Shannon, P., Guha, A., 2000. Expression of vascular endothelial growth factor by reactive astrocytes and associated neoangiogenesis. Brain Res. 883, 87–97.

Sanders, V.J., Pittman, C.A., White, M.G., Wang, G., Wiley, C.A., Achim, C.L., 1998. Chemokines and receptors in HIV encephalitis. AIDS 12, 1021–1026.

Sheng, J.G., Shirabe, S., Nishiyama, N., Schwartz, J.P., 1993. Alterations in striatal glial fibrillary acidic protein expression in response to 6-hydroxydopamine-induced denervation. Exp. Brain Res. 9, 450–456.

Shozuhara, H., Onodera, H., Katoh-Semba, R., Kato, K., Yamasaki, Y., Kogure, K., 1992. Temporal profiles of nerve growth factor beta-subunit level in rat brain regions after transient ischemia. J. Neurochem. 59, 175–180.

Simpson, J., Rezaie, P., Newcombe, J., Cuzner, M.L., Male, D., Woodroofe, M.N., 2000. Expression of the β-chemokine receptors CCR2, CCR3 and CCR5 in multiple sclerosis central nervous system tissue. J. Neuroimmunol. 108, 192–200.

Sköld, M., Culheim, S., Hammarberg, H., Piehl, F., Suneson, A., Lake, S., Sjögren, A.M., Walum, E., Risling, M., 2000. Induction of VEGF and VEGF receptors in the spinal cord after mechanical spinal injury and prostaglandin administration. Eur. J. Neurosci. 12, 3675–3686.

Smith, C., Berry, M., Clarke, W.E., Logan, A., 2001. Differential expression of fibroblast growth factor-2 and fibroblast growth factor receptor 1 in a scarring and nonscarring model of CNS injury in the rat. Eur. J. Neurosci. 13, 443–456.

Sondell, M., Sundler, F., Kanje, M., 2000. Vascular endothelial growth factor is a neurotrophic factor which stimulates axonal outgrowth through the flk-1 receptor. Eur. J. Neurosci. 12, 4243–4254.

Spirin, K.S., Ljubimov, A.V., Castellon, R., Wiedoeft, O., Marano, M., Sheppard, D., Kenney, M.C., Brown, D.J., 1999. Analysis of gene expression in human bullous keratopathy corneas containing limiting amounts of RNA. Invest. Ophthalmol. Vis. Sci. 40, 3108–3115.

Stadelmann, C., Kerschensteiner, M., Misgeld, T., Brück, W., Hohlfeld, R., Lassmann, H., 2002. BDNF and gp145trkB in multiple sclerosis brain lesions: neuroprotective interactions between immune and neuronal cells? Brain 125, 75–85.

Strack, A., Asensio, V.C., Campbell, I.L., Schlüter, D., Deckert, M., 2002. Chemokines are differentially expressed by astrocytes, microglia, and inflammatory leukocytes in Toxoplasma encephalitis and critically regulated by interferon-γ. Acta Neuropathol. 103, 458–468.

Strauss, S., Otten, U., Joggerst, B., Pluss, K., Volk, B., 1994. Increased level of nerve growth factor (NGF) protein and mRNA and reactive gliosis following kainic acid injection into the rat striatum. Neurosci. Lett. 168, 193–196.

Strömberg, I., Björklund, L., Johansson, M., Tomac, A., Collins, F., Olson, L., Hoffer, B., Humpel, C., 1993. Glial cell line-derived neurotrophic factor is expressed in the developing but not adult striatum and stimulates developing dopamine neurons *in vivo*. Exp. Neurol. 124, 401–412.

Wang, E., Miller, L.D., Ohnmacht, G.A., Liu, E.T., Marincola, F.M., 2000. High-fidelity mRNA amplification for gene profiling. Nat. Biotech. 18, 457–459.

Westmoreland, S.V., Rottman, J.B., Williams, K.C., Lackner, A.A., Sasseville, V.G., 1998. Chemokine receptor expression on resident and inflammatory cells in the brain of macaques with Simian Immunodeficiency Virus encephalitis. Am. J. Pathol. 152, 659–665.

Wick, A., Wick, W., Waltenberger, J., Weller, M., Dichgans, J., Schulz, J.B., 2002. Neuroprotection by hypoxic preconditioning requires sequential activation of vascular endothelial growth factor receptor and Akt. J. Neurosci. 22, 6401–6407.

Widenfalk, J., Lundströmer, K., Jubran, M., Brené, S., Olson, L., 2001. Neurotrophic factors and receptors in the immature and adult spinal cord after mechanical injury or kainic acid. J. Neurosci. 21, 3457–3475.

Winter, C.G., Saotome, Y., Levison, S.W., Hirsh, D., 1995. A role for ciliary neurotrophic factor as an inducer of reactive gliosis, the glial response to central nervous system injury. Proc. Natl Acad. Sci. USA 92, 5865–5869.

Xia, M.Q., Bacskai, B.J., Knowles, R.B., Qin, S.X., Hyman, B.T., 2000. Expression of the chemokine receptor CXCR3 on neurons and the elevated expression of its ligand IP-10 in reactive astrocytes: *in vitro* ERK1/2 activation and role in Alzheimer's disease. J. Neuroimmunol. 108, 227–235.

Zhao, B., Schwartz, J.P., 1998. Involvement of cytokines in normal CNS development and neurological diseases: recent progress and perspectives. J. Neurosci. Res. 52, 7–16.

Zhou, M., Schools, G.P., Kimelberg, H.K., 2000. GFAP mRNA positive glia acutely isolated from rat hippocampus predominantly show complex current patterns. Mol. Brain Res. 76, 121–131.

Advances in
Molecular and
Cell Biology

The nitric oxide/cyclic GMP pathway in CNS glial cells

Agustina García[a,*] and María Antonia Baltrons[b]

[a]*Instituto de Biotecnología y Biomedicina V. Villar Palasí, Universidad Autónoma de Barcelona, 08193 Bellaterra (Cerdanyola del Vallés), Barcelona, Spain*

[*]*Correspondence address: E-mail: agustina.garcia@uab.es(A.G.)*

[b]*Instituto de Biotecnología y Biomedicina V. Villar Palasí and Departamento de Bioquímica y Biología Molecular, Universidad Autónoma de Barcelona, 08193 Bellaterra (Cerdanyola del Vallés), Barcelona, Spain*

Contents

1. Introduction
2. NO formation in glial cells
   2.1. Expression and regulation of calcium-dependent NOS isoforms
   2.2. Induction and regulation of NOS2
3. Cyclic GMP synthesis
   3.1. Guanylyl cyclases
   3.2. NO-sensitive guanylyl cyclase expression in glial cells
   3.3. Regulation of NO-sensitive guanylyl cyclase in astroglial cells
4. Cyclic GMP inactivation in astroglial cells
   4.1. Cyclic GMP phosphodiesterases
   4.2. Cyclic GMP efflux
5. Targets and actions of cGMP in glial cells
6. Concluding remarks

The NO-cGMP signaling cascade participates in essential CNS functions. NO has been also recognized as a neuropathological agent mediating excitotoxic cell death and neuroinflammatory cell damage. All CNS parenchymal cells have the capacity to synthesize NO but only neurons and astroglial cells appear to express all the molecular components required for NO-cGMP-PKG signaling. Recent evidence implicates this pathway in the regulation of important aspects of astroglial physiology—calcium homeostasis, gene expression and survival—that are relevant for neuronal function.

1. Introduction

Shortly after the identification of NO as the endothelium-derived relaxing factor (Ignarro et al., 1987; Palmer et al., 1987), it was reported that this free radical gas was also

Advances in Molecular and Cell Biology, vol. 31, pages 575–593

ISBN: 0-444-51451-1

a signaling molecule in the CNS. The observation that stimulation of cyclic GMP (cGMP) formation by excitatory amino acids in brain cells was mediated by NO (Garthwaite et al., 1988) lead to a plethora of studies that have established the role of NO as an atypical neurotransmitter in both CNS and PNS, as well an important modulator of synaptic plasticity processes underlying memory formation and behavior, brain development, visual and sensory processing, neuroendocrine secretion and cerebral blood flow. Moreover, NO has been recognized as a neuropathological agent responsible for excitotoxic cell death and neuroinflammatory cell damage in conditions such as epilepsy, stroke and neurodegenerative disorders (for reviews see Zhang and Snyder, 1995; Szabó, 1996; Bredt, 1999; Bolaños and Almeida, 1999; Murphy, 2000; Garthwaite, 2000).

NO signaling begins with its synthesis by a nitric oxide synthase (NOS). Three NOS isoforms (types 1–3) encoded by distinct genes and several splice variants have been identified. In their active form NOS are homodimers that catalyze the formation of NO and L-citrulline from L-arginine, using O_2 and NADPH as co-substrates, and FMN, FAD, heme and tetrahydrobiopterin as cofactors. NOS1 and NOS3, also known as nNOS and eNOS because they were first identified in neurons and endothelial cells, are constitutively expressed and activated by binding calmodulin in a reversible calcium-dependent manner. In contrast, NOS2 or iNOS, originally identified in macrophages, is regulated by transcriptional induction and does not require calcium increase for activity, since it has tightly bound calmodulin (for reviews see Förstermann and Kleinert, 1995; Nathan, 1997; Alderton et al., 2001). The three NOS isoforms can be expressed in CNS parenchymal cells. In the normal brain, NOS1, predominantly expressed in discrete populations of neurons, is the main NOS responsible for controlled, transient generation of low concentrations of NO. However, in neuropathological conditions associated with inflammation, glial cells become an important source of sustained and high production of NO by the action of NOS2.

As in the cardiovascular system, the major physiological target for NO in the CNS is a NO-sensitive guanylyl cyclase that catalyzes the conversion of GTP into cGMP. This nucleotide mediates most of the low concentration effects of NO through interaction with cGMP-dependent protein kinases (PKGs), cGMP-regulated phosphodiesterases (PDEs) and cGMP-regulated ion channels (Schmidt et al., 1993; Garthwaite, 2000; Lucas et al., 2000). In contrast to the potentially ubiquitous NO synthesis in CNS cells, NO-dependent cGMP formation appears to occur mainly in neurons and astrocytes. The role of NO as a signaling molecule in neuronal cells and as a pathological agent in CNS disorders has been extensively reviewed. Here we review evidence on the expression and regulation of components of the NO-cGMP signaling pathway in CNS glial cells and their implication in the regulation of glial and neuronal activity.

2. NO formation in glial cells

2.1. *Expression and regulation of calcium-dependent NOS isoforms*

Most of the NOS1 constitutively expressed in brain cells is found in discrete populations of interneurons in different brain nuclei (Zhang and Snyder, 1995; Bredt, 1999). Of the three splice variants of NOS1 (α, β and γ) expressed in rodent brain,

NOS1α is the most abundant and the predominant form in neurons. NOS1α contains exon 2 that carries a PDZ domain that is absent in the β and γ variants. The PDZ domain interacts with postsynaptic density protein PSD-95 that also binds to glutamate NMDA receptors, thus allowing an efficient stimulation of NOS1 by calcium entering through the NMDA channel (reviewed by Tomita et al., 2001). The presence of the other calcium-dependent isoform NOS3 in populations of neurons has been also reported (Dinerman et al., 1994). Although initially controversial there is now solid evidence that NOS1 and NOS3 are also expressed in populations of astrocytes. In contrast, calcium-dependent NOS activities have not been detected in microglia or oligodendrocytes (Agulló et al., 1995; Keilhoff et al., 1998).

The first indication of the expression of calcium-dependent NOS activities in astrocytes was obtained using primary cultures from rat brain (reviewed in Murphy et al., 1993). In our laboratory, results from a comparative study in cultured forebrain neurons and astrocytes, using a series of agonists known to stimulate cGMP formation in different CNS preparations, showed that while several agonists stimulated NO-dependent cGMP formation in astrocytes, only glutamate (acting on NMDA receptors) gave significant responses in neurons (Agulló and García, 1992). Noradrenaline acting on α_1-adrenoceptors was the most effective agonist in astrocytes from different brain regions and the largest response was observed in cerebellar cells (Agulló et al., 1995). Since noradrenaline had no effect on neuronal cultures we suggested astrocytes as the site of the α_1-adrenoceptor-mediated stimulation of cGMP formation well documented in cerebellar slices (Wood, 1991). We later showed that glutamate acting on AMPA receptors and endothelin acting through ET_A receptors also elicited large responses in cerebellar astroglia (Baltrons and García, 1997; Saadoun and García, 1999). Immunoreactivity towards different anti-NOS1 antibodies as well as the histochemical reaction for NADPH diaphorase was demonstrated in the cerebellar astroglial cultures fixed with low concentration of aldehyde (Arbonés et al., 1996). Immunohistochemical studies using higher aldehyde concentrations showed weak NOS1 immunoreactivity in astroglial cells in different brain regions of rodents and humans (Schmidt et al., 1992; Wendland et al., 1994; Egberongbe et al., 1994; Lüth, 1997). More convincing evidence for the expression of NOS1 in astrocytes in situ was reported by Kugler and Drenckhahn (1996). These authors, using freeze-dried tissue sections to minimize diffusion artifacts and avoid loss of antigenicity due to fixation with aldehydes, demonstrated strong NOS1 immunostaining in rat hippocampal and cerebellar astrocytes and Bergmann glia.

The NOS1 expressed in rat cerebellar astroglial cells has the same biochemical characteristics (Km for L-arginine, EC_{50} for calcium, inhibitory potency of L-arginine analogs) as the isoform expressed in cerebellar granule cells, but the astroglial activity is totally cytosolic, while a significant fraction of the neuronal activity is membrane-bound (Arbonés et al., 1996). The more labile association of the astroglial enzyme to membranous structures could explain its higher sensitivity to fixation conditions, and it suggests that it may be a variant lacking the PDZ motif. This suggestion is supported by a recent report showing increased expression of the PDZ-lacking NOS1 splice variants β and γ in astrocytes of amyotrophic lateral sclerosis (ALS) patients (Catania et al., 2001). This and other studies additionally suggest a differential regulation of the neuronal and astroglial NOS1 variants and an implication of the astroglial enzyme in brain pathologies. In ALS patients (Catania et al., 2001) and in a transgenic mice model of ALS, overexpressing a human SOD-1 mutation (Cha et al., 1998), NOS1 is up-regulated in reactive astrocytes in the spinal cord but not in neurons. Intense NOS1 immunopositive reactive astrocytes have also been observed around β-amyloid plaques in hippocampus and entorhinal cortex of brains from Alzheimer disease patients (Simic et al., 2000), and increased NOS1 expression has been documented in astrocytes after cortical spreading depression (Caggiano and Kraig, 1998; Shen and Gundlach, 1999). In this context, NO produced by astroglial NOS1 may be involved in immunomodulatory actions of neurotransmitters, as suggested by reports showing that it inhibits expression of MHC class-II (Heuschling, 1995) and activation of the redox-sensitive transcription factor NFκB (Togashi et al., 1997).

In recent studies, NOS3 immunoreactive astrocytes have been detected in different brain regions of rat and primate species (Gabbot and Bacon, 1996; Wiencken and Casagrande, 1999). As for NOS1 (Kugler and Drenckhahn, 1996), astrocyte processes immunopositive for NOS3 were observed in close contact with the external surface of capillary walls, suggesting that NO generated in astroglia in response to neurotransmitters may couple neuronal activity to changes in blood flow. However, a role for astroglial NOS3 in brain pathologies is also suggested by the observation of increased NOS3 expression in rodent brain astrocytes following a neurotropic viral infection and after intraperitoneal injection of cytokines or bacterial endotoxin (lipopolysaccharide; LPS) (Barna et al., 1996; Iwase et al., 2000), as well as in brains of victims of Alzheimer's disease and other neurodegenerative conditions (Sohn et al., 1999).

2.2. *Induction and regulation of NOS2*

Since the initial demonstration ten years ago (reviewed by Murphy et al., 1993), that LPS and combinations of proinflammatory cytokines (IL-1β, TNF-α, INF-γ) induced NOS2 expression in rodent microglial and astroglial cultures, and that glial-derived NO was toxic to neurons and oligodendrocytes, a large amount of literature has accumulated showing expression of NOS2 in several inflammatory and degenerative diseases of the CNS. The controversial issue of the implication of NO produced by NOS2 in neurotoxicity/neuroprotection continues to stimulate interest in the investigation of the cellular source and factors that regulate NOS2 expression.

A large amount of evidence gathered both in vitro and in vivo indicates that in rodent brain microglia and astroglia are the main cellular site of NOS2 expression. Studies in cultured cells have shown that apart from LPS and cytokines other substances that participate in pathogenic processes leading to neurodegeneration, such as β-amyloid peptides (Rossi and Bianchini, 1996; Hu et al., 1998), protein S100β (Hu et al., 1996), HIV coat proteins (Koka et al., 1995; Hori et al., 1999), chromogranin A (Taupenot et al., 1996) and prion protein fragments (Fabrizi et al., 2001) cause transcriptional activation of NOS2. Additionally, astroglial and/or microglial NOS2 induction has been detected in brains of animals with acute and chronic viral infection (Sun et al., 1995; Grzybicki et al., 1997), following traumatic brain injury (Grzybicki et al., 1998), global and focal cerebral ischemia (Endoh et al., 1994; Loihl et al., 1999), excitotoxic damage (Calka et al., 1996; Acarin et al., 2002), or experimental allergic encephalitis (Okuda et al., 1997) and in transgenic mouse models of familial ALS (Almer et al., 1999) and Alzheimer's disease (Lüth et al., 2001). In contrast, the predominant site of NOS2 expression in humans appears to be astrocytes rather than microglia. NOS2 induction in response to IL-1β and INF-γ or TNF-α is readily observed in fetal and adult human astrocytes (Lee et al., 1993; Hu et al., 1995; Zhao et al., 1998), but only occasionally demonstrated in microglia (Colasanti et al., 1995; Ding et al., 1997). Astroglial NOS2 immunoreactivity has been reported in postmortem brain tissue of patients suffering from Multiple Sclerosis (Bö et al., 1994), Alzheimer's disease (Wallace et al., 1997), Parkinsonism (Hunot et al., 1996) and Krabbe's disease (Giri et al., 2002) and in the optic nerve head of human glaucomatous eyes (Liu and Neufeld, 2000). There is increasing evidence from in vitro and in vivo studies that NOS2 can also be expressed in immunostimulated neurons in rodents and humans (review by Heneka and Feinstein, 2001). However, expression of NOS2 in oligodendrocytes has been demonstrated only in cultured rodent cells (Merrill et al., 1997; Molina-Holgado et al., 2001).

Expression of glial NOS2 is mainly regulated at the transcriptional level, although changes in mRNA stability in response to some agents have also been described (review in Murphy, 2000; Colasanti and Persichini, 2000; Baltrons and García, 2001). As found for other cell types, mitogen-activated protein kinase (MAPK)-mediated signaling appears to play a prominent role in endotoxin and cytokine transcriptional activation of glial NOS2. The promoter regions of NOS2, as well as other proinflammatory molecules, contain consensus binding sites for numerous transcription factors such as NFκB, AP-1, CRE and C/EBP (Eberhardt et al., 1996) that are directly or indirectly phosphorylated by MAPKs. NFκB, whose activation has been described during neurological disease and trauma (Mattson and Camandola, 2001), plays a key role in the transcriptional activation of NOS2 expression. The activation of NFκB involves its release from a complex with IκB followed by translocation to the nucleus. Phosphorylation of IκB by a cytokine-activated kinase marks it for destruction by the 20S proteasome. Several factors prevent NOS2 induction by interfering with NFκB activation or binding to the gene promoter. In astrocytes, as in other cells, exogenous NO has been shown to reduce NOS2 transcriptional activation by both of these mechanisms, suggesting an explanation for the transient expression of NOS2 (review in Colasanti and Persichini, 2000). The heat shock response induced by hyperthermia or treatment with the fungal-derived antibiotics ansamycins blocks LPS and cytokine-dependent activation of NFκB and expression of NOS2 in cultured astroglia and in vivo by increasing IκBα (Feinstein et al., 1996; Heneka et al., 2000; Murphy et al., 2002). The same mechanism appears to be responsible for suppression of NFκB activation and NOS2 induction by other anti-inflammatory compounds like glucocorticoids (Quan et al., 2000). Furthermore, the neurotransmitter noradrenaline, that blocks NOS2 expression in astrocytes via the cAMP/PKA pathway, increases IκB expression (Galea and Feinstein, 1999). Conversely, depletion of central noradrenaline levels by lesioning of the locus ceruleus reduces IκB levels and potentiates inflammatory responses to LPS (Gavrilyuk et al., 2002).

3. Cyclic GMP synthesis

3.1. Guanylyl cyclases

Cyclic GMP is formed by an ubiquitous family of guanylyl cyclases (GCs) that comprises peptide-regulated and NO-sensitive isoforms. Peptide-regulated GCs or particulate GCs (pGCs) are homodimers or homotetramers of single membrane-spanning subunits that contain a N-terminal extracellular ligand binding domain and intracellular regulatory and catalytic domains (for reviews see Garbers, 1999; Lucas et al., 2000). The pGCs expressed in the CNS belong to the natriuretic peptide receptor group. Using immunocytochemical techniques, accumulation of cGMP in response to atrial natriuretic peptides has been observed in astrocytes in restricted areas of the adult rat brain, but only in a few neuronal structures (De Vente and Steinbusch, 2000), and not in oligodendrocytes or microglial cells (Tanaka et al., 1997).

The NO-sensitive GC, a major target for NO, was until very recently (see below) considered a cytosolic enzyme and usually referred to as soluble GC. For the sake of simplicity we will use the abbreviation sGC for the NO-sensitive GC throughout this article. sGCs are largely heterodimers of α and β subunits containing one protoporphyrin IX type heme that is the binding site for NO. Each subunit contains a N-terminal domain involved in heme binding, a central domain thought to be implicated in dimerization and a C-terminal catalytic domain homologous to pGC and adenylyl cyclase catalytic domains. The α1 and α2 subunits can form functional heterodimers with the β1 subunit that are functionally indistinguishable, and both α1β1 and α2β1 have been found at the protein level (for reviews see Denninger and Marletta, 1999; Russwurm and Koesling, 2002). The β2 subunit, primarily expressed in kidney, presents activity in the absence of other subunits (Koglin et al., 2001).

Recent reports question the concept of NO-activated GC as a soluble enzyme. It has been shown that the α2/β1 isoform associates with brain membranes by interaction of the α2 subunit C-terminal amino acids with a PDZ domain of PSD-95 (Russwurm et al., 2001). This interaction will position the enzyme close to the NMDA receptor-activated NOS1 (see Section 2.1) leading to an efficient stimulation of NO-dependent cGMP formation by glutamate. However, NOS1 associated sGC may be a minor portion of brain sGC, since immunocytochemical studies have generally found that co-localization of NOS1 and NO-stimulated cGMP accumulation is minor, whereas independent but complementary staining is abundant throughout the brain (reviewed by De Vente and Steinbusch, 2000). The α1β1 dimer was also reported to be partially associated with membranes in different tissues, including brain, in a state sensitized to NO. Activation of the enzyme increases its translocation to the membrane in a manner dependent on increased intracellular calcium (Zabel et al., 2002). This regulatory mechanism contrasts with the reported inhibition of NO-stimulated sGC activity by calcium (Parkinson et al., 1999).

3.2. NO-sensitive guanylyl cyclase expression in glial cells

In situ hybridization studies in adult (Matsuoka et al., 1992; Furuyama, 1993) and postnatal (Gibb and Garthwaite, 2001) rat brain demonstrate a widespread expression of the sGC β1 subunit and a more limited distribution of the α1 and α2 subunits, indicating that different regions predominantly express either the α1β1 or the α2β1 isoform. A differential expression of α1 and β1 subunits in regions of the human brain was also reported (Zabel et al., 1998). Early immunohistochemical studies with antibodies against purified sGC demonstrated immunoreactivity in rat brain neurons and astrocytes (Ariano et al., 1982; Nakane et al., 1983). Furthermore, differential lesioning of cell types in slice preparations (Garthwaite and Garthwaite, 1987) and measurement of GC activity in acutely dissociated cells from rat cerebellum (Bunn et al., 1986) indicated that astrocytes are a major site of NO-dependent cGMP accumulation. In agreement with this, cGMP immunoreactivity was demonstrated in astrocytes in different regions of adult and immature rat brain slices stimulated with NO donors, more prominently in hippocampal and cerebellar astrocytes and in Bergmann glia (reviewed by De Vente and Steinbusch, 2000). Recently, the β1 subunit of sGC was shown to co-localize with cGMP inmunoreactivity in hippocampal astrocytes (Teunissen et al., 2001). In contrast, cGMP immunoreactivity in oligodendrocytes is only observed in immature but not adult rat brain and microglia does not appear to accumulate cGMP in response to NO (Tanaka et al., 1997). Our studies in primary cultures also demonstrate regional differences in NO-dependent cGMP formation in rat astrocytes and the lack of sGC activity in microglia (Agulló et al., 1995). In humans, information about cGMP formation in glial cells is scant. In one report fetal cortical astrocytes in culture were shown to accumulate low levels of cGMP after induction of NOS2 (Ding et al., 1997). Our recent results using a similar preparation (T. Sardón, M.A. Baltrons and A. García, unpublished) show NO donor-stimulated cGMP accumulation in human astrocytes in the presence of a phosphodiesterase inhibitor (Fig. 1A and B).

3.3. Regulation of NO-sensitive guanylyl cyclase in astroglial cells

Despite the recognition of sGC as a major receptor for NO, mediating numerous of its physiological functions there is little information about the mechanisms that regulate its activity in living cells. Studies in rat cerebellar astrocytes show that sGC located within cells is activated by NO with higher potency and faster activation and deactivation than the purified enzyme. Furthermore, within seconds of adding NO, desensitization rapidly occurs in cells but not with the purified enzyme, suggesting the existence of regulatory factors (Bellamy et al., 2000; Bellamy and Garthwaite, 2001a). These characteristics indicate that in the cellular environment sGC behaves like a neurotransmitter receptor. As occurs after chronic exposure of receptors to their agonists, continuous exposure to NO donors has been shown to down-regulate sGC in different cell types (Ujiie et al., 1994; Papapetropoulos et al., 1996). In smooth muscle cells, this phenomenon that is thought to contribute to nitrovasodilator-induced tolerance, apparently results from a decrease in subunit mRNA stability (Filippov et al., 1997). A similar effect is produced by endogenous NO generated by NOS2 in cells treated with proinflammatory cytokines (Takata et al., 2001).

In contrast, in rat astroglial cultures we have shown that chronic treatment with LPS or IL-1β decreases sGC activity and β1 subunit levels in a NO-independent manner

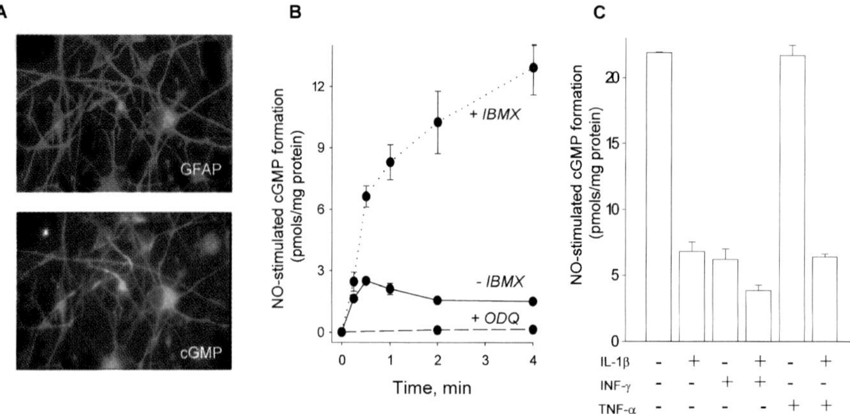

Fig. 1. NO-stimulated cGMP accumulation in human fetal cortical astrocyte cultures. (A) Cultures stimulated with sodium nitroprusside (SNP; 100 μM, 10 min) in the presence of the phosphodiesterase inhibitor 3-isobutyl-1-methylxanthine (IBMX; 1 mM) were double immunofluorescence labeled with sheep anti-formaldehyde fixed cGMP (1:4000) and mouse anti-glial fibrillary acidic protein (GFAP; 1:350), visualized with fluorescein-conjugated anti-sheep and rhodamine-conjugated anti-mouse IgGs, respectively; magnification 630 × in the negative. (B) Time course of cGMP accumulation in response to diethylamine/NO (2 μM) in the presence or absence of IBMX (1 mM) or the guanylyl cyclase inhibitor 1H-[1,2,4]oxodiazolo[4,3-a]quinoxalin-1-one (ODQ; 1 μM) added 10 min before. (C) Cell cultures were treated or not with human recombinant IL-1β (100 U/ml), TNF-α (100 U/ml) or INF-γ (200 U/ml) for 72 h, washed and stimulated with SNP in the presence of IBMX.

(Baltrons and García, 1999; Pedraza et al., 2002). Down-regulation of the β1 subunit is due to a decrease in the half-life of the protein and requires transcription and protein synthesis, suggesting that the inflammatory agents induce the expression of a protein that is directly or indirectly responsible for sGC degradation. Additionally, LPS and IL-1β induce a decrease in sGC α1 and β1 subunit mRNA levels that is NO-dependent. The latter mechanism will contribute to maintain sGC levels low under conditions of prolonged reactive gliosis, associated with high levels of NO production (Fig. 2), and it may explain the long time required for recovery of sGC activity after removal of the inflammatory compound (Baltrons and García, 1999). Treatment with β-amyloid peptides that induce astroglial reactivity similarly down-regulates sGC (Baltrons et al., 2002). The decrease in sGC protein and mRNA also occurs in rat brain after intracerebral administration of LPS, IL-1β or β-amyloid peptides (Baltrons et al., 2002; Pedraza et al., 2002). Inflammatory agents may exert a similar effect in human brain since sGC activity was reported to be decreased in brain of Alzheimer's disease patients (Bonkale et al., 1995). In agreement with this observation, our recent results show that IL-1β and INF-γ, but not TNF-α decrease NO-stimulated cGMP formation in human foetal astroglia-enriched cultures (Fig. 1C). The relevance of impaired NO-dependent cGMP formation in reactive astroglia during neuroinflammation is difficult to ascertain at present.

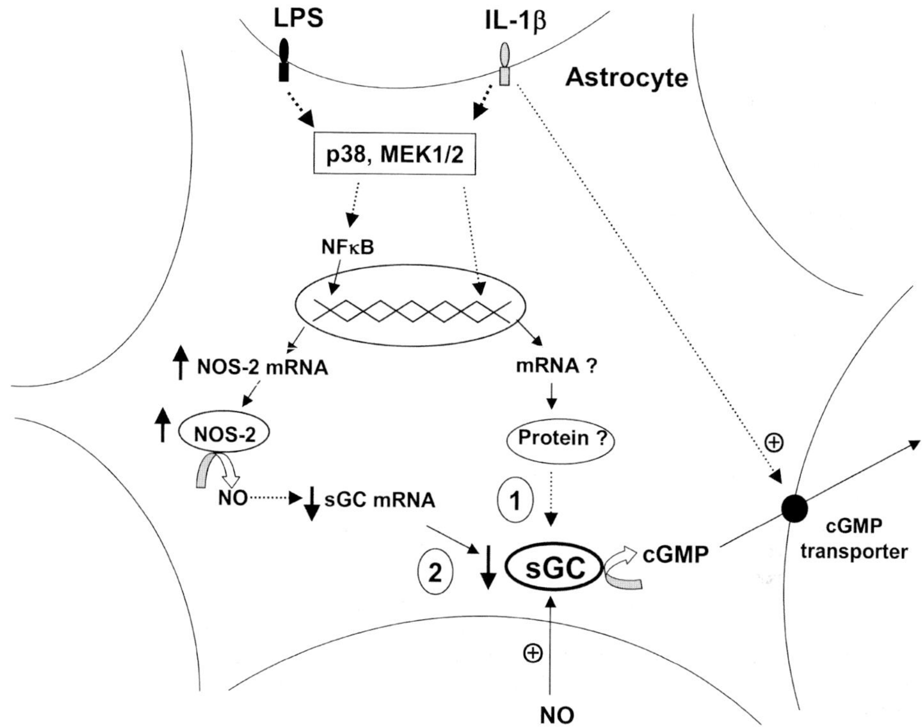

Fig. 2. Schematic representation of LPS and IL-1β effects on NO-dependent cGMP accumulation in astroglial cells. Treatment of rat brain astrocytes with LPS or IL-1β increases NOS2 expression and simultaneously down-regulates sGC by two mechanisms: (1) a NO-independent decrease of the half-life of sGC protein that requires transcription and translation; (2) a NO-dependent decrease in sGC mRNA levels that will contribute to maintain sGC levels low during prolonged periods of time. Additionally, IL-1β increases transporter-mediated efflux of cGMP.

4. Cyclic GMP inactivation in astroglial cells

4.1. Cyclic GMP phosphodiesterases

Cyclic nucleotide phosphodiesterase hydrolysis of cGMP is the major mechanism underlying the clearance of the nucleotide. Eleven families of PDEs have been identified that comprise more than 50 isoforms encoded by 21 genes (Soderling and Beavo, 2000; Lucas et al., 2000). PDEs differ in their primary structure, affinity for cyclic nucleotides, sensitivity to calcium and other inhibitors and tissue distribution. According to the specificity for the nucleotide determined in cell free systems, PDEs are divided in three groups: PDEs hydrolyzing cAMP and cGMP (PDE1, PDE2, PDE3, PDE10 and PDE11), PDEs hydrolyzing cAMP (PDE4, PDE7 and PDE8) and PDEs highly specific for cGMP. Among the latter PDE6 is unique to photoreceptors and plays a crucial role in the visual transduction cascade, while PDE5 and PDE9 are differentially expressed in rat brain regions. In situ hybridization and immunocytochemical studies indicate that both are strongly expressed in the Purkinje cell layer of the rat cerebellum (Andreeva et al., 2001; Kotera et al., 1997), where local administration of the PDE-5/PDE-9 inhibitor zaprinast facilitates cGMP regulated long-term depression (Hartell, 1996). PDE9A, the isoform that shows the highest affinity for cGMP, is also expressed in the forebrain, hippocampus and olfactory bulb, with a distribution similar but not identical to NOS1 and sGC (Andreeva et al., 2001). Additionally, early studies indicated that the predominant cGMP degrading activity in brain is a highly active calcium/calmodulin-dependent PDE1 (Shenolikar et al., 1985; Mayer et al., 1992), and various isoforms of this group have been localized in neuronal structures (Kincaid et al., 1987; Yan et al., 1994; Furuyama et al., 1994). No mention was made in any of these reports about a glial localization of PDE isoforms. However, immunoreactivity for cGMP is potentiated by the non-specific PDE inhibitor IBMX in neurons and astrocytes in different rat brain regions, indicating that both cells types contain cGMP-PDE activities (De Vente and Steinbusch, 2000). Potentiation by IBMX of NO-stimulated cGMP accumulation is also observed in cultured astrocytes from rat (Baltrons et al., 1997) and human brain (Fig. 1A and B).

In rat cerebellar granule cells and astrocytes we observed that agents that increase intracellular calcium decrease cGMP accumulation in a calmodulin-dependent manner (Baltrons et al., 1997). Activity measurements in homogenates confirmed that a calcium/calmodulin-dependent PDE or PDE1 of similar biochemical and pharmacological characteristics was present in both cell types (Agulló and García, 1997). The calcium concentration required for half-maximal stimulation of PDE1 and NOS1 was similar, suggesting that larger NO-stimulated cGMP accumulations will occur in cell compartments different from those where NO is generated in a calcium-dependent manner. In agreement with this hypothesis, immunocytochemical studies in cerebellar slices show that stimulation of neuronal NO production by NMDA increases cGMP more in astrocytes than in neurons, whereas NO donors produce more generalized increases in cGMP (De Vente and Steinbusch, 1992). In contrast to our results, Bellamy and Garthwaite (2001b) using acutely dissociated astrocytes from 8-day-old rat cerebellum showed by pharmacological means that cGMP was predominantly hydrolyzed by PDE5, with a minor contribution from PDE4 and no detectable calcium-dependent PDE activity. Since PDEs are developmentally regulated (Billingsley et al., 1990; Kotera et al., 1997), a difference in the developmental stage of the two cell preparations may explain the discrepancy. Nevertheless, our recent results also indicate that a PDE5 is present in rat cerebellar astroglial cultures. An increase in this activity appears to be responsible for the decrease in NO-dependent cGMP accumulation that is induced by long-term treatment with the HIV-1 virus coat protein gp120 (Navarra et al., 2002).

4.2. *Cyclic GMP efflux*

Release of cGMP after stimulation of GC by NO or natriuretic peptides has been shown in several tissues, including brain (Kapoor and Krishna, 1977; Tjörnhammar et al., 1986; Schultz et al., 1998). This efflux has been generally considered a mechanism that in conjunction with PDEs would return intracellular cGMP to basal levels. Additionally, extracellular cGMP may play a role in cell–cell communication by regulating the activity of membrane proteins through interaction with extracellular sites. For instance, in cerebellar granule neurons extracellular cGMP inhibits kainate receptors (Pouloupoulou and Nowak, 1998) and protects against glutamate-induced toxicity (Montoliu et al., 1999). In rat cortical astrocytes, cGMP released after stimulation with natriuretic peptides was reported to inhibit a Na^+/H^+ exchanger leading to a decrease in intracellular pH, an effect that could modulate important astroglial functions such as K^+ conductance or cell proliferation (Touyz et al., 1997). We have found that treatment of cerebellar astroglial cells with IL-1β induces efflux of cGMP formed after induction of NO release by the cytokine or by LPS (Pedraza et al., 2001). Like the primary active transport system for cGMP described in erythrocyte membranes (Schultz et al., 1998), cGMP extrusion in astrocytes was blocked by inhibitors of organic anion transporters (Touyz et al., 1997; Pedraza et al., 2001). Recently, cGMP was shown to be a high affinity substrate for the multidrug resistance protein isoform, MRP5 (Jedlitschky et al., 2000), and this organic anion transporter is expressed in astrocytes (Hirrlinger et al., 2002). Thus it is tempting to speculate that IL-1β may regulate the activity/expression of MRP5 or a similar transporter in astrocytes, and it will be interesting to investigate possible actions of extracellular cGMP in the context of neuroinflammation.

5. Targets and actions of cGMP in glial cells

All three recognized molecular targets for cGMP, i.e., PKGs, cGMP-regulated PDEs and cGMP-regulated ion channels, have been found in CNS structures, but only in the case of PKGs is there any information about its expression and function in glial cells. The two isoforms of PKG, PKGI (splice variants Iα and Iβ) and PKGII are expressed in brain cells. PKGI, a cytosolic enzyme, is particularly concentrated in Purkinje cells of the cerebellum and in sensory neurons (El-Husseini et al., 1999), while PKGII, a membrane bound enzyme, shows a widespread distribution in brain, often co-localizing with NO-stimulated cGMP formation, and is associated with neurons, astrocytes and oligodendrocytes (De Vente et al., 2001). Although PKGs are thought to be the major intracellular receptor proteins for cGMP and mediate many of the actions of the nucleotide in brain cells very few specific substrates have been demonstrated (Wang and Robinson, 1997; Lucas et al., 2000).

In astrocytes the NO-cGMP-PKG pathway has been implicated in the generation of calcium waves (see chapter by Cornell-Bell et al.). Addition of NO to forebrain astrocytes was shown to increase intracellular calcium by mobilization from ryanodine-sensitive stores (see chapter by Scapignini et al.) and calcium influx and to produce intercellular calcium waves via cGMP and PKG. Mechanical stress of individual astrocytes also resulted in intercellular calcium waves which were NO, sGC and PKG dependent (Willmott et al., 2000). Amplification of the NO effect by activation of astroglial calcium-dependent NOS could contribute to the propagation of calcium waves. This mechanism may be relevant for the bi-directional communication between neurons and astroglia and the active participation of astroglial cells in synaptic integration and processing of information. It is now evident that neuronal activity can trigger calcium increases in astrocytes that are followed by exocytotic release of glutamate which feeds back to neuronal targets (reviewed in Vesce et al., 2001; Araque et al., 2001). This is thought to be due to synaptically released neurotransmitters, such as glutamate or noradrenaline, and these transmitters can both stimulate calcium-dependent NO and cGMP formation in astrocytes (Agulló et al., 1995; Baltrons and García, 1997). Interestingly, in astrocytes of the visual cortex it was shown that the frequency of calcium oscillations was altered by glutamate in a NO-dependent manner, but the production of NO by neurons was not necessary, suggesting that astroglial NO was involved (Pasti et al., 1995). On the other hand, a receptor-independent, but NO-dependent, calcium influx has recently been demonstrated in Bergmann glia after stimulation of parallel fibers, an effect that could have implications for the induction of long-term depression in the cerebellum (Matyash et al., 2001).

Synergistic effects of calcium and NO on gene transcription may be involved in the regulation of neuronal differentiation and survival and synaptic plasticity (Peunova and Enikolopov, 1995; Gudi et al., 1999). Recently it was reported that the NO and cGMP induced activation of the fos promoter mediated by membrane-bound, extranuclear PKGII, requires cell-specific factors that are found in cells of neuronal and astroglial origin (Gudi et al., 1999). The NO/cGMP pathway in astrocytes may also affect neuronal development by regulating NGF release (Xiong et al., 1999).

Compared to the large amount of studies on the mechanisms of NO-induced cell death, much less is known about the implication of cGMP in neurotoxicity/neuroprotection mechanisms. Recent evidence suggests a protective role of cGMP in neural cells (Kim et al., 1999). In cortical astrocytes cGMP via PKG has been shown to prevent apoptosis by inhibiting the mitochondrial transition pore (Takuma et al., 2001). In an immortalized oligodendrocyte cell line derived from O-2A progenitors, cGMP via PKG also protected from kainate-induced cell death by decreasing calcium influx (Yoshioka et al., 2000).

6. Concluding remarks

Existing evidence indicates that besides neurons astroglial cells are the only CNS parenchymal cell type where all the molecular components required for NO-cGMP signaling are expressed. NO can be formed in astrocytes by the three known NOS isoforms and NO can stimulate sCG, resulting in increased intracellular cGMP and PKG-dependent phosphorylation of cell proteins. In contrast, only NOS2 expression has been documented in microglial cells and, as in their peripheral counterparts the macrophages, NO hence produced seems to be implicated in defense mechanisms and neuroinflammation associated cell death. On the other hand, information about the NO/cGMP system in oligodendrocytes is scarce and controversial in some respects. Although no direct evidence has so far been reported, regulation of astroglial function by NO and cGMP is likely to contribute to some of the important neuromodulatory actions in which NO has been implicated. Three important aspects of astroglial physiology—calcium homeostasis, gene expression and survival—that are relevant for neuronal function, are regulated by the NO-cGMP-PKG pathway. Additionally, cGMP extruded from astroglial cells may affect the activity of neuronal membrane proteins. In the next few years we will probably witness significant advances in the understanding of how the NO-cGMP pathway participates in astroglial regulation of neuronal development and in synaptic integration and processing of information. Tools like the transgenic mice expressing GFAP promoter-controlled green fluorescent protein (Nolte et al., 2001) will probably be very valuable to investigate this fundamental aspect of glial cell function under normal and pathological conditions.

Acknowledgements

We want to thank Carlos E. Pedraza, Teresa Sardón and Michele Navarra who generated part of the results of our laboratory reviewed here and Francisca García and Annabel Segura for technical assistance. We also thank Dr Nuria Toran for providing human brain tissue from 20 to 22-week-old therapeutically aborted fetuses, obtained by a protocol approved by the Ethical Committee of the Hospital de la Vall d'Hebron (Barcelona, Spain). The sheep anti-formaldehyde fixed cGMP antibody was kindly provided by Dr J. de Vente (European Graduate School of Neuroscience, Maastricht University, The Netherlands). This work was supported by SAF2001-2540, Fundació La Marató TV3 1008/97 and SGR2001-212 grants.

References

Acarin, L., Peluffo, H., Gonzalez, B., Castellano, B., 2002. Expression of inducible nitric oxide synthase and cyclooxygenase-2 after excitotoxic damage to the immature rat brain. J. Neurosci. Res. 68, 745–754.

Agulló, L., Baltrons, M.A., García, A., 1995. Calcium-dependent nitric oxide formation in glial cells. Brain Res. 686, 160–168.

Agulló, L., García, A., 1992. Different receptors mediate stimulation of nitric oxide-dependent cyclic GMP formation in neurons and astrocytes in culture. Biochem. Biophys. Res. Commun. 182, 1362–1368.

Agulló, L., García, A., 1997. Ca^{2+}/calmodulin dependent cyclic GMP phosphodiesterase activity in granule neurons and astrocytes from cerebellum. Eur. J. Pharmacol. 323, 119–125.

Alderton, W.K., Cooper, C.E., Knowles, R.G., 2001. Nitric oxide synthases: structure, function and inhibition. Biochem. J. 357, 593–615.

Almer, G., Vukosavic, S., Romero, N., Przedborski, S., 1999. Inducible nitric oxide synthase up-regulation in a transgenic mouse model of familial amyotrophic lateral sclerosis. J. Neurochem. 72, 2415–2425.

Andreeva, S.G., Dikkes, P., Epstein, P.M., Rosenberg, P.A., 2001. Expression of cGMP-specific phosphodiesterase 9A mRNA in the rat brain. J. Neurosci. 21, 9068–9076.

Araque, A., Carmignoto, G., Haydon, P.G., 2001. Dynamic signaling between astrocytes and neurons. Annu. Rev. Physiol. 63, 795–813.

Arbonés, M.L., Ribera, J., Agulló, L., Baltrons, M.A., Casanovas, A., Riveros-Moreno, V., García, A., 1996. Characteristics of nitric oxide synthase type I of rat cerebellar astrocytes. Glia 18, 224–232.

Ariano, M.A., Lewicki, J.A., Brandwein, H.J., Murad, F., 1982. Immunohistochemical localization of guanylate cyclase within neurons of rat brain. Proc. Natl Acad. Sci. USA 79, 1316–1320.

Baltrons, M.A., García, A., 1997. AMPA receptors are couple to the nitric oxide/cyclic GMP pathway in cerebellar astroglial cells. Eur. J. Neurosci. 9, 2497–2501.

Baltrons, M.A., García, A., 1999. Nitric oxide-independent down-regulation of soluble guanylyl cyclase by bacterial endotoxin in astroglial cells. J. Neurochem. 73, 2149–2157.

Baltrons, M.A., García, A., 2001. The nitric oxide/cyclic GMP system in astroglial cells. Progr. Brain Res. 132, 335–347.

Baltrons, M.A., Pedraza, C.E., Heneka, M.T., García, A., 2002. β-Amyloid peptides decrease soluble guanylyl cyclase expression in astroglial cells. Neurobiol. Dis. 10, 139–149.

Baltrons, M.A., Saadoun, S., Agulló, L., García, A., 1997. Regulation by calcium of the nitric oxide/cyclic GMP system in cerebellar granule cells and astroglia in culture. J. Neurosci. Res. 49, 333–341.

Barna, M., Komatsu, T., Reiss, C.S., 1996. Activation of type III nitric oxide synthase in astrocytes following a neurotropic viral infection. Virology 223, 331–343.

Bellamy, T.C., Garthwaite, J., 2001. Sub-second kinetics of the nitric oxide receptor, soluble guanylyl cyclase, in intact cerebellar cells. J. Biol. Chem. 276, 4287–4292.

Bellamy, T.C., Garthwaite, J., 2001. "cAMP-specific" phosphodiesterase contributes to cGMP degradation in cerebellar cells exposed to nitric oxide. Mol. Pharmacol. 59, 54–61.

Bellamy, T.C., Wood, J., Goodwin, D.A., Garthwaite, J., 2000. Rapid desensitization of the nitric oxide receptor, soluble guanylyl cyclase, underlies diversity of cellular cGMP responses. Proc. Natl Acad. Sci. USA 97, 2928–2933.

Billingsley, M.L., Polli, J.W., Balaban, C.D., Kincaid, R.L., 1990. Developmental expression of calmodulin-dependent cyclic nucleotide phosphodiesterase in rat brain. Brain Res. Dev. Brain Res. 53, 253–263.

Bö, L., Dawson, T.M., Wesselingh, S., Mork, S., Choi, S., Kong, P.A., Hanley, D., Trapp, B.D., 1994. Induction of nitric oxide synthase in demyelinating regions of multiple sclerosis brains. Ann. Neurol. 36, 778–786.

Bolaños, J.P., Almeida, A., 1999. Roles of nitric oxide in brain hypoxia–ischemia. Biochim. Biophys. Acta 1411, 415–436.

Bonkale, W.L., Winblad, B., Ravid, R., Cowburn, R.F., 1995. Reduced nitric oxide responsive soluble guanylyl cyclase activity in the superior temporal cortex of patients with Alzheimer's disease. Neurosci. Lett. 187, 5–8.

Bredt, D.S., 1999. Endogenous nitric oxide synthesis: biological functions and pathophysiology. Free Radic. Res. 31, 577–696.

Bunn, St. J., Garthwaite, J., Wilkin, G.P., 1986. Guanylate cyclase activities in enriched preparations of neurones, astroglia and a synaptic complex isolated from rat cerebellum. Neurochem. Int. 8, 179–185.

Caggiano, A.O., Kraig, R.P., 1998. Neuronal nitric oxide synthase expression is induced in neocortical astrocytes after spreading depression. J. Cereb. Blood Flow Metab. 18, 75–87.

Calka, J., Wolf, G., Schmidt, W., 1996. Induction of cytosolic NADPH-diaphorase/nitric oxide synthase in reactive microglia/macrophages after quinolinic acid lesions in the rat striatum: an electron and light microscopical study. Histochem. Cell Biol. 105, 81–89.

Catania, M.V., Aronica, E., Yankaya, B., Troost, D., 2001. Increased expression of neuronal nitric oxide synthase spliced variants in reactive astrocytes of amylotrophic lateral sclerosis human spinal cord. J. Neurosci. 21, RC148(1-5).

Cha, C.I., Kim, J.M., Shin, D.H., Kim, Y.S., Kim, J., Gurney, M.E., Lee, K.W., 1998. Reactive astrocytes express nitric oxide synthase in the spinal cord of transgenic mice expressing a human Cu/Zn SOD mutation. Neuroreport 9, 1503–1506.

Colasanti, M., Persichini, T., 2000. Nitric oxide: an inhibitor of NFκB/Rel system in glial cells. Brain Res. Bull. 52, 155–161.

Colasanti, M., Persichini, T., Di Pucchio, T., Gremo, F., Lauro, G.M., 1995. Human ramified microglial cells produce nitric oxide upon *Escherichia coli* lipopolysaccharide and tumor necrosis factor alpha stimulation. Neurosci. Lett. 200, 144–146.

Denninger, J.V., Marletta, M.A., 1999. Guanylate cyclase and the NO/cGMP signaling pathway. Biochim. Biophys. Acta 1411, 334–350.

De Vente, J., Asan, E., Gambaryan, S., Markerink-van Ittersum, M., Axer, H., Gallatz, K., Lohmann, S.M., Palkovits, M., 2001. Localization of cGMP-dependent protein kinase type II in rat brain. Neuroscience 108, 27–49.

De Vente, J., Steinbusch, H.W.M., 1992. On the stimulation of soluble and particulate guanylate cyclase in the rat brain and the involvement of nitric oxide as studied by cGMP immuno-cytochemistry. Acta Histochem. 92, 13–38.

De Vente, J., Steinbusch, H.W.M., 2000. Nitric oxide-cGMP signaling in the rat brain. In: Steinbusch, H.W.M., De Vente, J., Vincent, S.R., (Eds.), Handbook of Chemical Neuroanatomy. Elsevier, Amsterdam, pp. 355–415.

Dinerman, J.L., Dawson, T.M., Schell, M.J., Snowman, A., Snyder, S.H., 1994. Endothelial nitric oxide synthase localized to hippocampal pyramidal cells: implications for synaptic plasticity. Proc. Natl Acad. Sci. USA 91, 4214–4218.

Ding, M.Z., Stpierre, B.A., Parkinson, J.F., Medberry, P., Wong, J.L., Rogers, N.E., Ignarro, L.J., Merrill, J.E., 1997. Inducible nitric-oxide synthase and nitric oxide production in human fetal astrocytes and microglia. J. Biol. Chem. 272, 11327–11335.

Eberhardt, W., Kunz, D., Hummel, R., Pfeilschifter, J., 1996. Molecular cloning of the rat inducible nitric oxide synthase gene promoter. Biochem. Biophys. Res. Commun. 223, 752–756.

Egberongbe, Y.I., Gentleman, S.M., Falkai, P., Bogerts, B., Polak, J.M., Roberts, G.W., 1994. The distribution of nitric oxide synthase immunoreactivity in the human brain. Neuroscience 59, 561–578.

El-Husseini, A.E., Williams, J., Reiner, P.B., Pelech, S., Vincent, S.R., 1999. Localization of the cGMP-dependent protein kinases in relation to nitric oxide synthase in the brain. J. Chem. Neuroanat. 17, 45–55.

Endoh, M., Maiese, K., Wagner, J., 1994. Expression of the inducible form of nitric oxide synthase by reactive astrocytes after transient global ischemia. Brain Res. 651, 92–100.

Fabrizi, C., Silei, V., Menegazzi, M., Salmona, M., Bugiani, O., Tagliavini, F., Suzuki, H., Lauro, G.M., 2001. The stimulation of inducible nitric-oxide synthase by the prion protein fragment 106–126 in human microglia is tumor necrosis factor-alpha-dependent and involves p38 mitogen-activated protein kinase. J. Biol. Chem. 276, 25692–25696.

Feinstein, D.L., Galea, E., Aquino, D.A., Li, G.C., Xu, H., Reis, D.J., 1996. Heat shock protein 70 suppresses astroglial iNOS expression by decreasing NFkB activation. J. Biol. Chem. 271, 17724–17732.

Filippov, G., Bloch, D.B., Bloch, K.D., 1997. Nitric oxide decreases stability of mRNAs encoding soluble guanylate cyclase subunits in rat pulmonary artery smooth muscle cells. J. Clin. Invest. 100, 942–948.

Förstermann, U., Kleinert, H., 1995. Nitric oxide synthase: expression and expressional control of the three isoforms. Naunyn-Schmiedeberg's Arch. Pharmacol. 356, 351–364.

Furuyama, T., Inagaki, S., Takagi, H., 1993. Localizations of $\alpha 1$ and $\beta 1$ subunits of soluble guanylate cyclase in the rat brain. Mol. Brain Res. 20, 335–344.

Furuyama, T., Iwahashi, Tano, Y., Takagi, H., Inagaki, S., 1994. Localization of 63-kDa calmodulin-stimulated phosphodiesterase mRNA in the rat brain by in situ hybridization histochemistry. Mol. Brain Res. 26, 331–336.

Gabbot, P.L.A., Bacon, S., 1996. Localisation of NADPH diaphorase activity and NOS immunoreactivity in astroglia in normal adult rat brain. Brain Res. 714, 135–144.

Galea, E., Feinstein, D.L., 1999. Regulation of the expression of the inflammatory nitric oxide synthase (NOS2) by cyclic AMP. FASEB J. 13, 2125–2137.

Garbers, D.L., 1999. The guanylyl cyclase receptors. Methods 19, 477–484.

Garthwaite, J., 2000. The physiological roles of nitric oxide in the central nervous system. In: Mayer, B. (Ed.), Nitric Oxide. Springer, Berlin, pp. 259–275.

Garthwaite, J., Charles, S.L., Chess-Williams, R., 1988. Endothelium-derived relaxing factor release on activation of NMDA receptors suggests role as intracellular messenger in the brain. Nature 336, 385–388.

Garthwaite, J., Garthwaite, G., 1987. Cellular origins of cyclic GMP responses to excitatory amino acid receptor agonists in rat cerebellum in vitro. J. Neurochem. 48, 29–39.

Gavrilyuk, V., Dello Russo, C., Gavrilyuk, V., Heneka, M.T., Pelligrino, D., Weinberg, G., Feinstein, D.L., 2002. Norepinephrine increases IKBα expression in astrocytes. J. Biol. Chem. 277, 29662–29668.

Gibb, B.J., Garthwaite, J., 2001. Subunits of the nitric oxide receptor, soluble guanylyl cyclase, expressed in rat brain. Eur. J. Neurosci. 13, 539–544.

Giri, S., Jatana, M., Rattan, R., Won, J.S., Singh, I., Singh, A.K., 2002. Galactosylsphingosine (psychosine)-induced expression of cytokine-mediated inducible nitric oxide synthases via AP-1 and C/EBP: implications for Krabbe disease. FASEB J. 16, 661–672.

Grzybicki, D., Kwack, K.B., Perlman, S., Murphy, S., 1997. Nitric oxide synthase type II expression by different cell types in MHV-JHM encephalitis suggests distinct roles for nitric oxide in acute versus persistent virus infection. J. Neuroimmunol. 73, 15–27.

Grzybicki, D., Moore, S.A., Schelper, R., Glabinski, A., Ransohoff, R.M., Murphy, S., 1998. Expression of MCP-1 and nitric oxide synthase-2 following cerebral trauma. Acta Neurophathol. 95, 98–103.

Gudi, T., Hong, G.K., Vaandrager, A.B., Lohmann, S.M., Pilz, R.B., 1999. Nitric oxide and cGMP regulate gene expression in neuronal and glial cells by activating type II cGMP-dependent protein kinase. FASEB J. 13, 2143–2152.

Hartell, N.A., 1996. Inhibition of cGMP breakdown promotes the induction of cerebellar long-term depression. J. Neurosci. 16, 2881–2890.

Heneka, M.T., Feinstein, D.L., 2001. Expression and function of inducible nitric oxide synthase in neurons. J. Neuroimmunol. 114, 8–18.

Heneka, M.T., Sharp, A., Klockgether, T., Gavrilyuk, V., Feinstein, D.L., 2000. The heat shock response inhibits NF-kB activation, nitric oxide synthase type 2 expression, and macrophage/microglial activation in brain. J. Cereb. Blood Flow Metab. 20, 800–811.

Heuschling, P., 1995. Nitric oxide modulates γ-interferon-induced MHC class II antigen expression on rat astrocytes. J. Neuroimmunol. 57, 63–69.

Hirrlinger, J., Konig, J., Dringen, R., 2002. Expression of mRNAs of multidrug resistance proteins (Mrps) in cultured rat astrocytes, oligodendrocytes, microglial cells and neurones. J. Neurochem. 82, 716–719.

Hori, K., Burd, P.R., Furuke, K., Kutza, J., Weih, K.A., Clouse, K.A., 1999. Human immunodeficiency virus-1-infected macrophages induce inducible nitric oxide (NO) production in astrocytes: astrocytic NO as a possible mediator of neural damage in acquired immunodeficiency syndrome. Blood 93, 1843–1850.

Hu, J., Akama, K.T., Krafft, G.A., Chromy, B.A., Van Eldik, L.J., 1998. Amyloid-β peptide activates cultured astrocytes: morphological alterations, cytokine induction and nitric oxide release. Brain Res. 785, 195–206.

Hu, J., Castets, F., Guevara, J.L., Van Eldik, L.J., 1996. S100 beta stimulates inducible nitric oxide synthase activity and mRNA levels in rat cortical astrocytes. J. Biol. Chem. 271, 2543–2547.

Hu, S.X., Sheng, W.S., Peterson, P.K., Chao, C.C., 1995. Differential regulation by cytokines of human astrocyte nitric oxide production. Glia 15, 491–494.

Hunot, S., Boissiere, F., Faucheux, B., Brugg, B., Mouattprigent, A., Agid, Y., Hirsch, E.C., 1996. Nitric oxide synthase and neuronal vulnerability in Parkinson's disease. Neuroscience 72, 355–363.

Ignarro, L.J., Buga, G.M., Wood, K.S., Byrns, R.E., Chaudhuri, G., 1987. Endothelium-derived relaxing factor produced and released from artery and vein is nitric oxide. Proc. Natl Acad. Sci. USA 84, 9265–9269.

Iwase, K., Miyanaka, K., Shimizu, A., Nagasaki, A., Gotoh, T., Mori, M., Takiguchi, M., 2000. Induction of endothelial nitric oxide synthase in rat brain astrocytes by systemic lipopolysaccharide treatment. J. Biol. Chem. 275, 11929–11933.

Jedlitschky, G., Burchell, B., Keppler, D., 2000. The multidrug resistance protein 5 functions as an ATP-dependent export pump for cyclic nucleotides. J. Biol. Chem. 275, 30069–30074.

Kapoor, C.L., Krishna, G., 1977. Hormone-induced cyclic guanosine monophosphate secretion from guinea pig pancreatic lobules. Science 196, 1003–1005.

Keilhoff, G., Seidel, B., Wolf, G., 1998. Absence of nitric oxide synthase in rat oligodendrocytes: a light and electron microscopic study. Acta Histochem. 100, 409–417.

Kim, Y.-M., Chung, H.-T., Kim, S.-S., Han, J.-A., Yoo, Y.-M., Kim, K.-M., Lee, G.-H., Yun, H.-Y., Green, A., Li, J., Simmons, R.L., Billiar, T.R., 1999. Nitric oxide protects PC12 cells from serum deprivation-induced apoptosis by cGMP-dependent inhibition of caspase signaling. J. Neurosci. 19, 6740–6747.

Kincaid, R.L., Balaban, C.D., Billingsley, M.L., 1987. Differential localization of calmodulin-dependent enzymes in rat brain: evidence for selective expression of cyclic nucleotide phosphodiesterase in specific neurons. Proc. Natl Acad. Sci. USA 84, 1118–1122.

Koglin, M., Vehse, K., Budaeus, L., Scholz, H., Behrends, S., 2001. Nitric oxide activates the β2 subunit of soluble guanylyl cyclase in the absence of a second subunit. J. Biol. Chem. 276, 30737–30743.

Koka, P., He, K., Zack, J.A., Kitchen, S., Peacock, W., Fried, I., Tran, T., Yashar, S.S., Merril, J.E., 1995. Human immunodeficiency virus 1 envelope proteins induce interleukin 1β, tumor necrosis factor-α, and nitric oxide in glial cultures derived from fetal, neonatal, and adult human brain. J. Exp. Med. 182, 941–952.

Kotera, J., Yanaka, N., Fujishige, K., Imai, Y., Akatsuka, H., Ishizuka, T., Kawashima, K., Omori, K., 1997. Expression of rat cGMP-binding cGMP-specific phosphodiesterase mRNA in Purkinje cell layers during postnatal neuronal development. Eur. J. Biochem. 249, 434–442.

Kugler, P., Drenckhahn, D., 1996. Astrocytes and Bergmann glia as an important site of nitric oxide synthase I. Glia 16, 165–173.

Lee, S.C., Dickson, D.W., Liu, W., Brosnan, C.F., 1993. Induction of nitric oxide synthase activity in human astrocyte by interleukin-1β and interferon-γ. J. Neuroimmunol. 46, 19–24.

Liu, B., Neufeld, A.H., 2000. Expression of nitric oxide synthase-2 (NOS-2) in reactive astrocytes of the human glaucomatous optic nerve head. Glia 30, 178–186.

Loihl, A.K., Campbell, I., Murphy, S., 1999. The expression of NOS-2 following permanent focal ischemia and the role of NO in infarct generation in male, female and NOS-2 deficient mice. Brain Res. 830, 155–164.

Lucas, K.A., Pitari, G.M., Kazerounian, S., Ruiz-Stewart, I., Park, J., Schulz, S., Chepenik, K.P., Waldman, S.A., 2000. Guanylyl cyclases and signaling by cyclic GMP. Pharmacol. Rev. 52, 375–414.

Lüth, H.J., 1997. Ultrastructural demonstration of constitutive nitric oxide synthase (cNOS) in neocortical glial cells and glial perisynaptic sheaths. Anat. Anz. 179, 221–225.

Lüth, H.J., Holzer, M., Gartner, U., Staufenbiel, M., Arendt, T., 2001. Expression of endothelial and inducible NOS-isoforms is increased in Alzheimer's disease, in APP23 transgenic mice and after experimental brain lesion in rat: evidence for an induction by amyloid pathology. Brain Res. 913, 57–67.

Mattson, M.P., Camandola, S., 2001. NF-kB in neuronal plasticity and neurodegenerative disorders. J. Clin. Invest. 107, 247–254.

Matsuoka, I., Giuli, D., Poyard, M., Stengel, D., Parma, J., Guellaen, G., Hanoune, J., 1992. Localization of adenylyl and guanylyl cyclase in rat brain by in situ hybridization: comparison with calmodulin mRNA distribution. J. Neurosci. 12, 3350–3360.

Matyash, V., Filippov, V., Mohrhagen, K., Kettenmann, H., 2001. Nitric oxide signals parallel fiber activity to Bergmann glial cells in the mouse cerebellar slice. Mol. Cell. Neurosci. 18, 664–670.

Mayer, B., Klatt, P., Böhme, E., Schmidt, K., 1992. Regulation of neuronal nitric oxide and cyclic GMP formation by Ca^{2+}. J. Neurochem. 59, 2024–2029.

Merrill, J.E., Murphy, S.P., Mitrovic, B., Mackenzie-Graham, A., Dopp, J.C., Ding, M., Griscavage, J., Ignarro, L.J., Lowenstein, C.J., 1997. Inducible nitric oxide synthase and nitric oxide production by oligodendrocytes. J. Neurosci. Res. 48, 372–384.

Molina-Holgado, E., Vela, J.M., Arevalo-Martin, A., Guaza, C., 2001. LPS/IFN-γ cytotoxicity in oligodendroglial cells: role of nitric oxide and protection by the anti-inflammatory cytokine IL-10. Eur. J. Neurosci. 13, 493–502.

Montoliu, C., Llansola, M., Kosenko, E., Corbalan, R., Felipo, V., 1999. Role of cyclic GMP in glutamate neurotoxicity in primary cultures of cerebellar neurons. Neuropharmacolgy 38, 1883–1891.

Murphy, S., 2000. Production of nitric oxide by glial cells: regulation and potential roles in the CNS. Glia 29, 1–14.

Murphy, P., Sharp, A., Shin, J., Gavrilyuk, V., Dello Russo, C., Weinberg, G., Sharp, F.R., Lu, A., Heneka, M.T., Feinstein, D.L., 2002. Suppressive effects of ansamycins on inducible nitric oxide synthase expression and the development of experimental autoimmune encephalomyelitis. J. Neurosci. Res. 67, 461–470.

Murphy, S., Simmons, M.L., Agullo, L., Garcia, A., Feinstein, D.L., Galea, E., Reis, D.J., Minc-Golomb, D., Schwartz, J.P., 1993. Synthesis of nitric oxide in CNS glial cells. Trends Neurosci. 16, 323–328.

Nakane, M., Ichikawa, M., Deguchi, T., 1983. Light and electron microscopic demonstration of guanylate cyclase in rat brain. Brain Res. 273, 9–15.

Nathan, C., 1997. Inducible nitric oxide synthase. J. Clin. Invest. 100, 2417–2423.

Navarra, M., Pedraza, C.E., Sardón, T., Baltrons, M.A., García, A., 2002. HIV-1 coat protein GP120 decreases NO-dependent cGMP accumulation in rat brain astrocytes by increasing phosphodiesterase 5 activity. Glia S1, S28.

Nolte, C., Matyash, M., Pivneva, T., Schipke, C.G., Ohlemeyer, C., Hanisch, U.-K., Kirchhoff, F., Kettenmann, H., 2001. GFAP promoter-controlled EGFP-expressing transgenic mice: a tool to visualize astrocytes and astrogliosis in living brain tissue. Glia 33, 72–86.

Okuda, Y., Sakoda, S., Fujimura, H., Yanagihara, T., 1997. NO via an inducible isoform of NOS is a possible factor to eliminate inflammatory cells from the CNS of mice with EAE. J. Neuroimmunol. 73, 107–166.

Palmer, R.M.J., Ferrige, A.G., Moncada, S., 1987. Nitric oxide release accounts for the biological activity of endothelium-derived relaxing factor. Nature 327, 524–526.

Papapetropoulos, A., Go, C., Murad, F., Catravas, J.D., 1996. Mechanisms of tolerance to sodium nitroprusside in cultured rat aortic smooth muscle cells. Br. J. Pharmacol. 117, 147–155.

Parkinson, S.J., Jovanovic, A., Jovanovic, S., Wagner, F., Terzic, A., Waldman, S.A., 1999. Regulation of nitric oxide-responsive recombinant soluble guanylyl cyclase by calcium. Biochemistry 38, 6441–6448.

Pasti, L., Pozzan, T., Carmignoto, G., 1995. Long-lasting changes of calcium oscillations in astrocytes. A new form of glutamate-mediated plasticity. J. Biol. Chem. 270, 15203–15210.

Pedraza, C.E., Baltrons, M.A., García, A., 2001. Interleukin-1β stimulates cyclic GMP efflux in brain astrocytes. FEBS Lett. 507, 303–306.

Pedraza, C.E., Baltrons, M.A., García, A., 2002. Pro-inflammatory cytokines decrease soluble guanylyl cyclase in astroglial cells. Glia S1, S60.

Peunova, N., Enikolopov, G., 1995. Nitric oxide triggers a switch to growth arrest during differentiation of neuronal cells. Nature 373, 68–73.

Pouloupoulou, C., Nowak, L., 1998. Extracellular 3′,5′ cyclic guanosine monophosphate inhibits kainate-activated responses in cultured mouse cerebellar neurons. J. Pharmacol. Exp. Ther. 286, 99–109.

Quan, N., He, L., Lai, W., Shen, T., Herkenham, M., 2000. Induction of IkBα mRNA expression in the brain by glucocorticoids: a negative feedback mechanism for immune-to-brain signaling. J. Neurosci. 20, 6473–6477.

Rossi, F., Bianchini, E., 1996. Synergistic induction of nitric oxide by β-amyloide and cytokines in astrocytes. Biochem. Biophys. Res. Commun. 225, 474–478.

Russwurm, M., Koesling, D., 2002. Isoforms of NO-sensitive guanylyl cyclase. Mol. Cell. Biochem. 230, 159–164.

Russwurm, M., Wittau, N., Koesling, D., 2001. Guanylyl cyclase/PSD-95 interaction: targeting of the nitric oxide-sensitive α2/β1 guanylyl cyclase to synaptic membranes. J. Biol. Chem. 276, 44647–44652.

Saadoun, S., García, A., 1999. Endothelin stimulates nitric oxide-dependent cyclic GMP formation in rat cerebellar astroglia. Neuroreport 10, 1–4.

Schmidt, H.H.H.W., Gagne, G.D., Nakane, M., Pollock, J.S., Miller, M.F., Murad, F., 1992. Mapping of neural nitric oxide synthase in the rat suggests frequent co-localization with NADPH diaphorase but not with soluble guanylyl cyclase, and novel paraneural functions for nitrinergic signal transduction. J. Histochem. Cytochem. 40, 1439–1456.

Schmidt, H.H.H.W., Lohmann, S.M., Walter, U., 1993. The nitric oxide and cGMP signal transduction system: regulation and mechanism of action. Biochem. Biophys. Acta 1178, 153–175.

Schultz, C., Vaskinn, S., Kildalsen, H., Sager, G., 1998. Cyclic AMP stimulates the cyclic GMP egression pump in human erythrocytes: effects of probenecid, verapamil, progesterone, theophylline, IBMX, forskolin, and cyclic AMP on cyclic GMP uptake and association to inside-out vesicles. Biochemistry 37, 1161–1166.

Shen, P.J., Gundlach, A.L., 1999. Prolonged induction of neuronal NOS expression and activity following cortical spreading depression (SD): implications for SD- and NO-mediated neuroprotection. Exp. Neurol. 160, 317–332.

Shenolikar, S., Thompson, W.J., Strada, S.J., 1985. Characterization of a Ca^{2+}-calmodulin-stimulated cyclic GMP phosphodiesterase from bovine brain. Biochemistry 24, 672–677.

Simic, G., Lucassen, P.J., Krsnik, Z., Kruslin, B., Kostovic, I., Winblad, B., Bogdanovi, N., 2000. nNOS expression in reactive astrocytes correlates with increased cell death related DNA damage in the hippocampus and entorhinal cortex in Alzheimer's disease. Exp. Neurol. 165, 12–26.

Soderling, S.H., Beavo, J.A., 2000. Regulation of cAMP and cGMP signaling: new phosphodiesterases and new functions. Curr. Opin. Cell Biol. 12, 174–179.

Sohn, Y.K., Ganju, N., Bloch, K.D., Wands, J.R., de la Monte, S.M., 1999. Neuritic sprouting with aberrant expression of the nitric oxide synthase III gene in neurodegenerative diseases. J. Neurol. Sci. 162, 133–151.

Sun, N., Grzybicki, D., Castro, R.F., Murphy, S., Perlman, S., 1995. Activation of astrocytes in the spinal cord of mice chronically infected with a neurotropic coronavirus. Virology 213, 482–493.

Szabó, C., 1996. Physiological and pathological roles of nitric oxide in the central nervous system. Brain Res. Bull. 41, 131–141.

Takata, M., Filippov, G., Liu, H., Ichinose, F., Janssens, S., Bloch, D.B., Bloch, K.D., 2001. Cytokines decrease sGC in pulmonary artery smooth muscle cells via NO-dependent and NO-independent mechanisms. Am. J. Physiol. Lung Cell. Mol. Physiol. 280, L272–L278.

Takuma, K., Phuagphong, P., Lee, E., Mori, K., Baba, A., Matsuda, T., 2001. Anti-apoptotic effect of cGMP in cultured astrocytes: inhibition by cGMP-dependent protein kinase of mitochondrial permeable transition pore. J. Biol. Chem. 276, 48093–48099.

Tanaka, J., Markerink-Van Ittersum, M., Steinbusch, H.W.M., De Vente, J., 1997. Nitric oxide-mediated cGMP synthesis in oligodendrocytes in the developing rat brain. Glia 19, 286–297.

Taupenot, L., Ciesielski-Treska, J., Ulrich, G., Chasserot-Golaz, S., Aunis, D., Bader, M.F., 1996. Chromogranin A triggers a phenotypic transformation and the generation of nitric oxide in brain microglial cells. Neuroscience 72, 377–389.

Teunissen, C., Steinbusch, H., Markerink-van Ittersum, M., Koesling, D., de Vente, J., 2001. Presence of soluble and particulate guanylyl cyclase in the same hippocampal astrocytes. Brain Res. 891, 206–212.

Tjörnhammar, M.L., Lazaridis, G., Bartfay, T., 1986. Efflux of cyclic guanosine 3′,5′-monophosphate from cerebellar slices stimulated by L-glutamate or high K + or *N*-methyl-*N*′-nitro-*N*-nitrosoguanidine. Neurosci. Lett. 68, 95–99.

Togashi, H., Sasaki, M., Frohman, E., Taira, E., Ratan, R.R., Dawson, T.M., Dawson, V.L., 1997. Neuronal (type I) nitric oxide synthase regulates NFkB activity and immunologic (type II) nitric oxide synthase expression. Proc. Natl Acad. Sci. USA 94, 2676–2680.

Tomita, S., Nicoll, R.A., Bredt, D.S., 2001. PDZ protein interactions regulating glutamate receptor function and plasticity. J. Cell Biol. 153, F19–F24.

Touyz, R.M., Picard, S., Schffrin, E.L., Deschepper, C.F., Cyclic, G.M.P., 1997. inhibits a pharmacologically distinct Na^+/H^+ exchanger variant in cultured rat astrocytes via an extracellular site of action. J. Neurochem. 68, 1451–1461.

Ujiie, K., Hogarth, L., Danziger, R., Drewett, J.G., Yuen, P.S.T., Pang, I.-H., Star, R.A., 1994. Homologous and heterologous desensitization of a guanylyl cyclase-linked nitric oxide receptor in cultured rat medullary interstitial cells. J. Pharmacol. Exp. Ther. 270, 761–767.

Vesce, S., Bezzi, P., Volterra, A., 2001. Synaptic transmission with the glia. News Physiol. Sci. 16, 178–184.

Wallace, M.N., Geddes, J.G., Farquhar, D.A., Masson, M.R., 1997. Nitric oxide synthase in reactive astrocytes adjacent to β-amyloid plaques. Exp. Neurol. 144, 266–272.

Wang, X., Robinson, P.J., 1997. Cyclic GMP-dependent protein kinase and cellular signalling in the nervous system. J. Neurochem. 68, 443–456.

Wendland, B., Schweizer, F.E., Ryan, T.A., Nakane, M., Murad, F., Scheller, R.H., Tsien, R.W., 1994. Existence of nitric oxide synthase in rat hippocampal pyramidal cells. Proc. Natl Acad. Sci. USA 91, 2151–2155.

Wiencken, A.E., Casagrande, V.A., 1999. Endothelial nitric oxide synthetase (eNOS) in astrocytes: another source of nitric oxide in neocortex. Glia 26, 280–290.

Willmott, N.J., Wong, K., Strong, A.J., 2000. A fundamental role for the nitric oxide-G-kinase signaling pathway in mediating intercellular Ca^{2+} waves in glia. J. Neurosci. 20, 1767–1779.

Wood, P.L., 1991. Pharmacology of the second messenger, cyclic guanosine 3′,5′-monophosphate, in the cerebellum. Pharmacol. Rev. 43, 1–25.

Xiong, H., Yamada, K., Jourdi, H., Kawamura, M., Takei, N., Han, D., Nabeshima, T., Nawa, H., 1999. Regulation of nerve growth factor release by nitric oxide through cyclic GMP pathway in cortical glial cells. Mol. Pharmacol. 56, 339–347.

Yan, C., Bentley, J.K., Sonnenburg, W.K., Beavo, J.A., 1994. Differential expression of the 61 kDa and 63 kDa calmodulin-dependent phosphodiesterases in the mouse brain. J. Neurosci. 14, 973–984.

Yoshioka, A., Yamaya, Y., Saiki, S., Kanemoto, M., Hirose, G., Pleasure, D., 2000. Cyclic GMP/cyclic GMP-dependent protein kinase system prevents excitotoxicity in an immortalized oligodendroglial cell line. J. Neurochem. 74, 633–640.

Zabel, U., Kleinschnitz, C., Oh, P., Nedvetsky, P., Smolenski, A., Muller, H., Kronich, P., Kugler, P., Walter, U., Schnitzer, J.E., Schmidt, H.H., 2002. Calcium-dependent membrane association sensitizes soluble guanylyl cyclase to nitric oxide. Nat. Cell Biol. 4, 307–311.

Zabel, U., Weeger, M., La, M., Schmidt, H.H.W., 1998. Human soluble guanylate cyclase: functional expression and revised isoenzyme family. Biochem. J. 335, 51–57.

Zhang, J., Snyder, S.H., 1995. Nitric oxide in the nervous system. Annu. Rev. Pharmacol. Toxicol. 35, 213–233.

Zhao, M.-L., Liu, J.S.H., He, D., Dickson, D.W., Lee, S.C., 1998. Inducible nitric oxide synthase expression is selectively induced in astrocytes isolated from adult human brain. Brain Res. 813, 402–405.

Advances in
Molecular and
Cell Biology

Potassium homeostasis in the brain at the organ and cell level

Wolfgang Walz

Department of Physiology, University of Saskatchewan, 107 Wiggins Road, Saskatoon, Sask., Canada S7N 5E5
Correspondence address: Tel.: +1-306-966-6535; fax: +1-306-966-6532.
E-mail: walz@sask.usask.ca(W. W.)

Contents

1. Introduction
2. Blood–brain barrier
3. Choroid plexus
4. Potassium levels in the extracellular space
5. Effect of excess extracellular potassium on neuronal processing
6. Astrocytes as the site of potassium regulatory mechanisms
7. Astrocytes as spatial buffers
8. Astrocytes as transient storage sites for potassium
9. Reactive astrocytes
10. Concluding remarks

Active neurons release potassium ions into the extracellular space. The resulting increase in extracellular potassium concentration is high enough to interfere with ion channel gating and therefore information transfer in neurons. The excess potassium is not released across the blood–brain barrier into the blood. Astrocytes are now acknowledged to be the major site of the clearance of excess extracellular potassium. The major astrocytic mechanism is uptake of KCl via Na/K ATPase and electroneutral Na,K,2Cl carrier. Spatial buffering by potassium currents across astrocytic membranes works only over short distances due to the small length constant and does therefore not contribute significantly. The only exception might be the periphery of the retina. After injury, reactive astrocytes express chloride conductance(s), which allow KCl accumulation via Donnan forces in addition to the carrier-mediated accumulation. Thus, KCl accumulation into astrocytes by active and passive (after injury) means is now a well established cellular mechanism in the nervous system.

Advances in Molecular and Cell Biology, Vol. 31, pages 595–609

ISBN: 0-444-51451-1

1. Introduction

The potassium ion has a crucial role in neuronal excitability. Its distribution across neuronal membranes determines the resting membrane potential features, transmitter release and the kinetics of voltage-gated ion channels. Thus, small absolute changes of the normally low extracellular but not necessarily of the higher intracellular concentration of potassium can play havoc with excitability and neuronal information transfer. Therefore, mechanisms exist which control potassium concentrations in the brain extracellular space (ECS) and cerebrospinal fluid (CSF). There are two sources of potassium instability: (i) potassium concentration in the plasma varies due to dietary intake or other factors, and the brain as an organ must be sheltered from these changes; and (ii) neurones release potassium to the ECS during action and synaptic potentials and due to the low extracellular K^+ concentration and small dimension of the ECS accumulation of released K^+ can lead to large relative changes. The homeostatic mechanisms of the neurones alone are not sufficient to prevent activity-induced increases in extracellular K^+ concentration, and therefore non-neuronal elements bear most of the responsibility for potassium homeostasis.

2. Blood–brain barrier

Increases or decreases of the potassium concentration in the blood plasma have no effect on the corresponding concentration in the ECS or CSF (Jones and Keep, 1987). This applies to both, acute and chronic changes (Stummer et al., 1995). Since the total content of brain potassium is not changed during plasma concentration changes, it is unlikely that astrocytes play a role in regulation of whole brain K^+, when the systemic potassium concentration is changed, because the potassium content of the astrocytic compartment is not altered. Rather it must be the blood–brain barrier itself, which is responsible for this homeostatic mechanism (Keep, 2002). The regulation is so effective, that even an experimental breach of the blood–brain barrier did not raise extracellular K^+ in brain parenchyma, despite the fact that this breach led to dye penetration (Somjen et al., 1991). Raising blood plasma potassium concentration from its normal level of 5 up to 17 mM with a breached blood–brain barrier caused waves of spreading depression with associated changes of the potassium concentration, but no long term increase in extracellular potassium.

The Na/K ATPase activity of the endothelial cell is polarized: there is a much higher activity on the interstitial than on the luminal surface (Somjen, 2002). There are also indications of different isoforms of Na/K ATPase existing on luminal versus abluminal surfaces (Keep, 2002). The turnover rate of brain potassium across the blood–brain barrier is approx. 2%/h. This compares with 35%/h for sodium and 20%/h for chloride (Keep, 2002). These 10–20 times lower permeability rates mean that the barrier for potassium is very effective in shutting plasma potassium out of the brain. Whatever transport exists is out of the brain via the Na/K ATPase, using the ATP created by the high density of mitochondria in the endothelial cells (Oldendorf et al., 1977).

Another question is if the brain releases excess local extracellular potassium into the plasma either during normal operation or during pathological events. There is no evidence for transport of potassium out into the blood during normal brain function. The situation is, however, different in brain injury. During focal ischemia, when extracellular K^+ may reach very high levels (see below), the Na/K ATPase seems to lead to a translocation of potassium from brain parenchyma to blood, coupled with an import of sodium into the parenchyma (Betz et al., 1989, 1994). In this context it is of interest that the Na/K ATPase of the endothelial cells has sufficiently high affinity for K^+ to be stimulated by increased potassium concentrations (Schielke et al., 1990).

3. Choroid plexus

The choroid plexus is responsible for production of CSF and its secretion into the ventricles (see chapter by Weaver et al.). Unlike the situation in most other epithelia, the Na/K ATPase is located on the apical side, the side that faces the CSF, as is a Na,K,2Cl cotransporter. The most comprehensive study to date about the potassium homeostatic role of this epithelium was carried out by Husted and Reed (1976). They found only small increases of potassium transport from blood to CSF when blood serum potassium was raised. However, increases in CSF K^+ were effectively regulated by (i) a reduction of potassium content of the CSF fluid secreted without any effect on secretion rate and (ii) an increased K^+ re-absorption rate out of the CSF. Wu et al. (1998) confirmed that Na/K ATPase and a Na,K,2Cl cotransporter on the CSF side are responsible for removing potassium from the CSF. It is unclear which transport mechanism is responsible for potassium transport across the basolateral membrane back into blood. Wu et al. found evidence for a barium-sensitive potassium conductance without locating it. Hung et al. (1993) located potassium channels in the apical membrane, but did not investigate the basolateral side. A KCC3 (KCl) cotransporter was however localized by Pearson et al. (2001) in the basolateral membrane. Its role was not investigated. Whatever the mechanism, although the choroid plexus is responsible for secretion of CSF into the ventricles, it seems to have the ability to extract excess potassium from the CSF without reducing its capacity for CSF production. In contrast elevated K^+ in blood only increases potassium transport slightly.

4. Potassium levels in the extracellular space

The resting level of K^+ in the ECS of the central nervous system is lower than that in blood or other organs. It varies between 2.7 and 3.5 mM (Somjen, 2002). The most frequently used method of estimating increases in the external potassium concentration is the use of potassium-sensitive double-barreled microelectrodes. This method is accurate enough for the estimation of widespread or massive potassium release. However, release by point sources will be underestimated. This is due to the creation of a dead space around the tip of the electrode, whose diameter is several times the width of the extracellular space. Therefore any limited amount of potassium released (as the one during the passage of a single action potential) will be severely underestimated. In these cases one will get more reliable estimates using indirect methods like the amplitude of the after-hyperpolarization of the action potential, which in many axons is a phase of exclusive potassium permeability of the membrane (Frankenhaeuser and Hodgkin, 1956). Using this kind of estimate, most authors agree that there is a transient increase in the extracellular potassium concentration after the passage of a *single* action potential that amounts to nearly 1 mM above the resting level of 3 mM (Adelman and Fitzhugh, 1975). The magnitude of the increase in extracellular potassium depends upon stimulation frequency and number of neuronal elements that contribute to the release of K^+ (Sykova, 1991). Both action potentials and synaptic potentials are the source of this activity-dependent increase in extracellular potassium. Another factor contributing to the increase of extracellular K^+ is a decrease in the volume of the extracellular space under most conditions of increased activity (Ransom et al., 1986). Accordingly, during artificial, intensive stimulation of neuronal pathways extracellular K^+ may increase by as much as nearly 5 mM above the resting level, but it does not exceed this level. During seizures, the accumulation of extracellular potassium is further enhanced, but again there seems to be a ceiling level of 12 mM (Heinemann and Lux, 1977). Only during injury (hypoxia/ischemia, trauma, hypoglycaemia) is this ceiling level disrupted and concentrations of up to 25 mM are reached (Hansen, 1985). An extreme case is waves of spreading depression, which result in transient elevations of the extracellular potassium concentration of 30–80 mM in the intact nervous tissue (Irwin and Walz, 1999).

5. Effect of excess extracellular potassium on neuronal processing

There is only a need for potassium homeostasis if values of excess potassium in the range encountered above are able to significantly change the excitability of neurons. That this is the case has repeatedly been shown. These concentrations affect transmitter release (Gage and Quastel, 1965; Erulkar and Weight, 1977) and electrical properties of axons (Malenka et al., 1981). More specifically, it was shown that increases of the extracellular potassium concentration to 5 mM change the action potential threshold of hippocampal neurons, leading to hyperexcitability (Voskuyl and Ter Keurs, 1981; Balestrino et al., 1986; Kreisman and Smith, 1993). Increases in the extracellular potassium concentration, that exceed 5 mM, increase, in addition, the efficacy of synaptic transmission in hippocampal neurons (Balestrino et al., 1986; Rausche et al., 1990). The underlying mechanisms seem to be a combination of potassium evoked changes in the membrane potential and a direct gating effects of the potassium ion on some of the participating ion channels (Leech and Stanfield, 1981).

6. Astrocytes as the site of potassium regulatory mechanisms

One would expect that immediate re-uptake of K^+ into neurons and diffusion in the extracellular space would be sufficient to prevent a build-up of excess extracellular potassium released during neuronal activity. There must, however, be potassium removal sites, which do not resident in neurones, because iontophoretically applied extracellular potassium is removed as efficiently as the potassium that is released from neurons. Only in the latter case is there a simultaneous accumulation of sodium ions into neurones that stimulate the neuronal Na/K pump. Kinetic analysis of potassium transients and their sensitivity to temperature changes and strophantidin, led Ransom et al. (2000) to conclude that a fast initial removal is based on glial Na/K ATPases, and a slower, sustained process is due to stimulation of the neuronal enzyme. In addition, the existence of a ceiling level of 12 mM under normal conditions as well as the observed undershoot below the normal resting potassium concentration after a transient phase of extracellular potassium accumulation are strong arguments in favor of active removal processes situated in non-neuronal cells (Heinemann and Lux, 1977; Heinemann et al., 1983). If K^+ is iontophoretically applied into the extracellular space, the diffusion curves appeared normal, but the volume fraction routinely exceeds 100% of the ECS (Rice and Nicholson, 1991). This is physically impossible and the only explanation is that the exit of potassium from the extracellular space through adjacent cell membranes is facilitated by auxiliary mechanisms.

The existence of two different potassium clearance mechanisms residing in astrocytes was suggested a long time ago. The first mechanism is removal of excess potassium by uptake and accumulation into adjacent astrocytes, including its transient storage and subsequent release back into the extracellular space (Hertz, 1965).

The second mechanism is re-distribution of excess potassium by current loops through the astrocytic syncytium ('spatial buffering'), which would not lead to potassium accumulation in astrocytes, since for each potassium ion entering another one would leave the syncytium at a site distant from the active neurons (Orkand et al., 1966). For some time these two mechanisms were seen as exclusive alternatives. However, this does not need to be so, since they could co-exist. Accordingly the question arises whether they might be differently distributed and which one is likely to play the dominant role in gray matter.

Recently, a study in the dentate gyrus employing 8 mM of potassium did not find much evidence for spatial buffering, but suggested that Na/K ATPase-mediated removal plays a major role (Xiong and Stringer, 2000). Along similar lines D'Ambrosio et al. (2002) used selective blockers and controlled neuronal stimulation in slices of CA3 stratum pyramidale. They concluded that spatial buffering did not affect potassium clearance rates but that Na/K ATPases are responsible for determining the rate of recovery. Thus, these two recent studies support a potassium accumulation rather than spatial buffering as the major mechanism of potassium clearance. However, a differentiation between neuronal and glial Na/K ATPases was not possible in these studies. In the retina, where spatial buffering has been most convincingly demonstrated (see below) K^+ uptake by spatial buffering in Muller cells seems to be restricted to the plexiform layers. In the central retina active uptake seems to dominate (Skatchkov et al., 1999). The spatial buffer concept will be discussed in considerable detail below, since it played a dominant role during several decades, since spatial buffering does seem to be important for K^+ clearance in the retina and occurs over short distances in brain and spinal cord, and since estimates of its contribution to K^+ homeostasis at the cellular level of the CNS can only be evaluated when the underlying principles are understood.

7. Astrocytes as spatial buffers

The spatial buffer concept (Orkand et al., 1966) assumes a glial syncytium, coupled by gap junctions (see chapter by Scemes and Spray), in which extracellular potassium is selectively increased in one region. In a syncytium, the membrane potential of neighboring cells has a tendency to attain a similar value. Therefore the region experiencing an increased extracellular potassium concentration will have a potassium equilibrium potential that is less negative than the membrane potential. This will lead to an inward driving force for potassium, and since the membrane is highly permeable to potassium, it will enter the cell via a passive current. The current loop has to be closed and the potassium-mediated current will immediately reach the other parts of the syncytium via the gap junctions. In regions further removed from the area exposed to high extracellular potassium concentrations, the potassium equilibrium potential is more negative than the membrane potential due to the tendency of the syncytium to remain isopotential. Therefore at these areas there is an outwardly directed driving force for potassium. Thus the current into the cells (at regions with high extracellular potassium), inside the syncytium and out of the cells (at regions distant from local elevations of potassium), is almost completely carried by potassium as the dominant intracellular ion and an ion which can pass through gap junctions. The loop is closed by a return current in the extracellular space. This current is carried mainly by sodium and chloride ions (the dominant ions in the ECS). Therefore, despite the closed current loop, potassium does not cycle but is passively transported away from an extracellular location with high potassium concentrations and dispersed to extracellular areas with lower potassium. The original concept allowed for flexibility, since any part of the syncytium could be the source of the potassium current, depending solely on the location of the elevated extracellular potassium concentration.

In order for astrocytes to act as spatial buffers, the following criteria must be met: (i) When neuronal elements are active, neighboring astrocytes must always be depolarized. (ii) A large enough population of astrocytes has to undergo that depolarization at the same time. (iii) There has to be electrotonic continuity of spatially extended astrocytes and/or the syncytium from the active to the inactive region. (iv) The length constant (see below) of the glial syncytium has to be compatible with the spread of current.

Criterion (i) seems to present no problem. In general, astrocytes have a 20 mV more negative membrane resting potential than neighboring neurons, i.e., at least -90 mV. This is mainly due to the relatively lower sodium permeability of astrocytes (Somjen, 1995). Whether or not the astrocytic membrane has a significant chloride permeability in situ has been recently reviewed (Walz, 2002) and will not be repeated here. Normal astrocytes seem to have no chloride permeability, but activation into a reactive sub-type induces such a permeability. An almost selective potassium permeability is the reason why astrocytes are more sensitive to increases in extracellular potassium than neurons (Somjen, 2002). The relationship between extracellular K^+ concentration and membrane potential is under most conditions somewhat sub-Nernstian, i.e., the amplitude is less than predicted from the Nernst equation. This can be explained by the simultaneous potassium uptake leading to an increased intracellular potassium concentration rather than a stationary intracellular K^+, by the modulatory effects of electrogenic uptake and by ionotropic responses of transmitters (see below). In situ, all transmitter agonists, which cause changes in ionic conductance in astrocytes, lead to a depolarization. Examples are Kainate, GABA and dopamine (Murdick-Donnon et al., 1993; Jabs et al., 1997; Bekar et al., 1999). The depolarization is due to three factors: changes in ion equilibrium potentials and in ion channel conductances (Jabs et al., 1997) as well as electrogenic transmitter uptake (Bergles and Jahr, 1997). Thus, substances other than potassium released from active neurons could contribute to astrocytic membrane potential changes, and astrocytes always react with a depolarization to neuronal activity.

Criterion (ii) does also not present a large challenge. Orthodromic stimulation in a variety of preparations leads to a depolarization of astrocytes that persists longer than the stimulation or the activity of neurons, including synaptic potentials (Somjen, 1975). Lothman and Somjen (1975) studied the relationship of stimulation-induced extracellular K^+ increases and glial membrane potentials in the spinal cord. They found potassium increases up to 8 mM with corresponding glial depolarizations of 25 mV. An amplitude of 25–30 mV seems to be the maximal astrocytic depolarization achievable by stimulation of neuronal pathways in vivo and in brain slices (Ballanyi et al., 1987). If more physiological stimuli are used, like light flashes, glial depolarizations are much smaller (8 mV) as are the simultaneous rises in extracellular potassium, but there is never any hyperpolarization of glial cells (Karwoski and Proenza, 1977; Kelly and Van Essen, 1974). During seizure-like events, the glial depolarizations may reach 35 mV, which corresponds to the 10–12 mM ceiling level (Heinemann and Lux, 1977). Studies using membrane potential-sensitive dyes in glial cells in vivo are lacking; thus, it is not possible to estimate the extent of the direct depolarizations of glial cells by axon bundles or groups of neurons. Presumably, the extent of the direct depolarization is limited to the region that contains raised potassium levels.

Criterion (iii) is easily fulfilled since there undoubtedly is extensive dye- and electrical coupling of astrocytes. A dye can appear in as many as 100 astrocytes, when injected into one (Ransom, 1995). The spread of the dye seems to be spherical outward from the injected cell. There is every reason to assume that astrocytes that are dye-coupled are also coupled electrically (Dermietzel et al., 1991).

In contrast, criterion (iv) is the major problem for efficient distribution of spatial buffering to potassium homeostasis at the cellular level. The length constant is the distance along a process to the site, where a voltage amplitude has decayed to 37% of its value due to the leakage of current across the cell membrane. It is obvious that a low specific resistance will decrease the length constant and therefore the length that a significant amount of current can travel in the processes of the glial syncytium. Since astrocytes have a relatively low membrane resistance, the issue of a restricted length constant is the most serious argument against the spatial buffer function (Heinemann and Walz, 1998). There are now several studies that estimated the length constant and found it to be too small in all types of astrocytes or astrocyte-like cells tested with the possible exception of those in the retinal periphery. Barres et al. (1990) measured the length constant of an acutely isolated astrocyte from rat optic nerve as 100 μm, whereas the whole length of the cell was 400 μm. Skatchkov et al. (1999) found 70 μm in Muller cells of the frog retina. This was not long enough for long central cells, but sufficient for short peripheral cells. Using data from rat cerebral cortex, Chen and Nicholson (2000) concluded that spatial buffering, as constrained by the leaky cable properties, does not appear to be able to operate over long distances. Thus a major deficiency of the spatial buffer theory seems to be that spatial buffering can only operate across short distances.

Evidence from in situ systems support the concept of spatial buffering. Nicholson and Phillips (1981) studied diffusion in rat cerebellum using ion-selective microelectrodes and iontophoretic point sources. Assuming an extracellular space of 20% they found for a variety of anions and cations a close correlation of their movement with the laws of macroscopic diffusion. However, there was one ion that did not fit into the scheme: the movement of potassium in an electrical gradient is more keeping with an extracellular space that occupies more than 100% of the brain volume. The anomalous nature of the potassium migration can be explained by assuming that the ion does not remain in the extracellular space, but is in fact, the major current carrier across cell membranes. Hounsgaard and Nicholson (1983) analyzed the possibility in more detail. They measured changes in the extracellular potassium concentration in the vicinity of Purkinje cells in guinea pig cerebellar slices. No extracellular potassium changes were seen when the Purkinje cells were hyperpolarized with current passage or during sub-threshold depolarizations. Only during spike activity was there a rise in extracellular potassium levels. In the vicinity of glial cells, however, a hyperpolarizing current injected into glia reduced the outside potassium concentration in a symmetrical manner, while depolarizing current injection induced a rise in the outside potassium levels. These results demonstrate that the potassium ions are crossing the glial cell membrane during their movement in an electrical gradient. They also provide additional evidence supporting the view that glial cells function as spatial buffer for potassium (see also chapter by Laming).

Thanks to its easy accessibility and layered structure, the retina is the part of the CNS that is best characterized in terms of spatial buffering. Light stimulation results in external potassium increases in the two plexiform layers and decreases in the sub-retinal space (Reichenbach et al., 1998—see also chapter by Bringmann et al.). In order to buffer the extracellular K^+ changes, potassium currents enter the Muller cells in the two plexiform layers, where external potassium is high, whereas potassium efflux occurs into the vitreous humor at the endfoot and into the sub-retinal space. In a vascularized retina further efflux can occur at the blood vessels. These currents, K(IR), use inwardly rectifying potassium channels, which open at -70 to -90 mV (Chao et al., 1994), mediate both inward and outward currents (Skatchkov et al., 1995) and increase their conductance when external potassium is raised (Newman, 1993). The Muller cells have high densities of K(IR) at regions, which act as sinks for buffering currents on the endfoot (Brew et al., 1986). It has been shown, that blockade of these channels with barium raises external potassium amplitudes and increases clearance time (Oakley et al., 1992). Newman (1986) showed with single Muller cells that these cells are capable of carrying potassium currents and releasing substantial amounts at their endfeet. However, the aforementioned work by Skatchkov et al. (1999) seems to show that the length constant permits efficient K^+ clearance by spatial buffering only in the periphery, where the distances are shorter, and that Na/K pump mediated processes must be involved in long distance clearance of K^+.

Whether a similar mechanism as in Muller cells operates in cerebral astrocytes is still not clear. There are indications from hippocampal astrocytes that only a sub-population of the cells has K^+ inward rectifier channels (D'Ambrosio et al., 1998; Zhou and Kimelberg, 2000). These studies support the view that different astrocytes are responsible for inward and outward currents. Also there seems to exist a dynamic coupling and uncoupling of astrocytes, caused by unknown factors, which would add another complexity to such a buffering system (McKhann et al., 1997). However, one should keep in mind the previous mentioned estimates, that potassium buffering in the cortex is only suited for operation over very short distances (Chen and Nicholson, 2000) and that compared with the Na/K ATPase mediated removal of potassium, spatial buffering plays a minor role in K^+ homeostasis (Xiong and Stringer, 2000; D'Ambrosio et al., 2002). Thus, one has to conclude that spatial buffering can take place locally but not in an extended area.

8. Astrocytes as transient storage sites for potassium

The spatial buffer current does not mediate cellular storage of potassium ions: for every potassium ion entering the syncytium, one is leaving at the same time, although at a different location. Thus no significant accumulation of potassium inside astrocytes will take place, and any observed accumulation can be seen as an indication of a mechanism other than operation of a spatial buffer. In all preparations tested (cultured astrocytes: Walz et al., 1984; cultured oligodendrocytes: Kettenmann, 1987; brain slices: Ballanyi et al., 1987; drone retina: Coles and Orkand, 1983; leech CNS: Wuttke, 1990; reactive astrocytes in situ: Walz and Wuttke, 1999) glial cells accumulate potassium ions, when the extracellular potassium concentration increases. They release potassium again as soon as the extracellular concentration is lowered. This transient accumulation or storage was observed regardless whether the extracellular potassium concentration was increased artificially in a uniform way (a situation where spatial buffer currents would not arise) or whether the activity of neighboring neurons was stimulated by physiological or non-physiological means. It was found in the optic nerve (Ransom et al., 2000; MacVicar et al., 2002) and hippocampal slices (Xiong and Stringer, 2000; D'Ambrosio et al., 2002) that active uptake via a transporter is more efficient than spatial buffering. This conclusion is strengthened by the aforementioned observations by various authors that the amount of accumulated potassium is similar when the neurons are stimulated and when the external potassium is experimentally increased to levels observed during the neuronal stimulation. This is important since neuronal stimulation increases not only the internal sodium and external potassium concentrations but also significantly changes calcium and proton concentrations as well as transmitter and metabolite levels (Ballanyi, 1995; Ransom, 1992). The increase in glial intracellular potassium is therefore exclusively a response to the increased extracellular potassium levels; the other changes might modify this response but do not cause it.

The estimated maximal increase in the glial intracellular potassium concentration during neuronal activity amounts to approximately 10 mM in the experiments mentioned above. If the extracellular potassium concentration is increased artificially to the levels observed during neuronal stimulation, the increase in the glial K^+ concentration is higher (20–25 mM). The difference is probably due to the uniform increase around the glial syncytium in the latter case as compared to a more localized increase during stimulation. The storage of potassium in astrocytes is transient and depends on the extracellular potassium levels: as soon as the extracellular potassium concentration decreases, the glial cells start releasing the accumulated potassium. If the external levels again reach the normal value, no storage of excess potassium occurs. The release mechanism is unknown. In cell culture the increase in terms of concentration can amount to a doubling if all the external sodium ions are replaced by potassium ions (Walz, 1987). Since at this stage massive swelling occurs, the absolute amount of potassium taken up is about three times the content at resting levels. Presumably such levels play a role following anoxia/hypoxia and spreading depression in vivo (Erecinska and Silver, 1994). Two alternate uptake mechanisms exist, which can complement each other and work in parallel (Walz and Wuttke, 1999). They are discussed below.

The intracellular sodium and potassium concentrations are controlled by the activity of the Na/K pump or Na/K ATPase (Walz and Wuttke, 1999; Walz et al., 1993). However, although part of the potassium uptake is sensitive to ouabain, the prototypical Na/K pump blocker, at high uptake rates for potassium, there would simply not be enough sodium in the cells to account for a 1:1, let alone a 3:2 exchange of sodium with potassium. The internal sodium concentration of astrocytes is around 10–15 mM (Rose and Ransom, 1996) and does not change during potassium uptake (Walz and Hertz, 1983). This can be explained by operation of a transmembrane sodium cycle suggested by Walz and Hinks (1986): potassium is moved inward by the Na/K pump in exchange with the sodium ion, and a decline in intracellular sodium concentration is prevented by a simultaneous stimulation of the electroneutral bumetanide-sensitive Na–K–Cl cotransporter, which passively follows the combined driving forces of all three ions, leading to an inward flow of these ions. Sodium is extruded by the Na/K pump and replenished by the Na–K–Cl carrier. The net effect is KCl accumulation and swelling due to the osmolyte increase (Walz, 1987). The evidence for such a mechanism is based on radiotracer analysis in cultured astrocytes and the fact that inhibition of either pathway was not additive, but showed considerable overlap between both. In addition, there is no evidence for different kinetic properties of the glial and neuronal Na/K pump (Sweadner, 1995—see, however, also chapter by Peng et al.). This is a puzzling finding, which could mean that the glial Na/K pump is actually stimulated by intracellular sodium, which enters through the Na–K–Cl cotransporter. This would explain the close involvement of an ouabain-sensitive component despite the fact that the main properties of the pump seem to show no kinetic differences between astrocytes and neurons. This view was supported by findings by Rose and Ransom (1996) that the bumetanide-sensitive cotransporter was an efficient mechanism to replenish sodium ions that were pumped out during potassium uptake into astrocytes. A kinetic investigation in the colon (Payne et al., 1995) showed that the Na–K–Cl cotransporter is well suited for potassium uptake due to the combined strong inward driving force of all participating ions and a high affinity for extracellular potassium. That this kinetic arrangement is indeed responsible for astrocytic potassium uptake also in situ was confirmed by MacVicar et al. (2002).

Another alternative for K^+ uptake in astrocytes is passive uptake of KCl through ion channels. The driving force for such an uptake is Donnan forces and it will continue until the equilibrium potentials for both chloride and potassium are identical and have the same value as the membrane potential. This accumulation, however, is crucially dependent upon a significant chloride permeability of the cell membrane. It therefore contradicts the paradigms of the spatial buffer mechanisms, which assumes selective potassium permeability. The issue was recently reviewed (Walz, 2002) and it was concluded that there is little evidence for such a Donnan-mediated KCl mechanism in astrocytes in *normal tissue*. There is a porin in the cell membrane of cultured astrocytes, similar to those found in mitochondrial membranes (Dermietzel et al., 1994). At positive membrane potentials these porins would activate their large unitary conductance. However, their exact activation mechanism and the occurrence of this porin in situ still has to be confirmed. It is of interest, however, that during spreading depression a large anion conductance is transiently mediating chloride uptake into cells (Phillips and Nicholson, 1979).

9. Reactive astrocytes

High chloride conductances exist in the cell membrane of astrocytes in pathologically altered tissue. These could assist in Donnan-mediated KCl uptake. There is evidence from cell culture studies that transformation of the astrocytes into a more reactive shape and state is activating various chloride conductances (Lascola et al., 1998; Fava et al., 2001). In reactive gliosis in situ such chloride conductances mediate KCl uptake in parallel with the carrier-assisted mechanism (Walz and Wuttke, 1999). In case of a blockade of the carriers the Donnan-mediated KCl uptake can substitute fully. Thus, it is suggested that this channel-mediated KCl uptake comes into play in reactive astrocytes but not in normal cells. Moreover, voltage-gated potassium channels are altered in reactive astrocytes in situ (Jabs et al., 1997; Francke et al., 1997; D'Ambrosio et al., 1999; Bordey et al., 2001). All these studies seem to confirm a down-regulation of glial inward (and some outward potassium) currents. This seems to interfere with spatial buffering, but apparently not with Donnan-mediated KCl uptake (Walz and Wuttke, 1999).

10. Concluding remarks

The notion that astrocytes are the major site of potassium homeostasis in the brain seems to have matured to the status of a scientific fact more than 35 years after it was first suggested by Hertz (1965). Moreover, all in situ studies designed to test the importance of buffering versus active uptake for K^+ homeostasis at the cellular level of the CNS constantly reach the conclusion that the carrier-mediated uptake is by far the most important component at all locations with the possible exception of the peripheral retina. After injury astrocytes undergo changes in their mechanism(s) for ion homeostasis, but they are still fully capable of potassium clearance, which maybe now becomes even more important than in healthy tissue.

Acknowledgement

Supported by operating funds from the Heart and Stroke Foundation of Saskatchewan.

References

Adelman, W.J., Fitzhugh, R., 1975. Solutions of the Hodgkin–Huxley equations modified for potassium accumulation in periaxonal spaces. Fedn Proc. 34, 1322–1329.

Balestrino, M., Aitken, P.G., Somjen, G.G., 1986. The effects of moderate changes of extracellular potassium and calcium on synaptic and neuronal function in the CA1 region of the hippocampal slices. Brain Res. 377, 229–239.

Ballanyi, K., 1995. Modulation of glial potassium, sodium and chloride activities by the extracellular milieu. In: Kettenmann, H., Ransom, B.R. (Eds.), Neuroglia. Oxford University Press, New York, pp. 289–298.

Ballanyi, K., Grafe, P., Ten Bruggencate, G., 1987. Ion activities and potassium uptake mechanisms of glial cells in guinea pig olfactory cortex slices. J. Physiol. 382, 159–174.

Barres, B.A., Koroshetz, W.J., Chun, L.L.Y., Corey, D.P., 1990. Ion channel expression by white matter glia: the type-1 astrocyte. Neuron 4, 507–524.

Bekar, L.K., Jabs, R., Walz, W., 1999. $GABA_A$ receptor agonists modulate potassium currents in adult hippocampal glial cells in situ. Glia 26, 129–138.

Bergles, D.E., Jahr, C.E., 1997. Synaptic activation of glutamate transporters in hippocampal astrocytes. Neuron 19, 1297–1308.

Betz, A.L., Iannotti, F., Hoff, J.T., 1989. Brain edema: a classification based on blood–brain barrier integrity. Cerebrovasc. Brain Metab. Rev. 1, 133–154.

Betz, A.L., Keep, R.F., Beer, M.E., Ren, X.D., 1994. Blood–brain-barrier permeability and brain concentration of sodium, potassium, and chloride during focal ischemia. J. Cereb. Blood Flow Metab. 14, 29–37.

Bordey, A., Lyons, S.A., Hablitz, J.J., Sontheimer, H., 2001. Electrophysiological characteristics of reactive astrocytes in experimental cortical dysplasia. J. Neurophysiol. 85, 1719–1731.

Brew, H., Gray, P.T., Mobbs, P., Attwell, D., 1986. Endfeet of retinal glial cells have higher densities of ion channels that mediate K^+ buffering. Nature 324, 466–468.

Chao, T.I., Henke, A., Reichelt, W., Eberhardt, W., Reinhardt-Maelicke, S., Reichenbach, A., 1994. Three distinct types of voltage-dependent K^+ channels are expressed by Muller (glial) cells of the rabbit retina. Pflugers Arch. 426, 51–60.

Chen, K.C., Nicholson, C., 2000. Spatial buffering of potassium ions in brain extracellular space. Biophys. J. 78, 2776–2797.

Coles, J.A., Orkand, R.K., 1983. Modification of potassium movement through the retina of the drone (*Apis mellifera* male) by glial uptake. J. Physiol. 340, 157–174.

D'Ambrosio, R., Gordon, D.S., Winn, H.R., 2002. Differential role of KIR channel and Na/K pump in the regulation of extracellular K^+ in rat hippocampus. J. Neurophysiol. 87, 87–102.

D'Ambrosio, R., Maris, D.O., Grady, M.S., Winn, H.R., Janigro, D., 1999. Impaired K^+ homeostasis and altered electrophysiological properties of post-traumatic hippocampal glia. J. Neurosci. 19, 8152–8162.

D'Ambrosio, R., Wenzel, J., Schwartzkroin, P.A., McKhann, G.M., Janigro, D., 1998. Functional specialization and topographic segregation of hippocampal astrocytes. J. Neurosci. 18, 4425–4438.

Dermietzel, R., Hertzberg, E.L., Kessler, J.A., Spray, D.C., 1991. Gap junctions between cultured astrocytes: immunocytochemical, molecular, and electrophysiological analysis. J. Neurosci. 11, 1421–1432.

Dermietzel, R., Hwang, T.K., Buettner, R., Hofer, A., Dotzler, E., Kremer, M., Deutzmann, R., Thinnes, F.P., Fishman, G.I., Spray, D.C., Siemen, D., 1994. Cloning and in situ localization of a brain-derived porin that constitutes a large-conductance anion channel in astrocytic plasma membranes. Proc. Natl Acad. Sci. USA 91, 499–503.

Erecinska, M., Silver, I.A., 1994. Ions and energy in mammalian brain. Prog. Neurobiol. 43, 37–71.

Erulkar, S.D., Weight, F.F., 1977. Extracellular potassium and transmitter release at the giant synapse of the squid. J. Physiol. 226, 209–218.

Fava, M., Ferroni, S., Nobile, M., 2001. Osmosensitivity of an inwardly rectifying chloride current revealed by whole-cell and perforated-patch recordings in cultured rat cortical astrocytes. FEBS Lett. 492, 78–83.

Francke, F., Pannicke, T., Biedermann, B., Faude, F., Wiedemann, P., Reichenbach, A., Reichelt, W., 1997. Loss of inwardly rectifying potassium currents by human retinal glial cells during diseases of the eye. Glia 20, 210–218.

Frankenhaeuser, B., Hodgkin, A.L., 1956. The after effects of impulses in the giant nerve fibers of Loligo. J. Physiol. 131, 341–376.

Gage, P.W., Quastel, D.M.J., 1965. Dual effect of potassium on transmitter release. Nature 206, 625–626.

Hansen, A.J., 1985. Effect of anoxia on ion distribution in the brain. Physiol. Rev. 65, 101–148.

Heinemann, U., Lux, H.D., 1977. Ceiling of stimulus-induced rises of extracellular potassium concentration in the cerebral cortex of cat. Brain Res. 120, 231–249.

Heinemann, U., Neuhaus, S., Dietzel, I., 1983. Aspects of potassium regulation in normal and gliotic brain tissue. In: Baldy-Moulinier, M., Ingvar, D., Beldrum, B. (Eds.), Current Problems in Epilepsy. John Libbey, London, pp. 271–277.

Heinemann, U., Walz, W., 1998. Contributions of potassium currents and glia to slow potential shifts. In: Laming, P., Sykova, E., Reichenbach, A., Hatton, G., Bauer, H. (Eds.), Glial Cells: Their Role in Behaviour. Cambridge University Press, Cambridge, pp. 197–209.

Hertz, L., 1965. Possible role of neuroglia: a potassium-mediated neuronal–neuroglial–neuronal impulse transmission system. Nature 206, 1091–1094.

Hounsgaard, J., Nicholson, C., 1983. Potassium accumulation around individual Purkinje cells in cerebellar slices from guinea pig. J. Physiol. 340, 359–388.

Hung, B.C., Loo, D.D., Wright, E.M., 1993. Regulation of mouse plexus apical Cl^- and K^+ channels by serotonin. Brain Res. 617, 285–295.

Husted, R.F., Reed, D.J., 1976. Regulation of cerebrospinal fluid potassium by the cat choroid plexus. J. Physiol. 259, 213–221.

Irwin, A., Walz, W., 1999. Spreading depression waves as mediators of secondary injury and of protective mechanisms. In: Walz, W. (Ed.), Cerebral Ischemia. Humana Press, Totowa, pp. 35–44.

Jabs, R., Paterson, I.A., Walz, W., 1997. Qualitative analysis of membrane currents in glial cells from normal and gliotic tissue in situ: down-regulation of Na^+ current and lack of P2 purinergic responses. Neuroscience 81, 847–860.

Jones, H.C., Keep, R.F., 1987. The control of potassium concentration in the cerebrospinal fluid and brain interstitial fluid of developing rats. J. Physiol. 383, 441–453.

Karwoski, C.J., Proenza, L.M., 1977. Relationship between Muller cell responses, a local transretinal potential, and potassium flux. J. Neurophysiol. 40, 244–259.

Keep, R.F., 2002. The blood–brain-barrier. In: Walz, W. (Ed.), The Neuronal Environment. Humana Press, Totowa, pp. 277–308.

Kelly, J.P., Van Essen, D.C., 1974. Cell structure and function in the visual cortex of the cat. J. Physiol. 238, 515–547.

Kettenmann, H., 1987. Potassium and chloride uptake by cultured oligodendrocytes. Can. J. Physiol. Pharmacol. 65, 1033–1037.

Kreisman, N.R., Smith, M.L., 1993. Potassium-induced changes in excitability in the hippocampal CAI region of immature and adult rats. Der. Brain Res. 76, 67–73.

Lascola, C.D., Nelson, D.J., Kraig, R.P., 1998. Cytoskeletal actin gates a chloride channel in neocortical astrocytes. J. Neurosci. 18, 1679–1692.

Leech, C.A., Stanfield, P.R., 1981. Inward rectification in frog skeletal muscle fibers and its dependence on membrane potential and external potassium. J. Physiol. 319, 295–309.

Lothman, E.W., Somjen, G.G., 1975. Extracellular potassium activity, intracellular and extracellular potential responses in the spinal cord. J. Physiol. 252, 115–136.

MacVicar, B.A., Feighan, D., Brown, A., Ransom, B., 2002. Intrinsic optical signals in the rat optic nerve: role for K uptake via NKCC1 and swelling of astrocytes. Glia 37, 114–123.

Malenka, R.C., Kocsis, J.D., Ransom, B.R., Waxman, S.G., 1981. Modulation of parallel fiber excitability by postsynaptically mediated changes in extracellular potassium. Science 214, 339–341.

McKhann, G.M. 2nd, D'Ambrosio, R., Janigro, D., 1997. Heterogeneity of astrocyte resting membrane potentials and intercellular coupling revealed by whole-cell and gramicidin-perforated patch recordings from cultured neocortical and hippocampal slice astrocytes. J. Neurosci. 17, 6850–6863.

Murdick-Donnon, L.A., Williams, P.J., Pittman, Q.J., MacVicar, B.A., 1993. Postsynaptic potentials mediated by GABA and dopamine evoked in stellate glial cells of the pituitary pars intermedia. J. Neurosci. 13, 4660–4668.

Newman, E.A., 1986. High potassium conductance in astrocyte endfeet. Science 233, 453–454.

Newman, E.A., 1993. Inward-rectifying potassium channels in retinal glial cells. J. Neurosc., 13, 3333–3345.

Nicholson, C., Phillips, J.M., 1981. Ion diffusion modified by tortuosity and volume fraction in the extracellular microenvironment of the rat cerebellum. J. Physiol. 321, 225–257.

Oakley, B., Katz, B.J., Xu, Z., Zheng, J., 1992. Spatial buffering of extracellular potassium by Muller (glial) cells in the toad retina. Exp. Eye Res. 55, 539–550.

Oldendorf, W.H., Cornford, M.E., Brown, W.J., 1977. The large apparent work capability of the blood–brain-barrier: a study of the mitochondrial content of capillary endothelial cells in brain and other tissues of the rat. Ann. Neurol. 1, 409–417.

Orkand, R.K., Nicholls, J.G., Kuffler, S.W., 1966. The effect of nerve impulses on the membrane potential of glial cells in the CNS of amphibian. J. Neurophysiol. 29, 788–806.

Payne, J.A., Xu, J.C., Haas, M., Lytle, C.Y., Ward, D., Forbush, D., 1995. Primary structure, functional expression, and chromosomal location of a bumetanide-sensitive Na–K–Cl cotransporter in human colon. J. Biol. Chem. 270, 17977–17985.

Pearson, M.M., Lu, J., Mount, D.B., Delpire, E., 2001. Localization of the KCl cotransporter KCC3 in the central and peripheral nervous systems: expression in the choroid plexus, large neurons and white matter tracts. Neuroscience 103, 481–491.

Phillips, J.M., Nicholson, C., 1979. Anion permeability in spreading depression investigated with ion-sensitive microelectrodes. Brain Res. 173, 567–571.

Ransom, B.R., 1992. Glial modulation of neuronal excitability mediated by extracellular pH: a hypothesis. Prog. Brain Res. 94, 37–46.

Ransom, B.R., Carlini, W.G., Connors, B., 1986. Brain extracellular space: developmental studies in rat optic nerve. Ann. N. Y. Acad. Sci. 481, 78–105.

Ransom, C.B., Ransom, B.R., Sontheimer, H., 2000. Activity-dependent extracellular K accumulation in rat optic nerve: the role of glial and axonal Na^+ pumps. J. Physiol. 522, 427–442.

Rausche, G., Igelmund, P., Heinemann, U., 1990. Effects of changes in extracellular potassium, magnesium and calcium concentration on synaptic transmission in area CA1 and the dentate gyrus of rat hippocampal slices. Pflugers Arch. 415, 588–593.

Reichenbach, A., Skatchkov, S.N., Reichelt, W., 1998. The retina as a model of glial function in the brain. In: Laming, P., Sykova, E., Reichenbach, A., Hatton, G., Bauer, H. (Eds.), Glial Cells: Their Role in Behaviour. Cambridge University Press, Cambridge, pp. 63–82.

Rice, M.E., Nicholson, C., 1991. Diffusion characteristics and extracellular volume fraction during normoxia and hypoxia in slices of rat neostriatum. J. Neurophysiol. 65, 264–272.

Rose, C.R., Ransom, B.R., 1996. Intracellular sodium homeostasis in rat hippocampal astrocytes. J. Physiol. 491, 291–305.

Schielke, G.P., Moises, H.C., Betz, A.L., 1990. Potassium activation of the Na, K pump in isolated brain microvessels and synaptosomes. Brain Res. 524, 291–296.

Skatchkov, S.N., Krusek, J., Reichenbach, A., Orkand, R.K., 1999. Potassium buffering by Muller cells isolated from the center and periphery of the frog retina. Glia 27, 171–180.

Skatchkov, S.N., Vyklicky, L., Orkand, R.K., 1995. Potassium currents in endfeet of isolated Muller cells from the frog retina. Glia 15, 54–64.

Somjen, G.G., 1975. Electrophysiology of neuroglia. Annu. Rev. Physiol. 37, 163–190.

Somjen, G.G., 1995. Electrophysiology of mammalian glial cells in situ. In: Kettenmann, H., Ransom, B.R. (Eds.), Neuroglia. Oxford University Press, New York, pp. 319–331.

Somjen, G.G., 2002. Ion regulation in the brain: implications for pathophysiology. The Neuroscientist 8, 254–267.

Somjen, G.G., Segal, M.B., Herreras, O., 1991. Osmotic-hypertensive opening of the blood brain barrier in rats does not necessarily provide access to: potassium to cerebral interstitial fluid. Exp. Physiol. 76, 507–514.

Stummer, W., Betz, A.L., Keep, R.F., 1995. Mechanisms of brain ion homeostasis during acute and chronic variations of plasma potassium. J. Cereb. Blood Flow Metab. 15, 336–344.

Sweadner, K.J., 1995. Neuroglia. In: Kettenmann, H., Ransom, B.R. (Eds.). Oxford University Press, New York, pp. 259–272.

Sykova, E., 1991. Activity-related ionic and volume changes in neuronal microenvironment. In: Fuxe, K., Agnati, L.F. (Eds.), Volume Transmission in the Brain. Raven Press, New York, pp. 217–236.

Voskuyl, R.A., Ter Keurs, H.E.D.J., 1981. Modification of neuronal activity in olfactory cortex slices by extracellular potassium. Brain Res. 230, 372–377.

Walz, W., 1987. Swelling and potassium uptake in cultured astrocytes. Can. J. Physiol. Pharmacol. 65, 1051–1057.

Walz, W., 2002. Chloride/anion channels in glial cell membranes. Glia 40, 1–10.

Walz, W., Hertz, L., 1983. Intracellular ion changes of astrocytes in response to extracellular potassium. J. Neurosci. Res. 10, 411–423.

Walz, W., Hinks, E.C., 1986. A transmembrane sodium cycle in astrocytes. Brain Res. 368, 226–232.

Walz, W., Klimaszewski, A., Paterson, I.A., 1993. Glial swelling in ischemia. Dev. Neurosci. 15, 216–225.

Walz, W., Wuttke, W., 1999. Independent mechanisms of potassium clearance by astrocytes in gliotic tissue. J. Neurosci. Res. 56, 595–603.

Walz, W., Wuttke, W., Hertz, L., 1984. Astrocytes in primary cultures: membrane potential characteristics reveal exclusive potassium conductance and potassium accumulator properties. Brain Res. 292, 367–374.

Wu, Q., Delpire, E., Hebert, S.C., Strange, K., 1998. Functional demonstration of Na^+–K^+–$2Cl^-$ cotransporter activity in isolated, polarized choroid plexus cells. Am. J. Physiol. 275, C1565–C1572.

Wuttke, W., 1990. Mechanism of potassium uptake in neuropile glial cells in the CNS of the leech. J. Neurophysiol. 63, 1089–1097.

Xiong, Z.Q., Stringer, J.L., 2000. Sodium pump activity, not spatial buffering, clears potassium after epileptiform activity in the dentate gyrus. J. Neurophysiol. 83, 1443–1451.

Zhou, M., Kimelberg, H.K., 2000. Freshly isolated astrocytes from rat hippocampus show two distinct current patterns and different K uptake capabilities. J. Neurophysiol. 84, 2746–2757.

Advances in
Molecular and
Cell Biology

Potassium and glia-derived slow potential shifts in relation to behaviour

Peter R. Laming

Medical Biology Centre, School of Biology and Biochemistry, Queen's University of Belfast, 97, Lisburn Road, Belfast BT9 7BL, Northern Ireland, UK
Correspondence address: Tel.: +44-1232-272269; fax: +44-1232-236505
E-mail: p.laming@qub.ac.uk(P.R.L.)

Contents

1. Introduction
2. Spatial buffering of potassium by glial cells
3. Slow potential shift origins in intact animals
4. Slow potential shifts and behaviour
5. SPS responses during arousal and attention
6. SPS responses associated with changes in motivation
7. Effects of potassium dynamics on motivation
8. The effect of imposed DC shifts on neuronal responsivity and behaviour
    8.1. Neuronal responsivity
    8.2. Behaviour
9. Changes in the SPS during habituation
10. Concluding remarks

This chapter examines evidence that in fish and amphibians glial cells respond to changes in extracellular potassium ($[K^+]_e$) in ways that contribute to modulation of neuronal activity and thereby behaviour. Glial cells spatially (and probably directionally) redistribute potassium from regions of increasing concentration to those with a lesser concentration. This redistribution is largely responsible for slow potential shifts associated with behavioural responses of animals. These slow shifts are related in amplitude to the level of 'arousal' of an animal, and its motivational state. In addition, glia, especially astrocytes, respond to changes in $[K^+]_e$. Simulating these effects by imposing potassium loads and by DC stimulation interacts with previous motivational state to alter neuronal responses and behaviour. The responses of glia to changes in extracellular potassium after neuronal activity have been associated with at least some forms of learning, including habituation.

The implication of these effects of potassium signalling in the brain is that there is considerable involvement of glia in a number of processes crucial to neuronal activity.

Advances in Molecular and Cell Biology, Vol. 31, pages 611–633

ISBN: 0-444-51451-1

Glia may also form another route for information distribution in the brain that is at least bi-directional, though less specific than its neuronal counterparts. It is evident that the Neuroscience of the future will have to incorporate much more study of neuron–glial interactions than hitherto.

1. Introduction

It was in the 1960s that the first definitive evidence was produced to reveal that glial cells in vivo were depolarised by activity of associated neurones; firstly in the leech (Nicholls and Kuffler, 1964, 1965) and subsequently in the optic nerve of the amphibian, *Necturus* (Orkand et al., 1966).

At about the same time, there were indications that glial cells were uniquely sensitive to extracellular potassium ($[K^+]_e$). These findings led Galambos (1961) to speculate that glia may modulate neuronal activity by controlling the extracellular milieu. However, it was Hertz (1965) that suggested that glia respond to elevated $[K^+]_e$ by acting as a sensing and conduction medium for information derived from neurones and by depolarising contribute to slow potential shifts (SPSs), associated with behaviour.

Since that time, evidence for these suggestions has emerged from a variety of species. Here we will concentrate on that derived from fish and amphibians with especial emphasis on the contributions of glia to SPSs that are associated with behaviour and its modulation.

2. Spatial buffering of potassium by glial cells

Activation of neurones releases potassium (K^+), causing astrocytic or ependymal cell depolarisation (Orkand et al., 1966; Kuffler et al., 1966). In contrast, glial hyperpolarisation results from activation of the electrogenic Na^+/K^+ pump (Walz and Kimelberg, 1985).

There is now considerable evidence that extracellular potassium concentration in mammals in vivo may fluctuate from its 'normal' level of ca. 3.5 mM to up to 9^+ mM in conditions of high neuronal activity (Syková, 1983, 1992). In the amphibian tectum $[K^+]_e$ rises from a baseline of 3 mM to 4 mM after a brief light flash has been reported, corresponding to an SPS of 1 mV recorded extracellularly (Roitbak et al., 1992). Elevated $[K^+]_e$ is not restricted to the immediate region of neuronal activity, as glial cells are connected by gap junctions to form a functional syncytium that allows spatial buffering of ions. Thus, a negative SPS induced by electrical stimulation of the tectum or a photic stimulus to a frog *Rana esculenta* (Roitbak et al., 1992, Fig. 4) or by electrical stimulation of the tectum of a fish, *Carassius auratus* (Nicol et al., 1993), induces a surface negative SPS that declines with depth and inverts at ca. 400 μm. Since astrocytes are relatively sparse in the tectum of amphibians and teleost fish (Roots and Laming, 1998) it is probable that potassium redistribution and the consequent SPS occurs through radial ependymoglia (Fig. 2, Lazar, 1989).

The concept of the spatial buffer function for potassium redistribution assumes a glial syncytium with $[K^+]_e$ increased at one region. The region experiencing increased $[K^+]_e$ will have a more positive K^+ equilibrium potential than the membrane potential, providing an inward driving force for K^+. Since the membrane is highly permeable to K^+, it will enter the cell via a passive current to be distributed to the other parts of the syncytium via gap junctions. In regions distant from the high K^+ region the K^+ equilibrium potential therefore becomes more negative than the membrane potential, and there is a driving force for K^+ out of the cell. Thus the current flows into the cells at high K^+ regions, through the syncytium, and out of the cells at regions distant from high K^+ locations. This intracellular current is almost completely carried by K^+. The loop is closed mainly by a return current in the extracellular space (ECS) carried by Na^+ and Cl^-.

A problem with the concept that glia generate the SPS by spatial buffering of potassium lies in measures of the length constant of glial cells. This is the distance along a process to the site, where a voltage amplitude has decayed to 37% of its value, due to leakage of current across the cell membrane. Barres et al. (1990) measured the length constant of an isolated astrocyte of the rat optic nerve for a single process as 100 μm. The overall length of the process is 400 μm, meaning that no significant portion of a current could cross into neighbouring glial cells. However, the length of the glial cell would be sufficient to redistribute K^+ from a site of maximal K^+ accumulation to a remote less active area. Indirect measurements based on K^+ transport induced by an electric field estimated a length constant of around 200 μm for the rat cerebral cortex (Gardner-Medwin, 1983). In the tectum of the frog (Fig. 2), *Rana esculenta*, the SPS and associated $[K^+]_e$ declined to 37% in 300 μm (Roitbak et al., 1992), whilst in the goldfish the SPS decline to this magnitude was about 250 μm (Nicol et al., 1993). In these animals this is well within distances from surface to depth that could influence the activity of intrinsic tectal neurons, involved in the processing of visual information.

Nicholson and Phillips (1981) studied ion diffusion in rat cerebellum using ion-selective micropipettes and ionophoretic point sources. They found that the movement of potassium in an electrical gradient behaves as if the ECS volume was more than 100%. This anomaly can be solved by assuming that the ion does not remain in the ECS, but is in fact the major current carrier across cell membranes. Hounsgaard and Nicholson (1983) measured changes in extracellular K^+ concentration in the vicinity of Purkinje cells in guinea-pig cerebellar slices. No extracellular K^+ changes were seen when the cells were hyperpolarised with current passage or during sub-threshold depolarisations. Only during spike activity was there a rise in extracellular K^+. In the vicinity of glial cells, however, a hyperpolarising current injected into glia reduced extracellular K^+ concentration in a symmetrical manner, while depolarising current injection induced a rise in $[K^+]_e$. These experiments show that the K^+ ion is mainly using the glial cell membrane during its movement in an electrical gradient.

Elevations of $[K^+]_e$ are never restricted to just one layer in a cortical-type structure. Thus, the problem of the restricted length constant of a glial cell syncytium is less severe than just discussed. For example, a number of glial cells may be depolarised at the point of maximal potassium accumulation, so that the depolarisation spreads over a short distance. Potassium released from active neurones will thus be redistributed to the less active surround. There, more spatially extended glial or coupled glial cells could redistribute the potassium further into the surround. The different local currents gradients will then add and contribute to generation of slow potentials over the whole structure.

To summarise, changes in the $[K^+]_e$ due to release on neuronal activity causes depolarisation of the glial membrane and K^+ uptake. Electrotonic current flow then occurs through glia by spatial buffering, and it may be recorded extracellularly as an SPS. Near the source of neuronal excitation the K^+ potential reflects glial depolarisation and the SPS in mammalian cortex (Roitbak et al., 1987) and in the frog brain (Roitbak et al., 1992).

3. Slow potential shift origins in intact animals

Electrical stimulation of the mammalian cortex with a short (0.2–0.8 ms) square wave pulse induces a brief (20–30 ms) negative wave at the cortical surface. This is known as the dendritic response (Chang, 1951), as it represents the excitatory post-synaptic potentials (EPSPs) of apical dendrites. Such responses can be induced regardless of the polarity of stimulation (Roitbak and Fanardjian, 1981). As the strength of the stimulating current is increased, the amplitude of the dendritic response increases proportionally, and a second, longer duration (250 ms) negativity develops (Chang, 1951; Goldring et al., 1959). The intensity required to provoke the secondary negativity is close to that required to induce depolarisation of deeply situated glial cells identified by their high resting potential and their lack of either spike activity or activity within the EEG range (Roitbak and Fanardjian, 1981). Increasing the frequency of the applied pulses to the cortical surface causes prolonged negativity, associated with summation of the secondary negativity (O'Leary and Goldring, 1964) and glial depolarisation (Karahashi and Goldring, 1966; Ransom and Goldring, 1973a,b; Roitbak and Fanardjian, 1981; Roitbak et al., 1987, Fig. 1). The latter paper suggested that the SPS can be almost entirely accounted for by glial depolarisation, except for the first 300 ms, to which inhibitory post-synaptic potentials (IPSPs) make a major contribution. It is probable that events occurring in the first 500–700 ms after stimulation include elements of both neuronal and glial, event-related depolarisations, whereas later events are mainly or exclusively glial.

Generation of SPSs also occurs with stimulation of the optic tectum of conscious but immobilised anuran amphibia (frogs and toads). The optic tectum is the main integrative region of the brains of these animals and its electrical stimulation by single stimuli provokes dendritic responses (Fig. 1A). The initial negativity of these dendritic responses declines on repeated stimulation, and is replaced by an SPS (Fig. 1B, Laming et al., 1992), which, as in mammals, outlasts the stimulus by seconds (Roitbak and Fanardjian, 1981). The amplitude of SPSs so generated reflects the 'strength' of the stimulus, whether this be measured by pulse duration, current or frequency.

4. Slow potential shifts and behaviour

To record SPSs in relation to behaviour, natural stimuli and unanaesthetised preparations are required. In unrestrained animals, movement can produce artefactual contamination of recordings, or SPSs induced by the central motor command centres themselves may render interpretation of the sensory response component difficult. Averaging techniques may overcome some of these problems if the response is resistant to habituation (Roughan and Laming, 1998). In spite of these difficulties there has developed a quite substantial literature on mammalian 'reactive' SPS responses to sensory stimuli

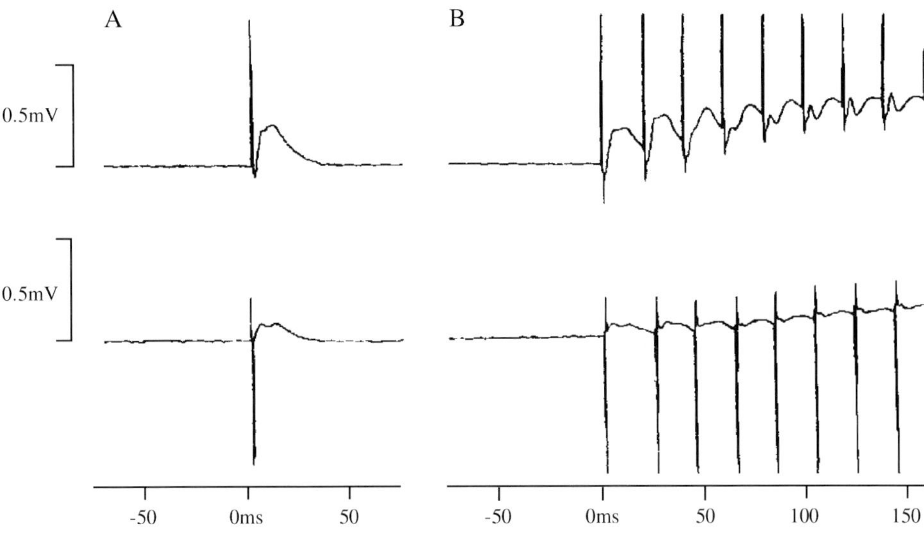

Fig. 1. (A) Dendritic response to a 600 μs, 100 μA pulse to the surface of the optic tectum of a frog. Note that the artefact changes polarity with polarity of the stimulus (upper compared to lower trace). (B) With repetition of a 500 μs, 100 μA stimulus at 50 Hz for 1 s, the dendritic responses decline, superimposed on a generated SPS (after Laming et al., 1992).

(Rowland et al., 1985). These responses may have a general distribution in the cortex or may be localised.

Lickey and Fox (1966), using a muscle relaxant to immobilise the animal, similarly found surface negative SPS responses to visual, auditory and peripheral electroshock stimuli, localised in the primary sensory region for the stimulus modality used. The occurrence of SPS responses in the mammalian cortex has led many to assume that it is a strictly cortical phenomenon, yet SPSs in response to visual stimulation have been recorded from the hyperstriatum of pigeons, a region lacking a laminated cortical structure (Durkovic and Cohen, 1966, 1968). This led to the proposal that a radially organised structure was not necessary for the expression of SPSs. This concept might be supported by recordings of SPSs from apparently non-radially organised or laminated regions of the fish brain, such as the telencephalon and medulla (Nicol and Laming, 1992). Both fish and amphibians are useful 'model' vertebrates for the study of relationships between sensory neuronal activity and SPS responses, as most visual sensory experience is integrated in the midbrain optic tectum (Fig. 3) and auditory information in the underlying torus semicircularis. In anuran amphibians (frogs and toads), the retinal ganglion cell output projects via the optic nerve to the surface of the 9-layered contralateral tectum (Fig. 2), where information is processed from surface to depth. Microelectrodes can be used to simultaneously record both neuronal and SPS responses from the tectum (Fig. 3), and these can be monitored separately by selective filtration and amplification of the recorded signal. Similarly, the electrode can be moved to maximise the amplitude of the signal of one neuron (a unit) whilst smaller signals are filtered out. Recordings of this nature have

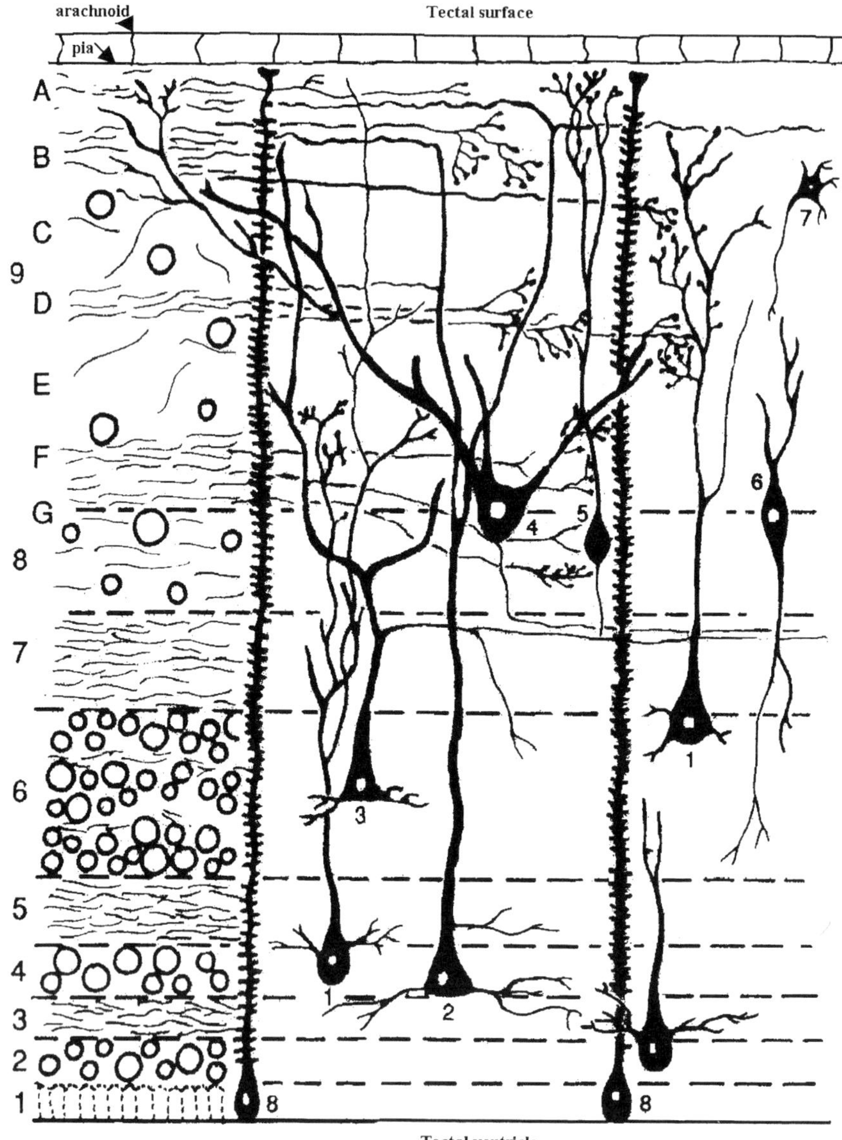

Fig. 2. Schematic representation of the layers and types of cell in the optic tectum of an anuran. Together, the depth of all the layers amounts to approximately 0.5 mm. The left illustrates the tectal cytoarchitecture and fibre patterns. Numbers on the extreme left indicate layers; A to G are sub-divisions of layer 9. Unmyelinated fibres in sub-layers A, C, E are not drawn. Arachnoid and pia are meninges of these names. 1,2—large pear-shaped neuron; 3—large pyramidal neuron; 4—large ganglionic neuron; 5—small pear-shaped neuron; 6—bipolar neuron; 7—stellate neuron; 8—ependymo-glial cell (modified from Roitbak et al., 1992).

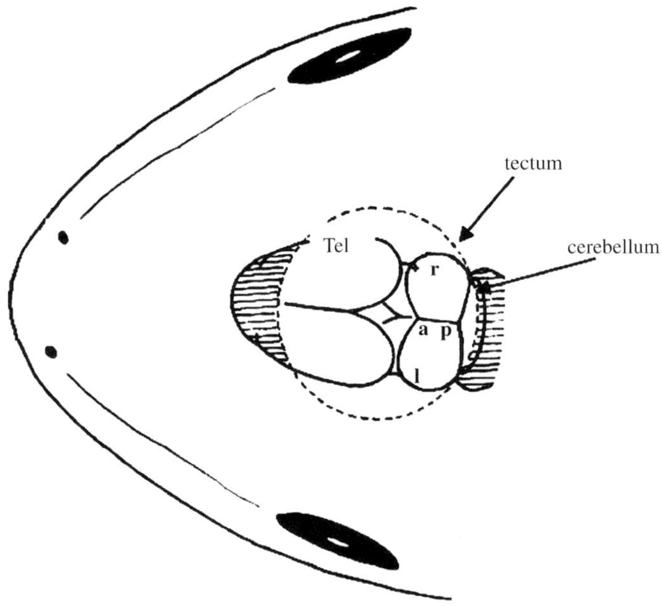

Fig. 3. Schematic drawing of an anuran brain to illustrate positions of recording and stimulating electrodes described in Sections 8.1 and 8.2. Tel = telencephalon; electrodes used in immobilised animals; + = recording microelectrode, a, p: anterior and posterior ball electrodes for both DC stimulation and SPS recording. Stimulating electrode positions for recording behavioural responses; r = right, l = left.

allowed neuronal units to be classified into types based both on the responses they make to a range of visual stimuli presented to the animal and on the size of their receptive field.

The main target for retinal afferents is the midbrain optic tectum which has a laminated cortex-like structure (Fig. 2). The innermost layer (layer 1) is composed of cell bodies of ependymoglial cells. Layers 2–9 are alternately cellular and fibrous layers, the outermost ones of which (layers 8 and 9) receive retinal input. Retinal fibres, exclusively from the contralateral eye, terminate on the tectum with a retinotopic distribution. Four types of retinal cell input to the tectum have been identified on the basis of their response properties to visual stimuli. All are responsive to moving visual stimuli with circular receptive fields (with sizes of 2–4, 8 and 10†15 degrees of visual angle) and 'on, on–off and off' responses to illumination change. Although these units respond to moving visual stimuli, their activity does not show the configurational selectivity of prey-catching behaviour, so the basis for this must lie in cellular elements beyond the retina.

Although the tectum has a laminated appearance, it is composed of cells, neurons and glia, which have a column-like organisation, with their processes oriented perpendicular to the surface (Fig. 2). These intrinsic tectal units have been identified on the basis of their physiological responses to moving visual stimuli. Seven major classes of tectal (T) neurons have been identified in this manner, though some have been sub-categorised. Of these units, the T5 units and their sub-classes, $T5_1$, $T5_2$ and $T5_3$, have provoked most interest because they show sensitivity and, in the case of $T5_2$, selectivity for 'worm-like' stimulus properties mirroring those which provoke prey-catching behaviour. T5 units are sensitive to the area of a moving stimulus, $T5_1$ units to its extension in the direction of movement and $T5_3$ units to its extension perpendicular to its direction of movement. Most significantly, $T5_2$ units are sensitive to both so that they are selective for a worm-like stimulus. They have a response pattern of worm preference which reflects the probability that the stimulus will be treated as 'prey' (Ewert, 1989).

Evidence has accumulated to emphasise that the SPS recorded intracranially is a general expression of sensory activity in vertebrates as a whole. In response to direct electrical stimulation, or sensory stimuli, the negative SPS reflects the strength of the stimulus or the activity of local neurons, respectively. Thus, in anurans, the SPS amplitude is correlated with visual unit activity at the tectal surface, where it is presumably the retinal ganglion cell input (neuron #4 in Fig. 2) that releases K^+ and thus generates the SPS (Laming and Ewert, 1984). In response to the onset of illumination or to electrical stimulation of the tectal surface SPS also closely reflects changes in $[K^+]_e$ (Roitbak et al., 1992, Fig. 4). The SPS at the surface is probably a reflection of excitation

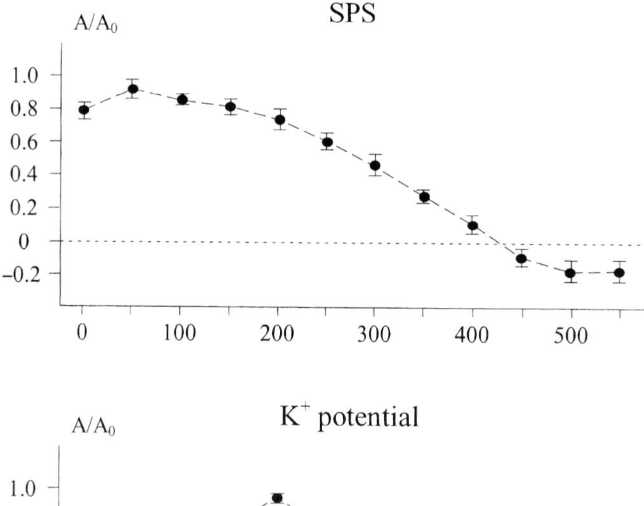

Fig. 4. Plot of mean ± SEM (n = 10) of changes in SPS and K^+ potential (potential recorded with a potassium selective electrode) to electrical stimulation of the frog tectal surface. Note that the SPS is maximal at 50 μm, the K^+ potential at 200 μm (after Roitbak et al., 1992). A/A_0 = the proportion of a value (A) compared to the maximum value obtained (A_0).

of unmyelinated retinal afferent terminals in layer 9. These and deeper intrinsic tectal neurons contribute to the rise in $[K^+]_e$ which is maximal at a depth of between 100 and 200 μm. In mammals also, the reactive SPS to sensory stimulation is fairly closely correlated with neuronal activity (Rowland, 1968). Reactive SPSs are always negative at the source of their generation, but may be reflected elsewhere in the brain by either positivity or negativity. In the teleost, *Carassius auratus* (goldfish), a negative SPS is evoked in the midbrain in response to visual or pressure-wave stimuli, whilst in the fore- and hindbrain regions, the SPS is initially positive (Quick and Laming, 1990). In the common toad, *Bufo bufo*, SPS responses were first recorded in conscious but immobilised animals in response to stimuli, which emulated natural prey objects (Laming and Ewert, 1984). In the optic tectum, these stimuli caused activation of visual units, an increase in EEG frequency and amplitude and a monophasic negative SPS, these being recorded with the same electrode. At the tectal surface, the region of retinal fibre input, the unit activity preceded EEG changes and the SPS, but in deeper layers of the sensory processing system, the SPS marginally preceded the EEG change but significantly pre-empted the activity of local units. This led to the suggestion illustrated in Fig. 5, that the radial glial potassium buffering currents were acting to translocate K^+ to deep layers of the sensory processing system prior to the onset of activity in the neurones themselves (Laming, 1989a). A system of this nature is made possible by the lack of synaptic delay in the glial syncytium, and would be adaptive, in that it would 'prime' neurones likely to be imminently in receipt of visual input. Spatial buffering would make this sensitisation mechanism passive, but it could be more dramatic as at least mammalian glia have active uptake of K^+ (see chapter by Walz) and participate in glutamate metabolism (see chapter by Schousboe and Waagepetersen). In sensory systems that are organised topographically to represent spatial sensory experience, like the visual system, neurones in areas close to the part of the sensory map being stimulated would also be sensitised, and responses to moving objects would be enhanced. The release of $[K^+]_e$ in response to sensory stimulation, as implied by this hypothesis, is evidenced by SPSs and changes in $[K^+]_e$ recorded in response to a visual stimulus in frogs. These have a similar depth profile as that induced by tectal electrical stimulation (Roitbak et al., 1992, Fig. 4). Thus, in anurans at least, the column of neurones responding to a particular retinotopic projection would seem to be associated also with a parallel oriented radial glial cell or cells, participating in the responses of the neural information processing system, much as envisioned by Hertz (1965).

The toad tectum derives much of its ability to discriminate prey (e.g., an elongate piece of black card moving along its long axis against a contrasting background, mimicking a worm) from non-prey (other configurations of the black card) due to inhibitory projections arising from the pretectal thalamus that are activated by non-prey stimuli. If appropriately positioned small lesions are made in the pretectal thalamus, the toad will treat almost any moving object as prey (Ewert, 1989). In an immobilised animal, a similar lesion causes enhanced neuronal responses to visual stimuli, and the selectivity of responding units for the configuration of the visual stimulus is lost. In spite of the greatly enhanced neuronal activity in the tectum, the SPS is considerably attenuated (Laming and Ewert, 1983). In contrast, telencephalic ligature causes declines in both unit and SPS responses

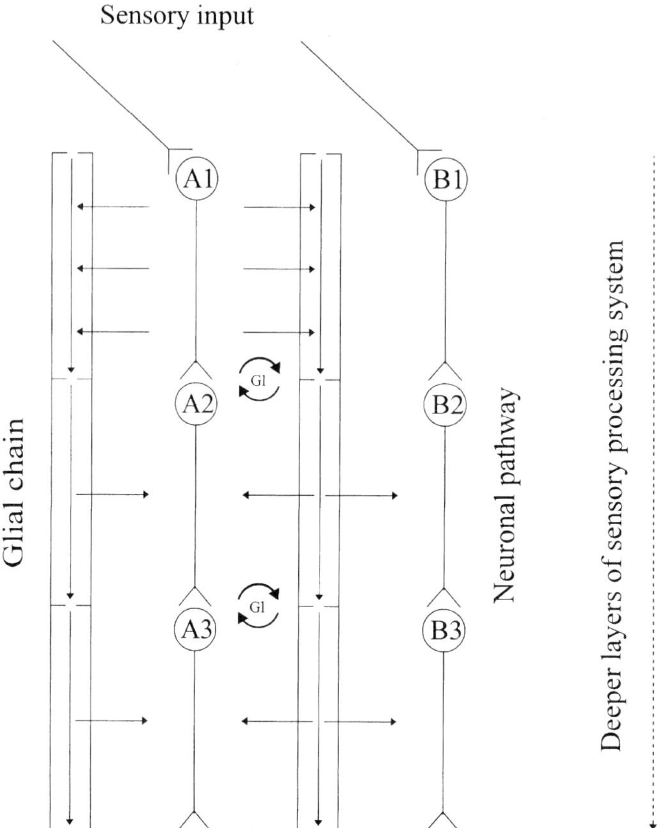

Fig. 5. A mechanism by which glia may sensitise neurones by relocation of potassium in a radially organised sensory system, like the toad tectum. On arrival of the sensory input, active neurones, e.g., A1, release K^+, which is translocated by glia to the regions of A2 and A3, which are therefore partially depolarised prior to being reached by the specific neuronal activity through the A1/A2 and A2/A3 synapses. Spatial buffering would make this sensitisation mechanism passive, but it could be more dramatic as glia have active uptake of K^+ and participate in glutamate metabolism. In sensory systems that are organised topographically to represent spatial sensory experience, like the visual system, neurones in areas close to the part of the sensory map being stimulated would also be sensitised (B1, B2, B3) and responses to moving objects would be enhanced (after Laming, 1989a).

(Laming et al., 1984b). It would thus seem that there are experimentally contrived conditions when the SPS is not a simple mirror of local neuronal activity in toads.

In behaving toads, both the frequency of tectal visual unit responses and the amplitude of the SPS response to a visual stimulus is greater than in immobilised animals, and it is associated with prey-catching behaviour. However, if simultaneous defensive behaviour is elicited by a tactile stimulus, then visual unit activity is minimal, whilst the SPS recorded from the same electrode is larger than from visual stimulation alone (Laming et al., 1984a). In this case the SPS may derive from deeper midbrain regions responding to the tactile stimulus, emphasising that an important aspect of ionic currents through glia is that they enable conduction in any direction in which they are connected by gap junctions.

Thus, deep activation could potentially act to sensitise more superficial regions, a role previously ascribed to reticular formation projections. Presuming the SPS to be due to spatial buffering currents derived from glia, these results suggest that glia may act to integrate signals derived from a number of stimuli or stimulus modalities.

When a number of SPS recording electrodes are employed to examine responses of behaving animals to a sensory event, quite complex waveforms may emerge. Thus, toads respond to prey-like objects in their frontal visual field with an initially negative SPS in the corresponding retinal projection (anterio-lateral) region of the tectum. The response in the posterior tectum, which is not directly activated by this visual stimulus, is initially positive. In the behaving animal, observation of the prey leads to orienting (see below) and approach behaviour (Ewert, 1989), during which time (ca. 4 s) the polarity of the regional waves becomes reversed. If the same animals are immobilised, they still exhibit anterior negativity and posterior positivity in tectal SPS responses to the prey stimulus. However, under such conditions these responses are monophasic, suggesting that some aspect of the behaviour itself might explain 'rebound' in SPS polarity. One explanation is that the movement of the toad causes visual input to the whole tectum to be activated and thus causes many different sources and sinks to be generated (Laming et al., 1995).

Of perhaps greater interest was the finding that in immobilised animals with no visual stimulation of the posterior tectal projection region, but visual stimulation of the anterior tectum, a positive SPS was recorded in the posterior tectal projection region, which mirrored the negative SPS in the stimulated region of the anterior tectum. This implies that sinks and sources for current may be present across the surface of the tectum as well as through its depth. If the conduction pathway through the depth of the optic tectum is via radially oriented glia, then that across the surface may be via tangentially oriented glia. Glial processes are often tangentially oriented at the brain surface, which itself is comprised of a sheet of ependymal glial cells or radial glial end feet as part of the glia limitans, which separates cerebrospinal fluid (CSF) of the sub-arachnoid space from the fluid of the ECS of the brain. The apparent separations between sources and sinks (a distance of 1 mm) found in these immobilised animals, would suggest a low resistance pathway for current spread, potentially available through the CSF, along the glia limitans, which express high density of gap junctions, or possibly even via meningeal (pial) gap junction-coupled cells covering the surface of the tectum as shown in Fig. 2 (see also chapter by Mercier and Hatton). It would be interesting to determine whether the movement of ions, especially K^+, between ECS and CSF has the buffering properties ascribed to similar K^+ movements in the retina, between vitreous humour and Muller cells (Newman, 1986—see also chapter by Bringmann et al. and by Scemes and Spray) and whether such properties might form part of a mechanism for controlling ECS constituents.

5. SPS responses during arousal and attention

The reactive responses of animals to sensory input, as described above, vary according to their behavioural state. This may provide clues to the origin of that state.

Animals are not in a constant state of alertness, and even when they are awake, there are periods when neural activity is increased and the behaviour of the animal is activated. The animal may be described as being 'aroused'. External factors, which induce arousal, are stimuli, which have biological relevance, because they are novel, or stimuli which are relevant, either due to an endogenous trigger (food, mate), or because they have acquired relevance through experience. The response to such relevant stimuli is the 'orientation reaction' (Sokolov, 1963), initially comprising a non-directed alerting response (arousal), associated with generalised increases in neuronal activity, changes in EEG frequencies and amplitudes, a general reduction in sensory thresholds, development of SPSs and changes in measures of peripheral physiology, such as heart and ventilatory rates. This initial generalised response is brief, especially in mammals, and it is followed by a more directed behaviour towards the source of the stimulus and a reduction in the expression of the responses in those regions of the brain, which are not associated with further assessment of the stimulus. This secondary phase may be described as 'attention' and includes behavioural orienting towards the stimulus source (Laming, 1989a,b). The magnitude of the SPS (in terms of amplitude and/or duration) appears to reflect the level of arousal or activation of the brain in mammals (Rowland, 1968).

In fish, a transient bradycardia (reduction in heart rate) provides a quantifiable measure of the arousal response to a novel stimulus and of the decline of that response with repetition of the stimulus (Laming and Savage, 1980). Situations that evoke a bradycardia, like the onset of increased illumination, also induce SPSs, recorded with implanted electrodes in the midbrain, hindbrain and forebrain (Quick and Laming, 1990). In response to a moving visual stimulus, the onset of illumination, or a tap to the side of the aquarium, the resulting 4–8 s SPSs were predominantly negative in the midbrain, positive in the forebrain and mildly positive in the hindbrain. During early presentations of the tap stimulus, the cardiac arousal responses were large and strongly related, as evaluated by regression analysis, to the amplitude of the midbrain SPS response to the tap stimulus.

Interest in this relationship provoked further studies of cardiac arousal and SPS responses, this time with Ag/AgCl electrodes on the telencephalic, anterior tectal, cerebellar and medullary surfaces (Nicol and Laming, 1992). Initial presentation of the onset of illumination to fish in a darkened enclosure evoked a bradycardiac arousal response, accompanied by a predominantly positive SPS on the medullary surface, a predominantly negative SPS on the cerebellum, and an SPS that was initially negative (1–2 s), then positive and eventually returning to negativity, on the anterior tectal and the telencephalic surfaces (Nicol and Laming, 1992). An SPS response to an arousing stimulus can thus be recorded in many regions of the fish brain, not only those most directly involved in processing the primary stimulus input, though such regions are associated with SPS responses of greatest amplitude.

During arousal, all brain responses are considered to be amplified, whereas during attention there is selective response inhibition (Lynn, 1966). In rabbits, visual evoked potentials are enhanced in amplitude if preceded by cortical negativity, leading to the conclusion that the amplitude of the evoked potentials could be changed by reinforcement (Richter et al., 1992). We have performed studies on fish to determine if evoked potentials in response to a relatively neutral stimulus (sound) were affected by the prior presentation of an arousing, SPS-evoking stimulus. Initial presentation of a non-acoustic stimulus to fish may cause an increase or a decrease in the acoustic evoked potential (AEP) response in the brainstem to subsequently presented 'click' stimuli (Laming and Brooks, 1985). When more localised recordings were made, it was found that it is the peak to peak amplitude that is most affected by the preceding priming stimulus (Laming and Bullock, 1991). Although tenuous relationships have been found between the broad-band increase in EEG power in response to the non-acoustic stimulus and the change in the AEP, these were stimulus-specific and could not alone account for the change in sensory evoked activity (Laming et al., 1991a). Sustained potential shift responses and changes in AEP amplitude were thus simultaneously recorded in the midbrain tectum and torus semicircularis in response to water or saline applied to the flank of a carp. A 4 s negative SPS wave was followed by a positive wave in response to either type of priming stimulus alone (Laming et al., 1991b). Trains of six clicks delivered in the absence of a priming stimulus showed that the highest amplitude AEP was that evoked by the first click in the train. With a priming stimulus, this AEP was the most attenuated, indicating an attentional rather than an arousal response type (since arousal as mentioned above is associated with increased evoked potential amplitudes). These changes in AEP amplitude did not relate to the SPS in the acoustic-response region in which these were recorded. Rather, they showed relationships with the SPS in the other region from which recordings were made (i.e., torus or tectum).

Although these studies suggested links between mechanisms generating the SPS and mechanisms causing changes in sensory evoked responses, they were not shown to be causally related and were coincidentally timed so that the SPS was going from negative to positive at a time when the AEP was attenuated. Further experiments were therefore performed, in which the fish was subjected to continuous background auditory stimulation (clicks delivered at 1 click/s), with averaging of AEP response changes, subsequent to the induction of SPSs on the telencephalic, posterior tectal and medullary surfaces in response to the onset of illumination. The posterior midbrain response was similar to that of the cerebellum reported previously, i.e., a largely monophasic, 10 s long negative wave; the medulla showed a small positive wave and the wave on the telencephalic surface was smaller still, and positive (Nicol and Laming, 1993). Attenuation of AEPs was recorded in response to illumination onset as the priming stimulus, again suggesting an attentional response. However, least attenuation was associated with large cardiac responses, indicating that both facilitatory and inhibitory (arousal and attention) mechanisms might be simultaneously active. Again, the amplitudes of these SPSs were closely related across the different regions. Correlations were also found between stimulus-evoked changes in AEP amplitude and simultaneously recorded SPSs during initial presentations of the stimulus. In the telencephalon SPS amplitude and changes in AEP amplitude were significantly correlated, and changes in midbrain AEP amplitude were significantly correlated with SPS amplitudes in the telencephalon and the medulla. These studies have revealed that many modalities of naturalistic stimuli induce SPS responses in most regions of the brains of fish. These responses are often related to cardiac (arousal) responses but they also share features of attention in that they accompany regional attenuation of sensory evoked neuronal activity.

6. SPS responses associated with changes in motivation

In toads, the feeding motivational state of the animal can be tested and quantified by the number of times the animal will perform the behavioural components of the prey catching sequence, when presented with a simulated prey object. The prey is an elongate piece of black card moving along its long axis against a contrasting background; i.e., 'a worm'. The evoked behaviours are categorised as 'orient', 'approach', 'fixate' and 'snap' (Ewert, 1989). After testing for the motivational state, animals were prepared for recording SPSs during subsequent stimulus presentation. Ag/AgCl 0.5 mm ball electrodes were placed bilaterally on the anterior, mid and posterior tectal surfaces through small apertures in the cranium and fixed in place before leaving the animal 24 h to recover. Testing with the simulated prey showed that orientation to prey was associated with a negative shift on the anterior tectal surface, the retinal projection region of the frontal visual field. This negative shift was followed by a positive wave after ca. 4 s. On the surface of the posterior tectum, the reverse occurred, i.e., a positive shift was followed by negativity (Laming et al., 1995). The amplitude of these shifts was related to the prey catching motivation prior to the operation. Animals were then re-anaesthetised and immobilised by a muscle relaxant and retested to the 'worm' stimulus. The SPS response now was a monophasic negative wave in the anterior tectum and a monophasic positive wave in the posterior tectum. Again their amplitudes were related to the previously recorded motivational state. In order to determine if the motivational state had been affected by the operation and subsequent testing, animals were again anaesthetised to allow them to recover from the muscle relaxant and they were then tested again. They demonstrated no change in prey catching activity from that exhibited prior to the experiment (Laming et al., 1995). Thus, it can be concluded that the motivational state of the animal measured by prey-catching behaviour, influences the amplitude of SPSs.

7. Effects of potassium dynamics on motivation

Section 6 showed that changes in neuronal responsiveness and SPSs are associated with changes in motivation as measured by behaviour in anurans. To test the hypothesis that these changes in motivation were related to changes in potassium dynamics, toads tested for motivation to feed were prepared for tectal recordings of visual units, overall neuronal activity and SPSs in response to application of isotonic artificial CSF to the tectum and subsequent administration of visual stimuli. Though only a few animals were motivated to feed, all were sexually active, which might account for the fact that most of the units found post-operatively responded best to moving square stimuli ($T5_1$ unit), which are more likely to represent potential mates than prey. This might account for the high level of overall neuronal responses to this type of stimulus before any solution addition, especially in non-hungry toads. It might also account for the significantly larger overall neuronal and SPS responses made by these animals compared to those of hungry animals. However, all animals showed significant unit, overall neuronal activity and SPS responses to presentation of the moving square stimulus alone at the start of experiments (Laming and Laming, 2003).

With the exception of 41 mM K^+ solution, which was applied last, all lower concentrations were applied in random order and a summary of the effects of solution addition alone follows. Addition of 0 and 4 mM K^+ isotonic solution to the pia-covered brain surface produced no significant changes in unit or massed unit activity or any SPS. This may be because this potassium concentration is similar or less than that of 'normal' extracellular space. Although 7 mM K^+ solution addition showed no overall effect, there was an earlier increase in neuronal activity and an earlier SPS in hungry toads than in non-hungry animals. This is the first indication that the different motivational characteristics of the two groups may be reflected in the manner that the brain (or meninges) deal with a potassium load, marginally reduced sodium or physical trauma. With the addition of 11 and 17 mM K^+ there were significant rises in unit activity and an SPS. With 17 mM K^+, there was also an increase in massed unit activity and the SPS was more evident in hungry toads. Addition of 27 mM K^+ solution only produced significant unit activity, and the final (non-randomised) addition of 41 mM K^+ only elicited a significant SPS, though at this point in the experiment, all responses appeared erratic, with a high level of massed unit activity. Overall, the concentration range from 7 to 17 mM K^+ produced the most significant responses to solution addition and revealed differences in those responses between the hungry and non-hungry groups. The differential responsiveness of the two groups and the fact that responses were only obtained at these concentrations suggests fairly free permeability of the pia to the ionic constituents. The fact that only these concentrations induced responses might also suggest that the physical trauma of solution addition may not be responsible for the responses themselves, but the lack of relationship between concentration and response would suggest that responses were not entirely due to diffusion characteristics of raised K^+ but may also involve the slightly lowered Na^+ concentration.

In response to the visual stimulus, there was an increase in overall neuronal activity at all except 7 mM K^+ concentrations. In all cases it was the non-hungry toads that demonstrated the largest responses. This effect was reflected also in the single unit responses at 7 mM K^+, although a significant unit response was obtained at all concentrations.

Orienting behaviour prior to the experiment was positively regressed on overall neuronal activity in response to solution addition and negatively on overall neuronal activity in response to the visual stimulus at both 4 and 7 mM K^+ concentrations. Thus, hungry animals showed more neuronal activity on solution addition and less than non-hungry animals to the visual stimulus. This followed a general trend in that non-hungry animals responded best to the moving square stimulus, but were less obviously sensitive to the addition of solutions. These differences in responsivity were most evident in the concentration range of 4–17 mM K^+. What emerges is the probability that the hungry and non-hungry animals all respond to both solution applications to the brain and to a visual stimulus, but in different ways. Non-hungry animals were more sensitive to solution addition, but slower, and more responsive towards the (non-prey) visual stimulus. The latter response suggests that their motivation might be sexual rather than food related. Their later, but more sensitive response to solution addition might reflect a different pial or cellular permeability to potassium. The converse explanation might obtain for the hungry toads that show a reduced responsivity to the non-prey stimulus, but respond faster to solution addition, perhaps because of relatively enhanced pial or cellular permeability to potassium and/or sodium. The possibility that permeability changes could contribute to changes in responsiveness and motivation is worthy of greater examination.

8. The effect of imposed DC shifts on neuronal responsivity and behaviour

8.1. Neuronal responsivity

One way to emulate the effects of the SPS recorded in the tectum is to generate an SPS by DC stimulation and record it together with its associated neural responses both with and without additional visual stimuli. Although there is considerable evidence that a prolonged (days, weeks) DC stimulation damages neural tissue (Hurlbert et al., 1993), there is no evidence that DC stimulation at levels that mimic those reflected normally as sustained or SPSs has any adverse effect. Nor did the behavioural studies (see below) suggest that any damage had occurred.

Toads (*Bufo bufo*) were tested for their prey catching motivation prior to experiments as described above. All animals showed prey-catching responses of varying degrees and motivation was measured by the number of orienting responses. Experiments involved recording SPS responses from platinum/iridium ball electrodes on the medial anterior and posterior left tectum and neuronal unit responses and SPS from a microelectrode in the anteriolateral right tectum, corresponding to the frontal visual field (Fig. 3, Sterritt et al., 2003). Prior to any DC stimulation, regression analysis showed that the neuronal responses of toads to a prey-like visual stimulus reflected their motivational tendency prior to operations, with a maximum response (~ 10 spikes/s) after 4 s (Fig. 6a). Positive DC stimulation in the proximity of an electrode recording from the toad tectum enhanced and accelerated neuronal responses recorded by that electrode to a visual stimulus (Fig. 6b). Negative DC stimulation inhibited those neuronal responses. It would appear from regression analysis that the positive stimulation reinforces the motivational tendency of an individual animal. Although the data showed links between positivity and the neuronal response, this may not be the only interpretation as the visual stimulus was actually presented after the electrical stimulus, during the period when the polarity rebound was occurring, i.e., when the polarity was negative. Thus it can be concluded that DC stimulation interacted with the behavioural measures of motivational tendency to produce enhanced neuronal responses, whilst the potential was going from positive to negative.

Recently, experiments have been performed using transcranial DC electrical stimulation in humans, that have revealed that a polarity dependent activation or suppression of cortical activity can be achieved with anodal or cathodal stimulation, respectively (Nitsche et al., 2002). Although those authors recognise the potential therapeutic use of such shifts they do not seem to consider the likelihood of a glial contribution to naturally occurring DC shifts, even though a glial origin of similar shifts has been demonstrated in cats (Roitbak et al., 1987) and amphibians (Roitbak et al., 1992), where the shifts mirror movements of extracellular potassium. It has been suggested that these potassium concentration changes may be a mechanism for modulation of neural excitability (Laming, 2000; Laming et al., 2000). The experiments described in Section 7 are consistent with this concept.

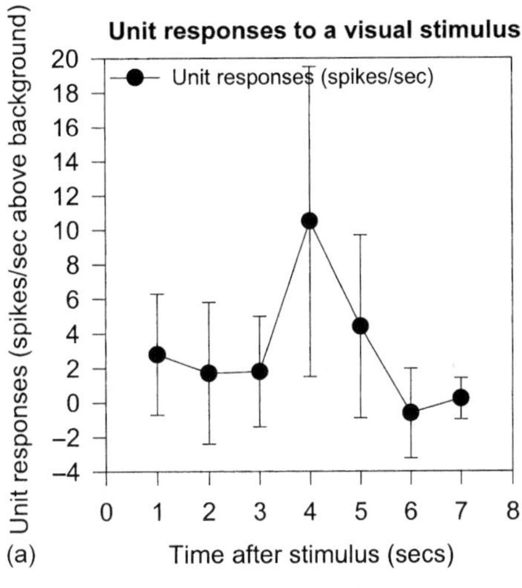

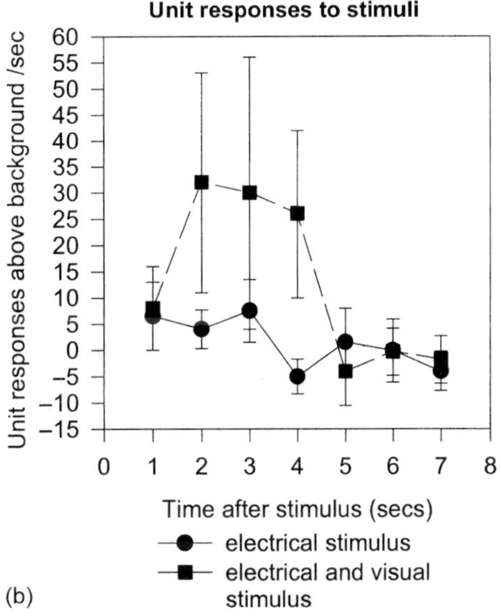

Fig. 6. Mean ± SEM of unit responses to the presentation of a moving worm-like stimulus, either (a) alone or (b) after stimulating the anterior tectal ball electrode (see Fig. 3) at +1.5 mV for 1 s against the negative posterior electrode, compared to DC stimulation alone (after Sterritt et al., 2003).

8.2. Behaviour

The observation that DC currents and a thus imposed DC shift affect neuronal and SPS responses led to the question whether DC shifts also might have a modulatory effect on motivation. Thus, experiments were carried out to examine how DC or equivalent current AC stimulation of the toad tectum might affect the prey catching or avoidance responses exhibited by toads exposed to a simulated prey object. As before, toads were tested for their prey catching motivation. Under anaesthesia an operation was then performed to place a stimulating electrode on the anterior surface of each tectum (Fig. 3), and the animal was left for 24 h to recover before stimulating at various, randomly ordered, current strengths of both polarities (Sterritt and Laming, 2003).

Prey-like visual stimuli, remotely operated, were presented to the toad in both clockwise and anti-clockwise directions, and the behavioural responses were monitored. The imposition of a DC current to the tectal surface (Fig. 3) caused a current strength-related increase in behavioural responses to a visual prey-like stimulus (Fig. 7). In all the experiments there was a large increase in prey-catching behaviour at currents above 50 μA, compared to the control without DC stimulation or the initial DC stimulation of 0.1 μA, and prey-catching behaviour increased as a function of increasing current, regardless of whether a toad started the experiment hungry or not. For orient and crouch

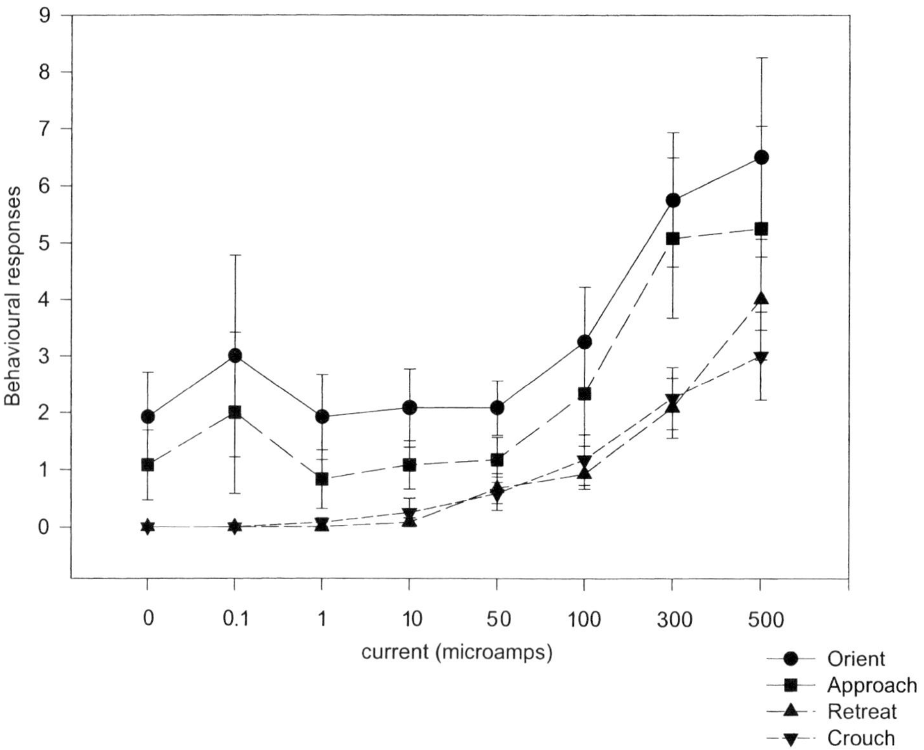

Fig. 7. Mean ± SEM of orient, approach, (representing prey catching) retreat and crouch (representing avoidance) behaviours in response to 10 s transtectal DC stimulation followed by presentation of a remotely operated worm-like stimulus (3 × 0.75 cm^2 black card moving along its longitudinal axis at 2 cm/s) (after Sterritt and Laming, 2003).

there was an interaction between right tectal negativity and anti-clockwise response preference, suggesting that there might be an additive effect of right tectal negativity with the negativity biologically generated as an SPS as the worm-like stimulus passes the left eye. In general, best responses for orient, approach and crouch behaviours were when the right tectum was negative, the left tectum positive. Analogously, Bauer et al. (1989) found that their human subjects exposed to epicranial DC responded faster to positive currents than negative. Based on these results they proposed that neurons depolarized to near the firing threshold would be activated or firing delayed, depending on the polarity of the current and spatial arrangement of the cells.

At the higher currents visual stimuli in addition elicited defensive behaviours, like crouch and retreat, as well as inappropriate responses like delayed orient or orient in an inappropriate direction. It would seem that at these higher currents the tectal stimulation of the visual system was not always simply additive to the visual stimulus, but it was causing the toad problems in its interpretation of the visual stimulus. In some cases these problems were transient, as shown by the observation that during DC stimulation at higher currents some of the toads crouched and retreated for the first 2–3 s of the 10 s exposure to the DC stimulation, but then they recovered and orientated and approached before the worm-like visual stimulus had passed the window.

Ablation of the optic tectum abolishes all visually guided prey-catching and avoidance movements (Comer and Grobstein, 1981), while AC electrical stimulation of the same structures elicits normal orienting responses (Ewert, 1970). It was therefore interesting to compare the effectiveness of AC versus DC stimulation. It was found that AC stimulation at the current strengths used was particularly ineffective. There are of course no direct comparisons as the energy transfer of DC stimulation is much higher than at an equivalent AC current.

9. Changes in the SPS during habituation

During habituation to a variety of arousing stimuli, the decline in the cardiac arousal response of fish to all types of stimuli used has been found to be related to a decline in the positivity of SPS responses in the telencephalon (Quick and Laming, 1990). This is of interest because of the considerable body of evidence linking telencephalic regions (posterior dorso-central) with habituation of arousal responses in fish (see Rooney and Laming, 1986, 1988).

Nicol and Laming (1992) found in goldfish that during the period of habituation of the cardiac arousal response to the onset of illumination, SPSs recorded from the telencephalic, tectal and medullary surfaces in response to the same acoustic stimulus also declined in amplitude. In a further study (Nicol and Laming, 1993), it was found that such changes in the amplitudes of SPS responses to an unaltered stimulus are related to concurrent changes in the modulatory effects of the presentation of the arousing stimuli on the amplitudes of AEPs evoked by continuous background auditory stimulation. This was particularly apparent in the telencephalon. In this region, over successive presentations of the visual stimulus, the visually induced attenuation of AEPs declined, and this decline was related to the change, over the same series of presentations, in the amplitudes of SPSs recorded from the same site.

In toads (*Bufo bufo*), intrinsic tectal visual unit activity frequency and burst duration habituate on repeated presentation of a prey-like object to immobilised but conscious animals. This habituation may reflect the unobtainability of the simulated prey. SPS amplitude and duration as well as the duration of characteristics associated with arousal in the EEG recorded with the same electrode, similarly decline (Laming and Ewert, 1984). However, the relationship is not a simple one. Over 20 presentations of the stimulus, unit frequency and burst duration declined to 50% of the original values, a similar decline as that of the duration of SPS and EEG responses. The decline in SPS amplitude was much larger, in the order of 70%, suggesting its habituation may be independent of the decline in unit responses. During habituation, the SPS duration was reduced to closely match that of the bursts of the local intrinsic tectal unit, suggesting that initially it was deriving some of its source $[K^+]_e$ from more superficial (retinal input) regions, perhaps by spatial buffering. It would seem that the spatial buffering currents decline in magnitude to a greater degree than the decline in neuronal activity. This may be due to passive build up of $[K^+]_e$ over the repeated stimuli, or to an exponential decline in potentially sensitising, spatial buffering currents. If the former, then it would appear that 1 min is insufficient time for potassium to equilibrate (the time between successive stimulus presentations), if the latter, then it would appear that glia may be actively involved in the process of habituation.

10. Concluding remarks

It is apparent that glial cells, especially astrocytes and ependymoglia are highly responsive to elevated extracellular potassium, which triggers its own redistribution by these cells, influencing all cells within range of spatial buffering currents. The extracellular currents, largely caused by spatial buffering through glia, can be recorded as SPSs, associated with behavioural arousal and simple forms of learning in fish and amphibians. The motivational state of an animal is reflected in SPS amplitudes. Experimental changes in extracellular potassium or imposed DC shifts in the brains of amphibia interact with the animal's motivational state in a way that suggests that ionic dynamics between neurons and glia may contribute to motivation.

DC stimulation of the tectum of the common toad, *Bufo bufo*, enhances prey catching behaviour, though at higher currents it also generates avoidance. The finding that the polarity of the stimulus is relevant both to these behavioural responses and to neuronal activation, suggests that directional ionic movements such as those involved in spatial buffering and active uptake by glia may contribute to neuromodulation and behaviour.

References

Barres, B.A., Koroshetz, W.J., Chun, L.L.Y., Corey, D.P., 1990. Ion channel expression by white matter glia: the type-1 astrocyte. Neuron 5, 527–544.

Bauer, H., Muhr, R., Korunka, C., Leodolter, M., 1989. Possible contribution of neuroglia to the DC-potential level and Slow Potential Shifts, Proceedings of EPIC IX, Noordwiyk, Holland.

Chang, H.T., 1951. Dendritic potentials of cortical neurons produced by direct electrical stimulation of the cerebral cortex. J. Neurophysiol. 14, 1–21.

Comer, C., Grobstein, P., 1981. Tactually elicited prey acquisition behaviour in the frog *Rana pipens*, and a comparison with visually elicited behaviour. J. Comp. Physiol. 142, 141–150.

Durkovic, R.O., Cohen, D.H., 1966. DC potential activity in a nervous system lacking neocortex: the pigeon telencephalon. Anat. Rec. 154, 341.

Durkovic, R.O., Cohen, D.H., 1968. Spontaneous evoked and defensively conditioned steady potential changes in the pigeon telencephalon. Electroencephalogr. Clin. Neurophysiol. 24, 474–481.

Ewert, J., 1970. Neural mechanism of prey-catching and avoidance behaviour in the toad (*Bufo bufo* L.). Brain Behav. Evol. 3, 36–56.

Ewert, J.P., 1989. The release of visual behaviour in toads: stages of parallel/hierarchical information processing. In: Ewert, J.P., Arbib, M.A. (Eds.), Visuomotor Coordination: Amphibians, Comparisons Models and Robots, Plenum Press, New York, pp. 39–120.

Galambos, R., 1961. A glial–neuronal theory of brain function. Proc. Natl Acad. Sci. 47, 129–136.

Gardner-Medwin, A.R., 1983. Analysis of potassium dynamics in mammalian brain tissue. J. Physiol. (Lond.) 335, 393–426.

Goldring, S., O'Leary, J.L., Winter, D.L., Pearlman, A.L., 1959. Identification of a prolonged post-synaptic potential of cerebral cortex. Proc. Soc. Exp. Biol. Med. 100, 429–431.

Hertz, L., 1965. Possible role of neuroglia: a potassium-mediated neuronal–neuroglial–neuronal impulse transmission system. Nature 4989, 1091–1094.

Hounsgaard, J., Nicholson, C., 1983. Potassium accumulation around individual Purkinje cells in cerebellar slices from guinea-pig. J. Physiol. 340, 359–388.

Hurlbert, R.J., Tator, C.H., Therialt, E.T., 1993. Dose-response study of the pathological effects of chronically applied direct current stimulation on the normal rat spinal cord. Neurosurgery 79, 905–916.

Karahashi, Y., Goldring, S., 1966. Intracellular potentials from 'idle' cells in cerebral cortex of cat. Electroencephalogr. Clin. Neurophysiol. 20, 600–607.

Kuffler, S.W., Nicholls, J.G., Orkand, R.K., 1966. Physiological properties of glial cells in the central nervous system of amphibia. J. Neurophysiol. 29, 768–787.

Laming, P.R., 1989. Central representation of arousal. In: Ewert, J.-P., Arbib, M.A. (Eds.), Visuomotor Coordination: Amphibians, Comparisons, Models and Robots, Plenum Press, New York, pp. 693–727.

Laming, P.R., 1989. Do glia contribute to behaviour? A neuromodulatory review. J. Comp. Biochem. Physiol. 94, 555–568.

Laming, P.R., 2000. Potassium signalling in the brain: its role in behaviour. Neurochem. Int. 36, 271–290.

Laming, P.R., Borchers, H.W., Ewert, J.P., 1984. Visual unit, EEG and sustained potential shift responses in the brain of toads (*Bufo bufo*) during alert and defensive behaviour. Physiol. Behav. 32, 463–468.

Laming, P.R., Brooks, M., 1985. Effects of visual, chemical and tactile stimuli on the auditory evoked response of the teleost, *Rutilus rutilus*. Comp. Biochem. Physiol. 82A(3), 667–673.

Laming, P.R., Bullock, T.H., 1991. Changes in early acoustic-evoked potentials by mildly arousing priming stimuli in carp (*Cyprinus carpio*). J. Comp. Biochem. Physiol. 99A(4), 567–575.

Laming, P.R., Bullock, T.H., McClune, M., 1991. Changes in EEG power, acoustic evoked potentials and heart rate after mildly arousing non-acoustic priming stimulus in the carp, (*Cyprinus carpio*). J. Comp. Biochem. Physiol. 100A(1), 81–93.

Laming, P.R., Bullock, T.H., McClune, M., 1991. Sustained potential shifts and changes in acoustic evoked potentials after presentation of a non-acoustic stimulus to carp (*Cyprinus carpio*). J. Comp. Biochem. Physiol. 100A(1), 95–104.

Laming, P.R., Ewert, J.P., 1983. The effects of pretectal lesions on neuronal, sustained potential shift and electroencephalagraphic responses of the toad tectum to presentation of a visual stimulus. J. Comp. Biochem. Physiol. 154, 89–101.

Laming, P.R., Ewert, J.P., 1984. Visual unit, EEG and sustained potential shift responses to biologically significant stimuli in the brain of toads (*Bufo bufo*). J. Comp. Physiol. 154, 89–101.

Laming, P.R., Ewert, J.P., Borchers, H.W., 1984. The effects of telencephalic ablation on unit, EEG and sustained potential shift responses of the toad tectum to a visual stimulus. Behav. Neurosci. 98(1), 118–124.

Laming, P.R., Kimelberg, H., Robinson, S., Salm, A., Hawrylak, N., Muller, C., Roots, B., Ng, K., 2000. Neuronal–glial interactions and behaviour. Neurosci. Biobehav. Rev. 24, 295–340.

Laming, P.R., Laming, G.E., 2003. Tectal responses to potassium loads and subsequent visual stimuli are affected by motivational state in the toad, *Bufo bufo*. Comp. Biochem, Physiol. Submitted for publication.
Laming, P.R., Ocherashvili, I.V., Nicol, A.U., 1992. Dendritic and sustained shifts in potential to electrical stimulation of the anuran tectal surface. J. Comp. Biochem. Physiol. 101A, 91–96.
Laming, P.R., Ocherashvili, I.V., Nicol, A.U., Roughan, J.V., Laming, B.A., 1995. Sustained potential shifts in the toad tectum reflect prey catching and avoidance behaviour. Behav. Neurosci. 109, 150–160.
Laming, P.R., Savage, G.E., 1980. Physiological changes observed in the goldfish (*Carassius auratus*) during behavioral arousal and fright. Behav. Neural Biol. 29, 255–275.
Lazar, G., 1989. Cellular architecture and connectivity of the frog's optic tectum and pretectum. In: Ewert, J.P., Arbib, M.A. (Eds.), Visuomotor Coordination: Amphibians, Comparisons, Models and Robots, Plenum Press, New York, pp. 175–199.
Lickey, M.E., Fox, S.S., 1966. Localisation and habituation of sensory evoked DC responses in cat cortex. Exp. Neurol. 15, 437–454.
Lynn, R., 1966. Attention, Arousal and the Orientation Reaction. Pergamon Press, Oxford.
Newman, E.A., 1986. High potassium conductance in astrocyte endfeet. Science 233, 453–454.
Nicholls, J.G., Kuffler, S.W., 1964. Extracellular space as a pathway for exchange between blood and neurons in the central nervous system of the leech: ionic composition of glial cells and neurons. J. Neurophysiol. 27, 645–673.
Nicholls, J.G., Kuffler, S.W., 1965. Na and K content of glial cells and neurons determined by flame photometry in the central nervous system of the leech. J. Neurophysiol. 28, 519–525.
Nicholson, C., Phillips, J.M., 1981. Ion diffusion modified by tortuosity and volume fraction in the extracellular microenvironment of the rat cerebellum. J. Physiol. 321, 225–257.
Nicol, A.U., Laming, P.R., 1992. Sustained potential shift responses and their relationship to the ECG response during arousal in the goldfish (*Carassius auratus*). J. Comp. Biochem. Physiol. 101A(3), 517–532.
Nicol, A.U., Laming, P.R., 1993. Sustained potential shifts, alterations in acoustic evoked potential amplitude and bradycardic responses to onset of illumination in the goldfish (*Carassius auratus*). J. Comp. Physiol. 173A, 353–362.
Nicol, A.U., Savage, U., Laming, P.R., 1993. The depth profile of electrically induced tectal SPS responses in the goldfish, (*Carassius auratus*). Behav. Neural Biol. 59, 58–161.
Nitsche, M.A., Leibetanz, D., Tergau, F., Paulus, W.T.I., 2002. Modulation of cortical excitability in man using transcranial direct current stimulation. Nervenarzt, 332–335.
O'Leary, J.L., Goldring, S., 1964. DC potentials in the brain. Physiol. Rev. 44, 91–125.
Orkand, R.K., Nicholls, J.G., Kuffler, S.W., 1966. The effect of nerve impulses on the membrane potential of glial cells in the central nervous system of amphibia. J. Neurophysiol. 29, 788–806.
Quick, I.A., Laming, P.R., 1990. Relationship between ECG, EEG and SPS responses during arousal in the goldfish (*Carassius auratus*). J. Comp. Biochem. Physiol. 95A(3), 459–471.
Ransom, B.R., Goldring, S., 1973. Slow depolarisation in cells presumed to be glia in cerebral cortex of cat. J. Neurophysiol. 36, 879–892.
Ransom, B.R., Goldring, S., 1973. Ionic determinants of membrane potentials presumed to be glia in cerebral cortex of cat. J. Neurophysiol. 36, 855–868.
Richter, F., Wicher, C., Schmidt, D., Leichsenring, A., Haschke, W., 1992. Activation state of the cortex could be changed by reinforcement of low-amplitude visual evoked potentials in rabbits. Neurosci. Lett. 135, 133–135.
Roitbak, A.I., Fanardjian, V.V., 1981. Depolarization of cortical glial cells in response to electrical stimulation of the cortical surface. Neuroscience 6, 2529–2537.
Roitbak, A.I., Fanardjian, V.V., Melkonyan, D.S., Melkonyan, A.A., 1987. Contribution of glia and neurons to the surface-negative potentials of the cerebral cortex during its electrical stimulation. Neuroscience 20, 1057–1067.
Roitbak, A.I., Ocherashvili, I.V., Laming, P.R., Roitbak, T.A., 1992. Stimulus-evoked sustained potential shifts and changes in $[K^+]$: of the frog optic tectum. J. Comp. Physiol. 170A, 327–333.
Rooney, D.J., Laming, P.R., 1986. Localisation of telencephalic regions concerned with habituation of cardiac and ventilatory responses associated with arousal in the goldfish, (*Carassius auratus*). Behav. Neurosci. 100(1), 45–50.

Rooney, D.J., Laming, P.R., 1988. Effects of ablation of the goldfish telencephalon on short-term (within session) and long-term (between session) habituation of arousal responses. Behav. Neural Biol. 49, 83–96.

Roots, B., Laming, P.R., 1998. The phylogeny of glial–neuronal relationships and behaviour. In Laming, P.R., Reichenbach, A., Sykova, E., Bauer, H., Hatton, G., (Eds), Glial Cells: Their Role in Behaviour. Cambridge University Press, Cambridge, pp. 22–44.

Roughan, J.V., Laming, P.R., 1998. Epicortical slow potential shifts and sensory evoked potentials are related to seizure propensity in gerbils. J. Comp. Physiol. A 182, 827–838.

Rowland, V., 1968. Cortical steady potential (direct current potential) in reinforcement and learning. In: Stellar, E., Sprague, J.M. (Eds.), Progress in Physiological Psychology, Academic Press, New York, pp. 1–77.

Rowland, V., Gluck, H., Sumergrad, S., Dines, G., 1985. Slow and multiple unit potentials in trace and temporal conditioning controlled by electrical reward in the rat. Electroencephalogr. Clin. Neurophysiol. 61, 559–568.

Sokolov, E.N., 1963. Higher nervous functions: the orienting reflex. Ann. Rev. Physiol. 25, 545–580.

Sterritt, L., Laming, P.R., 2003. DC stimulation enhances prey catching behaviour in the toad, *Bufo bufo*. In preparation.

Sterritt, L., Laming, G.E., Laming, P.R., 2003. Tectal responses to DC stimulation are affected by motivational state in the toad, *Bufo bufo*. J. Comp. Physiol. Submitted for publication.

Syková, E., 1983. Extracellular K^+ accumulation in the central nervous system. Prog. Biophys. Mol. Biol. 42, 135–189.

Syková, E., 1992. Ion-selective electrodes Monitoring Neuronal Activity: A Practical Approach. In: Stamford, J. (Ed.). Oxford University Press, New York, pp. 261–282.

Walz, W., Kimelberg, H.K., 1985. Differences in cation transport properties of primary astrocyte cultures from mouse and rat brain. Brain Res. 340, 333–340.

Advances in
Molecular and
Cell Biology

Regulation of Ca^{2+} stores in glial cells

Giovanni Scapagnini,[a,b,*] Thomas J. Nelson[a] and Daniel L. Alkon[a]

[a]*Blanchette Rockefeller Neurosciences Institute, West Virginia University, Rockville, MD 20850, USA*
[*]*Correspondence address: Blanchette Rockefeller Neurosciences Institute, JHU, Academic and Research Building, 9601 Medical Center Drive, Rm. 351, Rockville, MD 20850-3332, Tel.: (301) 294-7191; fax: (301) 294-7007*
E-mail: gscapag@brni-jhu.org
[b]*Institute of Neurological Sciences, CNR, 95123 Catania, Italy*

Contents

1. Calcium homeostasis in glial cells
2. Ca^{2+} storage organelles and intracellular Ca^{2+} release
   2.1. Sarco(endo)plasmic reticulum Ca^{2+}-ATPases
   2.2. Ca^{2+}-binding proteins
   2.3. Calcium release
   2.4. Structural complexity and compartmentalization of ER in glial cells
   2.5. Mitochondria
3. Capacitative calcium entry
4. Intracellular calcium sensors and effectors
   4.1. S-100
   4.2. Calcium-binding proteins in Alzheimer's disease and apoptosis
5. Concluding remarks

Virtually all cell functions are regulated by changes in the cytosolic free Ca^{2+} concentration. Ca^{2+} signals within cells can be local or global, can involve waves, oscillations, or even more-complex patterns, and can be modulated in terms of both amplitude and frequency. In astrocytes, calcium signaling is not restricted to single cells, but can cross cell borders via gap junctions, resulting both in intracellular Ca^{2+} waves, traveling from one glial cell to the next, and in induction of Ca^{2+} responses in neurons. In mammals, many glial cells contain an elaborate endoplasmic reticulum. Associated with this membrane network are pumps for Ca^{2+} uptake (the so-called SERCAs), that, as in other cells, are inhibited by thapsigargin and cyclopiazonic acid. The membranous network also contains Ca^{2+}-binding proteins for Ca^{2+} storage and channels for Ca^{2+}-induced calcium release. These include the ubiquitous IP_3 receptors and, in many cells, the ryanodine receptors. The ryanodine receptors are regulated by a variety of accessory proteins and are sensitive to the classical modulators ryanodine and caffeine. In glia,

Advances in Molecular and Cell Biology, Vol. 31, pages 635–660

ISBN: 0-444-51451-1

however, the major mechanism for Ca^{2+} release from internal stores involves activation of inositol 1,4,5-trisphosphate (IP_3)-gated Ca^{2+} release channels (IP_3 receptors). Regulation of IP_3 receptors is complex with alternative splicing of at least three different isoforms, posttranslational modification of IP_3R by phosphorylation, and interaction with adenine nucleotides, calcium, and immunophilins or FK506 binding proteins. The mitochondria and calcium storage proteins also function as high-capacity storage sites for calcium. Other calcium binding proteins play important roles in cell signaling, maintenance of the cytoskeleton, and apoptosis.

1. Calcium homeostasis in glial cells

Virtually all cell functions, from cell birth to cell death, are directly or indirectly regulated by changes in the intracellular free Ca^{2+} concentration $[Ca^{2+}]_i$, that act as an eclectic second messenger system. Cells have developed specialized machinery to control the spatial and temporal characteristics of these Ca^{2+} signals. These include transmembrane Ca^{2+} transporters and Ca^{2+}-permeable channels, cytoplasmic buffers, and intracellular organelles that are able to accumulate, store, and release Ca^{2+}. The fundamental importance of Ca^{2+} for signaling within cells has been established for a multitude of cell types and intracellular compartments. Glial cells express a complex set of molecules controlling Ca^{2+} signaling, including voltage-gated Ca^{2+} channels and Ca^{2+}-release channels from intracellular pools. Moreover, it has become evident that the different types of glial cells, such as cortical astrocytes, oligodendrocytes or microglia, are quite distinct with respect to their repertoire of Ca^{2+}-signalling mechanisms. This machinery allows these cells to sense, integrate and respond to external environment.

Some aspects of Ca^{2+}-signaling depend directly upon Ca^{2+} entry from the extracellular fluid. Within the outer cell membrane, a broad range of channels allows the influx of Ca^{2+} in response to various stimuli; indeed glial cells express a large diversity of membrane receptors coupled to the phosphoinositide breakdown pathway as well as ionotropic receptors and voltage-gated calcium channels (MacVicar, 1984; Berger et al., 1992). Within the membranes of intracellular organelles, such as the endoplasmic reticulum (ER), Golgi apparatus and mitochondria, are additional response elements that control Ca^{2+} uptake from or release into cytosolic domains (or both). In glia, similar to most cells, ER Ca^{2+} storage and release contribute crucially to the Ca^{2+} signals. Net entry of Ca^{2+} through plasma membrane channels is usually much more limited in magnitude and duration in comparison to the larger, more sustained elevations caused by release from the ER. The ER, in fact, is important not only for buffering Ca^{2+}, but also for local Ca^{2+} signaling and for rapidly transmitting Ca^{2+} signals to the cell interior.

Because multiple processes in cells can be influenced simultaneously by changes in $[Ca^{2+}]_i$, the spatial and temporal patterns of Ca^{2+} signals is a critical point (Alkon et al., 1998). Indeed, the Ca^{2+} signals within cells can be local or global, can involve waves, oscillations, or even more-complex patterns, and can be modulated in terms of both amplitude and frequency (see chapter by Shuai et al.). These transient rises of $[Ca^{2+}]_i$ in turn trigger or regulate various intracellular events, including metabolic processes (see chapter by Hertz, Peng et al.), gene expression, and ion transport systems (Bootman et al.,

1997; Dolmetsch et al., 1998). Moreover, in astrocytes, calcium signaling is not restricted to single cells, but can cross cell borders via gap junctions, resulting in intracellular Ca^{2+} waves traveling from one glial cell to the next, or the induction of Ca^{2+} responses in neurons. Calcium signaling might thus be a form of glial excitability, enabling these cells to integrate extracellular signals, communicate with each other and exchange information with neurons. Moreover, since it is apparent that glial Ca^{2+} signaling is an important pathway of glial–glial and neuronal–glial cross-talk, it is important to understand its role during pathological conditions.

2. Ca^{2+} storage organelles and intracellular Ca^{2+} release

In glial cells (Simpson and Russel, 1997; Deitmer et al., 1998), as in all other types of animal cells, stimulation causes an elevation of the $[Ca^{2+}]_i$, which triggers a large spectrum of physiological responses. Although some of this 'signal calcium' may come directly from the extracellular fluid through different types of channels (Verkhratsky et al., 1998), much of the signal Ca^{2+} comes from the intracellular Ca^{2+} stores, primarily the ER (Deitmer et al., 1998; Verkhratsky et al., 1998). The role of ER in Ca^{2+} sequestration was first demonstrated in neurons by Ca^{2+} flux and electron microprobe studies. Shortly thereafter, the study of Ca^{2+} signaling was revolutionized by the development of Ca^{2+}-sensitive fluorescent probes for use in intact, living cells. Application of these dyes revealed a near-universal distribution of ER Ca^{2+}-concentrating and Ca^{2+}-release mechanisms in cells, including glial cells. Nevertheless, to date, in spite of major advances during the last years, little is known about the precise structural and functional organization of the ER Ca^{2+} store. Direct visualization of ER using electron microscopy and fluorescent staining with lipophilic carbocyanine dyes suggests that the ER is a continuous, interconnecting network of tubules and cisterns (Pozzan et al., 1994; Terasaki et al., 1994).

Aplysia glial cells were found to have an unusual analog of ER Ca^{2+} stores, which may retain an enormously high (up to 50–100 mM) Ca^{2+} concentration. These glial cells surrounding the identified giant nerve cell bodies R2 or LP1 of *Aplysia punctata* were studied by quantitative electron microscopy and found to contain specific, electron-dense but non-osmiophilic membrane-bound granules, approximately 0.3 μm in diameter, called 'gliagrana' (Keicher et al., 1991). The density of these gliagrana varies with fluctuations in extracellular Ca^{2+} ($[Ca^{2+}]_o$). Increases or decreases in glial calcium depending on $[Ca^{2+}]_o$ suggest the possible involvement of glia in the regulation of $[Ca^{2+}]_o$. Similar glial granules are more often found in marine than in freshwater molluscs, possibly because they represent a calcium store used to compensate excess Na^+ in the extracellular milieu of marine species and to regulate perineuronal calcium concentration (Keicher et al., 1992). In agreement with this hypothesis, the abundance of gliagrana (= number of glial granules per square micron) is found to be higher in animals adapted to low Ca^{2+} artificial sea water than in animals kept in high Ca^{2+} (or low Na^+) conditions. This finding is not observed after 1 week but after 2 weeks of adaptation. Similar stores have been described in frog ependymal glia (Gambetti et al., 1975).

In mammals, many glial cells contain an elaborate ER (Privat et al., 1995). Associated with this membrane network are pumps for Ca^{2+} uptake (the so-called SERCAs), that, as in other cells, are inhibited by thapsigargin and cyclopiazonic acid (CPA). The membranous network also contains Ca^{2+}-binding proteins for Ca^{2+} storage and channels for Ca^{2+} release, the ubiquitous inositol triphosphate receptors (IP_3R) and, in many cells, also the ryanodine receptors (RyR).

2.1. Sarco(endo)plasmic reticulum Ca^{2+}-ATPases

In order for the cell to utilize Ca^{2+} as a signaling molecule, Ca^{2+} gradients across the plasma membrane and Ca^{2+} stores within intracellular organelles must be maintained. This is accomplished primarily by the activity of several dozen Ca^{2+}-transporting ATPases encoded by alternatively spliced transcripts from as many as eight different genes. These include three distinct sarco(endo)plasmic reticulum Ca^{2+}-ATPases (SERCA1–3) (MacLennan et al., 1997) (human gene nomenclature, *ATP2A1–3*), four distinct plasma membrane Ca^{2+}-ATPases (PMCA1–4) (human gene nomenclature, *ATP2B1–4*), and a putative mammalian secretory pathway Ca^{2+}-ATPase, (human gene nomenclature, *ATP2C1*).

Ca^{2+} storage in the ER is mediated by a Ca^{2+}-dependent ATPase that is blocked by agents such as thapsigargin and CPA (Inesi and Sagara, 1994). The discovery of these potent and relatively selective SERCA inhibitors has enabled the widespread study of the roles of SERCA pumps in Ca^{2+} signaling (Simpson and Russel, 1997). In many cell types, SERCA inhibition leads to elevation of cytosolic $[Ca^{2+}]$ secondary to leakage of Ca^{2+} from stores. Both thapsigargin and CPA have previously been reported to be effective in inhibiting SERCA pumps in cultured glia (Blaustein and Golovina, 2001). Some SERCAs (or perhaps non-SERCA Ca^{2+} pumps), however, are quite resistant to these agents (Tanaka and Tashjian, 1993; Golovina and Blaustein, 2000). The inhibitor-resistant Ca^{2+} pumps appear to be associated with Ca^{2+} stores that respond to caffeine and ryanodine, but not to inositol 1,4,5-trisphosphate (IP_3) (Golovina and Blaustein, 2000).

Three mammalian genes (SERCA1, -2 and -3) encode at least six SERCA isoforms as a result of alternative splicing (Shull, 2000). These gene products are differently expressed in various tissues and exhibit functional differences.

SERCA1 consists of two C-terminal variants, both of which are restricted to fast-twitch skeletal muscle. The SERCA2 gene encodes both SERCA2a, the cardiac SR Ca^{2+} pump, and SERCA2b, the major intracellular Ca^{2+} pump of smooth muscle and non-muscle tissues. The two variants arise as a result of alternative splicing of sequences encoding their extreme C-termini. SERCA2b is similar to SERCA2a throughout most of its length, but its C-terminus contains an alternative 49- or 50-amino-acid sequence, depending on the species, in place of the last four amino acids found in SERCA2a (Shull, 2000). The extra sequence in SERCA2b contains an additional transmembrane domain, which places the C-terminus in the lumen of the ER, where it apparently interacts with, and is modulated by, calreticulin. The ubiquitous expression of SERCA2b and the relatively limited cell-type distribution of the other isoforms suggests that SERCA2b is the major Ca^{2+} pump serving ER Ca^{2+} stores in most tissues, included glial cells. Although some cases of thapsigargin-resistance have been observed, treatment of cultured cells with thapsigargin usually leads to cell cycle arrest or apoptosis. Thus, it seems likely that SERCA2b plays a housekeeping function that is critical for the long-term viability of most mammalian cell types. In contrast, SERCA2a exhibits a restricted tissue distribution and is expressed at very high levels in cardiac muscle, where it clearly plays an organ-specific function. Contraction of cardiac muscle is initiated when Ca^{2+} influx across the sarcolemma (SR) triggers release of much larger quantities of Ca^{2+} from the SR, and relaxation occurs as Ca^{2+} is resequestered within the SR by SERCA2a and extruded from the cell by the Na^{+}/Ca^{2+} exchanger and the plasma membrane Ca^{2+} pump. Although its importance in cardiac physiology is clear, the degree to which normal cardiac function and long-term health are dependent on the appropriate levels of SERCA2a in heart is uncertain. The function of SERCA3, which has a lower Ca^{2+} affinity and a higher pH optimum than the other isoforms, is poorly understood. Three C-terminal variants of the enzyme arise by alternative splicing of the primary transcript. In situ hybridization, Western blot, and Northern blot analyses show that SERCA3 is expressed in many tissues, but that its cell-type distribution is quite limited. In all of the cell types in which it is present, SERCA3 appears to be co-expressed with SERCA2b, suggesting that it might play a specialized role in Ca^{2+} signaling or provide some redundancy in Ca^{2+} sequestering activity.

Several different SERCA subtypes are known to be expressed together in some cells; for example, SERCA2a and SERCA3 are co-expressed with SERCA2b in cerebellar Purkinje neurons (Baba-Aissa et al., 1998). The functional significance of this co-expression, and the question of whether the different isoforms reside on the same or different ER components, remain to be elucidated. A further complication is that other, less well characterized, non-SERCA Ca^{2+}-ATPases (i.e., products of non-homologous genes) might also play a role in intracellular Ca^{2+} sequestration.

2.2. Ca^{2+}-binding proteins

ER Ca^{2+}-binding proteins provide a high capacity buffering mechanism which results in the lowering of the concentration of free Ca^{2+} in the ER, and thus a reduction in the gradient against which pumps must transport cytoplasmic Ca^{2+} into the store. They are also thought to be important in localizing Ca^{2+} to sites of release, and in modulating release activity via protein–protein interactions with release channels (Mackrill, 1999). The best described of these Ca^{2+}-binding proteins are calsequestrin and calreticulin (CRT). The first of these proteins to be identified was calsequestrin in the striated muscle sarcoplasmic reticulum 19. The very acidic C-terminal domain of calsequestrin binds Ca^{2+} with low-affinity (average $K_d = 1$ mM) and high capacity (-50 Ca^{2+}-binding sites per molecule). A similar Ca^{2+}-binding protein, CRT, predominates in most (if not all) non-muscle cells (Krause and Michalak, 1997). In addition to the numerous low-affinity sites of its C-terminal domain, CRT also includes a single high-affinity site positioned towards the middle of the molecule. Localization of calreticulin to $InsP_3R$-containing membrane vesicles has been reported in some cell types using density gradient techniques. The function of this co-expression, however, has remained controversial. Recent reports have indicated that calreticulin may play a role in regulating Ca^{2+} signals, including perhaps serving as a luminal sensor for Ca^{2+} store depletion. CRT is now recognized to be a multifunctional protein that is associated with cellular responses in many ways. CRT has been reported to interact with the subunit of the integrin receptors, and has been proven to be critical for regulating integrin-mediated cell adhesion in other systems. A recent study highlights the participation of CRT in integrin–ligand interaction in oligodendrocytes (Gudz et al., 2002). This work clearly showed the presence of CRT on the surface of oligodendrocytes. However, data on CRT association with integrins show that it interacts with the cytoplasmic tail of the integrin, and chelating intracellular Ca^{2+} removes CRT from the integrin–CRT complex, suggesting an intracellular localization. This dichotomy is unresolved, but it has been proposed that two forms of CRT may exist, an endoCRT molecule localized on the intracellular surface of the plasma membrane and an ectoCRT molecule localized on the extracellular surface of cells. Thus, further characterization of the CRT associated with this complex is needed. However, for CRT, extensive binding of, Ca^{2+} has been shown to take place both in vitro and within the ER of living cells. Such information is still missing for other ubiquitous lumenal proteins, in particular BiP, p94 (endoplasmin), Erp72, PDI and the membrane protein calnexin, all of which bind Ca^{2+} with low affinity when tested in vitro. Because these proteins do not possess an acidic C-terminal domain, their Ca^{2+} binding is believed to occur at doublets and triplets of acidic amino acids scattered throughout the molecule. This type of Ca^{2+} binding is not a general property of the lumenal Ca^{2+}-binding proteins of the ER. Other such proteins, e.g., reticulocalbin and p55 (calstorin, which is abundant especially in the brain), were shown to include typical EF-hand domains and may bind Ca^{2+} with higher affinity. Thus, Ca^{2+} buffering in the ER lumen depends on a host of Ca binding proteins with variable affinities for calcium.

2.3. Calcium release

On the ER membrane, two major receptors trigger the release of Ca^{2+} into the cytosol: the inositol-1,4,5-trisphosphate receptor (IP_3R) (Berridge, 1993) and the RyR (Sutko and Airey, 1996). Receptor-mediated activation of phospholipase C (PLC) at the outer cell membrane cleaves phosphatidylinositol bisphosphate (PIP_2) to generate 1,2-diacylglyerol (DAG) and IP_3, a second messenger which activates the IP_3R-mediated release of Ca^{2+} (IICR) from the ER. Endogenous signaling pathway(s) that activate the RyR-mediated release of Ca^{2+} from the ER have, until recently, been largely unknown. For levels of $[Ca^{2+}]_i \geq 10\ \mu M$ (Buratti et al., 1995), calmodulin (CaM) has been shown to block RyR function (Tripathy et al., 1995), and cyclic ADP ribose (cADPR) at unknown levels can activate the RyR in vitro (Lee et al., 1995). However, neither the endogenous levels nor initial signaling molecules for cADPR have been determined. Nonetheless, it is known that activation of the RyR depends on the levels of Ca^{2+} already present in the cytosol (Verkhratsky and Shmigol, 1996). There is evidence, however, that IP_3R activation also depends on $[Ca^{2+}]_i$ because $[Ca^{2+}]_i$ increases IP_3R sensitivity to IP_3 (Ehrlich, 1995). In any case, in muscle cells, the RyR does mediate Ca^{2+}-induced Ca^{2+} release (CICR) (Dousa et al., 1996). The role of RyR in CICR in glial cells (see below) has not been elucidated completely.

2.3.1. IICR

In glia the major mechanism for Ca^{2+} release from internal stores involves activation of IP_3-gated Ca^{2+} release channels. (Berridge, 1993; Verkhratsky et al., 1998). The production of IP_3, in turn, is achieved by the activation of PLC coupled via G proteins with numerous 'metabotropic' plasmalemmal receptors (see Fig. 1). The direct activation of IP_3R by photorelease of IP_3 from caged compound was shown in cultured astrocytes (Khodakhah and Ogden, 1993; Shao and McCarthy, 1995). Astrocytic IP_3 receptors appear to be substantially more sensitive to IP_3 than IP_3R in Purkinje neurons; the threshold IP_3 concentration for activation of the IP_3-gated channel in astrocytes was 0.2–0.5 μM, whereas in Purkinje neurons, it was 9 μM (Khodakhah and Ogden, 1993).

The IP_3R has been purified from rat cerebellum and localized primarily to the ER (Ross et al., 1989), though nuclear, plasma membrane and neurotransmitter vesicle localizations have also been described (Petersen, 1996). Molecular cloning reveals three different isoforms, encoded by different genes, IP_3R1, 2, and 3 (Ross et al., 1992). Regulation of IP_3Rs is complex with alternative splicing of at least IP_3R1 and posttranslational modification of IP_3R by phosphorylation. In addition, there is regulation by adenine nucleotides, calcium, and the immunophilin FK506 binding protein (Sharp et al., 1999). Different isoforms of the receptor may have different affinities for IP_3 and different forms of regulation. Receptors in a number of cell lines appear to be heterotetramers composed of more than one isoform (Nucifora et al., 1996), and there is evidence for homotetramers of the IP_3R3 isoform in some cultured cells (Nucifora et al., 1996), with the possibility that

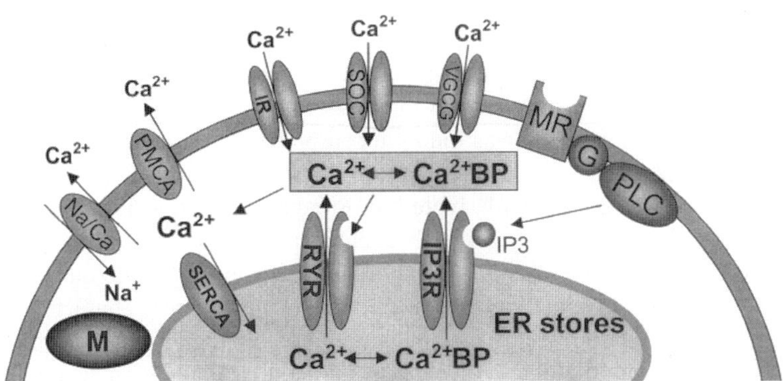

Fig. 1. *Components of Ca^{2+} signaling in glial cells.* Many glial cells are endowed with voltage gated Ca^{2+} channels (VGCG) and Ca^{2+}-permeable ionotropic receptors (IR), which lead to an increase in the intracellular concentration of Ca^{2+}. In glia the major mechanism for Ca^{2+} signaling involves the activation of G protein (G) coupled metabotropic receptors (MR), which, through the synthesis of inositol trisphosphate (IP_3) by PLC, activate Ca^{2+} release from the endoplasmic reticulum (ER), trough inositol trisphosphate receptors (IP_3R). In some populations of glial cells a calcium induced Ca^{2+} release through ryanodine receptors (RYR) has also been demonstrated. Depletion of the ER stores might trigger additional Ca^{2+} entry via capacitative mechanisms, which involve the opening of the store-operated calcium channels (SOC). Regulation of Ca^{2+} signaling depends also on the intracellular Ca^{2+} buffering systems that include plasmalemmal Ca^{2+} ATP-ases (PMCA), the Na^+/Ca^{2+} exchanger (Na/Ca) and the storage by intracellular organelles by sarco(endo)plasmatic reticulum ATPases (SERCA) in the ER, or Ca^{2+} uniporters in mitochondria (M). A differential expression or activation of all the above-mentioned components results in a large heterogeneity of Ca^{2+} responses.

homotetramers of other isoforms could also exist. Among tissues, IP_3R heterotetramers have been demonstrated so far only in liver. Single-channel studies using isolated channel proteins incorporated in lipid bilayers have shown that the kinetic properties of the individual channel subtypes are different (Hagar et al., 1998; Ramos-Franco et al., 1998). The characteristic biophysical properties of the individual IP_3R subtypes (Cardy et al., 1997; Miyakawa et al., 1999) expressed in a cell appears to determine the specific temporal and spatial characteristics of the Ca^{2+} signals that are elicited. Indeed, in the type 2-expressing cell, agonist-evoked $[Ca^{2+}]_i$ signals were oscillatory, the type 1- and 3-expressing cells showed more transient Ca^{2+} responses. This property is believed to be due to the lack of Ca^{2+}-dependent inactivation of IP_3 receptor subtype R2, which unlike the R1 and R3 subtypes are not inactivated at elevated Ca^{2+} concentrations.

The expression of IP_3R subtypes 1, 2, and 3 varies with development, and each subtype is expressed by cells in a tissue-specific manner IP_3 is highly expressed in brain, IP_3R2 apparently highly expressed in spinal cord, and IP_3R3 highly expressed in intestine (Ross et al., 1992; Sharp et al., 1999). Different isoforms also predominate in different cultured cell types. The nature of the IP_3Rs subtypes in different glial cells is not known in detail and remained for a while controversial. Astrocytes in culture express all three subtypes of IP_3R, but this is not necessarily true for astrocytes in situ. In a couple of studies, only type 3 but not type 1 and 2 IP_3Rs have been immunolocalized in rat cortical astrocytes, cerebellar Bergmann glial cells (Yamamoto-Hino et al., 1995) and astrocytes of suprachiasmatic nucleus (Hamada et al., 1999). The latter studies used an antibody raised against a peptide differing significantly from the rat sequence and the possibility exists that the IP_3R antibody cross-reacts with the rat IP_3R2 protein expressed by astrocytes (Sharp et al., 1999). Recently an extensive study has been conducted using more specific antibodies (Holtzclaw et al., 2002). Using dual indirect immunohistochemistry, it has been shown that astrocytes in several adult rat brain areas express only the type 2 isoform of IP_3R. In addition IP_3R2 labeling was found in cell bodies of oligodendrocyte lineage and in microglial cells. Moreover, in astrocytes receptor labeling appears in punctate patches and extends into the entire network, including fine hair-like branches of processes. The fact that these fine processes showed IP_3R2 labeling suggests that ER elements extend into them and that these structures participate in cellular signaling. The unique expression of the IP_3R2 ion channels in astrocytes in situ, because of the intrinsic characteristics of this isoform, may allow astrocytes to sustain stimulation at high frequency for extended periods. The patchy distribution of IP_3R2 in astrocyte cell bodies and processes may represent clusters of receptors, probably linked to other proteins and organelles involved in Ca^{2+} signaling, and constituting specialized Ca^{2+} release sites. Intercellular signaling between astrocytes and between astrocytes and neurons may be supported by this organization of signaling elements in specialized microdomains (see also chapter by Shuai et al.).

2.3.2. *CICR*

The other type of intracellular channel that mediates release of Ca^{2+} from intracellular stores, the Ca^{2+}-gated channel or RyR, is also comprised of a family of proteins that share sequence homology and structural similarities (Meissner et al., 1997). For example, channels of both families are assembled into heterotetrameric structures to form the functional ion pore. Each subunit consists of large proteins—300 kDa for IP_3Rs and 500 kDa for RyRs. Three genes for IP_3Rs and three genes (located on different human chromosomes) encode the three types of RyR isoforms (RyR1, RyR2 and RyR3).

RyR1, characterized originally in skeletal muscle, is important for excitation-contraction coupling. RyR1 is located on the SR of skeletal myocytes and triggers Ca^{2+} release following an action potential by its direct interaction with the dihydropyridine-sensitive, voltage-operated Ca^{2+} channel in the plasma membrane. Single-channel recordings, measurements of $[^{45}Ca^{2+}]$ efflux from SR vesicles, and measurements of binding of $[^3H]$ryanodine have shown that skeletal muscle RyR activity is affected by binding of cations to Ca^{2+} regulatory sites, as well as by binding of adenine nucleotides and possibly organic polycations (Meissner, 1994).

In contrast to RyR1, RyR2 activates excitation–contraction coupling principally in heart muscle cells in response to voltage-dependent Ca^{2+} influx [through dihydropyridine receptor (DHPR) channels] that accompanies depolarization of the cardiac SR and T-tubules. Here, the DHPR channels are not in physical contact but are in close proximity to the RyR channels. RyR2 amplifies the DHPR signal by means of calcium-induced calcium release (CICR) (Meissner, 1994). Unlike IP_3Rs, RyR2 conductance correlates positively with the level of Ca^{2+} already present in the cytosol, thus producing a positive feedback amplification for Ca^{2+} signaling. CICR is made possible in myocardial cells by local junctions between the plasma membrane and closely juxtaposed RyRs on the SR. Apparently, this RyR-membrane complex is important for the generation of local $[Ca^{2+}]_i$ oscillations.

RyR2 is the most abundant isoform throughout the brain, with the highest levels being found in hippocampal regions such as CA3 and the dentate gyrus, as well as in the cerebellum, olfactory bulb and in certain cortical areas. Although CICR is well documented in peripheral neurons, its function in central neurons has not been established. While CICR has not been as well characterized in neurons as in muscle cells, RyR2-plasma membrane juxtaposition within dendritic structures might contribute to local oscillations of $[Ca^{2+}]_i$, as in myocardial cells. While a number of observations using caffeine, voltage-clamp control of Ca^{2+} influx and ER depletion do suggest a role for CICR in neurons (Verkhratsky and Shmigol, 1996), a definitive demonstration awaits further study.

The RyR3 isoform was originally described in the brain, where it is localized predominantly in CA1 hippocampal neurons. All three RyR isoforms are co-expressed in some neurons. Distinct isoforms might be localized in distinct subcompartments of the ER and thereby provide different sensitivities to ligand- or voltage-based stimuli, or both, particularly within dendritic branches (Furuichi et al., 1994).

All three RyR isoforms consist of at least two functional elements: (i) a relatively small transmembrane domain located near the C-terminus region that appears to form a channel, and (ii) a large N-terminal segment that protrudes into the cytosol and is organized into different domains on which reside binding sites for putative modulatory substances. While the recently discovered second messenger, cADPR, does activate some in vitro RyR-mediated Ca^{2+} release (Lee et al., 1995), effective levels of cADPR and upstream transduction pathways have not been identified. In addition to CaM, which binds to the

C-terminal region of the RyR, another class of putative RyR-modulating proteins includes immunophilins. These low-molecular-weight proteins were characterized originally as receptors of immunosuppressant drugs, such as cyclosporin A, FK506 and rapamycin, and bind both RyR and IP_3R and regulate Ca^{2+} efflux from the ER (Gold, 1997), possibly by maintaining the configuration of the tetrameric structure. The immunophilins are also expressed at higher levels within the brain than in immunocompetent tissues.

The expression of RyR in glia has been poorly investigated and is still debatable. Functional Ca^{2+}-induced Ca^{2+} release, sensitive to the classical modulators ryanodine and caffeine and activated under physiological conditions, has been demonstrated initially for periaxonal Schwann cells (Lev Ram and Ellisman, 1995) and for freshly isolated Muller glial cells from salamander retina (Keirstead and Miller, 1995). In Bergmann glial cells, studied in cerebellar slices, caffeine and ryanodine triggered a moderate $[Ca^{2+}]_i$ elevation and attenuated $[Ca^{2+}]_i$ transients evoked by kainate. In astrocytes, data on CICR are controversial. Initial studies conducted in cultured and freshly isolated astrocytes failed to detect an obvious caffeine-triggered $[Ca^{2+}]_i$ effect (Charles et al., 1993). Other studies in cultured embryonic cortical astrocytes clearly showed that caffeine triggered a $[Ca^{2+}]_i$ increase (Golovina and Blaustein, 1997). A recent study demonstrated that cultured mouse astrocytes express only the RyR type 3 mRNA, but not Ryr1 or Ryr2 genes (Matyash et al., 2002). Immunolabeling with specific RYR3-antibodies confirmed the distribution of this channel in astrocyte cytoplasm, and its close relation to ER. Similar data were obtained in brain astrocytes in situ. Furthermore, RyR activation with the specific agonist 4-chloro-*m*-cresol triggers an elevation of intracellular Ca^{2+} in all astrocytes that have been investigated, indicating that astrocytes in the brain express functional RyRs. In the same study, using either pharmacological blockade of RyR by an antagonizing concentration of ryanodine (200 μM) or RyR3 knockout mice, it has been shown that functional RyR is fundamental for astrocyte motility and migration. RyR3 has not been detected in purified microglial cells. Another study identified RyRs by immunocytochemistry with specific antibodies in cultured oligodendrocytes, in type 2 astrocytes and in the bipotential precursor cells (O-2A progenitors) from which oligodendrocytes and type 2 astrocytes can develop (Simpson et al., 1998). Glia acutely isolated from rat brain or in situ in cortical sections were similarly found to express RyRs. Caffeine elicited Ca^{2+} responses in most cultured type 2 astrocytes and in approximately 50% of cultured oligodendrocytes. Increases of $[Ca^{2+}]_i$ elicited by caffeine were inhibited by pretreatment with ryanodine or thapsigargin, while ionotropic glutamate receptor activation by kainate increased the magnitude of Ca^{2+} elevation evoked by caffeine in all the cells evaluated.

2.4. *Structural complexity and compartmentalization of ER in glial cells*

Although the ER has been described as a 'continuous network' of tubules (Terasaki et al., 1994), some direct studies of Ca^{2+} store organization provide evidence for heterogeneity and compartmentalization (Golovina and Blaustein, 1997). Cell fractionation studies and immunocytochemical studies also indicate that neuronal ER is heterogeneous. Resolution of the ambiguity about ER compartmentalization requires detailed, direct measurements of the intra-ER Ca^{2+} concentration ($[Ca^{2+}]_{ER}$) with sufficiently high spatial resolution to visualize ER subcompartments. Several methods have been employed. Some involve the use of Ca^{2+}-sensitive fluorochromes such as Furaptra, which do not saturate at the ambient $[Ca^{2+}]_{ER}$. Others involve transfection of cells with Ca^{2+} reporter molecules (aequorin or CaM-linked green fluorescent protein) targeted to specific organelles, or the application of electron microprobe analysis. Anyway, all these methods have limitations, and a wide range of $[Ca^{2+}]_{ER}$ values (5 μM–5 mM) has been reported. Probably, to solve the question about compartmentalization, rather than focusing on absolute $[Ca^{2+}]_{ER}$ levels, relative changes in $[Ca^{2+}]_{ER}$ have to be considered.

Some indirect observations have led to the conclusion that the two ER membrane receptors (IP_3R and RyR) release Ca^{2+} from the same pool (Khodakhah and Armstrong, 1997). However, many other functional studies suggest that the IP_3-sensitive and Ry/caffeine-sensitive Ca^{2+} stores are independent (Kostyuk and Kirischuk, 1993). For example, imaging of intact snail neurons has revealed spatially distinct cytosolic elevations of $[Ca^{2+}]_i$ in response to caffeine (higher in the subplasmalemmal region) and to IP_3 injection (higher in the cell center). Similarly, in adrenal chromaffin cells, IP_3 and caffeine-evoked responses appear to be spatially distinct (Cheek et al., 1991). Moreover, in these cells, the two types of stores can be independently emptied: the response to caffeine is unaffected by prior, selective depletion of the IP_3-sensitive store in response to either IP_3 or thapsigargin. Similarly, studies conducted on cortical astrocytes revealed that ER Ca^{2+} stores are organized into functionally distinct subcompartments that can be unloaded and refilled independently (Golovina and Blaustein, 1997). The agonists glutamate and ATP released Ca^{2+} primarily from CPA-sensitive ER Ca^{2+} stores. Agonist-evoked release was abolished by prior treatment with CPA, but it was unaffected by prior depletion of caffeine/ryanodine (CAF/RY)-sensitive ER Ca^{2+} stores. Conversely, prior depletion of the CPA-sensitive stores did not interfere with Ca^{2+} release or reuptake in the CAF/RY-sensitive stores. These studies also addressed the question of whether the functionally independent CPA-sensitive (IP_3-sensitive) and CAF/RY-sensitive ER Ca^{2+} stores were spatially separate. To this end, the ER of living cells was loaded with Furaptra, and the changes in $[Ca^{2+}]_{ER}$ were studied with high spatial resolution imaging methods to visualize individual small elements of the ER. Image subtraction was then used to demonstrate that some components of the ER unload Ca^{2+} in the presence of CPA, while the $[Ca^{2+}]_{ER}$ in other components rises under these circumstances. The components of the ER that load with Ca^{2+} when CPA is added, are depleted by CAF. This implies that Ca^{2+} is sequestered in the CAF-sensitive stores by a CPA-insensitive Ca^{2+} pump. The effects of both CPA and CAF (on the specific stores) are reversible, and there is only a 10–18% overlap between these two types of stores. Moreover, when Ry (which has a relatively irreversible effect) was used to deplete the CAF/RY-sensitive stores in the presence of CPA, only the CPA-sensitive stores refilled after washout (Golovina and Blaustein, 1997). Thus, considering both the functional and spatial information presented in these studies, it is possible that there are two structurally separate (or independent) components of the ER Ca^{2+} stores in astrocytes.

The existence of two different types of ER Ca^{2+} release mechanisms has been studied also in oligodendrocytes (Haak et al., 2001). Both oligodendrocyte progenitors and myelinating oligodendrocytes are intimately associated with axons, suggesting the existence of neuronal signals affecting oligodendrocyte proliferation, migration, and differentiation (Barres and Raff, 1999). Although oligodendrocytes can differentiate without neurons, axons or axon-derived signals enhance myelin protein expression. Axonal signals may also be required for oligodendrocyte survival, and it has been suggested that neuronal electrical activity is linked to myelinogenesis, perhaps by stimulating the release of growth factors and neurotransmitters from axons or from astrocytes or other glial cells (Barres and Raff, 1999). In addition, neurotransmitter receptors are expressed by oligodendrocytes at several stages of differentiation, which suggests that they might participate in oligodendrocyte differentiation. In oligodendrocyte progenitor cells (OPs) activation of either IP_3R or RyRs produced kinetically distinct local Ca^{2+} release events. Spatial and temporal characteristics of Ca^{2+} signaling, from within microdomains to intracellular waves, have been studied in these particular glial cells that must migrate and proliferate before differentiating into myelinating cells. OPs express specific Ca^{2+} release channel subtypes: RyR3 and IP_3R2. These receptors are expressed in patches along OP processes. RyRs are co-expressed with IP_3Rs in some patches, but IP_3Rs are also found alone. This differential distribution pattern may underlie the differences in local and global Ca^{2+} signals mediated by these two receptors. Intracellular Ca^{2+} waves initiation seems to be dependent on the activation of IP_3Rs, while activation of the RyR3 in OPs appears to evoke highly localized Ca^{2+} signals. Local Ca^{2+} release from intracellular stores has been shown to be fundamental for cell guidance and migration during brain development (Simpson and Armstrong, 1999). Furthermore, in addition to their separate roles, IP_3R and RyRs appear to modulate each other, to tightly regulate Ca^{2+} release from the ER.

2.5. *Mitochondria*

Mitochondria are another capacious Ca^{2+} storage site (Pozzan and Rizzuto, 2000; Ganitkevich, 2003). Generally Ca^{2+} can enter mitochondria through two mechanisms, which have not yet been identified at the molecular level: a saturable low-affinity (10–20 μM) uniporter and a saturable rapid uptake mode (Gunter and Gunter, 2001). It is clear that most Ca^{2+} ions entering the mitochondrion are bound and only a very small portion remains free. The Ca^{2+}-binding ratio in mitochondria ($[Ca]_{total}/[Ca]_{free}$) was estimated to be in the range of ~4000 to 6000 in the steady-state (Kaftan et al., 2000). Mitochondria do not express significant amounts of Ca^{2+} binding proteins (e.g., like calreticulin or calsequestrin in ER), so Ca^{2+} is most likely to be bound to membrane phospholipids and/or to precipitate with phosphate ions (Pivovarova et al., 1999; Thayer et al., 2002). Large amounts of Ca^{2+} could also be stored in the mitochondrial matrix in the form of insoluble hydroxyapatite, if phosphate ions are available. Ca^{2+} leaves the mitochondrion mainly through two distinct saturable mechanisms. These are Na^+ dependent and Na^+ independent transport systems (Pfeiffer et al., 2001). In addition, Ca^{2+} could be released from mitochondria through a channel known as the permeability transition pore, which is activated with an excessive increase of free Ca^{2+} in the mitochondrial matrix and trigger cascades of cellular processes leading to cell death (Rizzuto et al., 2000). However, the role of mitochondria in $[Ca^{2+}]_i$ homeostasis in cells, and in particular in glia, is little understood. The dissipation of the mitochondrial electrochemical gradient by protonophores [carbonyl cyanide *m*-chlorophenylhydrazone (CCCP) and carbonyl cyanide *p*-trifluoromethoxyphenylhydrazone] triggers Ca^{2+} release in oligodendrocytes (Kirischuk et al., 1995a). CCCP treatment was not able to influence the kinetic parameters of the depolarization-triggered $[Ca^{2+}]_i$ transients (Kirischuk et al., 1995b). This suggests that mitochondrial Ca^{2+} accumulation does not play an important role in calcium signaling. However, protonophores are not totally selective for mithochondria, and probably do not represent an accurate method of investigation. Other studies highlighted the importance of mitochondrial location with respect to Ca^{2+} flux in the cytoplasm, to determinate their importance as Ca^{2+} sink (Collins et al., 2001). In a variety of cell types, the close apposition of mitochondria with ER has been documented (Hajnoczky et al., 2000). An interaction between ER and mitochondria has been implicated in mitochondrial Ca^{2+} uptake. During Ca^{2+} release from ER through the opening of IP_3-gated channels (Kaftan et al., 2000) or RyRs (Pacher et al., 2002), a high local Ca^{2+} concentration at sites of close contact between ER and mitochondria is believed to facilitate mitochondrial Ca^{2+} uptake (Rizzuto et al., 2000). It has been shown that mitochondria closely associated with ER tend to unload Ca^{2+} preferentially into microdomains, thus facilitating ER Ca^{2+} reloading (Arnaudeau et al., 2001). In oligodendrocyte processes, mitochondria were always found in association with sites where SERCA expression was elevated. This suggests that high concentrations of SERCA pumps together with other cellular specializations may be important in supporting elevated local Ca^{2+}-release kinetics in oligodendrocytes. This presumably allows mitochondria to play a specific role that depends on the spatial and temporal profile of cytosolic Ca^{2+} signals, that is, the mitochondrial contribution to Ca^{2+} regulation is location and stimulus specific. Therefore, it is not surprising that in some studies no apparent contribution was found of mitochondrial Ca^{2+} uptake during physiological stimulation.

3. Capacitative calcium entry

After stimulation, the temporal and spatial distribution of the increase in $[Ca^{2+}]_i$ in glial cells, especially astrocytes, is remarkably complex. At a given point, the time course of the increase may be a single spike, biphasic with an initial peak followed by a plateau, or oscillations (Verkhratsky and Kettenmann, 1995—see also chapters by Shuai et al., and by Cornell-Bell et al.). The plateau/oscillation phases requires extracellular Ca^{2+}. The Ca^{2+} influx pathway activated with metabotropic receptor stimulation is unclear. By analogy with other tissues, the most likely candidate is the store-operated Ca^{2+} channel, although the existence of this pathway in glial cells has not been completely understood. Store emptying generates a putative signal (Rzigalinski et al., 1999) that induces the opening of the store-operated calcium channel (SOC) at the level of the cell membrane, also known as the calcium release-activated calcium channel (CRAC) (Parekh and Penner, 1997), which allows calcium into the cells from the extracellular space. This channel has been identified as homologous to the transient receptor potential channels in *Drosophila* (Petersen et al., 1995). Calcium entry through the SOC/CRAC, activated by the depletion of the intracellular calcium stores, is also known as capacitative calcium entry (CCE). CCE has been identified in many cells (Parekh and Penner, 1997; Barritt, 1999; Putney, 1999), but the mechanism linking store-depletion and the increased Ca^{2+} permeability of the plasma membrane remains elusive. Several mechanisms have been proposed. In mast cells, an inward Ca^{2+} current (ICRAC, Ca^{2+} release-activated Ca^{2+} current) is seen following depletion of intracellular Ca^{2+} stores (Hoth and Penner, 1992), while, in mouse pancreatic acinar cells, CCE results from a ICRANC, Ca^{2+} release-activated non-selective cation current (Krause et al., 1996). It has been suggested that a Ca^{2+} influx factor, generated when stores are depleted, diffuses to the plasma membrane where the SOC is located, and activates CCE (Randriamampita and Tsien, 1993).

Alternatively, direct physical contact between the plasma membrane and the ER IP_3 receptor, which detects depletion of the intracellular Ca^{2+} stores, might be responsible for activating CCE (Irvine, 1990). Recently, two studies have provided evidence to support this latter hypothesis of an exocytosis-like mechanism of CCE activation (Yao et al., 1999; Patterson et al., 1999).

CCE is a mechanism whereby intracellular calcium stores are refilled (Berridge, 1995; Putney, 1999). The maintenance of the filled state of intracellular stores is involved in cell survival, as the prolonged depletion of calcium reservoirs causes cell death and apoptosis. Two hypotheses have been proposed to explain how calcium flowing through the SOC/CRAC during CCE can be captured and stored in the ER compartments and thus contribute to the spike or the plateau phase of the single calcium transient. Calcium, according to the preferential pathway hypotheses, could be directly sequestered within intracellular stores because of a physical association between the SOC/CRAC and the SERCA–ER complex (Berridge, 1995; Putney, 1999). Alternatively, calcium ions, once admitted into the cytosol by SOC/CRAC opening, might be quickly removed from the cytosol and accumulated in the ER by the SERCA (diffusion pathway) (Hofer et al., 1998). In addition to refilling the intracellular Ca^{2+} stores, the increased Ca^{2+} via SOC may also have other cellular functions. For example, Ca^{2+} in the microdomain of the channel mouth may activate enzymes or target proteins.

In glial cells CCE has not been fully characterized, and even the existence of SOC/CRAC channels has not been clearly shown. In a study performed in our lab we evaluated the relation between cytoskeleton rearrangement and CCE in cortical astrocytes (Grimaldi et al., 1999). Cultured conventional astrocytes (occasionally called type-1 astrocytes) express a wide array of second messenger-coupled receptors (Bhat, 1995; Verkhratsky et al., 1998). Among these receptors, P2Y, a subclass of purinergic receptors, activated by ATP (Shao and McCarthy, 1993) and bradykinin (BK) receptors (Chen et al., 1996) are coupled to PLC. Their activation causes calcium mobilization from intracellular calcium stores via PLC-induced IP_3 production (Berridge, 1995). In this study, we provided evidence that active and rapid rearrangement of astrocyte morphology induced by activation of the protein kinase A (PKA), is responsible for enhancement of IP_3-induced cytosolic calcium concentration ($[Ca^{2+}]_i$) elevation via an enhanced CCE. Enhanced CCE, in turn, is associated with reorganization of the spatial relationship between the outer cell membrane and the ER. In astrocytes, it is likely that the reorganization of the spatial relationship between ER structures and cell membrane may cause a change in the efficiency of calcium mobilization as a result of the improved refilling of the intracellular calcium stores. This would, in turn, result in larger calcium transients.

A common means of inducing morphological changes in astrocytes has been prolonged application of long lasting cAMP analogues. Astrocytes, which differentiate after prolonged exposure to cAMP analogues, changed from a flat polygonal form to a stellate process-bearing appearance after application of cAMP. Astrocytes differentiated by a long term exposure to cAMP showed changes in biochemical properties such as an increase in the production of IP_3 in response to both α_1-adrenergic agonists and BK. In addition, long-term cAMP-induced differentiation modifies membrane ionic conductance (Lascola and Kraig, 1996).

We assessed the arrangement of the ER in undifferentiated and differentiated cells. In undifferentiated astrocytes, ER structures were largely evident in the periphery of the cells where a tubular network of membrane-delimited structures was identifiable (see Fig. 2). In addition, ER structures and the Golgi complex could be identified in the perinuclear region of the astrocytes. In differentiated cells, perhaps as a consequence of the marked reorganization of the cell shape, the ER was so condensed that areas of the cytoplasm free of ER structures were virtually absent. Such a marked rearrangement may, in turn, increase the availability of calcium flowing from the extracellular space to provide for refilling of the stores during CCE, through a closer association of the outer cell membrane and ER structures.

Cell morphology has been previously shown to play a role in agonist-induced calcium mobilization in fibroblasts, in thapsigargin-induced store depletion, and in the associated

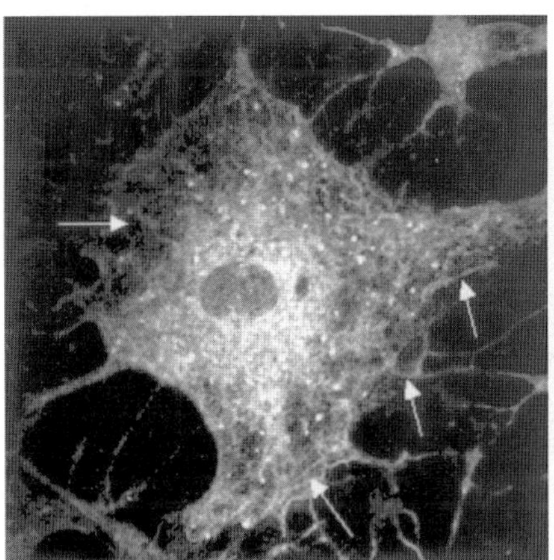

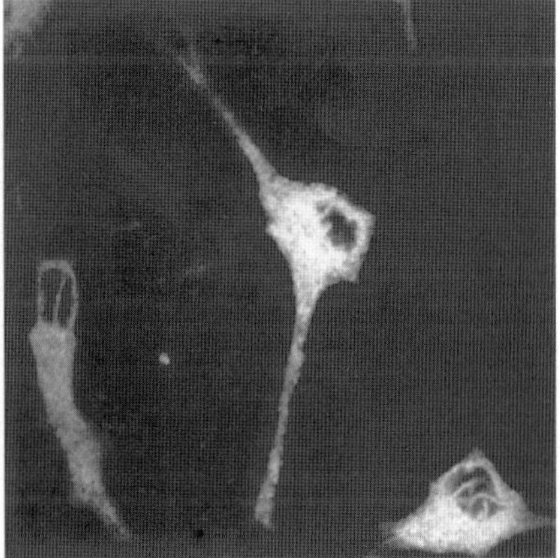

Fig. 2. Analysis of the ER distribution and association with cell membrane in control and differentiated astrocytes. ER was labeled in living cells by means of the ER tracker and analyzed with a confocal microscope, equipped with a 633 lens for ER-associated fluorescence. Images were then software zoomed to resolve single astrocytes. In undifferentiated astrocytes (left panel) the density of ER membranes is highest in the perinuclear area, whereas the density of the tubular network is decreased in the peripheral part of the cells. The arrowheads indicate some connections between the tubular ER structures and the cell membrane. Differentiated astrocytes (right panel) showed a more condensed ER organization without appreciable areas free of fluorescence signal across the cell.

CCE in endothelial cells (Holda and Blatter, 1997). Because cAMP-induced differentiation causes the rearrangement of the actin-formed stress fiber (Goldman and Chiu, 1984), we induced actin depolymerization with cytochalasin D (CytD) before exposing the cells to dibutyryl cAMP. As expected, in the presence of CytD, the cells did not reshape and maintained their flat polygonal appearance, despite the activation of PKA by cAMP. The prevention of the shape changes in dibutyryl cAMP-treated cells by CytD avoided the potentiation of ATP- and thapsigargin-induced $[Ca^{2+}]_i$ elevation (see Fig. 3). Both PKA activity in the cell cytosol and PKA localization at level of the cell membrane, where

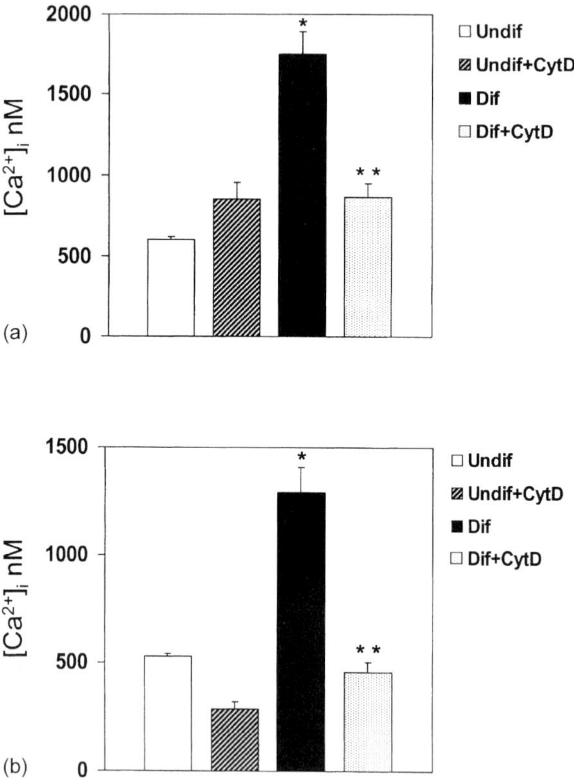

Fig. 3. Effect of stress fiber depolymerization on ATP- and thapsigargin-induced $[Ca^{2+}]_i$ elevation. The effect of stress fiber disassembly by means of 10 μM cytochalasin D (CytD) was studied. $[Ca^{2+}]_i$ values obtained at the peak of the response were averaged and graphed as a bar ± SEM. In panel a, ATP-induced intracellular calcium mobilization in undifferentiated and dibutyryl cAMP-treated astrocytes. Undifferentiated (open bar) and 10 μM CytD pretreated (hatched bar) astrocytes display a similar response to ATP. However, the potentiation of ATP stimulation in dibutyryl cAMP-differentiated astrocytes (solid bar versus open bar) was reversed by CytD pretreatment (square-filled bar). In panel b, the effect of CytD pretreatment on thapsigargin-induced intracellular calcium mobilization in undifferentiated and differentiated astrocytes is displayed. CytD pretreatment (square-filled bar) completely reversed the potentiation of thapsigargin response in differentiated astrocytes (solid bar). $*p < 0.05$ versus value in undifferentiated cells; $**p < 0.05$ versus value in dibutyryl cAMP-differentiated cells.

CRAC channels are localized, have been reported not to be affected by CytD pretreatment. Therefore, it seems likely that morphological changes involving the rearrangement of stress fibers play a critical role in the potentiation of the CCE.

Another study demonstrated that in rat cerebellar astrocytes, store depletion induced by thapsigargin activates CCE via SOCs (Lo et al., 2002). Furthermore, it has been shown that these channels are also permeable to Na^+ but not to Sr^{2+} and Ba^{2+} and that their activity is regulated by serine/threonine phosphorylation.

A study from our group has shown that 4-aminopyridine, a specific blocker of voltage-sensitive K^+ channels (Aronson, 1992), strongly potentiates CCE in astrocytes, but not in neurons (Grimaldi et al., 2001). Because the effect of 4-AP alone on CCE is not as large as when it is triggered by a large calcium mobilization, we believe that other mechanisms must be activated to uncover the potentiation of CCE that we observed. When ICS are depleted, either using an agonist able to cause a large production of IP_3, such as ATP or bradykinin or an agent able to completely discharge ICS, such as thapsigargin, a robust signal is generated that triggers the opening of CRAC. Such a signal has not been definitively identified and characterized. However, in cortical astrocytes a soluble factor, probably belonging to the family of the eicosanoid derivatives (Rzigalinski et al., 1999), may be responsible for the opening of CRAC channel. Alternatively, a physical association between SOC/CRAC and the IP_3 receptor may be involved in the opening of the CRAC channel after the emptying of ICS (Boulay et al., 1999). Regardless of the signal used to trigger the opening of the CRAC channels, 4-AP causes a considerably larger influx of calcium from the extracellular space than in control cells. CRAC and voltage-sensitive K^+ channels have some similarity in the amino acid sequence (Harteneck et al., 2000); therefore, it is conceivable that 4-AP interacts with the open CRAC channels in a similar manner as with K^+ channels, and thereby increases CCE. This action on CCE may explain some of the therapeutic effects of 4-AP in disorders in which impairment of neurotransmission is involved (Fujihara and Miyoshi, 1998; Andreani et al., 2000; Smith et al., 2000). Moreover, changes in calcium homeostasis induced by 4-AP in astrocytes might cause the release of trophic factors that are likely to support regrowth of neuronal extensions.

In microglial cells, a long-lasting activation of capacitative Ca^{2+} entry after maximal depletion of intracellular Ca^{2+} stores by stimulation with ATP in the absence of extracellular Ca^{2+} has been described recently (Toescu et al., 1998). Once activated, the capacitative Ca^{2+} entry pathway in microglial cells remained operative for tens of minutes, creating a steady-state $[Ca^{2+}]_i$ elevation that dramatically outlasted the period of agonist action.

4. Intracellular calcium sensors and effectors

After entering the cytoplasm, Ca^{2+} binds to a number of proteins that trigger various intracellular signal transduction pathways. Probably the best-known cytoplasmic Ca^{2+} sensor is CaM, which regulates the functional activity of at least three broad classes of enzymes, namely, CaM-dependent protein kinases, protein phosphatases, and adenylate cyclases. The latter either interact with cytoplasmic enzymes or transfer the signal further down to the nucleus, initiating other pathways responsible for gene expression. Among the many molecular targets of internal Ca^{2+} signaling, PKC isozymes α, β and γ are activated and translocated by combination(s) of Ca^{2+}, DAG and arachidonic acid (AA) (Nishizuka, 1984; Alkon et al., 1998). Elevated $[Ca^{2+}]_i$ also acts on Ca^{2+}/calmodulin-dependent (type II) kinase(s) (CaM kinases) that, in turn, can regulate voltage-dependent K^+ channels, cholinergic control of neuronal responsiveness, smooth muscle contraction and synaptic transmission. At higher Ca^{2+} levels (>10 μM), levels where CaM is known to block RyR function, CaM also regulates a mode of protein trafficking to the nucleus that is similar to, but independent of active nuclear transport regulated by the low-molecular-weight GTP-binding protein, ran (ran-mediated transport). Other evidence has also been found, suggesting an effect by CaM kinase(s) on transcription factors that in turn could influence protein synthesis (Gringhuis et al., 1997).

An alternative way by which cytoplasmic Ca^{2+} signals may influence gene expression is the pathway involving Ras proteins (small guanine nucleotide-binding proteins), which after being activated by Ca^{2+} trigger a cascade of phosphorylation events that lead to a modulation of gene expression. Finally, cytoplasmic Ca^{2+} signals may propagate to the nucleus, where they directly stimulate the synthesis of immediate early genes as well as structural genes. Unfortunately, little is known of the expression and role of these systems in glial cells; their characterization in glia is an important problem awaiting an experimental solution.

4.1. S-100

S-100 is a highly conserved, low molecular weight (10–12 kDa) multifunctional calcium binding protein synthesized in a variety of cells, including astroglia and Schwann cells. The protein is extremely stable and hydrophilic, and 95% of the cellular S-100 is found in the cytosol. The S-100 protein occurs as a homodimer, and it is one of a large family of low-MW EF-hand proteins that includes calcyclin, p9Ka, and the cystic fibrosis serum antigen MRP-8. A large number of functions have been attributed to S-100, including inhibition of protein phosphorylation, promotion of calcium-dependent microtubule dissociation (Fano et al., 1995), and maintenance of the cytoskeleton.

Like CaM, S-100 acts by binding to other proteins in a calcium-dependent fashion. For example, S-100 binds to glial fibrillary acidic protein (GFAP) and inhibits GFAP polymerization in a calcium-dependent manner. It also reportedly undergoes calcium-dependent binding to a number of targets, including fructose-1,6-bisphosphate aldolase and the RyR, which is involved in calcium-dependent calcium release from the ER, as described above. Inhibition of microtubule assembly was suggested to occur by binding of S-100 to tau proteins, preventing phosphorylation by protein kinase C (PKC).

The finding that the gene for S-100 is located in a gene cluster on chromosome 1q21 has suggested possible involvement with Alzheimer's disease and Down's syndrome. Indeed, elevated levels of S-100 immunoreactivity have been found in both Down's syndrome and Alzheimer's disease patients. In Alzheimer's disease and AIDS-related progressive neurodegeneration, elevated S-100 levels in extracellular fluids correlate with the extent of the pathological injury. Elevations of S-100 in hippocampus were also found in patients with Alzheimer's disease.

Other researchers have found that at high concentrations (0.5–2 μM), homologous dimers of S-100 produce pathological structural changes within minutes, leading to apoptosis. The effect is preceded by large increases in intracellular calcium. The protein synthesis inhibitor cycloheximide blocks S-100 induced apoptosis. Treatment of rat astrocytes with S100beta produces activation of inducible nitric oxide synthase, which produces nitric oxide resulting in cell death (Hu et al., 1997). However, mice in which S-100 is overexpressed show no behavioral abnormalities or pathological changes. Thus, it is still unclear whether the elevations in S-100 are a cause, an epiphenomenon, or part of a compensatory mechanism.

Low concentrations of S-100 that are secreted into the extracellular medium act as growth factors or cytokines, inducing proliferation. These effects are mediated by surface receptor RAGE (receptor for advanced glycation endproducts). S-100 applied extracellularly was found to hyperpolarize the resting potential and inhibit spontaneous discharge activity by interacting with potassium channels.

4.2. Calcium-binding proteins in Alzheimer's disease and apoptosis

Calsenilin is another EF-hand containing protein which, although mostly neuronal, is also found in non-neuronal cells. Calsenilin interacts with the C-terminal region of presenilins 1 and 2 (Buxbaum et al., 1998; Leissring et al., 2000). In human neuroglioma (H4) cells transfected with calsenilin, calsenilin expression correlated with the appearance of a proteolytic fragment of presenilin 2 created by the apoptotic protease caspase (Choi et al., 2001). However, calsenilin is not required for the gamma-secretase activity of presenilin 2 (Esler et al., 2002).

Calsenilin may also interact with the RyR, because calsenilin enhances apoptosis in the presence of thapsigargin, an inhibitor of Ca-ATP dependent calcium uptake in the ER, by promoting the release of calcium from intracellular stores (Lilliehook et al., 2002).

Calsenilin also interacts with voltage-gated potassium channels (Morohashi et al., 2002) and thus shares functional similarity with calexcitin, a learning-related calcium-binding protein found in invertebrates (Nelson et al., 1996). In addition to producing neuronal excitation by its direct inhibitory effect on voltage-dependent potassium (i_A) channels, calexcitin binds to and activates the RyR, producing rapid, transient increases in intracellular calcium levels that resemble oscillations (Nelson et al., 1999). Reduced levels of a calexcitin-like protein were also found in cultured cells from Alzheimer's disease patients (Kim et al., 1995), illustrating one of many similarities that have been noted between pathways involved in learning and Alzheimer's disease at the biochemical level (Alkon et al., 1998).

Although most research on apoptosis induced by the beta-amyloid peptide, the central figure in the amyloidogenic hypothesis of Alzheimer's disease, has concentrated on neurons, Alzheimer's disease is also accompanied by changes in white matter, loss of astrocytes and oligodendrocytes and cerebral vasculopathy. Vascular cell degeneration in patients with Alzheimer's disease complicated by cerebrovascular disease is associated with reduced levels of nitric oxide synthase 3 and increased levels of the apoptosis promoting protein p53. Vasculopathy is characterized by variable amyloid deposition and vascular smooth muscle cell apoptosis. Because apoptosis is accompanied by an increase in intracellular calcium, it is not surprising that calcium-buffering proteins such as calbindin-D28k and parvalbumin exert a significant protective effect (Wernyj et al., 1999; Tombal et al., 2002). Calcineurin, a calcium binding protein found in neurons, interacts with the apoptosis protection protein bcl-2. Similarly, neuronal apoptosis inhibitory protein binds to the calcium-binding protein hippocalcin (Lindholm et al., 2002). However, considering the central role played by calcium and calcium-activated proteases, such as calpains in apoptosis (Lu et al., 2002; Tombal et al., 2002), it is likely that any protein that interacts with calcium will either participate in apoptosis directly or exhibit a change in activity or binding behavior during apoptosis.

Calcium and hydrogen peroxide levels are inextricably related in the cell. Apoptosis can also be induced by calcium or by oxidative stress induced by hydrogen peroxide or lipid peroxidation products such as 4-hydroxynonenal (Kalinich et al., 2000). Hypoxia and ischemia can trigger changes in calcium homeostasis, allowing increased calcium influx, in a process that requires hydrogen peroxide production by beta-amyloid (Green et al., 2002). In the presence of copper or zinc, beta-amyloid catalytically generates hydrogen peroxide (Opazo et al., 2002), which may mediate the toxicity of beta-amyloid. Hydrogen peroxide and superoxide can alter intracellular calcium levels by their effects on the RyR (Okabe et al., 2000). Presenilin also activates CICR, activating PKC and causing apoptosis (Chan et al., 2002). Inhibition of PKC activity can also induce apoptosis by upregulating H_2O_2 (Liou et al., 2000). Pharmacological activation of PKC promotes the release of the soluble fragment of APP by activating alpha-secretase, reducing levels of beta-amyloid by depriving beta-secretase of substrate (McLaughlin and Breen, 1999). Although low concentrations of H_2O_2 activate PKC and the RyR, higher levels are inhibitory due to oxidation of essential sulfhydryl residues.

5. Concluding remarks

For a long time, glial cells were considered merely to provide structural and trophic support for neurons. However, this concept is now changing, since they have been shown to play a major role in the regulation of the extracellular pH (Deitmer, 1992—see also chapter by Bevensee and McAlear), K^+ levels (see chapter by Walz), CO_2 metabolism (see chapter by Hertz, Peng et al.), redox equilibrium and neurotransmitter uptake (Landis, 1994—see also chapter by Schousboe and Waagepetersen). Ca^{2+} signalling seems to play a pivotal role in the control of all this functions, and recently one of the major Ca^{2+} regulatory mechanisms, the capacitative Ca^{2+} influx, has been implicated in glial cell function (Grimaldi et al., 1999, 2001). Moreover, glial cells have been shown to exhibit an oscillatory Ca^{2+} response and a propagated Ca^{2+} wave, which can be transmitted across the gap junction to nearby neurons (Scemes, 2000). Since it is apparent that glial Ca^{2+} signalling is an important pathway of glia–glia and neuron–glia cross-talk (see chapters by Shuai et al., and by Cornell-Bell et al.), its role during brain pathological conditions has been highlighted. It is well known that injury or brain pathology leads to a complex reaction from the astrocytes and microglial cells (Landis, 1994). For example, elevation of $[Ca^{2+}]_i$ in astrocytes induced by ischemia has been related to astrocyte activation and release of growth factors (Duffy and MacVicar, 1996). In addition, it has been suggested that glial Ca^{2+} waves may be related to certain complex pathological conditions such as migraine headaches (Martins-Ferreira et al., 2000).

References

Alkon, D.L., Nelson, T.J., Zhao, W., Cavallaro, S., 1998. Time domains of neuronal Ca^{2+} signaling and associative memory: steps through a calexcitin, ryanodine receptor, K^+ channel cascade. Trends Neurosci. 21, 529–537.

Andreani, A., Leoni, A., Locatelli, A., Morigi, R., Rambaldi, M., Pietra, C., Villetti, G., 2000. 4-Aminopyridine derivatives with antiamnesic activity. Eur. J. Med. Chem. 35, 77–82.

Arnaudeau, S., Kelley, W.L., Walsh, J.V., Demaurex, N., 2001. Mitochondria recycle Ca^{2+} to the endoplasmic reticulum and prevent the depletion of neighbouring endoplasmic reticulum regions. J. Biol. Chem. 276, 29430–29439.

Aronson, J.K., 1992. Potassium channels in nervous tissue. Biochem. Pharmacol. 43, 11–14.

Baba-Aissa, F., Raeymaekers, L., Wuytack, F., Dode, L., Casteels, R., 1998. Distribution and isoform diversità of the organellar Ca^{2+} pumps in the brain. Mol. Chem. Neuropathol. 33, 199–208.

Barres, B.A., Raff, M.C., 1999. Axonal control of oligodendrocyte development. J. Cell Biol. 147, 1123–1128.

Barritt, G.J., 1999. Receptor-activated Ca^{2+} inflow in animal cells: a variety of pathways tailored to meet different intracellular Ca^{2+} signaling requirements. Biochem. J. 337, 153–169.

Berger, T., Walz, W., Schnitzer, J., Kettenmann, H., 1992. GABA- and glutamate-activated currents in glial cells of the mouse corpus callosum slice. J. Neurosci. Res. 4, 1277–1284.

Berridge, M.J., 1993. Inositol triphosphate and calcium signaling. Nature 361, 315–325.

Berridge, M.J., 1995. Capacitative calcium entry. Biochem. J. 312, 1–11.

Bhat, N.R., 1995. Signal transduction mechanisms in glial cells. Dev. Neurosci. 17, 267–284.

Blaustein, P., Golovina, V.A., 2001. Structural complexity and functional diversity of endoplasmic reticulum Ca^{2+} stores. Trends Neurosci. 24, 602–608.

Bootman, M.D., Berridge, M.J., Lipp, P., 1997. Cooking with calcium: the recipes for composing global signals from elementary events. Cell 91, 367–373.

Boulay, G., Brown, D.M., Qin, N., Jiang, M., Dietrich, A., Zhu, M.X., Chen, Z., Birnbaumer, M., Mikoshiba, K., Birnbaumer, L., 1999. Modulation of Ca^{2+} entry by polypeptides of the inositol 1,4,5-trisphosphate receptor (IP_3R) that bind transient receptor potential (TRP): evidence for roles of TRP and IP_3R in store depletion-activated Ca^{2+} entry. Proc. Natl Acad. Sci. USA 96, 14955–14960.

Buratti, R., Prestipino, G., Menegazzi, P., Treves, S., Zorzato, F., 1995. Calcium dependent activation of skeletal muscle Ca^{2+} release channel (ryanodine receptor) by calmodulin. Biochem. Biophys. Res. Commun. 213, 1082–1090.

Buxbaum, J.D., Choi, E.K., Luo, Y., Lilliehook, C., Crowley, A.C., Merriam, D.E., Wasco, W., 1998. Calsenilin: a calcium-binding protein that interacts with the presenilins and regulates the levels of a presenilin fragment. Nat. Med. 4, 1177–1181.

Cardy, T.J.A., Traynor, D., Taylor, C.W., 1997. Differential regulation of types-1 and -3 inositol trisphosphate receptors by cytosolic Ca^{2+}. Biochem. J. 328, 785–793.

Chan, S.L., Culmsee, C., Haughey, N., Klapper, W., Mattson, M.P., 2002. Presenilin-1 mutations sensitize neurons to DNA damage-induced death by a mechanism involving perturbed calcium homeostasis and activation of calpains and caspase-12. Neurobiol. Dis. 11, 2–19.

Charles, A.C., Dirksen, E.R., Merrill, J.E., Sanderson, M.J., 1993. Mechanisms of intercellular calcium signaling in glial cells studied with dantrolene and thapsigargin. Glia 7, 134–145.

Cheek, T.R., Barry, V.A., Berridge, M.J., Missiaen, L., 1991. Bovine adrenal chromaffin cells contain an inositol 1,4,5-trisphosphate-insensitive but caffeine-sensitive Ca^{2+} store that can be regulated by intraluminal free Ca^{2+}. Biochem. J. 275, 697–701.

Chen, C.C., Chen, W.C., 1996. ATP-evoked inositol phosphates formation through activation of P2U purinergic receptors in cultured astrocytes: regulation by PKC subtypes alpha, delta, and theta. Glia 17, 63–71.

Choi, E.K., Zaidi, N.F., Miller, J.S., Crowley, A.C., Merriam, D.E., Lilliehook, C., Buxbaum, J.D., Wasco, W.J., 2001. Calsenilin is a substrate for caspase-3 that preferentially interacts with the familial Alzheimer's disease-associated C-terminal fragment of presenilin 2. Biol. Chem. 276, 19197–19204.

Collins, T.J., Lipp, P., Berridge, M.J., Bootman, M.D., 2001. Mitochondrial Ca^{2+} uptake depends on the spatial and temporal profile of cytosolic Ca^{2+} signals. J. Biol. Chem. 276, 26411–26420.

Deitmer, J.W., 1992. Bicarbonate-dependent changes of intracellular sodium and pH in identified leech glial cells. Pflugers Arch. 420, 584–589.

Deitmer, J.W., Verkhratsky, A.J., Lohr, C., 1998. Calcium signaling in glial cells. Cell Calcium 24, 405–416.

Dolmetsch, R.E., Xu, K., Lewis, R.S., 1998. Calcium oscillations increase the efficiency and specificity of gene expression. Nature 392, 933–936.

Dousa, T.P., Chini, E.N., Beers, K.W., 1996. Adenine nucleotide diphosphates: emerging second messengers acting via intracellular Ca^{2+} release. Am. J. Physiol. 271, C1007–C1024.

Duffy, S., MacVicar, B.A., 1996. In vitro ischemia promotes calcium influx and intracellular calcium release in hippocampal astrocytes. J. Neurosci. 16, 71–81.

Ehrlich, B.E., 1995. Functional properties of intracellular calcium-release channels. Curr. Opin. Neurobiol. 5, 304–309.

Esler, W.P., Kimberly, W.T., Ostaszewski, B.L., Ye, W., Diehl, T.S., Selkoe, D.J., Wolfe, M.S., 2002. Activity-dependent isolation of the presenilin-gamma-secretase complex reveals nicastrin and a gamma substrate. Proc. Natl Acad. Sci. USA 99, 2720–2725.

Fano, G., Biocca, S., Fulle, S., Mariggio, M.A., Belia, S., Calissano, 1995. The S-100: a protein family in search of a function. Prog. Neurobiol. 46, 71–82.

Fujihara, K., Miyoshi, T., 1998. The effects of 4-aminopyridine on motor evoked potentials in multiple sclerosis. J. Neurol. Sci. 159, 102–106.

Furuichi, T., Furutama, D., Hakamata, Y., Nakai, J., Takeshima, H., Mikoshiba, K., 1994. Multiple types of ryanodine receptor/Ca^{2+} release channels are differentially expressed in rabbit brain. J. Neurosci. 14, 4794–4805.

Gambetti, P., Erulkar, S.E., Somlyo, A.P., Gontas, N.K., 1975. Calcium-containing structures in vertebrates glial cells. Ultrastructural and microprobe analysis. J. Cell Biol. 64, 322–330.

Ganitkevich, V.Y., 2003. The role of mitochondria in cytoplasmic Ca^{2+} cycling. Exp. Physiol. 88, 91–97.

Gold, B.G., 1997. FK506 and the role of immunophilins in nerve regeneration. Mol Neurobiol. 15, 285–306.

Goldman, J.E., Chiu, F.C., 1984. Dibutyryl cyclic AMP causes intermediate filament accumulation and actin reorganization in astrocytes. Brain Res. 306, 85–95.

Golovina, V.A., Blaustein, M.P., 1997. Spatially and functionally distinct Ca^{2+} stores in sarcoplasmic and endoplasmic reticulum. Science 275, 1643–1648.

Golovina, V.A., Blaustein, M.P., 2000. Unloading and refilling of two classes of spatially resolved endoplasmic reticulum Ca^{2+} stores in astrocytes. Glia 31, 15–28.

Green, K.N., Boyle, J.P., Peers, C., 2002. Hypoxia potentiates exocytosis and Ca^{2+} channels in PC12 cells via increased amyloid beta peptide formation and reactive oxygen species generation. J. Physiol. 541, 1013–1023.

Grimaldi, M., Atzori, M., Ray, P., Alkon, D.L., 2001. Mobilization of calcium from intracellular storse, potentiation of neurotransmitter-induced calcium transients, and capacitative calcium entry by 4-aminopyridine. J. Neurosci. 21, 3135–3143.

Grimaldi, M., Favit, A., Alkon, D.L., 1999. cAMP-induced cytoskeleton rearrangement increases calcium transients through the enhancement of capacitative calcium entry. J. Biol. Chem. 274, 33557–33564.

Gringhuis, S.I., de Leij, L.F., Wayman, G.A., Tokumitsu, H., Vellenga, E., 1997. The Ca^{2+}/calmodulin-dependent kinase type IV is involved in the CD5-mediated signaling pathway in human T lymphocytes. J. Biol. Chem. 272, 31809–31820.

Gudz, T.I., Schneider, T.E., Haas, T.A., Macklin, W.B., 2002. Myelin proteolipid protein forms a complex with integrins and may participate in integrin receptor signaling in oligodendrocytes. J. Neurosci. 22, 7398–7407.

Gunter, T.E., Gunter, K.K., 2001. Uptake of calcium by mitochondria: transport and possible function. IUBMB Life 52, 197–204.

Haak, L.H., Song, L.S., Molinski, T.F., Pessah, I.N., Cheng, H., Russell, J., 2001. Sparks and puffs in oligodendrocyte progenitors: cross talk between ryanodine receptors and inositol trisphosphate receptors. J. Neurosci. 21, 3860–3870.

Hagar, R.E., Butgstahler, A.D., Nathanson, M.H., Ehrlich, B.E., 1998. Type III InsP3 receptor channel stays open in the presence of increased calcium. Nature 396, 81–84.

Hajnoczky, G., Csordas, G., Madesh, M., Pacher, P., 2000. The machinery of local Ca^{2+} signalling between sarco-endoplasmic reticulum and mitochondria. J. Physiol. 529, 69–81.

Hamada, T., Niki, T., Ziging, P., Sugiyama, T., Watanabe, S., Mikoshiba, K., Ishida, N., 1999. Differential expression patterns of inositol trisphosphate receptor types 1 and 3 in the rat suprachiasmatic nucleus. J. Brain. Res., 838.

Harteneck, C., Plant, T.D., Schultz, G., 2000. From worm to man: three subfamilies of TRP channels. Trends Neurosci. 23, 159–166.

Hofer, A.M., Landolfi, B., Debellis, L., Pozzan, T., Curci, S., 1998. Free [Ca^{2+}] dynamics measured in agonist-sensitive stores of single living intact cells: a new look at the refilling process. EMBO J. 17, 1986–1995.

Holda, J.R., Blatter, L.A., 1997. Capacitative calcium entry is inhibited in vascular endothelial cells by disruption of cytoskeletal microfilaments. FEBS Lett. 403, 191–196.

Holtzclaw, L.A., Pandhit, S., Bare, D.J., Mignery, G.A., Russel, J.T., 2002. Astrocytes in adult rat brain express type 2 inositol 1,4,5-triphosphate receptors. Glia 39, 69–84.

Hoth, M., Penner, R., 1992. Depletion of intracellular calcium stores activates a calcium current in mast cells. Nature 355, 353–356.

Hu, J., Ferreira, A., Van Eldik, L.J., 1997. S100beta induces neuronal cell death through nitric oxide release from astrocytes. J. Neurochem. 69, 2294–2301.

Inesi, G., Sagara, Y., 1994. Specific inhibitors of intracellular Ca^{2+} transport ATPases. J. Membr. Biol. 141, 1–6.

Irvine, R.F., 1990. "Quantal" Ca^{2+} release and the control of Ca^{2+} entry by inositol phosphates—a possible mechanism. FEBS Lett. 263, 5–9.

Kaftan, E.J., Xu, T., Abercrombie, R.F., Hille, B., 2000. Mitochondria shape hormonally induced cytoplasmic calcium oscillations and modulate exocytosis. J. Biol. Chem. 275, 25465–25470.

Kalinich, J.F., Ramakrishnan, R., McClain, D.E., Ramakrishnan, N., 2000. 4-Hydroxynonenal, an end-product of lipid peroxidation, induces apoptosis in human leukemic T- and B-cell lines. Free Radic. Res. 33, 349–358.

Keicher, E., Maggio, K., Hernandez-Nicaise, M.L., Nicaise, G., 1991. The lacunar glial zone at the periphery of *Aplysia* giant neuron: volume of extracellular space and total calcium content of gliagrana. Neuroscience 42, 593–601.

Keicher, E., Maggio, K., Hernandez-Nicaise, M.L., Nicaise, G., 1992. The abundance of *Aplysia* gliagrana depends on Ca 2 and/or Na/concentrations in seawater. Glia 5, 131–138.

Keirstead, S.A., Miller, R.F., 1995. Calcium waves in dissociated retinal glial (Muller) cells are evoked by release of calcium from intracellular stores. Glia 14, 14–22.

Khodakhah, K., Armstrong, C.M., 1997. Induction of long-term depression and rebound potentiation by inositol trisphosphate in cerebellar Purkinje neurons. Proc. Natl Acad. Sci. USA 94, 14009–14014.

Khodakhah, K., Odgen, D., 1993. Functional heterogeneity of calcium release by inositol triphosphate in single Purkinje neurones, cultured cerebellar astrocytes and peripheral tissues. Proc. Natl Acad. Sci. USA 90, 4976–4980.

Kim, C.S., Han, Y.F., Etcheberrigaray, R., Nelson, T.J., Olds, J.L., Yoshioka, T., Alkon, D.L., 1995. Alzheimer and beta-amyloid-treated fibroblasts demonstrate a decrease in a memory-associated GTP-binding protein, Cp20. Proc. Natl Acad. Sci. USA 92, 3060–3064.

Kirischuk, S., Neuhaus, A., Verkhratsky, A., Kettenmann, H., 1995a. Preferential localization of active mitochondria in process tips of immature retinal oligodendrocytes. Neuroreport 6, 737–741.

Kirischuk, S., Scherer, J., Moller, T., Verkhratsky, A., Kettenmann, H., 1995b. Subcellular heterogeneity of voltage-gated Ca^{2+} channels in cells of the oligodendrocyte lineage. Glia 13, 1–12.

Kostyuk, P.G., Kirischuk, S.I., 1993. Spatial heterogeneity of caffeine- and inositol 1,4,5-trisphosphate-induced Ca^{2+} transients in isolated snail neurons. Neuroscience 53, 943–947.

Krause, K.H., Michalak, M., 1997. Calreticulin. Cell 88, 439–443.

Krause, E., Pfeiffer, F., Schmid, A., Schulz, I., 1996. Depletion of intracellular calcium stores activates a calcium conducting nonselective cation current in mouse pancreatic acinar cells. J. Biol. Chem. 271, 32523–32528.

Landis, D.M.D., 1994. The early reactions of non-neuronal cells to brain injury. Annu. Rev. Neurosci. 17, 133–151.

Lascola, C.D., Kraig, R.P., 1996. Whole-cell chloride currents in rat astrocytes accompany changes in cell morphology. J. Neurosci. 16, 2532–2545.

Lee, H.C., Aarhus, R., Graeff, R.M., 1995. Sensitization of calcium-induced calcium release by cyclic ADP-ribose and calmodulin. J. Biol. Chem. 270, 9060–9066.

Leissring, M.A., Yamasaki, T.R., Wasco, W., Buxbaum, J.D., Parker, I., LaFerla, F.M., 2000. Calsenilin reverses presenilin-mediated enhancement of calcium signaling. Proc. Natl Acad. Sci. USA 97, 8590–8593.

Lev Ram, V., Ellisman, M.H., 1995. Axonal activation-induced calcium transients in myelinating Schwann cells, sources, and mechanisms. J. Neurosci. 15, 2628–2637.

Lilliehook, C., Chan, S., Choi, E.K., Zaidi, N.F., Wasco, W., Mattson, M.P., Buxbaum, J.D., 2002. Calsenilin enhances apoptosis by altering endoplasmic reticulum calcium signaling. Mol. Cell. Neurosci. 19, 552–559.

Lindholm, D., Mercer, E.A., Yu, L.Y., Chen, Y., Kukkonen, J., Korhonen, L., Arumae, U., 2002. Neuronal apoptosis inhibitory protein: structural requirements for hippocalcin binding and effects on survival of NGF-dependent sympathetic neurons. Biochim. Biophys. Acta 1600, 138–147.

Liou, J.S., Chen, C.Y., Chen, J.S., Faller, D.V., 2000. Oncogenic ras mediates apoptosis in response to protein kinase C inhibition through the generation of reactive oxygen species. J. Biol. Chem. 275, 39001–39011.

Lo, K.J., Luk, H.N., Chin, T.Y., Chueh, S.H., 2002. Store depletion-induced calcium influx in rat cerebellar astrocytes. Br. J. Pharmacol. 135, 1383–1392.

Lu, T., Xu, Y., Mericle, M.T., Mellgren, R.L., 2002. Participation of the conventional calpains in apoptosis. Biochim. Biophys. Acta 1590, 16–26.

Mackrill, J.J., 1999. Protein–protein interactions in intracellular Ca^{2+}-release channel function. Biochem. J. 337, 345–361.

MacLennan, D.H., Rice, W.J., Green, N.M., 1997. The mechanism of Ca^{2+} transport by sarco(endo)plasmic reticulum Ca^{2+}-ATPases. J. Biol. Chem. 272, 28815–28818.

MacVicar, B.A., 1984. Voltage-dependent calcium channels in glial cells. Science 226, 1345–1347.

Martins-Ferreira, H., Nedergaard, M., Nicholson, C., 2000. Perspectives on spreading depression. Brain Res. Rev. 32, 203–214.

Matyash, M., Matyash, V., Nolte, C., Sorrentino, V., Kettenmann, H., 2002. Requirement of functional ryanodine receptor type 3 for astrocyte migration. FASEB J. 16, 84–86.

McLaughlin, M., Breen, K.C., 1999. Protein kinase C activation potentiates the rapid secretion of the amyloid precursor protein from rat cortical synaptosomes. J. Neurochem. 72, 273–281.

Meissner, G., 1994. Ryanodine receptor/Ca^{2+} release channels and their regulation by endogenous effectors. Annu. Rev. Physiol. 56, 485–508.

Meissner, G., Rios, E., Tripathy, A., Pasek, D.A., 1997. Regulation of skeletal muscle Ca^{2+} release channel (ryanodine receptor) by Ca^{2+} and monovalent cations and anions. J. Biol. Chem. 272, 1628–1638.

Miyakawa, T., Maeda, A., Yamazawa, T., Hirose, K., Kurosaki, T., Lino, M., 1999. Encoding of Ca^{2+}-signals by differential expression of IP_3 receptor subtypes. EMBO J. 18, 1303–1308.

Morohashi, Y., Hatano, N., Ohya, S., Takikawa, R., Watabiki, T., Takasugi, N., Imaizumi, Y., Tomita, T., Iwatsubo, T., 2002. Molecular cloning and characterization of CALP/KChIP4, a novel EF-hand protein interacting with presenilin 2 and voltage-gated potassium channel subunit Kv4. J. Biol. Chem. 277, 14965–14975.

Nelson, T.J., Cavallaro, S., Yi, C.L., McPhie, D., Schreurs, B.G., Gusev, P.A., Favit, A., Zohar, O., Kim, J., Beushausen, S., Ascoli, G., Olds, J., Neve, R., Alkon, D.L., 1996. Calexcitin: a signaling protein that binds calcium and GTP, inhibits potassium channels, and enhances membrane excitability. Proc. Natl Acad. Sci. USA 93, 13808–13813.

Nelson, T.J., Zhao, W.Q., Yuan, S., Favit, A., Pozzo-Miller, L., Alkon, D.L., 1999. Calexcitin interaction with neuronal ryanodine receptors. Biochem. J. 341, 423–433.

Nishizuka, Y., 1984. Turnover of inositol phospholipids and signal transduction. Science 225, 1365–1370.

Nucifora, F.C. Jr., Sharp, A.H., Milgram, S.L., Ross, C.A., 1996. Inositol 1,4,5,-trisphosphate receptors in endocrine cells: localization and association in hetero- and homotetramers. Mol. Biol. Cell 7, 949–960.

Okabe, E., Tsujimoto, Y., Kobayashi, Y., 2000. Calmodulin and cyclic ADP-ribose interaction in Ca^{2+} signaling related to cardiac sarcoplasmic reticulum: superoxide anion radical-triggered Ca^{2+} release. Antioxid. Redox Signal. 2, 47–54.

Opazo, C., Huang, X., Cherny, R.A., Moir, R.D., Roher, A.E., White, A.R., Cappai, R., Masters, C.L., Tanzi, R.E., Inestrosa, N.C., Bush, A.I., 2002. Metalloenzyme-like activity of Alzheimer's disease beta-amyloid.

Cu-dependent catalytic conversion of dopamine, cholesterol, and biological reducing agents to neurotoxic H_2O_2. J. Biol. Chem. 277, 40302–40308.

Pacher, P., Thomas, A.P., Hajnoczky, G., 2002. Ca^{2+} marks: miniature calcium signals in single mitochondria driven by ryanodine receptors. Proc. Natl Acad. Sci. USA 99, 2380–2385.

Parekh, A.B., Penner, R., 1997. Store depletion and calcium influx. Physiol. Rev. 77, 901–930.

Patterson, R.L., Van Rossum, D.B., Gill, D.L., 1999. Store-operated Ca^{2+} entry: evidence for a secretion-like coupling model. Cell 98, 487–499.

Petersen, O.H., 1996. Can Ca^{2+} be released from secretory granules or synaptic vescicles? Trends Neurosci. 19, 411–413.

Petersen, C.C., Berridge, M.J., Borgese, M.F., Bennett, D.L., 1995. Putative capacitative calcium entry channels: expression of *Drosophila* trp and evidence for the existence of vertebrate homologues. Biochem. J. 311, 41–44.

Pfeiffer, D.R., Gunter, T.E., Eliseev, R., Broekemeier, K.M., Gunter, K.K., 2001. Release of Ca^{2+} from mitochondria via the saturable mechanisms and the permeability transition. IUBMB Life 52, 205–212.

Pivovarova, N.B., Hongpaisan, J., Andrews, S.B., Friel, D.D., 1999. Depolarization-induced mitochondrial Ca accumulation in sympathetic neurons: spatial and temporal characteristics. J. Neurosci. 19, 6372–6384.

Pozzan, T., Rizzuto, R., 2000. The renaissance of mitochondrial calcium transport. Eur. J. Biochem. 267, 5269–5273.

Pozzan, T., Rizzuto, R., Meldolesi, J., 1994. Molecular and cellular physiology of intracellular calcium stores. Physiol. Rev. 74, 595–636.

Privat, A., Gimenez-Ribotta, M., Ridet, J.L., 1995. Morphology of astrocytes. In: Kettenmann, H., Ransom, B.R. (Eds.), Neuroglia. Oxford University Press, New York, pp. 3–22.

Putney, J.W. Jr., 1999. "Kissin cousins": intimate plasma membrane-ER interactions underlie capacitative calcium entry. Cell 99, 5–8.

Ramos-Franco, J., Fill, M., Mignery, G.A., 1998. Isoform-specific function of a single inositol 1,4,5,-trisphosphate receptor channels. Biophys. J. 75, 834–839.

Randriamampita, C., Tsien, R.Y., 1993. Emptying of intracellular Ca^{2+} stores releases a novel small messenger that stimulates Ca^{2+} influx. Nature 364, 809–814.

Rizzuto, R., Bernardi, P., Pozzan, T., 2000. Mitochondria as all-round players of the calcium game. J. Physiol. 529, 37–47.

Ross, C.A., Danoff, S.K., Schell, M.J., Snyder, S.H., Ullrich, A., 1992. Three additional inositol 1,4,5-triphosphate receptors: molecular cloning and differential localization in brain and peripheral tissues. Proc. Natl Acad. Sci. USA 89, 4265–4269.

Ross, C.A., Meldolesi, J., Milner, T.A., Satoh, T., Supattapone, S., Snyder, S.H., 1989. Inositol (1,4,5) triphosphate receptor localized to endoplasmic reticulum in cerebellar Purkinjie neurons. Nature 339, 468–470.

Rzigalinski, B.A., Willoughby, K.A., Hoffman, S.W., Falck, J.R., Ellis, E.F., 1999. Calcium influx factor, further evidence it is 5,6-epoxyeicosatrienoic acid. J. Biol. Chem. 274, 175–182.

Scemes, E., 2000. Components of astrocytic intercellular calcium signalling. Mol. Neurobiol. 22, 167–179.

Shao, Y., McCarthy, K.D., 1993. Regulation of astroglial responsiveness to neuroligands in primary culture. Neuroscience. 55, 991–1001.

Shao, Y., McCarthy, K.D., 1995. Receptor-mediated calcium signals in astroglia: multiple receptors, common stores and all or nothing responses. Cell Calcium 17, 187–196.

Sharp, A.H., Nucifora, F.C.J., Bondel, O.J., Sheppard, C.A., Zhang, C., Snyder, S.H., Russell, J.T., Ryugo, D.K., Ross, C.A., 1999. Differential cellular expression of isoforms of inositol 1,4,5-trisphosphate receptors in neurons and glia in brain. J. Comp. Neurol. 406, 207–220.

Shull, G.E., 2000. Gene knockout studies of Ca^{2+}-transporting ATPases. Eur. J. Biochem. 267, 5284–5290.

Simpson, P.B., Armstrong, R.A., 1999. Intracellular signals and cytoskeletal elements involved in olygodendrocyte progenitor migration. Glia 26, 22–35.

Simpson, P.B., Holtzclaw, L.A., Langley, D.B., Russell, J.T., 1998. Characterization of ryanodine receptors in oligodendrocytes, type 2 astrocytes, and O-2A progenitors. J. Neurosci. Res. 52, 468–482.

Simpson, B., Russel, J.T., 1997. Role of sarcoplasmic/endoplasmic-reticulum Ca^{2+}-ATPases in mediating Ca^{2+} waves and local Ca^{2+}-release microdomains in cultured glia. Biochem. J. 325, 239–247.

Smith, K.J., Felts, P.A., John, G.R., 2000. Effects of 4-aminopyridine on demyelinated axons, synapses and muscle tension. Brain 123, 171–184.

Sutko, J.L., Airey, J.A., 1996. Ryanodine receptor Ca^{2+} release channels: does diversity in form equal diversity in function? Physiol. Rev. 76, 1027–1071.

Tanaka, Y., Tashjian, A.H., 1993. Functional identification and quantitation of three intracellular calcium pools in GH_4C_1 cells: evidence that the caffeine-responsive pool is coupled to a thapsigargin-resistant, ATP-dependent process. Biochemistry 32, 12062–12073.

Terasaki, M., Slater, N.T., Fain, A., Schmidek, A., Reese, R.S., 1994. Continuous network of endoplasmic reticulum in cerebellar purkinje neurons. Proc. Natl Acad. Sci. USA 91, 7510–7514.

Thayer, S.A., Usachev, Y.M., Pottorf, W.J., 2002. Modulating Ca^{2+} clearance from neurons. Front. Biosci. 7, 1255–1279.

Toescu, E.C., Moller, T., Kettenmann, H., Verkhratsky, A., 1998. Long-term activation of capacitative Ca^{2+} entry in mouse microglial cells. Neuroscience 86, 925–935.

Tombal, B., Denmeade, S.R., Gillis, J.M., Isaacs, J.T., 2002. A supramicromolar elevation of intracellular free calcium $[Ca^{2+}]_i$ is consistently required to induce the execution phase of apoptosis. Cell Death Differ. 9, 561–573.

Tripathy, A., Xu, L., Mann, G., Meissner, G., 1995. Calmodulin activation and inhibition of skeletal muscle Ca^{2+} release channel (ryanodine receptor). Biophys. J. 69, 106–119.

Verkhratsky, A., Kettenmann, H., 1995. Calcium signalling in glial cells. Trends Neurosci. 19, 346–352.

Verkhratsky, A., Orkand, R.K., Kettenmann, H., 1998. Glial calcium: homeostasis and signaling function. Physiol. Rev. 78, 99–141.

Verkhratsky, A., Shmigol, A., 1996. Calcium-induced calcium release in neurones. Cell Calcium 19, 1–14.

Wernyj, R.P., Mattson, M.P., Christakos, S., 1999. Expression of calbindin-D28k in C6 glial cells stabilizes intracellular calcium levels and protects against apoptosis induced by calcium ionophore and amyloid beta-peptide. Mol. Brain. Res. 64, 69–79.

Yamamoto-Hino, M., Miyawaki, A., Kawano, H., Sugiyama, T., Furuichi, T., Hasegawa, M., Mikoshiba, K., 1995. Immunohistochemical study of inositol 1,4,5-trisphosphate receptor type 3 in rat central nervous system. Neuroreport 6, 273–276.

Yao, Y., Ferrer-Montiel, A.V., Montal, M., Tsien, R.Y., 1999. Activation of store-operated Ca^{2+} current in *Xenopus* oocytes requires SNAP-25 but not a diffusible messenger. Cell 98, 475–485.

Advances in Molecular and Cell Biology

Decoding calcium wave signaling

A. H. Cornell-Bell,* P. Jung and V. Trinkaus-Randall

Anscans Ivoryton, CT,
Ohio University Athens, O,
Boston University School of Medicine Boston, MA
**Correspondence address: E-mail: ahcb2002@yahoo.com*

Contents

1. Introduction
2. Glutamate application induces a receptor-mediated response
3. Mediators of intercellular astrocyte calcium wave propagation
   3.1. Gap junction blockade can inhibit calcium wave propagation
4. Extracellular ATP as mediators of calcium wave propagation
   4.1. Evidence of ATP-mediated propagation
   4.2. Purinergic receptors
   4.3. ATP release from astrocytes
5. Nitric oxide and PKG as mediators of calcium wave propagation
6. Role of endoplasmic reticulum and mitochondria in Ca^{2+} signaling
   6.1. Endoplasmic reticulum
   6.2. Mitochondria
7. Glutamate is involved in neuron-to-astrocyte signaling, but not in astrocytic propagation of calcium waves
   7.1. Effects of transmitter glutamate on astrocytes
   7.2. Glutamate release does not mediate interastrocytic calcium waves
   7.3. Glutamate as the mediator of astrocyte-to-neuron signaling
   7.4. Astrocytes can modulate neuronal responses and synaptic transmission
8. Pathology and Ca^{2+} waves
9. Concluding remarks

Intercellular calcium waves in astrocytes represent a phenomenon whereby a wave of increases in free cytosolic calcium concentration ($[Ca^{2+}]_i$) spreads from an initially stimulated cell across an astrocytic syncytium. Originally it was believed that the mechanism of the spread was transport of the second messenger inositol trisphosphate (IP_3) from cell to cell through connexin-mediated gap junctions, followed by an IP_3-mediated release of calcium from intracellular stores. Although such a mechanism might participate in some types of calcium waves, calcium wave propagation is not dependent

Advances in Molecular and Cell Biology, Vol. 31, pages 661–687

ISBN: 0-444-51451-1

upon this mechanism. Rather it depends in most cases upon an autocatalytic release of Ca^{2+} driven by release of ATP (which is partly Ca^{2+}-dependent and partly Ca^{2+} independent, but may be hemiconnexin-dependent) and subsequent autocrine/paracrine stimulation of purinergic P2 receptors, linked to release of Ca^{2+} from intracellular stores. In addition, NO-mediated signaling mechanisms are involved, but only in some types of calcium waves, especially those triggered by mechanical stimulation. Neuronal activity can induce astrocytic calcium waves by stimulation of metabotropic glutamate receptors on astrocytes. In turn, glutamate release from astrocytes sustaining a calcium wave is capable of triggering neuronal activity.

1. Introduction

Three key factors merged to trigger the discovery of calcium waves in the astrocyte syncytium. The first was finding that astrocytes possessed a full array of neurotransmitter receptors, which operated on a millisecond time frame similar to that expected by neurons (Bowman and Kimelberg, 1984; Sontheimer et al., 1988; Backus et al., 1989; Usowicz et al., 1989; reviewed by Cornell-Bell and Finkbeiner, 1991). The activation of these receptors resulted in a Ca^{2+} influx into astrocytes or intracellular calcium release with subsequent mobilization (Enkvist et al., 1989; Glaum et al., 1990; Inagaki et al., 1991). The second contribution was the many technological advances made in time-lapse video microscopy (Allen et al., 1981; Inoue, 1981, 1986), particularly involving the newly marketed confocal scanning laser microscope, which allowed longer recording periods free of phototoxicity (Brakenhoff et al., 1985; Amos et al., 1987; Inoue, 1995; Pawley, 1995). Lastly, most critical was the development of Ca^{2+} sensitive fluorescent probes that were ion-specific, which allowed precise quantification of changes in intracellular Ca^{2+} (Grynkiewicz et al., 1985; Tsien, 1988; Lipscombe et al., 1988; Kao et al., 1989; Harootunian et al., 1991).

Early imaging studies of calcium activity in astrocytes identified intracellular oscillations and intercellular Ca^{2+} waves as key elements in a complicated non-neuronal signaling system (Berridge and Gallione, 1988; Cornell-Bell et al., 1990; Jensen and Chiu, 1990; Charles et al., 1991; Teichberg, 1991). Compelling time-lapse sequences of astrocyte cultures responding to neurotransmitters stimulated speculation about an astrocytic long-range signaling system, which was postulated to be frequency encoded in the form of oscillations. Intracellular Ca^{2+} responses were thought to possibly synchronize neighboring astrocytes via a unified, long-lasting elevation of free cytosolic Ca^{2+} concentration ($[Ca^{2+}]_i$) that occurred during propagation of the long-distance waves over a region of cells. Oscillating cells were seen to dampen as the wave arrived, but then oscillations re-started following the passage of the wave. Initial experiments that defined the Ca^{2+} wave system recognized that either neurotransmitter receptor activation or mechanical stimulation of a single cell activated at least two separate Ca^{2+} elevation mechanisms (Cornell-Bell et al., 1990; Charles et al., 1991; Finkbeiner, 1993; Kim et al., 1994). These two distinct mechanisms were (i) Ca^{2+} entry from outside the cell, which was characteristic for kainate-induced Ca^{2+} waves (Kim et al., 1994); and (ii) IP_3-mediated release of Ca^{2+} from intracellular stores initiated by a mechanical stimulus or metabotropic glutamate receptor stimulation (Charles et al., 1991; Cornell-Bell and Finkbeiner, 1991; McCarthy and Salm, 1991; Venance et al., 1997). During the past decade, a large body of exciting work has unraveled second messenger systems involved in Ca^{2+} signaling processes. An understanding is emerging of the mechanism for entry of Ca^{2+} in response to glutamate and ATP, and aspects of what the signaling process actually may encode have emerged. Several laboratories have also investigated what happens when astrocytes are players in a pathological condition such as epilepsy. Different cell types have different signal initiating molecules, and different tissues have different requirements and mechanisms for eliciting Ca^{2+} waves. In this review we will touch on aspects of each of these topics during the process of dissecting what is currently known about the manner(s) in which glial cells and the neurons they face encode and respond to Ca^{2+} signals, which regulate responses in the nervous system.

2. Glutamate application induces a receptor-mediated response

Astrocytes expressing specific receptor systems linked to calcium regulation remained stable in vitro for several weeks even in the absence of neurons (McCarthy and Salm, 1991). One such receptor is glutamate. Selective agonists pharmacologically defined the glutamate subtypes on glial cells as kainate and quisqualate-preferring receptors, but in general not NMDA preferring receptors (Sontheimer et al., 1988; Usowicz et al., 1989; Backus et al., 1989; Cornell-Bell et al., 1990; Ahmed et al., 1990; Jensen and Chiu, 1990; Glaum et al., 1990; Cornell-Bell and Finkbeiner, 1991). Astrocytes produced Ca^{2+} spikes at glutamate concentrations of 100 nM or higher, with 50% of the cells in a culture responding at 300 nM (Cornell-Bell et al., 1990). Glutamate concentrations in the superfusate influenced whether oscillations occurred, the frequency of oscillations and whether both intracellular and intercellular waves were observed (Cornell-Bell et al., 1990). The average oscillation period decreased with increasing agonist concentrations. Coupling between oscillations in neighboring cells was not demonstrated, and oscillations within individual astrocytes showed either constant or monotonically decreasing frequency (Cornell-Bell and Finkbeiner, 1991; Finkbeiner, 1993—see also chapter by Shuai et al.).

In addition, the pattern of intracellular Ca^{2+} signaling was dependent upon the concentration of extracellular glutamate. For example, at concentrations of glutamate less than 1 μM small regions of the astrocyte cytoplasm flickered asynchronously, with propagation of the intracellular Ca^{2+} wave occurring in contained areas of the cytoplasm. If the extracellular glutamate concentrations were increased to between 1 and 10 μM, propagation of Ca^{2+} in waves was more common, and these waves traveled throughout the cell, often propagating into neighboring cells (Cornell-Bell et al., 1990). Analysis of single astrocytes showed that in more than 80% of the cells studied a gradient of intracellular calcium existed, and the locus from which the wave originated corresponded to the region of highest cytoplasmic Ca^{2+} concentration (Yagodin et al., 1994).

The locus of origin and the regions of calcium spikes with the highest amplitude remained invariant during the propagation of successive intracellular calcium waves

(Yagodin et al., 1994). Calcium waves were shown to be initiated at discrete regions of the cells where the calcium concentrations were highest, and they were propagated in a saltatory manner through the cytoplasm to other loci, where the rising calcium from the approaching wave front provoked a large Ca^{2+} increase. Models of intracellular wave propagation indicated that the propagation was dependent upon a Ca^{2+}-sensitive autocatalytic step (Berridge, 1993; Meyer and Stryer, 1991). Glutamate was shown to stimulate the release of Ca^{2+} from intracellular stores in astrocytes (Ahmed et al., 1990). One possible mechanism was that an increase in intracellular Ca^{2+} causes an increase in phospholipase C with resulting increase in IP_3 and Ca^{2+} (Monaghan et al., 1989; Lechleiter et al., 1991; Pearce et al., 1986). Another idea suggested that the calcium ion itself activates further release of calcium from stores by a calcium-induced calcium release (Backx et al., 1989; Charles et al., 1993—see also chapter by Scapagnini et al.).

3. Mediators of intercellular astrocyte calcium wave propagation

3.1. Gap junction blockade can inhibit calcium wave propagation

3.1.1. Gap junction-mediated propagation

Early applications of calcium imaging of cultured astrocytes revealed a striking new phenomenon: a cytosolic Ca^{2+} signal spreading from cell to cell and called a calcium wave (Cornell-Bell et al., 1990; Charles et al., 1991; Smith, 1992). Gap junction channels were identified as an important mediator of intercellular calcium waves (Bennet et al., 1991), since they provide a direct route between cells. Gap junctions are also permeable to calcium and IP_3 (Saez et al., 1983). Uncoupling gap junctions resulted in inhibition of the intercellular calcium wave (Anders, 1988; Dermietzel et al., 1991; Finkbeiner, 1992; Enkvist and McCarthy, 1992; Christ et al., 1992; Venance et al., 1995). When octanol and halothane were used to uncouple gap junctions in hippocampal astrocytes the incidence of Ca^{2+} waves propagating from cell to cell dropped from 50% occurrence to 0 and 5%, respectively (Finkbeiner, 1992). In contrast, the uncoupling did not dramatically affect the pattern or duration of intracellular calcium oscillations or intracellular waves (Finkbeiner, 1992). Normally, C6 glioma cells show no coupling to their neighbors. When C6 glioma cells were transfected with the gap junction protein, connexin43, calcium waves passed from one cell to the next, suggesting that the gap junction was a crucial functional component of the propagating calcium wave (Charles et al., 1992). Despite this concrete evidence of the involvement of gap junctions in wave regulation there were calcium waves propagating in a normal fashion in connexin43 (Cx43) knock-out (KO) mice (Scemes et al., 1998, 2000). Comparison of the strength of coupling between pairs of wildtype (WT) and Cx43 KO spinal cord astrocytes indicated that two-thirds of total coupling is attributable to channels formed by Cx43, with other connexins contributing the remaining one-third of junctional conductance (Scemes et al., 2000—see also chapter by Scemes and Spray). The brains of connexin43 null [Cx43 (−/−)] animals were shown to be macroscopically normal and to display patterns of cortical lamination remarkably similar to WT siblings (Dermietzel et al., 2000). Additionally the presence of Cx40 and Cx45 in brains and astrocytes cultured from both Cx43 (−/−) mice and WT littermates was confirmed. Cx30 mRNA was detected in long term (2 weeks) but not in fresh cultures of astrocytes. There is also the possibility that Cx26 may be present. These studies reveal that astrocyte gap junctions may be formed of multiple connexins (Dermietzel et al., 2000) providing the exciting possibility that specific roles for each of these proteins may exist in the regulation of calcium waves. It has recently been demonstrated in C6 glioma cells transfected with Cx32 or Cx43 that AMP and ADP and, especially, ATP pass much better through channels formed by Cx43, whereas the opposite applies to adenosine (Goldberg et al., 2000).

3.1.2. Modulation of gap junction-mediated propagation

Functional studies in brain slices and cultures have implicated neurotransmitters, cytokines, growth factors and other bioactive compounds in control of gap junctional coupling (Giaume and McCarthy, 1996; Spray et al., 1999). It has recently been demonstrated in various models of astrocyte and astrocyte/neuron cultures that gap junctional communication and expression of Cx43 in astrocytes can be up-regulated, and that the presence of neurons in the cultures enhances gap junction-mediated communication, Cx43 expression and the extent of calcium waves in astrocytes (Rouach et al., 2000). Interestingly, it was demonstrated in these experiments that plating of microglial cells on the astrocyte layer led to a decrease in gap junctional permeability and connexin43 expression, suggesting that several types of cell–cell interactions need to be considered in local control of gap junctions and their contribution to the propagating Ca^{2+} wave. Prolonged treatments (24–72 h) of striatal co-cultures of neurons and astrocytes with tetrodotoxin, picrotoxin, bicuculline and CNQX, an inhibitor of quisqualate-preferring glutamate receptors (Honore and Drejer, 1988; Yamada and Huzel, 1989), resulted in a significant reduction in gap junctional coupling without affecting the expression of connexins as determined by Western blot analysis (Rouach et al., 2000). It thus appears that neuron-induced up-regulation of astrocyte gap junctional coupling is dependent upon synaptic activity in striatal neurons, and it is also related to density and age of cultures (Rouach et al., 2000).

Inflammatory cytokines, including IL-1 β and TNF-α, have been shown to down-regulate gap junctional connectivity (John et al., 1999; Brosnan et al., 2001). Gap junctions form a molecular link for coordinated long-distance signaling via calcium waves, but there is also an important ATP-mediated extracellular pathway that can be influenced by cytokines (John et al., 1999), as will be described below. Communication in the astrocyte syncytium is sustained by a finely tuned interaction between gap junction-dependent and gap junction-independent mechanisms. When gap junctional coupling was reduced as in Cx43 knock out mice, an increase in the extracellular paracrine/autocrine communication dominated (John et al., 1999; Scemes et al., 2000; John et al., 2000). This interplay between the gap junctions and paracrine/autocrine signaling provides a high degree of plasticity for intercellular communication between astrocytes (Brosnan et al., 2001). The integral membrane components of gap junctions appear to be linked into a macromolecular complex that has been called the Nexus (Spray et al., 1999). Nexus components may vary such that the binding affinities within a Nexus containing an individual connexin may be altered by micro-environmental factors such as cytosolic pH, phosphorylation and changes in binding with other components. For the Cx43 Nexus, binding sites include src homology

(SH), PSD and zonula occludens (PDZ) binding domains, and the Cx43 molecule. Interactions of Cx43 with proteins containing PDZ, SH2 and SH3 domains is hypothesized to serve as the scaffold on which to assemble components of the intercellular signaling pathway into the multiprotein Nexus complex, which couples their activity to downstream signaling molecules (Brosnan et al., 2001). In the case of Cx43, a proline-rich region of the carboxyl terminus comprising amino acids 273–285 may bind to the SH3 domain, and a phosphorylated tyrosine at position 265 has been shown to interact with the SH2/SH3 domain proteins v-src and c-src (Swenson et al., 1990). Phosphorylation of Cx43 by v- and c-src is involved in decreased gap junctional conductance (Filson et al., 1990; Brosnan et al., 2001). Studies, such as these, place the Nexus into the framework of second messenger signaling cascades augmenting the importance of these proteins from a conduit between cells to an active regulatory mechanism.

4. Extracellular ATP as mediators of calcium wave propagation

4.1. Evidence of ATP-mediated propagation

Ca^{2+} waves in astrocytes can proceed by an extracellular pathway, as evidenced by the continued propagation of waves between astrocytes in culture even when cells are not in direct contact (Hassinger et al., 1996; Guthrie et al., 1999). Calcium waves cross cell-free gaps, and they are affected by the direction and strength of a perfusion bath (Hassinger et al., 1996). The most compelling evidence for a released extracellular factor is that subsequent calcium waves are produced if naïve cells are exposed to the medium collected from around cells that previously supported a propagating wave (Guthrie et al., 1999; Klepeis et al., 2001). Also, if gap junctional coupling is inhibited, propagation of Ca^{2+} waves is in several situations unaffected (Naus et al., 1997; Guan et al., 1997; John et al., 1999), at least partly dependent upon the concentration of the inhibitor (Cotrina et al., 1998a,b). Thus, when astrocytes were uncoupled from Muller cells, using octanol, the propagation of Ca^{2+} waves were maintained, highlighting the importance of the extracellular messenger in wave propagation between these two cell types (Zahs and Newman, 1997).

ATP is emerging as the extracellular signaling element that drives astrocytic Ca^{2+} waves. This applies not only to conditions when the calcium wave crosses an extracellular gap, but also to wave propagation between adjoining cells. Imaging techniques have detected ATP in the saline incubation medium surrounding astrocytes participating in a calcium wave (Guthrie et al., 1999; Stout et al., 2002; Coco et al., 2003). Using a combination of chemiluminescence and Fluo3AM calcium imaging at millisecond temporal resolution it was possible to show that extracellular ATP mediated intercellular calcium wave propagation (Wang et al., 2000).

4.2. Purinergic receptors

ATP acts on P2 receptors, whereas P1 receptors are adenosine receptors (see chapter by Hansson and Rönnbäck). Multiple subtypes of P2-purinergic receptors exist, as was first proposed by Burnstock (Burnstock and Kennedy, 1985; reviewed by Burnstock, 1997).

P2Y receptors function as G-protein coupled Ca^{2+}-mobilizing ATP receptors, operating via stimulation of phospholipase C and formation of inositol trisphosphate (IP_3) and diacylglycerol (DAG), the former of which causes release of Ca^{2+} from intracellular stores on the endoplasmic reticulum (see chapter by Scapagnini et al.); P2X-type receptors act as ligand-gated ion channels; and P2-Z receptors are associated with ATP-induced pore formation (Dubyak and el-Moatassim, 1993). Astrocytes express two subtypes of P2Y receptors, the P2Y1 and the P2Y2 receptor (Zhou and Kimelberg, 2001).

It is in support of the concept that ATP is the extracellular messenger for calcium wave propagation that P2 receptor antagonists inhibit Ca^{2+} wave propagation (Guan et al., 1997; Cotrina et al., 1998a,b; Guthrie et al., 1999; Fam et al., 2000). Also, suramin and PPADS, P2 receptor antagonists both inhibited Ca^{2+} wave propagation into Müller cells and astrocytes (Newman, 2001; Zanotti and Charles, 1997; Guan et al., 1997). In addition, wave propagation is reduced if ATP hydrolysis is enhanced with apyrase in cortical or retinal preparations (Guthrie et al., 1999; Cotrina et al., 1998a,b, 2000; Newman, 2001), but not in striatal preparations (Venance et al., 1997). Application of ATP initiated astrocyte calcium waves in cultures from suprachiasmatic nucleus cultures (van den Pol et al., 1992) and retinal cells (Newman and Zahs, 1997). Stimulation of the P2Y receptors on astrocytes was also shown by Kastritsis et al. (1992) to result in inositol phosphate formation and calcium mobilization. Using ^{45}Ca in the presence or absence of $LaCl_3$, an inhibitor of Ca^{2+} channels, Neary et al. (1998) showed, however, that ATP also stimulates the uptake of extracellular Ca^{2+} into cultured astrocytes. ATP-stimulated Ca^{2+} entry might be secondary to stimulation of a P2X receptor, as is the case in microglia (Verderio and Matteoli, 2001) and subsequent opening of voltage-sensitive Ca^{2+} channels and/or non-specific cation channels (Sun et al., 1999). It is in agreement with this suggestion that an ATP-induced increase in glial $[Ca^{2+}]_i$ in the acutely isolated rat optic nerve can be evoked not only by a P2Y-selective agonist but also by a P2X-selective agonist (James and Butt, 2001).

Calcium waves in other cell types are also propagated using ATP as the extracellular messenger (Osipchuk and Cahalan, 1992; Dubyak and el-Moatassim, 1993; Schlosser et al., 1996; Jorgensen et al., 1997; Frame and deFeijter, 1997; Schneider et al., 1994; Klepeis et al., 2001). On the other hand, activation of P2Y receptors on astrocytes have additional effects besides the ability to elicit calcium waves. Thus, P2Y receptors are coupled to the MAP kinases ERK by pathways which are distinct from that leading to an increase in $[Ca^{2+}]_i$ (Neary et al., 1999), a finding which explains how some astrocytes when exposed to ATP respond with increased intracellular Ca^{2+} and others respond trophically.

4.3. ATP release from astrocytes

As mentioned above, ATP is released from astrocytes during propagation of calcium waves (Guthrie et al., 1999; Wang et al., 2000; Stout et al., 2002; Coco et al., 2003). The ability of astrocytes to both release ATP and respond to ATP suggests that ATP may act as an autocrine or paracrine messenger between these glial cells (Quieroz et al., 1999). Two potential pathways exist for exit of ATP from the cell: (i) passage through anion channels formed by connexin hemichannels (Cotrina et al., 1998a,b; Stout et al., 2002); and (ii) Ca^{2+}-dependent and quantal release of vesicularly stored ATP (Coco et al., 2003).

If ATP release from astrocytes was a regenerative process, such that the stimulated cell as well as all subsequent cells rapidly released high concentrations of ATP an infinitely propagating wave would result (Guthrie et al., 1999). This is not the case, since propagating calcium waves only travel through approximately 20 cell diameters. Since not all astrocytes respond to ATP with the same dose dependence (Guthrie et al., 1999) or signaling pathways (Neary et al., 1999), and perhaps not all astrocytes release ATP, calcium waves are likely to be limited by extent (Guthrie et al., 1999). Although the study by Wang et al. (2000) showed that extracellular ATP mediated intercellular calcium wave propagation, the release and propagation of ATP was not calcium dependent. Use of the phospholipase C-inhibitor U-73122 blocked the ATP-stimulated Ca^{2+} wave, while BAPTA and thapsigargin (inhibitors of intracellular Ca^{2+} release—see chapter by Scapagnini et al.) did not, which implies that products of phospholipase C activity, IP_3 or DAG, are directly involved with ATP wave regulation (Wang et al., 2000). This suggestion is supported by observations that the use of caged-IP_3 followed by flash photolysis produces an ATP-mediated Ca^{2+} wave, whereas flash release of caged Ca^{2+} did not (Leybaert et al., 1998).

4.3.1. Release of cytosolic ATP

Ectopic expression of connexins was shown to cause an increase in ATP release with an increase in the radius of the propagating calcium wave (Cotrina et al., 2000). An intact cytoskeleton was needed for Ca^{2+} wave generation, since cytochalasin D reduced calcium signaling significantly (Cotrina et al., 1998b, 2000). Transjunctional currents between C6 and Cx43 cells were not altered by the inhibitor, confirming that gap junctions remained open (Cotrina et al., 1998a,b). The reason for the dependence on the cytoskeleton is unknown but an intact actin cytoskeleton is also necessary for calcium-dependent secretion in neurons and secretory cells (for review, see Trifaro and Vitale, 1993) and evidence is found that vesicular secretion of ATP contributes to ATP release (see below). Moreover, disruption of the actin cytoskeleton in cultured rat astrocytes suppresses ATP-induced calcium oscillations by reducing capacitative Ca^{2+} re-entry and store refilling (Sergeeva et al., 2000; see also chapter by Scapagnini et al.). Inhibition of the myosin light chain kinase also resulted in a suppressed Ca^{2+} signaling without effecting coupling or calcium mobilization (Cotrina et al., 1998b).

It is likely that connexin proteins supported purinergic-mediated Ca^{2+} signaling, not as a substrate for gap junctions, but rather as a facilitator of ATP release. This release may occur through ATP anion flow through connexin hemichannels, and it is enhanced by the removal of extracellular Ca^{2+} (Cotrina et al., 1998a). Cx-deficient cells can receive but not propagate Ca^{2+} signals (Cotrina et al., 2000). Recent experiments using real-time bioluminescence imaging of C6 glioma cells have demonstrated that the release of ATP is not uniform across a field, but is restricted to short bursts (Arcuino et al., 2002). The ATP bursts emanate from single cells with transient openings of non-selective channels resulting in a change in membrane permeability (Arcuino et al., 2002). Earlier studies of cystic-fibrosis airway epithelial cells (CFTR), using an atomic force microscope, showed that each affected cell studied contained distinct point sources that were dispersed across the epithelium of the cell surface in a random pattern (Schneider et al., 1994). The ATP that is released by the cell from a point source diffuses into the adjacent extracellular medium, or it is hydrolyzed by ecto-ATPases. More ATP molecules are in close proximity to the point source than at a distant location, with a concentration gradient developing as a function of the distance from the point source (Schneider et al., 1994). Increases in intracellular Ca^{2+} alone were not sufficient to initiate the changes in ATP membrane permeability (Arcuino et al., 2002). C6–Cx43 cells displayed repeated bursts of ATP release equivalent to the bursts seen for astrocytes. Mock transfected cells with null Cx showed no ATP bursts, further supporting that Cx expression mediates the release of ATP (Arcuino et al., 2002). It is tempting to conclude that ATP is released through functional hemichannels from the cell interior to its outside, secreting cells do express connexins, and in nearly all secretory epithelia intracellular, Ca^{2+}-activated ion conductances are stimulated by luminal ATP (and other nucleotides) to induce secretion of Cl−, K^+ or HCO_3^- (reviewed by Leipziger, 2003). ATP is highly concentrated in cytoplasm (approximately 2 mM) and in cultured astrocytes (Matz and Hertz, 1989), and efflux of ATP from a single cell potentially can activate several hundred neighboring cells (Arcuino et al., 2002). However, over-expression of Cx43 in C6–Cx43 cells is also associated with altered expression of other genes and with changes in volume regulation (Naus et al., 2000; Quist et al., 2000).

4.3.2. Release of vesicular ATP

Proteins generally involved in regulated exocytosis are expressed in cultured astrocytes (Madison et al., 1996), and they may provide a link between an extracellular component of calcium wave activity and the cytoskeleton (Cotrina et al., 1998b). It is in support of this concept that a recent study by Coco et al. (2003) has convincingly demonstrated vesicular release of ATP from primary cultures of hippocampal rat astrocytes. They reported that besides the described release of cytosolic calcium, which is enhanced during Ca^{2+} depletion, there is a gap junction-independent, Ca^{2+}-dependent release of ATP. This release is inhibited by bafilomycin, an inhibitor of ATP uptake into secretory granules, a process requiring an electrochemical proton gradient, maintained by a v-ATPase (for discussion of the v-ATPase, see chapter by Bevensee and McAlear). The vesicles containing ATP were functionally different from those releasing glutamate, as indicated by different inhibitor sensitivity, and the observation that activation of metabotropic glutamate receptors, which strongly evokes glutamate release, had only little effect on release of ATP.

5. Nitric oxide and PKG as mediators of calcium wave propagation

Additional signaling pathways mediate intercellular Ca^{2+} waves as indicated by the recent discovery that the nitric oxide/protein kinase G (PKG) pathway (see chapter by Garcia and Baltrons) has a fundamental role in regulating the initiation and propagation of Ca^{2+} waves following mechanical stimulation (Willmott et al., 2000). Molsidomine, a nitric oxide (NO) donor as well as an aqueous puff of NO itself produced an increase of $[Ca^{2+}]_i$ and propagating intercellular Ca^{2+} waves in primary mixed forebrain cultures.

Entry of extracellular Ca^{2+} was indicated by the observation that incubation in Ca^{2+}-free medium or application of an inhibitor of voltage-dependent Ca^{2+} channels reduced the Ca^{2+} response. NO is also an inducer of Ca^{2+} mobilization in several other cell types (Willmott et al., 2000), where Ca^{2+} was mobilized either through cGMP-dependent protein kinase-coupled activation of ADP-ribosyl cyclase, which produced ADP-ribose (Wilmott et al., 1996; Clementi et al., 1996), or via a direct nitrosylation of regulatory thiol groups of the ryanodine receptors (Stoyanovsky et al., 1997). Synthesis of NO by constitutive nitric oxide synthase is Ca^{2+}-dependent. It was hypothesized that a rise in intracellular Ca^{2+} would serve to amplify NO production, thus enhancing NO diffusion, which could contribute to propagation of intracellular and intercellular Ca^{2+} waves (Willmott et al., 2000). A small response was seen in Ca^{2+} free saline and no response was seen when just saline was puffed onto cells. Pretreating cells with an NO scavenger (PTIO) arrested the increase in Ca^{2+}, showing there was a direct relationship between the increase in Ca^{2+} and in NO. When NO was puffed onto cells treated with ryanodine no resulting $[Ca^{2+}]_i$ response was seen, indicating that Ca^{2+} is mobilized from intracellular stores (see chapter by Scapalgini et al.). Pretreating cells with a guanylate cyclase inhibitor abolished the NO-induced Ca^{2+} rise, whereas the PKG inhibitor Rp-8-pCPT-cGMP reduced the NO effect. Thus NO mobilizes Ca^{2+} from a ryanodine receptor-linked store through the cGMP-PKG signaling pathway in mechanical waves (Willmott et al., 2000). Studies of mechanically induced calcium waves indicated that they shared most essential features with those induced by NO, and an increase in intracellular NO could, indeed, be demonstrated in the calcium waves elicited by mechanical stimulation (Willmott et al., 2000).

In contrast, the increase in $[Ca^{2+}]_i$ induced by a bolus of ATP is not mediated via an increased production of NO in glial cells, and it is not dependent upon ryanodine receptor-linked Ca^{2+} release (Willmott et al., 2000). This is in direct contrast to mechanical wave studies (Simard et al., 1999; Guthrie et al., 1999). There was a significant inhibition, but no elimination, of the ATP-induced Ca^{2+} response after treatment with the phospholipase C inhibitor U 73122, suggesting that this response is dependent upon intracellular IP_3 generation and is independent of NO or ryanodine receptor-linked Ca^{2+} release (Centemeri et al., 1997; Willmott et al., 2000). It had previously been shown (Venance et al., 1997; Charles et al., 1993) that U 73122 alone was sufficient to completely eliminate the mechanically induced Ca^{2+} wave. These studies suggest there may be yet again another level of control of the glial Ca^{2+} signaling system, since the response to ATP presentation by direct application is driven by an IP_3-mediated receptor instead of the ryanodine receptor-linked signaling seen in mechanical waves. Evidence using the NO-specific fluoroprobe DAF-2 further confirmed that an ATP-induced Ca^{2+} rise was not dependent on an NO increase (Willmott et al., 2000).

6. Role of endoplasmic reticulum and mitochondria in Ca^{2+} signaling

6.1. Endoplasmic reticulum

Initial studies showed that norepinephrine-induced Ca^{2+} waves in cortical astrocyte cultures began at discrete initiation loci and from there propagated throughout the cytoplasm in a regenerative manner involving Ca^{2+} release sites (Sheppard et al., 1997).

A specific antibody to the type-2 IP_3 receptor subtype and specific staining of the endoplasmic reticulum showed domains of elevated Ca^{2+} response with kinetics defined as high amplitude and rapid rise rate, significantly correlated with high local intense staining of the IP_3 receptors (Holtzclaw et al., 2002). Further characterization of the sarcoplasmic–endoplasmic reticulum Ca^{2+} ATPase (SERCA) of cortical astrocytes and oligodendrocytes showed a slow onset Ca^{2+} response when SERCA was inhibited (Simpson and Russell, 1997). IP_3 receptors and SERCAs sensitive to thapsigargin and CPA (cyclopiazonic acid) are associated with a single Ca^{2+} store (see chapter by Scapagnini et al.). Inhibition of SERCA activates both Ca^{2+} release as a wave front and Ca^{2+} entry via store-operated channels (Simpson and Russell, 1997).

6.2. Mitochondria

Mitochondrial buffering of cytoplasm regulates the spread of astrocytic Ca^{2+} waves. Mitochondrial control of calcium uptake and release from cytoplasm has direct consequences on neuronal and glial Ca^{2+} responses (Simpson and Russell, 1996, 1998b). In oligodendrocytes, IP_3-dependent release of Ca^{2+} resulted in an elevated Ca^{2+} signal in the mitochondria, which modified cytosolic Ca^{2+} wave propagation (Simpson et al., 1997; Simpson and Russell, 1998a,b). Ca^{2+} release from stores regularly alternated with sites of removal from the cytosol (Laskey et al., 1998). Multiple mechanisms of Ca^{2+} removal from the cytosol contribute to the negative flux sites. Type-2 IP_3 receptors and calreticulin (a calcium binding protein) are expressed with high intensity at Ca^{2+} wave amplification sites along oligodendrocyte processes, when compared to other cell regions (Simpson et al., 1997). Stationary mitochondria were found at these specialized release sites in close association with high density ER proteins. These findings imply that the propagating Ca^{2+} wave can be modulated by special Ca^{2+}-release domains, involving both ER proteins and mitochondria (Simpson et al., 1997). The permeability transition pore forms the major Ca^{2+} efflux pathway from the mitochondria (Smaili et al., 2001). Ca^{2+} efflux from the mitochondrial matrix involves reversal of the uniporter and the inner membrane Na^+/Ca^{2+} exchanger. When mitochondrial function was blocked, there was a measurable decrease in wave speed and puff probability (Haak et al., 2000). An inhibitor of the electron transport chain (antimycin A) increased cytosolic Ca^{2+}, reduced agonist-evoked IP_3 production and enhanced PIP_2 binding to the Ca^{2+}-dependent protein, gelsolin (Haak et al., 2000). ATP application to cortical astrocytes mobilized intracellular Ca^{2+} stores with an increase in Ca^{2+} over the nucleus as well as cytosol, followed by a delayed increase in mitochondrial Ca^{2+}, that remained elevated for long periods after nuclear Ca^{2+} decrease (Boitier et al., 1999). The rise in the mitochondria appeared first at one end of the cell and then progressed in a wave-like pattern to the other end (Boitier et al., 1999). Using a mitochondrial uncoupling agent, FCCP, in conjunction with oligomycin, it was possible to dissipate the proton gradient across the inner mitochondrial membrane, while inhibiting mitochondrial consumption of ATP (Boitier et al., 1999). Collapse of the gradient by FCCP did not alter the peak amplitude of the ATP-induced Ca^{2+} transient over the nucleus, or the time required to reach half of the peak amplitude. However the decay phase was longer, suggesting that mitochondrial buffering plays a role in restoration of the nuclear and cytoplasmic concentrations to basal levels. The Ca^{2+} wave spreading across the cytoplasm traveled faster, when the mitochondrial gradient was inhibited, implying that mitochondrial uptake exerts a negative feedback action on the propagation rate of the Ca^{2+} wave (Boitier et al., 1999). These studies indicate that mitochondria could provide a significant regulatory function, modulating propagating intracellular and intercellular Ca^{2+} waves within the astrocyte syncytium.

7. Glutamate is involved in neuron-to-astrocyte signaling, but not in astrocytic propagation of calcium waves

7.1. Effects of transmitter glutamate on astrocytes

7.1.1. Ca^{2+}

Synaptically released neurotransmitters regulate astrocytic calcium levels, thus making astrocytes sensitive to neuronal signaling (Smith, 1992; Smith, 1994). In further studies, glial cells along axons in the optic nerve were shown to generate Ca^{2+} signals in response to applications of ATP or glutamate as well as to electrical stimulation of axons (Kriegler and Chiu, 1993). This study of white matter and an earlier study on gray matter (Dani et al., 1992) suggest that both synaptic and non-synaptic regions of a neuron can trigger dynamic glial calcium signaling. Glial Ca^{2+} spikes were induced by axonal activity in the absence of extracellular Ca^{2+} (Kriegler and Chiu, 1993). As a model, the release of glutamate from a synapse is vesicular and Ca^{2+} dependent, whereas release from axons occurs through reversal of a glutamate transporter, following activity-dependent alterations in Na^{+} and K^{+} gradients across the axon membrane. Released glutamate activates metabotropic receptors on glia leading to Ca^{2+} waves and spiking (Dani et al., 1992). The glial calcium signal may drive metabolic responses (see chapter by Hertz, Peng et al.), and ultimately glutamate receptor activation in astrocytes can induce release of other neurotransmitters (Gallo et al., 1991).

7.1.2. ATP

Activation of glutamate receptors on rat cortical astrocytes has also been shown to induce the release of ATP (Quieroz et al., 1999). The release is brought about by activation of any of the three ionotrophic glutamate receptor types, *N*-methyl-D-aspartate (NMDA), AMPA and kainate receptors (Queiroz et al., 1997). AMPA receptors seem to mediate at least a part of the effect of glutamate, but the additional involvement of the other subtype glutamate receptors cannot be ruled out. Two different mechanisms seem to be involved. The NMDA- and kainate-induced release of ATP requires an influx of calcium, it is not due to neuron-like exocytosis, and is reduced (by an unknown mechanism), but not abolished, by lithium. The AMPA-induced release does not require extracellular calcium, may be mediated by a transmembrane conductance regulator or a mechanism regulated by a transmembrane conductance regulator, and is abolished (by an unknown mechanism) by acute exposure to 1 mM lithium. Glia cells have also been shown to respond electrophysiologically to action potential generation in the giant axon by mechanisms involving amino acid neurotransmitter receptors (Lieberman and Hassan, 1988; Lieberman et al., 1989).

7.2. Glutamate release does not mediate interastrocytic calcium waves

Both neurons and glia had been shown to stimulate increased Ca^{2+} in the opposite cell type by release of glutamate (Hassinger et al., 1995; Nedergaard, 1994; Parpura et al., 1994—see also above). This made glutamate a strong candidate for the component likely to be released from astrocytes during the propagation of calcium waves. Historically, there was supporting evidence that astrocytes released glutamate and other neuroactive substances (Martin, 1992). Astrocytes synthesize, metabolize and release neurotransmitters in response to stimuli (Patel and Hunt, 1989; Nicholls and Atwell, 1990; Shain et al., 1986, 1989; Philbert et al., 1988). Pure astrocyte cultures stimulated with 55 mM KCl exhibit a calcium-independent release of D-aspartate, a non-metabolizable glutamate analog, after loading with labeled D-aspartate (Westergaard et al., 1991). The calcium-independence of this release suggests a non-vesicular release that may be due to the reversal of glutamate uptake systems (Szatkowski et al., 1990; Nicholls et al., 1987; Sanchez-Prieto et al., 1996). However, evidence continued to mount that astrocytes also released glutamate in a Ca^{2+}-dependent manner (Parpura et al., 1994; Nicholls, 1998). It was reasonable to assume that this released glutamate would play a role as the diffusible element along the path of the calcium wave, providing the regenerative element needed for propagation.

Evidence for glutamate providing the regenerative signal for astrocyte intercellular calcium waves, however, is lacking. As early as 1991, Andrew Charles made the critical observation, that the pattern of a mechanical wave initiated repetitively was not affected by the application of glutamate. The path the wave traveled was similar in the presence or absence of glutamate, and it involved a similar percentage of cells suggesting a minimal role for glutamate in the generation of the astrocyte wave process (Charles et al., 1991). Cortical astrocytes in the presence of glutamate receptor antagonists continued to propagate calcium waves between astrocytes, minimizing the role of glutamate might have as the astrocytic stimulus for wave propagation (Hassinger et al., 1995). Glutamate antagonists did however block the propagation of Ca^{2+} waves from astrocytes to neurons. Taken together, these results indicate that glutamate release was not necessary for the propagation of intercellular astrocyte waves, but had dramatic regulatory effects on neuronal activity.

7.3. Glutamate as the mediator of astrocyte-to-neuron signaling

Glutamate is not only released from neurons and stimulating glutamate receptors on astrocytes, but it is also released from astrocytes activating neuronal receptors, i.e., representing a bi-directional communication system between the two classes of cells (Dani et al., 1992; Porter and McCarthy, 1996; Harris-White et al., 1998; Parpura et al., 1994; Charles, 1994; Nedergaard, 1994; Hassinger et al., 1995; Pasti et al., 1997; Newman and Zahs, 1998). Using a NADH-based fluorescence system it was clearly demonstrated that glutamate is released in a regenerative manner from astrocytes upon arrival of the calcium wave, with subsequent cells that are involved in the wave releasing additional glutamate (Innocenti et al., 2000). The wave of glutamate release that underlies the NADH fluorescence propagated at a speed of approximately 26 μm/s, which correlated well with the rate of calcium wave progression (10–30 μm/s). The stimulus for this glutamate release was the increase in cytoplasmic calcium. Local accumulation of glutamate reached 1–100 μM (Innocenti et al., 2000). It is likely that this concentration range of extracellular glutamate can profoundly modulate the physiology of neighboring neurons (Innocenti et al., 2000). Extracellular glutamate levels in the order of 100 μM are high enough to be toxic (Finkbeiner and Stevens, 1988; Ward et al., 2000) and to have long-term desensitizing effects on other members of the glutamate receptor family (Zorumski et al., 1996; Trussel and Fishbach, 1989). It is provocative to imagine that stimulus-evoked release of glutamate at the synapse is enough to cause the regenerative release of the glutamate wave from astrocytes described by Innocenti et al. (2000) (see also chapter by Shuai et al.), which in turn can have a regulatory effect on neurons over large distances of cortical brain regions. In particular, the concept of long-distance Ca^{2+} waves regulating and coordinating large regions of cortex becomes appealing. This concept is supported by the demonstration that activity-dependent potentiation of inhibitory synaptic transmission in the hippocampus between interneurons and pyramidal neurons is critically dependent on glutamate release from astrocytes (Kang et al., 1998). Inhibitory interneurons are of great importance for brain function, being the controlling mechanism for modulation of the excitatory activity of neurons in the region.

7.4. Astrocytes can modulate neuronal responses and synaptic transmission

Glial contributions to excitatory neurotransmission were studied in single neuron micro-islands where neuronal autaptic and glial responses were recorded during excitatory synaptic events (Mennerick and Zorumski, 1994). The release of glutamate from astrocytes was recently shown to modulate synaptic transmission in cultured hippocampal neurons (Arague et al., 1998a,b), as well as in intact retina (Newman and Zahs, 1998).

Can astrocytes modulate electrical and transmission characteristics of neighboring neurons? To test this possibility, focally applied electrical field or laser stimulation was used to initiate a calcium response in astrocyte/neuronal cultures, which resembled waves evoked from other paradigms, including glutamate exposure (Nedergaard, 1994). Neurons lying on contiguous astrocyte monolayers responded to glial signaling with increased intracellular Ca^{2+}. In contrast, neurons cultured on fibronectin or on astrocytes not involved in a wave did not show corresponding increases in $[Ca^{2+}]_i$ and did not respond to field potentials (Nedergaard, 1994). Repeated stimulation of the same target neuron consistently provided the same results. To test whether the Ca^{2+} wave was transmitted by extracellular release and diffusion of an astrocyte-derived molecule the perfusion was changed to superfusion with a linear velocity vector opposite to and 20 times faster than the Ca^{2+} wave, a procedure which prevented any effects on signaling either between astrocyte or between astrocytes and neurons. Inhibition of synaptic mechanisms with tetrodotoxin had no effect on astrocyte to neuron signaling, and the source for the increased Ca^{2+} release appeared to be intracellular stores. Photobleach recovery experiments suggested that local neurons lack a significant degree of gap junction coupling to one another or to local astrocytes. However, gap junctions with unidirectional diffusion capabilities were suggested between heterologous cell types and the Ca^{2+} signaling from astrocytes to neurons was unidirectional and inhibited by the gap junction blocker octanol (Nedergaard, 1994). Further studies however (Murphy et al., 1993; Blank et al., 1998) did not corroborate the idea of electrotonic coupling between astrocytes and neurons, and glutamate has since been identified as the diffusible chemical messenger, which specifically stimulates neurons (Hassinger et al., 1996; Pasti et al., 2001).

Ca^{2+} oscillations in astrocytes may be the coordinating behavior, which regulates the release of glial glutamate (Pasti et al., 2001). HEK293 cells were transfected with the NMDA receptor to be used as a glutamate biosensor (Pasti et al., 2001). Oscillations of intracellular Ca^{2+} in astrocytes were shown to trigger synchronous and repetitive Ca^{2+} oscillations in the sensor HEK cells. Whole cell patch clamp recordings demonstrated activation of NMDA receptors in HEK cells, resulting in inward currents that have extremely fast kinetics, in the order of the NMDA receptor currents in postsynaptic neurons. Agents known to reduce neuronal exocytosis (i.e., tetanus toxin and bafilomycin A) reduced astrocyte Ca^{2+} oscillations. Ca^{2+} oscillations represent a frequency-encoded signaling system that controls a pulsatile release of glutamate from astrocytes (Pasti et al., 2001—see also chapter by Shuai et al.). This system is bi-directional since an increase in the firing rate of neuronal afferents results in an increased frequency of $[Ca^{2+}]_i$ oscillations in astrocytes (Pasti et al., 1997). The frequency of oscillations is under the dynamic control of neuronal activity raising the possibility that it represents a code for the transfer of information from neurons to astrocytes. It was shown that glial uptake removed synaptically released glutamate, thereby contributing to the termination of excitatory synaptic currents. If autaptically associated glial currents represent activation of electrogenic transporters, then under appropriate conditions reverse uptake should produce glial glutamate efflux. A depolarization of glia to $+50\,mV$ induced a slowly rising neuronal current and accompanying pharmacology showed that the neuronal response was mediated by NMDA receptors, responding to glutamate released from glia (Pasti et al., 2001). Astrocytes respond to neuronal activity through receptor-mediated events as well as electrical events making them important players in information processing in the brain. Other neurotransmitters including, gamma-aminobutyric acid (GABA) (Liu et al., 2000), growth factors (Beattie et al., 2002), amino acids (Baranano et al., 2001), neuroactive peptides (Blondel et al., 2000), and acetylcholine (at the neuromuscular synapse (Jahromi et al., 1992; Reist and Smith, 1992) may also direct astrocyte–neuronal signaling. However, whereas ATP mediates calcium waves between astrocytes, it is unable to trigger astrocytic-neuronal signaling (unpublished experiments).

Communication between neurons and glia can occur without the involvement of a synapse. Phosphorylation of myelin basic protein occurs following high frequency axonal firing and this activity-dependent signaling results in release of NO (Atkins et al., 1999). In the PNS, axonal firing was shown to release ATP from non-synaptic regions resulting in activation of the Schwann cell P2Y receptors and stimulation of a Ca^{2+}-activated signaling pathway (Stevens and Fields, 2000; Fields and Stevens-Graham, 2002).

8. Pathology and Ca^{2+} waves

Disease states are often characterized by chronic imbalances in signaling molecules, which directly affect activity of ion channels, neurotransmitters and transporters, compounding the progression of the pathology. A common feature to most of these conditions is excitotoxity, which is mediated by excessive activation of amino acid receptors with the resulting toxicity ultimately contributing to the degeneration of brain cells (Meldrum and Garthwaite, 1990; Rothstein et al., 1995—see also chapters by Barger, by Brown and Sassoon, by Ghorpade and Gendelman, and by Werner et al.). When concentrations of extracellular neurotransmitters reach excitotoxic levels there are profound effects on the astrocytes in the region of stress. Astrocytes cultured from resected medial temporal lobe epilepsy patients exhibited hyperexcitable responses to glutamate if they were derived from regions of hyperexcitable EEG activity, identified in the neurosurgery operating room (Cornell-Bell and Williamson, 1993). Changes in receptor expression often accompany disease states, as evidenced by up-regulated expression of group I and II metabotropic glutamate receptors in spinal cord of ALS patients (Aronica et al., 2001) and of mGlu receptors 2/3 and 5 in astrocytes from kainate-induced seizures in amygdala of rats (Ulas et al., 2000). The finding of mGlu receptors 2/3, 4 and 8 immunoreactive astrocytes in hippocampus suggested that these receptors are involved in gliosis at the seizure focus (Tang and Lee, 2001). An increase in astrocytic mGlu receptors may directly lead to CNS hyperexcitability as described in the chapter by Shuai et al. Astrocytes from epileptic focus exhibited increased gap junctional coupling (Lee et al., 1995) and altered expression of ion channels (Bordey and Sontheimer, 1998). The resting membrane potential of seizure focus astrocytes from mesial temporal lobe epilepsy were significantly depolarized (approximately -55 mV) compared with cortical astrocytes (-80 mV) and there was a much higher density of Na^+ channels, further leading to the excitability of astrocytes. The astroglial glutamate transporters were also impaired in epilepsy tissues, contributing to the elevation of the levels of extracellular glutamate (Ye et al., 1999; Gorter et al., 2002). Electrophysiological changes in astrocytes were also seen to accompany reactive gliosis following scar formation (MacFarlane and Sontheimer, 1997). Moreover hippocampal gene expression is altered following status epilepticus (Hendriksen et al., 2001). Increased levels of extracellular excitatory neurotransmitters including glutamate were directly measured in extracellular space using microdialysis (During and Spencer, 1993) with a loss of glutamate-stimulated GABA release that was secondary to a reduction in the number of GABA transporters (During et al., 1995). During ischemia, citrulline (by-product of NO synthesis), glutamate, glycine, and GABA concentrations were also shown to increase extracellularly (Tan et al., 1996—see also chapter by Håberg and Sonnewald). All of these conditions paint a picture of compromised balance. Once the extracellular environment is thrown out of equilibrium, glutamate levels can soar due to reversal of glutamate transporters (Atwell et al., 1993) and high levels of glutamate (in the millimolar range) can be released from the astrocytes to further injure neurons and escalate damage to glutamate homeostasis even more (Hertz et al., 1988; Szatkowski et al., 1990). Glioma cells release glutamate by cystine–glutamate exchange, which actively kills neurons in the vicinity of a tumor (Ye and Sontheimer, 1999; Ye et al., 1999). These are only a few representative studies of excitotoxic changes occurring in diseased tissues. How does the excitotoxic environment disrupt normal calcium wave communication in astrocytes? What other receptors may play a role in astrocyte signaling in diseased tissues and what effects can toxicants released from damaged cells play?

In excitotoxic disease states one striking characteristic of time-lapse movies is the distinct lack of long-lasting Ca^{2+} waves (Jung et al., 2001). Long-lasting signals are replaced by very short-lived elevations in Ca^{2+}, which rapidly die out and so travel only short distances. In time-lapse studies of calcium activity in a culture of astrocytes from a patient with Tuberous Sclerosis (TS) very short lived signals were noted in response to glutamate application (Jung et al., 2001). In a sequence of subtracted snapshots of calcium activity only local islands of calcium elevations were seen, lacking the large-scale organization observed in cultures of normal rat astrocytes. In these TS cultures periodically a drastic, large-scale organization of Ca^{2+} activity would develop in the form of a cylindrical wave, which would rapidly entrain nearly all of the cells in the field, indicating at least temporary strong coupling between the astrocytes, leading to a strong, fast wave. During these episodes the cells in the entire culture appear synchronized in their Ca^{2+} activity. After these events the synchrony is lost, and Ca^{2+} activity again becomes a local phenomenon. Using a time and space 'cluster' calculation as described in the chapter by Shuai et al., it was shown that the entropy for diseased astrocytes was below the entropy measured in normal rat astrocyte cultures. Entropy vanishes if the spatiotemporal patterns or 'excited states' belong to only one pattern. In the example of epileptic Ca^{2+} images, the spatiotemporal pattern prevails of many short lived waves that travel short distances and die out (Jung et al., 2001).

Time lapse-imaging studies of astrocytes from neocortical tumor cases and medial temporal lobe epilepsy cases exhibited profound increases in intracellular Ca^{2+} transients following exposure to glutamate (Cornell-Bell et al., 1992; Cornell-Bell and Williamson, 1993; Lee et al., 1995). Cultures from hyperexcitable cortex and parahippocampus showed significantly higher numbers of intercellular waves when compared to regions exhibiting normal EEG records. Close to 82% of the excitable cortex propagated intercellular waves compared to 19.0% of the normal cells or 16.4% of the hippocampal cells and 5.4% of the peritumoral astrocytes. There was some indication the velocity of these intercellular waves was higher and the lifetimes were shorter (Lee et al., 1995). Astrocytes from a tumor region rarely propagate intercellular waves, which may be an indication of their extreme excitotoxic environment (Ye and Sontheimer, 1999; Ye et al., 1999). There was significantly more gap junctional coupling between cells cultured from seizure foci as well (Lee et al., 1995). These studies of only the pathology of epilepsy clearly indicate that regulatory mechanisms affecting intracellular and intercellular Ca^{2+} signaling patterns are modified, sometimes rather drastically, by the pathophysiologic changes in the disease environment. Other pathologies including Leao's spreading depression and hypoxia are characterized by irregular intracellular Ca^{2+} homeostasis (Do Carmo and Somjen, 1994; Somjen et al., 1993) suggesting that study of the Ca^{2+} signaling in different diseases may help elucidate common irregularities that lend themselves to therapeutic intervention.

One organelle that is targeted by several disease states is the mitochondria (Boitier et al., 1999). Brain mitochondria have been found to be impaired in pathologies including stroke (Sims, 1995); Alzheimer's disease (Mattson and Fukuawa, 1996) and Parkinson's disease (Bowling and Beal, 1995). In such states, altered Ca^{2+} regulation by mitochondria will directly impinge on the local and long distance signaling throughout the astrocyte syncytium. Effects of diseased mitochondria on astrocyte signaling as it affects neuronal signaling are implied, but as of yet are not studied. In addition other Ca^{2+} regulating organelles such as endoplasmic reticulum and the nucleus may be targets for study in cultures of diseased tissues.

Other neurotransmitters besides those hitting amino acid receptors may be involved in disease states. Purinergic receptors especially warrant study, since ATP plays such an important regulatory role in astrocyte signaling, providing the extracellular mediator for Ca^{2+} waves. Extracellular nucleosides and nucleotides have been shown to mediate both proliferative and cytoskeletal alterations on astrocytes (Neary et al., 1996, 1999). This may be directly related to the response of astrocytes to brain injury, which commonly induce hypertrophic and hyperplastic responses (Eng et al., 1987; Norenberg, 1994—see also chapter by Kálmán). Changes in these astrocytes are characterized by elongation of cellular extensions, enhanced expression of GFAP and response to ATP or ATP analogs (Rathbone et al., 1992; Neary et al., 1994; Abbracchio et al., 1994; Bolego et al., 1997; Neary et al., 1999). Astrogliosis is also enhanced by infusion of adenosine analogs into brain (Hindley et al., 1994). Release of purines following injury such as hypoxia may contribute to the gliosis associated with a number of common neurological disorders such as stroke, trauma, degenerative and demyelinative disorders (Bergfeld and Forrester, 1992; Fredholm et al., 1997). Inflammatory mediators such as IL-1β have been shown to regulate Ca^{2+} wave propagation via P2 receptors and regulation of gap junction coupling (John et al., 1999). IL-1β induces the expression of proinflammatory genes (John et al., 1999; Brosnan et al., 2001). IL-1β also down regulates gap junction connectivity, which could also contribute to the disease state (Brosnan et al., 2001). Add to this growing list the role of astrocyte Ca^{2+} wave regulation and this opens up the whole field of pathologic changes to astrocyte–astrocyte and astrocyte–neuronal communication, which we are just beginning to appreciate in its complexity.

9. Concluding remarks

Few, if any, observations regarding astrocytes have created as much scientific interest and excitement as the demonstration of astrocytic calcium waves. We have now solid information about several basic characteristics of these waves, which can be used as tools in further investigations of the physiological (and pathological) role(s) of astrocytic calcium waves. The ability of neurons to trigger calcium waves in astrocytes, which in turn may activate not only additional astrocytes, but also other, non-synaptically connected, neurons across considerable distances enable astrocytes to play a role in signaling events in the nervous system. The importance of this signaling system may be crucial for CNS function, perhaps especially in connection with global phenomena like mood and attention. It will be an important task for future research to unravel the secrets of this neuronal–astrocytic-neuronal signaling system.

References

Abbracchio, M.P., Saffrey, M.J., Hopker, V., Burnstock, G., 1994. Modulation of astroglial cell proliferation by analogues of adenosine and ATP in primary cultures of rat striatum. Neuroscience 59, 67–76.

Ahmed, Z., Lewis, C.A., Faber, D.S., 1990. Glutamate stimulates release of Ca^{2+} from internal stores in astroglia. Brain Res. 516, 165–169.

Allen, R.D., Allen, N.S., Travis, J.L., 1981. Video-enhanced contrast, differential interference contrast (AVEC-DIC) microscopy: a new method capable of analyzing microtubule-related motility in the reticulopodial network of *Allogromia laticollaris*. Cell Motil. 1, 291–302.

Amos, W.B., White, J.G., Fordham, M., 1987. Use of confocal imaging in the study of biological structures. Appl. Opt. 26, 3239–3243.

Anders, J.J., 1988. Lactic acid inhibition of gap junctional intercellular communication in in vitro astrocytes as measured by fluorescence recovery after laser photobleaching. Glia 1, 371–379.

Arague, A., Parpura, V., Sanzgiri, R.P., Haydon, P.G., 1998a. Glutamate-dependent astrocyte modulation of synaptic transmission between cultured hippocampal neurons. Eur. J. Neurosci. 10, 2129–2142.

Arague, A., Sanzgiri, R.P., Papura, V., Haydon, P.G., 1998b. Calcium elevation in astrocytes causes an NMDA-receptor dependent increase in the frequency of miniature synaptic currents in cultured hippocampal neurons. J. Neurosci. 18, 16822–16829.

Arcuino, G., Lin, J.H., Takano, T., Liu, C., Jiang, L., Gao, Q., Kang, J., Nedergaard, M., 2002. Intercellular calcium signaling mediated by point-source burst release of ATP. Proc. Natl Acad. Sci. USA, 9840–9845.

Aronica, E., Yankaya, B., Janse, G.H., Keenstra, S., van Veelan, C.W.M., Gorter, J.A., Troost, D., 2001. Ionotropic and metabotropic glutamate receptor protein expression in glioneuronal tumors from patients with intractable epilepsy. Neuropathol. Appl. Neurobiol. 27, 223–237.

Atkins, C.M., Yon, M., Groome, N.P., Sweatt, J.D., 1999. Regulation of myelin basic protein phosphorylation by mitogen-activated protein kinase during increased action potential firing in the hippocampus. J. Neurochem. 73, 1090–1097.

Atwell, D., Barbour, B., Szatkowski, M., 1993. Nonvesicular release of neurotransmitter. Neuron 11, 401–407.

Backus, K.H., Kettenmann, H., Schachner, M., 1989. Pharmacological characterization of the glutamate receptor in cultured astrocytes. J. Neurosci. Res. 22, 274–282.

Backx, P.H., deTombe, P.P., Van Deen, J.H.K., Mulder, B.J.M., TerKeurs, H.E.D.J., 1989. A model of propagating calcium-induced calcium relelase mediated by calcium diffusion. J. Gen. Physiol. 93, 966–977.

Baranano, D.E., Ferris, C.D., Snyder, S.H., 2001. Atypical neural messengers. Trends Neurosci. 24, 99–106.

Beattie, M.S., Harrington, A.W., Lee, R., Kim, J.Y., Boyce, S.L., Longo, F.M., Bresnahan, J.C., Hempstead, B.L., Yoon, S.O., 2002. ProNGF induces p75-mediated death of oligodendrocytes following spinal cord injury. Neuron 36, 375–386.

Bennet, M.V.L., Barrio, T.A., Bargiello, T.A., Spray, D.C., Hetzberg, E., Saez, J.C., 1991. Gap junctions: new tools, new answers, new questions. Neuron 6, 305–320.

Bergfeld, G.R., Forrester, T., 1992. Release of ATP from human erythrocytes in response to a brief period of hypoxia and hypercapnia. Cardiovasc. Res. 26, 40–47.

Berridge, M.J., 1993. Inositol trisphosphate and calcium signaling. Nature 361, 315–325.

Berridge, M.J., Gallione, A., 1988. Cytosolic calcium oscillators. FASEB J. 2, 3074–3082.

Blank, E.M., Bruce-Keller, A.J., Mattson, M.P., 1998. Astrocytic gap junctional communication decreases neuronal vulnerability to oxidative stress-induced disruption of Ca^{2+} homeostasis and cell death. J. Neurochem. 70, 958–970.

Blondel, O., Collin, C., McCarran, W.J., Zhu, S., Zamostiano, R., Gozes, I., Brenneman, D.E., McKay, R.F., 2000. A glia-derived signal regulating neuronal differentiation. J. Neurosci. 20, 8012–8020.

Boitier, E., Rea, R., Duchen, M.R., 1999. Mitochondria exert a negative feedback on the propagation of intracellular Ca^{2+} waves in rat cortical astrocytes. J. Cell Biol. 145, 795–808.

Bolego, C., Ceruti, S., Brambilla, R., Puglisi, L., Cattabeni, F., Burnstock, G., Abbracchio, M.P., 1997. Characterization of the signalling pathways involved in ATP and basic fibroblast growth factor-induced astrogliosis. Br. J. Pharmacol. 121, 1692–1699.

Bordey, A., Sontheimer, H., 1998. Properties of humn glial cells associated with epileptic seizure foci. Epilepsy Res. 32, 286–303.

Bowling, A.C., Beal, M.F., 1995. Bioenergetic and oxidative stress in neurodegenerative diseases. Life Sci. 56, 1151–1171.
Bowman, C., Kimelberg, H., 1984. Excitatory amino acids directly depolarize rat brain astrocytes in primary culture. Nature 311, 656–659.
Brakenhoff, G.J., van der Voort, H.T.M., van Spronsen, E.A., Linnemans, W.A.M., Nanninga, N., 1985. Three dimensional chromatin in neuroblastoma nuclei shown by confocal scanning laser microscopy. Nature 317, 748–749.
Brosnan, C.F., Scemes, E., Spray, D.C., 2001. Cytokine regulation of gap junction connectivity: an open-and-shut case or changing partners at the Nexus? Am. J. Pathol. 158, 1565–1569.
Burnstock, G., 1997. The past, present and future of purine nucleotides as signaling molecules. Neuropharmacology 36, 1127–1139.
Burnstock, G., Kennedy, C., 1985. Is there as basis for distinguishing two types of P2-puirnoreceptor? J. Gen. Pharmacol. 16, 433–440.
Centemeri, C., Bolego, C., Abbracchio, M.P., Cattabeni, F., Puglisi, L., Burnstock, G., Nicosia, S., 1997. Characterization of the Ca^{2+} responses evoked by ATP and other nucleotides in mammalian brain astrocytes. Br. J. Pharmacol. 121, 1700–1706.
Charles, A.C., 1994. Glia–neuron intercellular calcium signaling. Dev. Neurosci. 16, 196–206.
Charles, A.C., Dirksen, E.R., Merril, J.E., Sanderson, M.J., 1993. Mechanisms of intercellular calcium signaling in glial cells studied with dantrolene and thapsigargin. Glia 7, 145–148.
Charles, A.C., Merrill, J.E., Dirksen, E.R., Sanderson, M.J., 1991. Intercellular signaling in glial cells: calcium waves and oscillations in response to mechanical stimulation and glutamate. Neuron 6, 983–992.
Charles, A.C., Naus, C.C.G., Zhu, D., Kidder, G.M., Dirksen, E.R., Sanderson, M.J., 1992. Intercellular calcium signaling via gap junctions in glioma cells. J. Cell Biol. 118, 195–201.
Christ, G.J., Moreno, A.P., Melman, A., Spray, D.C., 1992. Gap junction-mediated intercellular diffusion of Ca^{2+} in cultured human coporal smooth muscle cells. Am. J. Physiol. 263, C373–C383.
Clementi, E., Riccio, M., Sciorati, C., Nistico, G., Meldolesi, J., 1996. The type 2 ryanodine receptor of neurosecretory PC12 cells is activated by cyclic ADP-ribose. Role of the nitric oxide/cGMP pathway. J. Biol. Chem. 271, 17739–17745.
Coco, S., Calegari, F., Pravettoni, E., Pozzi, D., Taverna, E., Rosa, P., Matteoli, M., Verderio, C., 2003. Storage and release of ATP from astrocytes in culture. J. Biol. Chem. 278, 1354–1362.
Cornell-Bell, A., Finkbeiner, S.M., 1991. Ca^{2+} waves in astrocytes. Cell Calcium 12, 185–204.
Cornell-Bell, A.H., Finkbeiner, S.M., Cooper, M.S., Smith, S.J., 1990. Glutamate induces calcium wavs in cultured astrocytes: long-range glial signaling. Science 247, 470–473.
Cornell-Bell, A.H., Magge, S., During, M., 1992. Human cortical astrocytes from hyperexcitable epileptic foci are themselves hyperexcitable. In: Simon, R.P. (Ed.), Excitatory Amino Acids. Thieme Medical, New York, pp. 273–274.
Cornell-Bell, A.H., Williamson, A., 1993. Glutamate-induced hyperexcitability of astrocytes and neurons in epileptic human cortex associated with tumors. In: Federoff S., Burkholder, G, (Eds.), Biology and Pathology of Astrocyte–Neuron Interactions. Plenum Publishing, New York, pp. 51–65.
Cotrina, M.L., Lin, J.H., Alves-Rodrigues, A., Liu, S., Li, J., et al., 1998a. Connexins regulate calcium signaling by controlling ATP release. Proc. Natl Acad. Sci. 95, 15735–15740.
Cotrina, M.L., Lin, J.H., Lopez-Garcia, J.C., Naus, C.C., Nedergaard, M., 2000. ATP mediated glia signaling. J. Neurosci. 20, 2835–2844.
Cotrina, M.L., Lin, J.H.C., Nedergaard, M., 1998b. Cytoskeletal assembly and ATP release regulate astrocytic calcium signaling. J. Neurosci. 18, 8704–8804.
Dani, J.W., Chernjavsky, A., Smith, S.J., 1992. Neuronal activity triggers calcium waves in hippocampal astrocyte networks. Neuron 8, 429–440.
Dermeitzel, R., Hartzbert, E.L., Kessler, J.A., Spray, D.C., 1991. Gap junctions between cultured astrocytes: immunohistochemical, molecular and electrophysiological analysis. J. Neurosci. 11, 1421–1432.
Dermietzel, R., Kremer, M., Paputsoglu, G., Stang, A., Skerrett, I.M., Gomes, D., Srinivas, M., Janssen-Bienhold, U., Weiler, R., Nicholson, B.J., Bruzzone, R., Spray, D.C., 2000. Molecular and functional diversity of neural connexins in the retina. J. Neurosci. 20, 8331–8343.

Do Carmo, R.J., Somjen, G.G., 1994. Spreading depression of Leao: 50 years sine a seminal discovery. J. Neurophysiol. 72, 1–2.
Dubyak, G.R., el-Moatassim, C., 1993. Signal transduction via P2-purinergic receptors for extracellular ATP and other nucleotides. Am. J. Physiol. 265, C577–C606.
During, M.J., Ryder, K.M., Spencer, D.D., 1995. Hippocampal GABA transporter function in temporal-lobe epilepsy. Nature 376, 122–125.
During, M.J., Spencer, D.D., 1993. Extracellular hippocampal glutamate and spontaneous seizure in the conscious human brain. Lancet 341, 1607–1610.
Eng, L.F., Reier, P.J., Houle, J.D., 1987. Astrocyte activation and fibrous gliosis: glial fibrillary acidic protein immunostaining of astrocytes following intraspinal cord grafting of fetal CNS tissue. Prog. Brain Res. 71, 439–455.
Enkvist, M.O., Halopainen, I., Akerman, K.E., 1989. Glutamate-receptor linked changes in membrane potential and intracellular Ca^{2+} in primary rat astrocytes. Glia 2, 397–402.
Enkvist, M.O.K., McCarthy, K.D., 1992. Activation of protein kinase C blocks gap junction communication and inhibits spreads of calcium waves. J. Neurochem. 59, 519–526.
Fam, S.R., Gallagher, C.J., Salter, M.W., 2000. P2Y(1) purinorecptor-mediated Ca^{2+} signaling and Ca^{2+} wave propagation in dorsal spinal cord astrocytes. J. Neurosci. 20, 2800–2808.
Fields, R.D., Stevens-Graham, B., 2002. New insights into neuron–glia communication. Science 298, 556–562.
Filson, A.J., Azamia, R., Beyer, E.C., Lowenstein, W.R., Brugge, J.S., 1990. Tyrosine phosphorylation of gap junction protein correlates with inhibition of cell to cell communication. Cell Growth Differ. 1, 661–668.
Finkbeiner, S.M., 1992. Calcium waves in astrocytes-filling in the gaps. Neuron 8, 1101–1108.
Finkbeiner, S.M., 1993. Glial calcium. Glia 9, 83–104.
Finkbeiner, S., Stevens, C.F., 1988. Applications of quantitative measurements for assessing glutamate neurotoxicity. Proc. Natl Acad. Sci. USA 85, 4071–4074.
Frame, M.K., deFeijter, A.W., 1997. Propagation of mechanically induced intercellular calcium waves via gap junctions and ATP receptors in rat liver epithelial cells. Exp. Cell Res. 230, 197–207.
Fredholm, B.B., Abbracchio, M.P., Burnstock, G., Dubyak, G.R., Harden, T.K., Jacobson, K.A., Schwabe, U., Williams, M., 1997. Towards a revised nomenclature for P1 and P2 receptors. Trends Pharmacol. Sci. 18, 79–82.
Gallo, V., Patrizio, M., Levi, G., 1991. GABA release triggered by the activation of neuron-like non-NMDA receptors in cultured type 2 astrocytes is carrier mediated. Glia 4, 245–255.
Giaume, C., McCarthy, K.D., 1996. Control of junctional communication in astrocytic netweorks. TINS 19, 319–325.
Glaum, S.R., Holzwarth, J.A., Miller, R.J., 1990. Glutamate receptors activate Ca^{2+} mobilization and Ca^{2+} influx into astrocytes. Proc. Natl Acad. Sci. 87, 3454–3458.
Goldberg, G.S., Bechberger, J.F., Tajima, Y., Merritt, M., Omori, Y., Gawinowicz, M.A., Narayanan, R., Tan, Y., Sanai, Y., Yamasaki, H., Naus, C.C., Tsuda, H., Nicholson, B.J., 2000. Connexin43 suppresses MFG-E8 while inducing contact growth inhibition of glioma cells. Cancer Res. 60, 6018–6026.
Gorter, J.A., van Vliet, E.A., Lopes da Silva, F.H., Isom, L.L., Aronica, E., 2002. Sodium channel beta1-subunit expression is increased in reactive astrocytes in a rat model for mesial temporal lobe epilepsy. Eur. J. Neurosci. 16, 360–364.
Grynkiewicz, G., Poenie, M., Tsien, R., 1985. A new generation of Ca^{2+} indicators with greatly improved fluorescence properties. J. Biol. Chem. 260, 3440–3450.
Guan, X., Cravatt, B.F., Ehring, G.R., Hall, J.E., Boger, D.L., Lerner, R.A., Gilula, N.B., 1997. The sleep-inducing lipid oleamide deconvolutes gap junction communication and calcium wave transmission in glial cells. J. Cell Biol. 139, 1785–1792.
Guthrie, P.B., Knappenberger, J., Kegal, M., Bennett, M.V.L., Charles, A.C., Kater, S.B., 1999. ATP released from astrocytes mediates glial calcium waves. J. Neurosci. 189, 520–528.
Haak, L.L., Grimaldi, M., Russell, J.T., 2000. Mitochondria in myelinating cells: calcium signaling in oligodendrocyte precursor cells. Cell Calcium 28, 297–306.
Harootunian, A.T., Kao, J.P., Paranjape, S., Tsien, R.Y., 1991. Generation of calcium oscillations in fibroblasts by positive feedback between calcium and IP3. Science 251, 75–78.

Harris-White, M.E., Zanotti, S.A., Frautschy, S.A., Charles, A.C., 1998. Dpiral intercellular calcium waves in hippocampal slice cultures. J. Neurophysiol. 79, 1045–1052.

Hassinger, T.D., Atkinsoin, P.B., Strecker, G.J., Whalen, L.R., Dudek, F.E., Kossel, A.H., Kater, S.B., 1995. Evidence for glutamate-mediated activation of hippocampal neurons by glial calcium waves. J. Neurobiol. 28, 159–170.

Hassinger, T.D., Guthrie, P.B., Atkinson, P.B., Bennett, M.V.L., Kater, S.B., 1996. An extracellular signaling component in propagation of astrocytic calcium waves. Proc. Natl Acad. Sci. 93, 13268–13273.

Hendriksen, H., Datson, N.A., Ghijsen, W.E., van Vliet, E.A., da Silva, F.H., Gorter, J.A., Vreugdenhil, E., 2001. Altered hippocampal gene expression prior to the onset of spontaneous seizures in the rat post-status epilepticus model. Eur. J. Neurosci. 14, 1475–1484.

Hertz, L., Murthy, C.R.K., Schousboe, A., 1988. Metabolism of glutamate and related amino acids. In: Norenberg, M.D., Hertz, L., Schousboe, A. (Eds.), The Biochemical Pathology of Astrocytes, Alan Liss, New York, pp. 395–406.

Hindley, S., Herman, M.A., Rathbone, M.P., 1994. Stimulation of reactive astrogliosis in vivo by extracellular adenosine diphosphate or an adenosine A2 receptor agonist. J. Neurosci. Res. 38, 399–406.

Holtzclaw, L.A., Pandhit, S., Bare, D.J., Mignery, G.A., Russell, J.T., 2002. Astrocytes in adult rat brain express type 2 inositol 1,4,5-trisphosphate receptors. Glia 39, 69–84.

Honore, T., Drejer, J., 1988. Chaotropic ions affect the conformation of quisqualate receptors in rat cortical membranes. J. Neurochem. 51, 457–461.

Inagaki, N., Fukui, H., Ito, S., Wada, H., 1991. Type-2 astrocytes show intracellular Ca^{2+} elevation in response to various neuroactive substances. Neurosci. Lett. 128, 257–260.

Innocenti, B., Parpura, V., Haydon, P.G., 2000. Imaging extracellular waves of glutamate during calcium signaling in cultured astrocytes. J. Neurosci. 20, 1800–1808.

Inoue, S., 1981. Video image processing greatly enhances contrast, quality and speed in polarization-based microscopy. J. Cell Biol. 89, 346–356.

Inoue, S., 1986. Video Microscopy. Plenum Press, New York.

Inoue, S., 1995. Confocal imaging and light microscopy. In: Pawley, J.B. (Ed.), Handbook of Biological Confocal Microscopy. 2nd ed. Plenum Press, New York, pp. 1–14.

Jahromi, B.S., Robitaille, R., Charlton, M.P., 1992. Transmitter release increases intracellular calcium in perisynaptic Schwann cells in situ. Neuron 8, 1069–1077.

James, G., Butt, A.M., 2001. Changes in P2Y and P2X purinoceptors in reactive glia following axonal degeneration in the rat optic nerve. Neurosci. Lett. 312, 33–36.

Jensen, A.M., Chiu, S.Y., 1990. Fluorescence measurement of changes in intracellular calcium induced by excitatory amino acids in cultured cortical astrocytes. J. Neurosci. 10, 1165–1175.

John, G.R., Scemes, E., Suadicani, S.O., Liu, J.S.H., Charles, P.C., Lee, S.C., Spray, D.C., Brosnan, C.F., 1999. IL-1 B differentially regulates calcium wave propagation between primary human fetal astrocytes via pathways involving P2 receptors and gap junctional channels. Proc. Natl Acad. Sci. 96, 11613–11618.

John, G.R., Simpson, J.E., Woodroofe, M.N., Lee, S.C., Brosnan, C.F., 2001. Extracellular nucleotides differentially regulate interleukin-1beta signaling in primary human astrocytes: implications for inflammatory gene expression. J. Neurosci. 21, 4134–4142.

Jorgensen, N.R., Geist, S.T., Civitelli, R., Steinberg, T.H., 1997. ATP and gap junction-independent intercellular calcium signaling in osteoblastic cells. J. Cell Biol. 139, 497–506.

Jung, P., DeGrauw, A., Strawsberg, R., Cornell-Bell, A.H., Dreher, M., Trinkaus-Randall, V., 2001. Statistical Analysis and Modeling of calcium waves in healthy and pathological astrocytes syncitia. In: Moss, F., Gielen, S. (Eds.), Handbook of Biological Physics, vol. 4 Elsevier Science, New York, pp. 323–351.

Kang, J., Jiang, L., Goldman, S.A., Nedergaard, M., 1998. Astrocyte-mediated potentiation of inhibitory synaptic transmission. Nat. Neurosci. 1, 683–692.

Kao, J.P., Harootunian, A.T., Tsien, R.Y., 1989. Photochemically generated cytosolic calcium pulses and their detection by fluo-3. J. Biol. Chem. 264, 8179–8184.

Kastritsis, C.H.C., Salm, A.K., McCarthy, K., 1992. Stimulation of the P2Y receptor on type I astroglia results in inositol phosphate formation and calcium mobilization. J. Neurochem. 58, 1277–1284.

Kim, W.T., Rioult, M.G., Cornell-Bell, A.H., 1994. Glutmate-induced calcium signaling in astrocytes. Glia 11, 173–184.

Klepeis, V.E., Cornell-Bell, A., Trinkaus-Randall, V., 2001. Growth factors but not gap junctions play a role in injury-induced Ca^{2+} waves in epithelial cells. J. Cell Sci. 114, 4185–4195.

Kriegler, S., Chiu, S.Y., 1993. Calcium signaling of glial cells along mammalian axons. J. Neurosci. 13, 4229–4245.

Laskey, A.D., Roth, B.J., Simpson, P.B., Russell, J.T., 1998. Images of Ca^{2+} flux in astrocytes: evidence for spatially distinct sites of Ca^{2+} release and uptake. Cell Calcium 23, 423–432.

Lechleiter, J., Girrd, S., Peralta, E., Clapham, D., 1991. Spiral calcium wave propagation and annihilation in Xenopus laevis oocytes. Science 252, 123–126.

Lee, S.H., Magge, S., Spencer, D.D., Sontheimer, H., Cornell-Bell, A.H., 1995. Human epileptic astrocytes exhibit increased gap junction coupling. Glia 15, 195–202.

Leipziger, J., 2003. Control of epithelial transport via luminal P2 receptors. Am. J. Physiol. Renal Physiol. 284, F419–F432.

Leybaert, L., Paemeleire, K., Strahonja, A., Sanderson, M.J., 1998. Inositol-trisphosphate-dependent intercellular calcium signaling in and between astrocytes and endothelial cells. Glia 24, 398–407.

Lieberman, E.M., Abbott, N.J., Hassan, S., 1989. Evidence that glutamate mediates axon-Schwann cell signaling in the squid. Glia 2, 94–102.

Lieberman, E.M., Hassan, S., 1988. Studies of axon–glial cell interactions and periaxonal K^+ homeostasis. III. The effect of anisosmotic media and potassium on the relationship between the resistance in series with the axon membrane and the glial cell volume. Neuroscience 25, 971–981.

Lipscombe, D., Madison, D.V., Poenie, M., Reuter, H., Tsien, R.W., Tsien, R.Y., 1988. Imaging of cytosolic Ca^{2+} stores and Ca^{2+} channels in sympathetic neurons. Neuron 1, 355–363.

Liu, Q.Y., Schaffner, A.E., Chang, Y.H., Maric, D., Barker, J.L. 2000. Persistent activation of GABA(A) receptor/Cl(-) channels by astrocyte-derived GABA. J. Neurophysiol 3, 1392–1403.

Madison, D.L., Kruder, W.H., Kim, T., Pfeiffer, S.E., 1996. Differential expression of rab3 isoforms in oligodendrocytes and astrocytes. J. Neurosci. Res. 45, 258–268.

Martin, D.L., 1992. Synthesis and release of neuroactive substances by glial cells. Glia 5, 81–94.

Mattson, M.P., Fukuawa, K., 1996. Programmed cell life: anti-apoptotic signaling and therapeutic astrategies for neurodegenerative disorders. Restor. Neurol. Neurosci. 9, 191–205.

Matz, H., Hertz, L., 1989. Adenosine metabolism in neurons and astrocytes in primary cultures. J. Neurosci. Res. 2, 260–7.

MacFarlane, S.N., Sontheimer, H., 1997. Spinal cord astrocyte display a switch from TTX-sensitive to TTX-resistant sodium currents after injury-induced gliosis in vivo. J. Neurophysiol. 79, 2222–2226.

McCarthy, K.D., Salm, A.K., 1991. Pharmacologically distinct subsets of astroglia can be identified by their calcium response to neuroligands. Neuroscience 41, 325–333.

Meldrum, B., Garthwaite, J., 1990. Excitatory amino acid neurotoxicity and neurodegenerative disease. Trends Pharmacol. Sci. 11, 379–387.

Mennerick, S., Zorumski, C.F., 1994. Glial contributions to excitatory neurotransmission in cultured hippocampal cells. Nature 368, 59–62.

Meyer, T., Stryer, L., 1991. Calcium spiking. Annu. Rev. Biophys. Biophys. Chem. 20, 153–174.

Monaghan, D.T., Bridges, R.J., Cotman, C.W., 1989. The excitatory amino acid receptors: their classes, pharmacology, and distinct properties in the function of the central nervous system. Annu. Rev. Pharmacol. Toxicol. 29, 365–402.

Murphy, T.H., Blatter, L.A., Wier, W.G., Baraban, J.M., 1993. Rapid communication between neurons and astrocytes in primary cortical cultures. J. Neurosci. 13, 2672–2679.

Naus, C.C., Bechberger, J.F., Zhang, Y., Venance, L., Yamasaki, H., Juneja, S.C., Kidder, G.M., Giaume, C., 1997. Altered gap junctional communication, intercellular signaling, and growth in cultured astrocytes deficient in connexin43. J. Neurosci. Res. 49, 528–540.

Naus, C.C., Bond, S.L., Bechberger, J.F., Rushlow, W., 2000. Identification of genes differentially expressed in C6 glioma cells transfected with connexin43. Brain Res. Brain. Res. Rev. 32, 259–266.

Neary, J.T., van Breeman, C., Forster, E., Norenberg, L.O.B., Norenberg, M.D., 1998. ATP stimulates calcium influx in primary astrocyte cultures. Biochem. Biophys. Res. Commun. 157, 1410–1416.

Neary, J.T., McCarthy, M., Cornell-Bell, A., Kang, Y., 1999. Trophic signaling pathways activated by purinergic receptors in rat and human astroglia. Prog. Brain Res. 120, 323–332.

Neary, J.T., Whittemore, S.R., Zhu, Q., Norenberg, M.D., 1994. Destabilization of glial fibrillary acidic protein mRNA in astrocytes by ammonia and protection by extracellular ATP. J. Neurochem. 63, 2021–2027.
Neary, J.T., Zhu, Q., Kang, Y., Dash, P.K., 1996. Extracellular ATP induces formation of AP-1 complexes in astrocytes via P2 purinoceptors. Neuroreport 7, 2893–2896.
Nedergaard, M., 1994. Direct signaling from astrocytes to neurons in cultures of mammalian brain cells. Science 263, 1768–1771.
Newman, E.A., 2001. Propagation of intercellular calcium waves in retinal astrocytes and Muller cells. J. Neurosci. 21, 2215–2223.
Newman, E.A., Zahs, K.R., 1997. Calcium waves in retinal glial cells. Science 275, 844–847.
Newman, E.A., Zahs, K.R., 1998. Modulation of neuronal activity by glial cells in the retina. J. Neurosci. 18, 4022–4028.
Nicholls, D.G., 1998. Presynaptic modulation of glutamate release. Prog. Brain Res. 116, 15–22.
Nicholls, D., Atwell, D., 1990. The release and uptake of excitatory amino acids. Trends Pharmacol. Sci. 11, 462–468.
Nicholls, D.G., Sihra, T.S., Sanchez-Prieto, J., 1987. Calcium-dependent and -independent release of glutamate from synaptosomes monitored by continuous fluorometry. J. Neurochem. 49, 50–57.
Norenberg, M.D., 1994. Astrocyte responses to CNS injury. J. Neuropathol. Exp. Neurol. 53, 213–220.
O'Connor, E.R., Sontheimer, H., Spencer, D.D., de Lanerolle, N.C., 1998. Astrocytes from human hippocampal epileptogenic foci exhibit action potential-like responses. Epilepsia 39, 347–354.
Osipchuk, Y., Cahalan, M., 1992. Cell-to-cell spread of calcium signals mediated by ATP receptors in mast cells. Nature 17359, 241–244.
Parpura, V., Basarsky, T.A., Liu, F., Jeftinija, K., Jeftinija, A., Haydon, P.G., 1994. Gutamate-mediated astrocyte–neuron signaling. Nature 369, 744–747.
Pasti, L., Volterra, A., Pozzan, T., Carmignoto, G., 1997. Intracellular calcium oscillations in astrocytes: a highly plastic, bi-directional form of communication between neurons and astrocytes in situ. J. Neurosci. 17, 7817–7830.
Pasti, L., Zonta, M., Pozzan, T., Vicini, S., Carmignoto, G., 2001. Cytosolic calcium oscillations in astrocytes may regulate exocytotic release of glutamate. J. Neurosci. 2, 477–484.
Patel, A., Hunt, A., 1989. Regulation of production by primary cultures of rat forebrain astrocytes of a trophic factor important for the development of cholinergic neurons. Neurosci. Lett. 99, 223–228.
Pawley, J.B., 1995. Handbook of Biological Confocal Microscopy, 2nd ed. Plenum Press, New York.
Pearce, B.R., Morrow, C., Murphy, S., 1986. Receptor-mediated ionsitol phospholipid hydrolysis on astrocytes. Eur. J. Pharmacol. 121, 231–243.
Philbert, R.A., Rogers, K.L., Allen, A.J., Dutton, G.R., 1988. Dose-dependent K^+ stimulated efflux of endogenous taurine from primary astrocyte cultures is Ca^{2+} dependent. J. Neurochem. 51, 122–126.
Porter, J.T., McCarthy, K.D., 1996. Hippocampal astrocytes in situ respond to glutamate released from synaptic terminals. J. Neurosci. 16, 5073–5081.
Queiroz, G., Gebicke-Haerter, P.J., Schobert, A., Starke, K., von Kugelgen, I., 1997. Release of ATP from cultured rat astrocytes elicited by glutamate receptor activation. Neuroscience 4, 1203–1208.
Quieroz, G., Meyer, D.K., Meyer, A., Starke, K., Von Kugelgen, I., 1999. A study of the mechanism of release of ATP from rat cortical astroglial cells evoked by activation of glutamate receptors. Neuroscience 91, 1171–1181.
Quist, A.P., Rhee, S.K., Lin, H., Lal, R., 2000. Physiological role of gap-junctional hemichannels: extracellular calcium-dependent isosmotic volume regulation. J. Cell Biol. 148, 1063–1074.
Rathbone, M.P., Middlemiss, P.J., Kim, J.K., Gysbers, J.W., DeForge, S.P., Smith, R.W., Hughes, D.W., 1992. Adenosine and its nucleotides stimulate proliferation of chick astrocytes and human astrocytoma cells. Neurosci. Res. 13, 1–17.
Reist, N.E., Smith, S.J., 1992. Neurally evoked calcium transients in terminal Schwann cells at the neuromuscular junction. Proc. Natl Acad. Sci. 89, 7625–7629.
Rothstein, J.D., VanKammen, M., Levey, A.I., Martin, L.J., Kuncl, R.W., 1995. Selective loss of glial glutamate transporter GLT-1 in amyotrophic lateral sclerosis. Ann. Neurol. 38, 73–84.
Rouach, N., Glowinski, J., Giaume, C., 2000. Activity-dependent neuronal control of gap-junctional communication in astrocytes. J. Cell Biol. 149, 1513–1526.

Saez, J.C., Spray, D.C., Nairn, A.C., Hertzberg, E., Greengard, P., Bennet, V.L., 1983. cAMP increases junctional conductance and stimulates phosphorylation of the 27 kDa principal gap junction polypeptide. Proc. Natl Acad. Sci. 83, 2473–2477.

Sanchez-Prieto, J., Budd, D.C., Herrero, I., Vazquez, E., Nicholls, D.G., 1996. Presynaptic receptors and the control of glutamate exocytosis. Trends Neurosci. 19(6), 235–239.

Scemes, E., Dermeitzel, R., Spray, D.C., 1998. Calcium waves between astrocytes from Cx43 knockout mice. Glia 24, 65–73.

Scemes, E., Suadicani, S.O., Spray, D.C., 2000. Intercellular communication in spinal cord astrocytes: fine tuning between gap junctions and P2 nucleotide receptors in calcium wave propagation. J. Neurosci. 20, 1435–1445.

Schlosser, S.F., Burgstahler, A.D., Nathanson, M.H., 1996. Isolated rat hepatocytes can signal to other hepatocytes and bile duct cells by release of nucleotides. Proc. Natl Acad. Sci. USA 93, 9948–9993.

Schneider, H., Fallert, M., Wachsmuth, E.D., 1994. Kinetics of intracellular Ca^{2+} concentration changes and cell contraction of electrically stimulated cardiomyocytes as analysed by automated digital-imaging microscopy. J. Microsc. 175, 108–120.

Sergeeva, M., Ubl, J.J., Reiser, G., 2000. Disruption of actin cytoskeleton in cultured rat astrocytes suppresses ATP- and bradykinin-induced [Ca(2 +)](i) oscillations by reducing the coupling efficiency between Ca(2 +) release, capacitative Ca(2 +) entry, and store refilling. Neuroscience 97, 765–769.

Shain, W., Connor, J.A., Madelian, V., Martin, D.L., 1989. Spontaneous and beta-adrenergic receptor-mediated taurine release from astroglial cells are independent of manipulations of intracellular calcium. J. Neurosci. 9, 22306–22312.

Shain, W., Madelian, V., Martin, D.L., Kimelberg, H.K., Perrone, M., Lepore, R., 1986. Activation of beta-adrenergic receptors stimulates release of inhibitory transmitter from astrocytes. J. Neurochem. 46, 1298–1303.

Sheppard, C.A., Simpson, P.B., Sharp, A.H., Nucifora, F.C., Ross, C.A., Lange, G.D., Russell, J.T., 1997. Comparison of type 2 inositol 1,4,5-trisphosphate receptor distribution and subcellular Ca^{2+} release sites that support Ca^{2+} waves in cultured astrocytes. J. Neurochem. 68, 2317–2327.

Simard, M., Couldwell, W.T., Zhang, W., Song, H., Liu, S., Cotrina, M.L., Goldman, S., Nedergaard, M., 1999. Glucocorticoids-potent modulators of astrocytic calcium signaling. Glia. 1, 1–12.

Sims, N.R., 1995. Calcium, energy metabolism and the development of selective neuronal loss following short-term cerebral ischemia. Metab. Brain Dis. 10, 191–217.

Simpson, P.B., Russell, J.T., 1996. Mitochondria support inositol 1,4,5-trisphosphate-mediated Ca^{2+} waves in cultured oligodendrocytes. J. Biol. Chem. 271, 33493–33501.

Simpson, P.B., Russell, J.T., 1997. Role of sarcoplasmic/endoplasmic-reticulum Ca^{2+}-ATPases in mediating Ca^{2+} waves and local Ca^{2+}-release microdomains in cultured glia. Biochem. J. 325, 239–247.

Simpson, P.B., Mehotra, S., Lange, G.D., Russell, J.T., 1997. High density distribution of endoplasmic reticulum proteins and mitochondria at specialized Ca^{2+} release sites in oligodendrocyte processes. J. Biol. Chem. 272, 22654–22661.

Simpson, P.B., Russell, J.T., 1998a. Mitochondrial Ca^{2+} uptake and release influence metabotropic and ionotropic cytosolic Ca^{2+} responses in rat oligodendrocyte progenitors. J. Physiol. 508, 413–426.

Simpson, P.B., Russell, J.T., 1998b. Role of mitochondrial Ca^{2+} regulation in neuronal and glial cell signalling. Brain Res. Brain Res. Rev. 26, 72–81.

Smaili, S.S., Hsu, Y.T., Sanders, K.M., Russell, J.T., Youle, R.J., 2001. Bax translocation to mitochondria subsequent to a rapid loss of mitochondrial membrane potential. Cell Death Differ. 8, 909–920.

Smith, S.J., 1992. Do astrocytes process neural information? Prog. Brain Res. 94, 119–136.

Smith, S.J., 1994. Neural signaling. Neuromodulatory astrocytes. Curr. Biol. 4, 807–810.

Somjen, G.G., Aitken, P.G., Czeh, G., Jing, J., Young, J.N., 1993. Cellular physiology of hypoxia of the mammalian central nervous system. Res. Publ. Assoc. Res. Nerv. Ment. Dis. 71, 51–65.

Sontheimer, H., Kettenmann, H., Backus, K.H., Schachner, M., 1988. Glutamate opens Na^+/K^+ channels in cultured astrocytes. Glia 1, 328–336.

Spray, D.C., Duffy, H.S., Scemes, E., 1999. Gap junctions in glia. Types, roles, and plasticity. Adv. Exp. Med. Biol. 468, 339–359.

Stevens, B., Fields, R.D., 2000. Response of Schwann cells to action potentials in development. Science 287, 2267–2270.

Stout, C.E., Costantin, J.L., Naus, C.C., Charles, A.C., 2002. Intercellular calcium signaling in astrocytes via ATP release through connexin hemichannels. J. Biol. Chem. 277, 10482–10488.

Stoyanovsky, D., Murphy, T., Anno, P.R., Kim, Y.M., Salama, G., 1997. Nitric oxide activates skeletal and cardiac ryanodine receptors. Cell Calcium 21, 19–29.

Sun, H.Q., Yamamoto, M., Mejillano, M., Yin, H.L., 1999. Gelsolin, a multifunctional actin regulatory protein. J. Biol. Chem. 274, 33179–33182.

Swenson, K.I., Piwnica-Worms, H., MacNamee, H., Paul, D.L., 1990. Tyrosine-phosphorylation of the gap junction protein connexin43 is required for the pp60v-src-induced inhibition of communication. Cell Regul. 1, 998–1002.

Szatkowski, M., Barbour, B., Atwell, D., 1990. Non vesicular release of glutamate from glial cells by reversed electrogenic glutamate uptake. Nature 348, 443–446.

Tan, W.K., Williams, C.E., During, M.J., Mallard, C.E., Gunning, M.I., Gunn, A.J., Gluckman, P.D., 1996. Accumulation of cytotoxins during the development of seizures and edema after hypoxic–ischemic injury in late gestation fetal sheep. Pediatr. Res. 39, 791–797.

Tang, F.-R., Lee, W.-L., 2001. Expression of the group II and III metabotropic glutamate receptors in the hippocampus of patients with mesial temporal lobe epilepsy. J. Neurocytol. 30, 137–143.

Teichberg, V.I., 1991. Glial glutamate receptors: likely actors in brain signaling. FASEB J. 5, 3086–3091.

Trifaro, J.M., Vitale, M.L., 1993. Cytoskeleton dynamics during neurotransmitter release. Trends Neurosci. 16, 466–472.

Trussel, L.O., Fishbach, G.D., 1989. Glutamate receptor desensitization and its role in synaptic transmission. Neuron 3, 209–218.

Tsien, R.Y., 1988. Fluorescence measurement and photochemical manipulation of cytosolic free calcium. Trends Neurosci. 11, 419–424.

Ulas, J., Satou, T., Ivins, K.J., Kesslak, J.P., Cotman, C.W., Balazs, R., 2000. Expression of metabotropic glutamate receptor 5 is increased in astrocytes after kainate-induced epileptic seizures. Glia 30, 352–361.

Usowicz, M.M., Gallo, V., Cull-Candy, S.G., 1989. Multiple conductance channels in Type II cerebellar astrocytes activated by excitatory amino acids. Nature 339, 380–383.

van den Pol, A.N., Finkbeiner, M.S., Cornell-Bell, A.H., 1992. Calcium excitability and oscillations in Suprachiasmatic Nucleus neurons and glia in vitro. J. Neurosci. 12, 2648–2664.

Venance, L., Piomelli, D., Glowinski, J., Giaume, C., 1995. Inhibition by anandamide of gap junctions and intercellular calcium signaling in striatal astrocytes. Nature 376, 590–594.

Venance, L., Stella, N., Glowinski, J., Giaume, C., 1997. Mechasnism involved in initiation and propagation of receptor-linked intercellular calcium signaling in cultured rat astrocytes. J. Neurosci. 17, 1981–1992.

Verderio, C., Matteoli, M., 2001. ATP mediates calcium signaling between astrocytes and microglial cells: modulation by IFN-gamma. J. Immunol. 166, 6383–6391.

Wang, Z., Haydon, P.G., Yeung, E.S., 2000. Direct observation of calcium-dependent ATP signaling in astrocytes. Anal. Chem. 72, 2001–2007.

Ward, M.W., Rego, A.C., Frenguelli and Nicholls, D.G., 2000. Mitochondrial membrane potential and glutamate excitotoxicity in cultured cerbellar granulke cells. J. Neurosci. 20, 7208–7219.

Westergaard, N., Fosmark, H., Schousboe, A., 1991. Metabolism and release of glutamate in cerebellar granule cells cocultured with astrocytes from cerebellum or cerebral cortex. J. Neurochem. 56, 59–66.

Wilmott, N., Sethi, J.K., Walseth, T.F., Lee, H.C., White, A.M., Galione, A., 1996. Nitric oxide induces calcium mobilization via activation of the cyclic ADP-ribose pathway. J. Biol. Chem. 271, 3699–3705.

Willmott, N.J., Wong, K., Strong, A.J., 2000. A fundamental role for nitric oxide-G-kinase signaling pathway in mediating intercellular Ca^{2+} waves in glia. J. Neurosci. 20, 1767–1779.

Yagodin, S.V., Holtzclaw, L.A., Sheppard, C.A., Russell, J.T., 1994. Nonlinear propagation of agonist-induced cytoplasmic calcium waves in single astrocytes. J. Neurobiol. 25, 265–280.

Yamada, E.W., Huzel, N.J., 1989. Calcium-binding ATPase inhibitor protein of bovine heart mitochondria. Role in ATP synthesis and effect of Ca^{2+}. Biochemistry 28, 9714–9718.

Ye, Z., Rothstein, D., Sontheimer, H., 1999. Compromised glutamate transport in human glioma cells: reduction–mislocalization of sodium-dependent glutamate transporters and enhanced activity of cystine–glutamate exchange. J. Neurosci. 19, 10767–10777.

Ye, Z.-C., Sontheimer, H., 1999. Glioma cells release excitotoxic concentrations of glutamate. Cancer Res. 59, 4383–4391.

Zahs, K.R., Newman, E.A., 1997. Asymmetric gap junctional coupling between glial cells in the rat retina. Glia 20, 10–22.

Zanotti, S., Charles, A., 1997. Extracellular calcium sensing by glial cells: low extracellular calcium induces intracellular calcium release and intercellular signaling. J. Neurochem. 2, 594–602.

Zhou, M., Kimelberg, H.K., 2001. Freshly isolated hippocampal CA1 astrocytes comprise two populations differing in glutamate transporter and AMPA receptor expression. J. Neurosci. 21, 7901–7908.

Zorumski, C.F., Mennerick, S., Que, J., 1996. Modulation of excitatory synaptic transmission by low concentrations of glutamate in cultured rat hippocampal neurons. J. Physiol., 494.

Advances in
Molecular and
Cell Biology

Mathematical modeling of intracellular and intercellular calcium signaling

Jian-Wei Shuai,[a] Suhita Nadkarni,[a] Peter Jung,[a,*] Ann Cornell-Bell[b] and Vickery Trinkaus-Randall[c]

[a]*Department of Physics and Astronomy and Quantitative Biology Institute, Ohio University, Athens, OH 45701, USA*
[*]*Correspondence address: E-mail: jung@helios.phy.ohiou.edu*
[b]*ANSCANS, Ivoryton, CT 06442, USA*
[c]*School of Medicine, Boston University, Boston, MA 02118, USA*

Contents

1. Introduction
2. Modeling of intracellular Ca^{2+} signaling (IACW)
   2.1. Equations used
   2.2. Deterministic, non-phenomenological modeling
   2.3. Phenomenological modeling
   2.4. Stochastic modeling
3. Modeling of intercellular Ca^{2+} signaling (IRCW) in astrocytes
4. Bidirectional coupling between neurons and astrocytes
5. Concluding remarks

1. Introduction

The most complex system in the universe is probably the brain. Billions of neurons, interconnected to a large network, perform numerous cognitive and regulatory tasks. Most work on the modeling of brain functions is based on modeling of neuronal networks. The vast majority of cells in the brain, however, are non-neuronal cells or glial cells; about 90% of all brain cells are glial cells. Among several types of glial cells, the astrocytes are known to carry out many important functions, several of them in interactions with neurons.

Astrocytes, in contrast to most neuronal cells, do not fire action potentials due to insufficient Na^+ channel density (Bordeay and Sontheimer, 1998b). Documented exceptions are astrocytomas, where an enhanced expression of Na^+ channels allows the generation of action potentials (Bordeay and Sontheimer, 1998a). Astrocytes do not connect to other astrocytes or neurons via long processes. Therefore, for many years it has

Advances in Molecular and Cell Biology, Vol. 31, pages 689–706

ISBN: 0-444-51451-1

been believed that the role of the glia for brain function is to provide structural and chemical support for the neurons. Such support functions include for example uptake and recycling of neurotransmitters. Although it has been known for a long time that synaptic astrocytes respond with depolarization to neuronal action potentials, this was thought to be a passive response caused by the increased extracellular K^+ concentration. The discovery by Porter and McCarthy (1996), that astrocytes respond to neuronal action potentials by glutamate-mediated activation of metabotropic glutamate receptors, has changed the current thinking about the role of astrocytes dramatically. It is now clear that astrocytes listen to neuronal chatter at the synapses and in turn can modulate neuronal dynamics at the same synapse or over some distance.

As neurons fire, glutamate is released into the synaptic cleft, which is partially lined by the metabotropic glutamate receptors of the synaptic astrocytes. Upon binding of glutamate to the astrocyte, inositol 1,4,5-triphosphate (IP_3) is released into the intracellular space. IP_3 in turn binds to the IP_3 receptor of the endoplasmic reticulum (ER), and Ca^{2+} is released from the ER into the cytosol. As described in more detail below, such Ca^{2+} release can occur in forms of intracellular Ca^{2+} waves. The Ca^{2+} wave can propagate across the cell membrane, through the extracellular space into adjacent astrocytes. As will be discussed below, there are several mechanisms that may be involved in this intercellular signaling. Elevated Ca^{2+} concentrations in synaptic astrocytes generate extracellular glutamate that can modulate the neuronal synapse by generating additional inward currents. This feedback could of course also be inhibitory, if enhanced Ca^{2+} concentrations are activating inhibitory interneurons. In the remainder of this section, we will review recent progress in mathematical modeling of intra- and intercellular Ca^{2+} signaling in general and in the context of astrocytes and their control of synaptic plasticity in particular. We would also like to draw the attention of the reader to another recent review on the same topic by Schuster et al. (2002).

2. Modeling of intracellular Ca^{2+} signaling (IACW)

2.1. *Equations used*

Ca^{2+} is stored in internal stores, most notably the ER. It can be released into the cytosol through release channels and through passive leakage currents driven by the steep concentration gradient between the ER and the cytosol. The release channels, termed IP_3 receptor channels (IP_3Rs), have been modeled mathematically for more than ten years, and a number of models have been developed. Although the models differ in detail, most models have certain key-elements. Denoting the concentration of Ca^{2+} in the cytosol by c, conservation of Ca^{2+} is expressed by the equation of continuity

$$\frac{\partial c}{\partial t} + \nabla \mathbf{j}_c = \rho(c, \mathbf{x}, t), \tag{1}$$

with $\mathbf{j}_c$ denoting the Ca^{2+} flux density and $\rho(c, \mathbf{x}, t)$ the source density of Ca^{2+}. For not too large gradients in Ca^{2+} concentration Fick's law can be used to relate the flux density to the Ca^{2+} concentration, i.e.

$$\mathbf{j}_c = -D_{Ca}\nabla c, \tag{2}$$

with D_{Ca} denoting the diffusion coefficient for calcium. Inserting Eq. (2) into Eq. (1) yields

$$\frac{\partial c}{\partial t} = D_{Ca}\nabla^2 c + \rho_{Ca}(c, \mathbf{x}, t). \tag{3}$$

The source density describes the influx of Ca^{2+} into the cytosol per volume and unit time. In the absence of Ca^{2+} entry from the extracellular space, the Ca^{2+} concentration in the cytosol can change due to (i) Ca^{2+} entering from the ER through IP_3Rs, (ii) pump-mediated Ca^{2+} re-uptake by the ER, (iii) leakage of Ca^{2+} from the ER into the cytosol, (iv) Ca^{2+} release from mitochondria, and (v) Ca^{2+} re-uptake by mitochondria, i.e.

$$\rho_{Ca}(c, \mathbf{x}, t) = -\rho_{pump} + \rho_{channel} + \rho_{leak} - \rho_{mito}^{in} + \rho_{mito}^{out}. \tag{4}$$

The forms of the source density terms in Eq. (4) differ between the models. The pump term is of Hill-type with a Hill-coefficient of 2 in the two-pool model (Goldbeter et al., 1990)

$$\rho_{pump} = \frac{k_1 c^2}{k_2^2 + c^2}, \tag{5}$$

the single pool model (Somorgyi and Stucki, 2000; Dupont and Goldbeter, 1993), the De Young–Keizer model (De Young and Keizer, 1992) and the Li–Rinzel model (Li and Rinzel, 1994).

The mathematical form of the source density due to Ca^{2+} release from the ER through IP_3Rs, $\rho_{channel}$, differs between the models used for the IP_3 receptor. The single and two-pool models use a Hill-form for $\rho_{channel}$. In the De Young–Keizer model it is assumed that the IP_3 receptor has three independent subunits with three binding sites each: an activating binding site for Ca^{2+}, and inhibiting binding site for Ca^{2+}, and a binding site for IP_3. A subunit is activated if IP_3 is bound and the activating Ca^{2+} binding site is bound. The channel is open, if all three subunits are activated. We denote the state of each channel by a string 'abc' where a, b, and c can assume the values 0 or 1. The first letter indicates occupancy of the IP_3 binding site, the second one the occupancy of the activating Ca^{2+} binding site and the third letter the occupancy of the inactivating Ca^{2+} binding site. A value of '1' indicates that the binding site is occupied while a value '0' indicates that the binding site is unoccupied. The source density $\rho_{channel}$ is thus proportional to the fraction $x_{abc}^3 = x_{110}^3$ of channels with all three subunits bound with IP_3 and Ca^{2+}

$$\rho_{channel} = v_1 c_1 x_{110}^3 (c_{ER} - c), \tag{6}$$

where c_{ER} is the Ca^{2+} concentration in the ER. Thus, in this model, the Ca^{2+} flux through the IP_3Rs is driven by the Ca^{2+} concentration difference between the ER and the cytosol, and it is determined by the fraction of channels, x_{110}^3, with all subunits activated. The factor c_1 accounts for the ratio between the volume of the ER and the volume of the cytosol. The leak current is also driven by the concentration difference between the ER and the cytosol and it is assumed to be of the following form in most models

$$\rho_{\text{leak}} = \nu_2 c_1 (c_{\text{ER}} - c). \tag{7}$$

It is only recently that the effect of Ca^{2+} sequestering by mitochondria has been taken into account for modeling of intracellular Ca^{2+} signaling (Magnus and Keizer, 1997, 1998a,b). The uptake of Ca^{2+} by mitochondria through Ca^{2+} uniporters sets in only when the Ca^{2+} concentration exceeds a threshold. This threshold-like uptake is modeled by a Hill-type expression with a large Hill-coefficient, i.e. (Mahrl et al., 2000)

$$\rho_{\text{mito}}^{\text{in}} = k_{\text{in}} \frac{c^8}{k_2^8 + c^8}, \tag{8}$$

where k_{in} represents the maximum permeability of the Ca^{2+} uniporter and k_2 the half-saturation for Ca^{2+}. In other models, such as the one by Falcke et al. (1999), smaller Hill-coefficients have been used.

Ca^{2+} is released from the mitochondria through Na^{+}/Ca^{+} exchangers and through mitochondrial permeability transition pores. In Mahrl et al. (2000), these two fluxes have been combined into one term, i.e.

$$\rho_{\text{mito}}^{\text{out}} = \left(k_{\text{m}} + k_{\text{out}} \frac{c^2}{K_3^2 + c^2} \right) c_{\text{m}}, \tag{9}$$

where c_{m} is the Ca^{2+} concentration in the mitochondria. Other modelers (Falcke et al., 1999) use a more explicit model that includes the Na^{+} concentration in the cytosol and thus links Ca^{2+} signaling directly to transmembrane potentials (see, e.g., Mahrl et al., 1997).

Conservation of total Ca^{2+} requires two additional equations for the Ca^{2+} concentration in the ER, c_{ER}, and the Ca^{2+} concentration in the mitochondria c_{m}, i.e.

$$\frac{\mathrm{d}c_{\text{ER}}}{\mathrm{d}t} = \rho_{\text{Ca}}(c, \mathbf{x}, t) = \rho_{\text{pump}} - \rho_{\text{channel}} - \rho_{\text{leak}} + D_{\text{Ca}}^{(\text{ER})} \nabla^2 c_{\text{ER}}, \tag{10}$$

and

$$\frac{\mathrm{d}c_{\text{m}}}{\mathrm{d}t} = \rho_{\text{mito}}^{\text{in}} - \rho_{\text{mito}}^{\text{out}} + D_{\text{Ca}}^{(\text{mito})} \nabla^2 c_{\text{m}}. \tag{11}$$

The second messenger IP_3 is generated by binding of agonist and can diffuse relatively quickly through the cell, since it is less buffered than Ca^{2+}. The dynamics of IP_3 is thus described by a diffusion equation with a linear decay-term modeling the observed degradation of IP_3, and a production term $\alpha f(\mathbf{x}, t)$ i.e.

$$\frac{\partial c_{\text{IP}}}{\partial t} = \frac{1}{\tau_{\text{IP}}} (c_{\text{IP}}^* - c_{\text{IP}}) + \alpha f(\mathbf{x}, t) + D_{\text{IP}} \nabla^2 c_{\text{IP}}, \tag{12}$$

where c_{IP} is the concentration of IP_3, c_{IP}^* the equilibrium concentration, $1/\tau_{\text{IP}}$ the degradation rate of IP_3 (for recent values, see Wang et al., 1995), and D_{IP} the intracellular diffusion coefficient.

2.2. *Deterministic, non-phenomenological modeling*

Typically, these models predict Ca^{2+} oscillations or oscillatory spikes, if the concentration of IP_3 is within a certain interval. For the Li–Rinzel model (Li and Rinzel, 1994), e.g., where concentration gradients and mitochondria are neglected, the bifurcation diagram plotted in Fig. 1 is obtained.

For the original parameters in (Li and Rinzel, 1994) one finds a steady state concentration of Ca^{2+} at IP_3 concentrations below 0.345 μM and above 0.642 μM. In between the Ca^{2+} concentration oscillates between the minimum amplitude described by the lower branch in Fig. 1 and the maximum amplitude described by the upper branch.

If the Ca^{2+} concentration is spatially not uniform (the spatial derivatives in the equations above are now taken into account), the Ca^{2+} oscillations are organized in terms of non-linear waves. These waves can have the form of concentric rings (target waves), plane waves and rotating spiral waves. Such waves have been observed to exist in *Xenopus* oocyte (large cells) by Lechleiter and Clapham (1992) and by Lechleiter et al. (1991). Mathematical analysis of these waves have been performed using a two pool model (Dupont and Goldbeter, 1994) or piecewise linear models (Sneyd et al., 1993; Atri et al., 1993). The effect of Ca^{2+} buffers is reviewed, e.g., by Keener and Sneyd (1998). It has been shown by Smith et al. (1998) for cardiac myocytes that to obtain quantitative agreement between simulations and experiments the diffusible indicator dye, which also acts as a buffer, has to be taken into account. This gives rise to yet another pair of non-linear partial differential equations coupled to the equations above. The effect of mitochondria on intracellular Ca^{2+} waves is discussed by Falcke et al. (1999).

2.3. *Phenomenological modeling*

Phenomenological models that are less demanding in compute time than the stochastic models to be discussed below and also serve well in providing an intuitive picture have

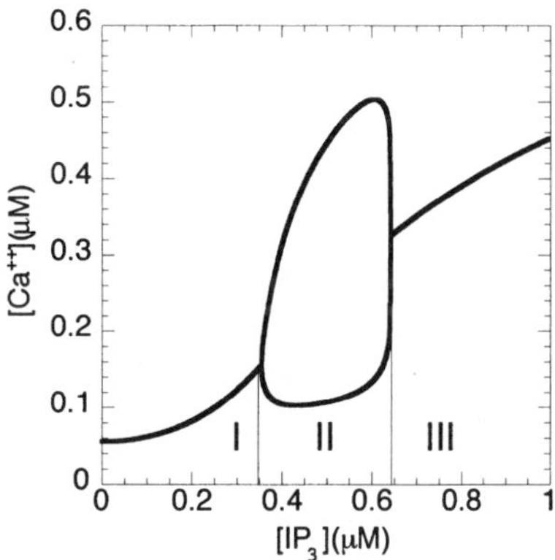

Fig. 1. Bifurcation diagram of the Li–Rinzel model (Li and Rinzel, 1994). The two branches in the interval $0.345 < [IP_3] < 0.642$ μM indicate the minimum and maximum amplitude of the Ca^{2+} oscillations. Below 0.345 μM and above 0.642 μM oscillations are absent.

been used to describe intracellular calcium waves with discrete sources. A simple, but very intuitive model is the fire-and-diffuse model (Keizer et al., 1998a; Dawson et al., 1999). In this one-dimensional model, clustered sources of Ca^{2+} are placed equidistantly on a line. Each source releases a fixed amount of Ca^{2+}, when the Ca^{2+} concentration c exceeds a threshold c_0. When Ca^{2+} is released from a site, it diffuses along the line, increases the Ca^{2+} concentrations at neighboring sites and—depending on the distance between the clusters and the amount of Ca^{2+} being released—can cause Ca^{2+} release at neighboring sites, that in turn can cause release of Ca^{2+} at their neighboring sites, and so on. The model yields analytic values for the speed of the calcium wave.

Another phenomenological model, similar to the fire-and-diffuse model is an excitable cellular-automaton model with discrete release sites (Jung et al., 1998). This model is a two-dimensional array of discrete release sites. Similar to the fire-and-diffuse model, each release site releases a fixed amount of Ca^{2+}, when the local Ca^{2+} concentration exceeds a threshold. According to the model, the released Ca^{2+} diffuses in the intracellular space and approaches instantaneously a stationary, Gaussian profile. In Falcke et al. (2000) and the fire-and-diffuse model, the profiles (although different) are obtained from the reaction-diffusion equations (1)–(12) (with various simplifications) and the one-dimensional diffusion equation, respectively. Qualitatively, the results are similar. If the coupling between the release clusters is small, the waves are abortive. Yet, fluctuations can generate and maintain local patterns spontaneously (Jung and Mayer-Kress, 1995; Jung et al., 1998; Falcke et al., 2000) and aid weak signals in generating a Ca^{2+} response. Characteristic for these noise-sustained patterns are power-law distributed lifetime and size distributions (Jung et al., 1998; Falcke et al., 2000).

2.4. *Stochastic modeling*

2.4.1. *Clustering of Ca^{2+} release channels*

Recently, high-resolution recordings in different types of cells have shed new light on the elementary intracellular Ca^{2+} release events. It has been observed that the Ca^{2+} release channels are spatially organized in clusters with only 20–50 release channels in each cluster, which has a size of about 100 nm. The calcium release through such small clusters is subject to random fluctuations due to thermal open–closed transitions of individual release channels. After Ca^{2+} is released, it rapidly diffuses within the cluster (within a few μs) and out into the cytosol. There, Ca^{2+} is absorbed by buffers and pumped back into the ER and out into the extracellular space (not considered for modeling in this paper) resulting in a spatially and temporally limited event that has been termed calcium puff or spark (Cheng et al., 1999; Callamaras et al., 1998; Melamed-Book et al., 1999; Gonzalez et al., 2000). Ca^{2+} blips arising from the opening of a single release channel have been observed as well (Bootman et al., 1997; Lipp and Niggli, 1998; Sun et al., 1998). Binding of IP_3 activates the calcium release channels and additional binding of Ca^{2+} opens the channels via Ca^{2+} induced Ca^{2+} release (Bezprovanny et al., 1991). Puffs remain spatially restricted at low concentration of IP_3 stimulus, whereas at high levels of IP_3 neighboring clusters become functionally coupled by Ca^{2+} diffusion and Ca^{2+}-induced Ca^{2+} release, so as to support intracellular

Ca^{2+} waves that propagate throughout the cell. Therefore, Ca^{2+} puffs serve as elementary building blocks of intracellular Ca^{2+} waves. Moreover, puffs can arise spontaneously before a wave is initiated and they can act as the triggers to initiate waves (Bootman et al., 1997). During the last three years, mathematical modeling of intracellular Ca^{2+} signaling starting at the elementary Ca^{2+} release of a single cluster have revealed that clustering of the release channels may have profound consequences for the cellular signaling capability (Shuai and Jung, 2003b).

2.4.2. *Stochastic models*

At the level of a single release cluster, stochastic (i.e., random) effects are dominating. The classification of the Ca^{2+} dynamics in terms of steady state and oscillatory becomes obsolete (Falcke et al., 2000; Shuai and Jung, 2002a,b). The calcium dynamics consists of a strongly random sequence of elementary calcium release events (see Fig. 2). To model intracellular Ca^{2+} dynamics according to a stochastic model of intracellular Ca^{2+} signaling with clustered release channels, the membrane of the ER is modeled as a mosaic of passive and active patches, where the active patches represent the release channel clusters and passive patches contain only pumps and leakage channels. The cell is usually assumed to be flat with a uniform Ca^{2+} concentration across its width in the ER, and another uniform Ca^{2+} concentration in the cytosol to allow two-dimensional modeling. The discreteness and small size of the Ca^{2+} release channels gives rise to novel intracellular Ca^{2+} patterns such as Ca^{2+} puffs, abortive waves (waves that propagate only a short distance and then die out) and tide waves (Falcke et al. 2000; Bär et al., 2001). Similar modeling studies for muscle cells, where the calcium release channels are Ryanodine receptors have been reported by Keizer and Smith (1998). One key issue in Keizer and Smith (1998) and Falcke et al. (2000) is the transition from local calcium sparks to propagating waves, i.e., the spark-to-wave transition. For IP_3 release channels

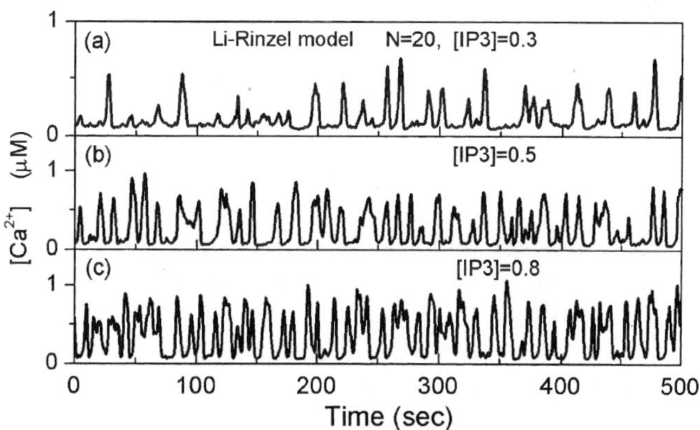

Fig. 2. Calcium released from a single release cluster with 20 release channels at various IP_3 concentrations generated with the stochastic version of the Li–Rinzel Model (Shuai and Jung, 2002a,b). The non-stochastic model predicts Ca^{2+} oscillations between 0.345 and 0.642 μM IP_3 and a steady state Ca^{2+} concentration anywhere else (see Fig. 1). The unit of the IP_3 concentration is μM.

it has been shown in Falcke et al. (2000) that as the intracellular IP_3 concentration is increased, starting from very small sub-threshold values, the calcium patterns change from puffs and abortive waves (sub-threshold) to propagating waves.

Figure 3 shows a sequence of space-time plots (Falcke et al., 2000), where the IP_3 concentration increases from the left panel to the right panel. In the second panel from the left, one can observe abortive waves that—initiated by spontaneous events—propagate through parts of the system and then abort spontaneously. As IP_3 increases, the waves—although still noisy—spread through the entire system and are repetitive, as an oscillatory model (the De Young–Keizer model) has been used.

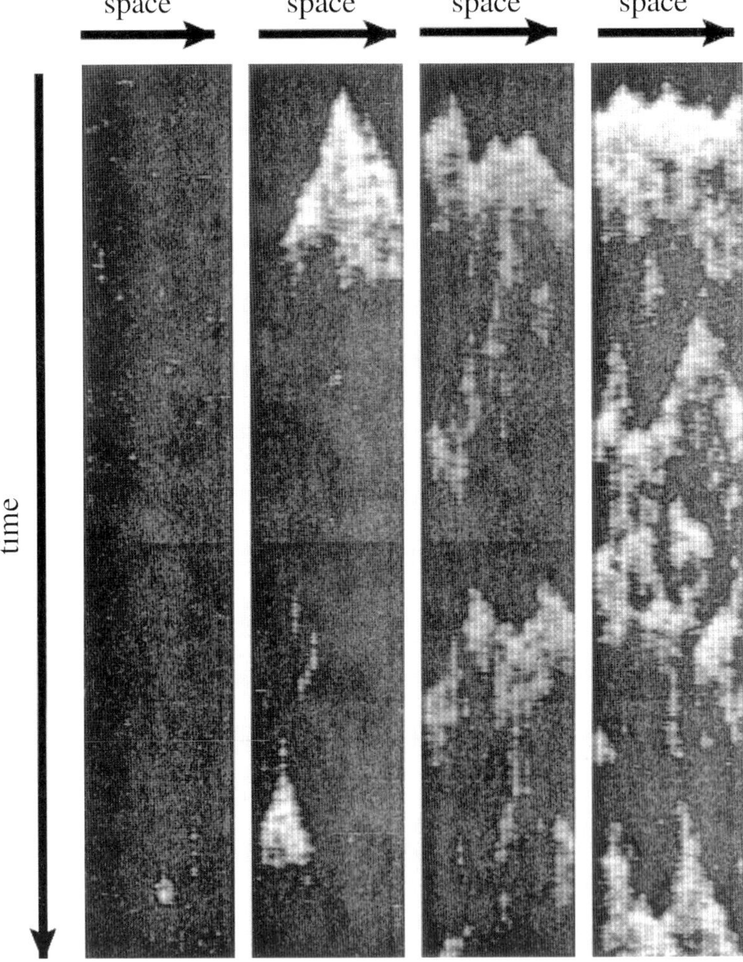

Fig. 3. Space-time plots obtained from a one-dimensional model for intracellular calcium waves with clustered release channels (Falcke et al., 2000). Lighter gray indicates increasing Ca^{2+} concentration. The IP_3 concentration increases from the left panel to the right panel.

A systematic classification of the firing patterns with discrete and stochastic Ca^{2+} release clusters, but neglecting buffers and Ca^{2+} handling by mitochondria (Shuai and Jung, 2003a) is shown in Fig. 4. The 'phase-diagram' of patterns (Ca^{2+} diffusion coefficient versus IP_3 concentration) presented in the figure revealed that essentially all types of observed intracellular Ca^{2+} release patterns could be reconstructed using Eqs. (1)–(12) supplemented with discrete sources of Ca^{2+}. In Fig. 5 snapshots can be seen of different Ca^{2+} transients simulated using this model (Shuai and Jung, 2003a).

2.4.3. *Benefit of clustered release channels*

Recently it has been suggested that the clustering of the IP_3Rs may enhance Ca^{2+} signaling capability (Shuai and Jung, 2003b). To this end, Eqs. (1)–(12), neglecting effects of mitochondria effects and slow buffers, have been supplemented with spatially discrete distributions of a fixed number of Ca^{2+} release channels with different spatial organization. The stochastic Li–Rinzel model was used to simulate local Ca^{2+} release through IP_3Rs. The concentration of IP_3 was kept below the threshold of Ca^{2+} oscillations observed in Fig. 1. Distributing the release channels homogeneously on the membrane of the ER yielded a cell-averaged low steady-state Ca^{2+} concentration—as one would expect. Clustering the channels at distances such that the total number of release channels is conserved (i.e., larger cluster-distance goes along with larger clusters), it was found that in a certain range of cluster distances (and corresponding sizes) the cell-averaged Ca^{2+} concentrations exhibited strong (almost noise-free) oscillations, coding a weak signal (small IP_3 concentration), that was not coded, when the clusters where homogeneously

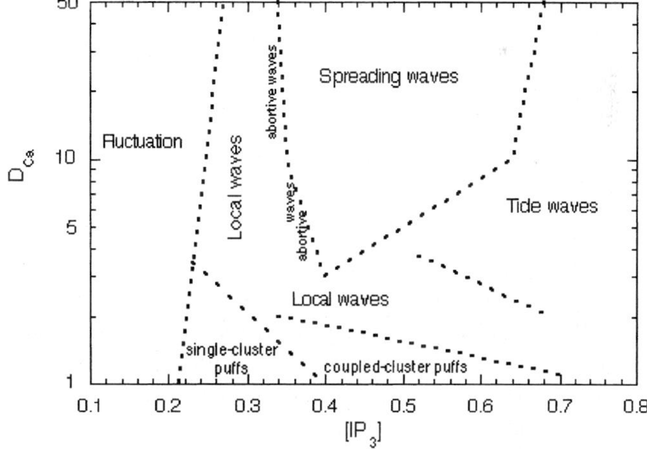

Fig. 4. Phase diagram of intracellular Ca^{2+} patterns from Shuai and Jung (2003a). The diffusion coefficient of Ca^{2+} (D_{Ca}) is indicated in $\mu m^2/s$ and the concentration of IP_3 in μM. The underlying model for the channel fluxes is the spatially extended, stochastic Li–Rinzel model with discrete and clustered Ca^{2+} release channels. The clusters contain 20 IP_3Rs each and are spaced at a distance of 2 μm. The size of the system is 60 $\mu m \times 60$ μm. In the notation used, abortive waves are local, non-propagating but almost propagating waves. Snapshots of the different types of Ca^{2+} transients are shown in Fig. 5.

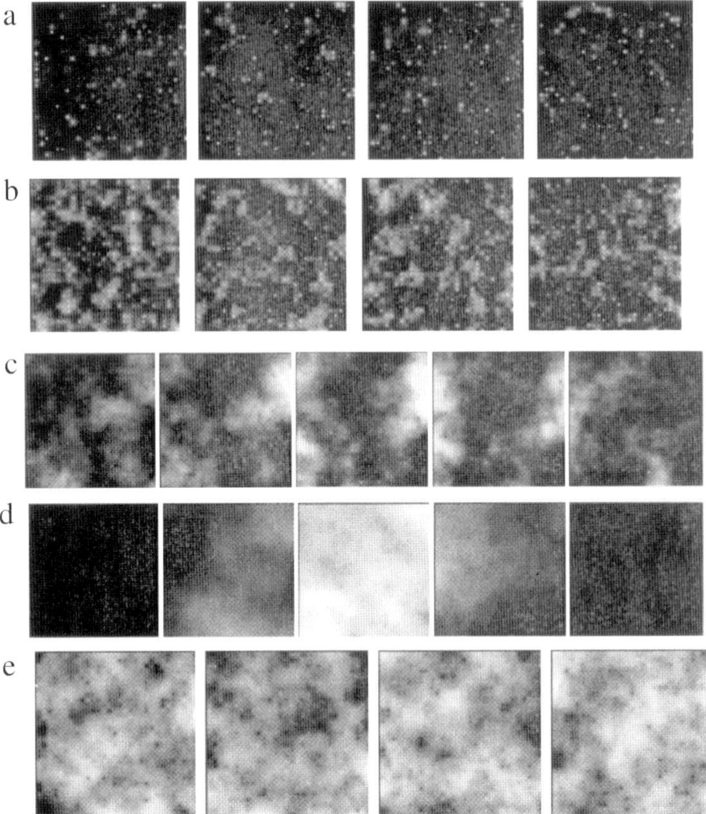

Fig. 5. Snapshots of simulated Ca^{2+} transients as a function of the diffusion coefficient of Ca^{2+} (D_{Ca}) and the IP_3 concentration. Lighter gray indicates increasing Ca^{2+} concentration. The system size is $60 \times 60\ \mu m^2$. (a) Ca^{2+} *puffs* at $D = 1\ \mu m^2/s$ and $[IP_3] = 0.3\ \mu M$ at a rate of 1 frame/3 s; (b) *local Ca^{2+} waves* at $D = 2\ \mu m^2/s$ and $[IP_3] = 0.5\ \mu M$ at a rate of 1 frame/3 s; (c) *abortive Ca^{2+} wave* at $D = 10\ \mu m^2/s$ and $[IP_3] = 0.35\ \mu M$ at a rate of 1 frame/1.4 s; (d) *spreading Ca^{2+} wave* at $D = 30\ \mu m^2/s$ and $[IP_3] = 0.5\ \mu M$ at a rate of 1 frame/2.4 s; (e) *Ca^{2+} tide wave* at $D = 10\ \mu m^2/s$ and $[IP_3] = 0.8\ \mu M$ at a rate of 1 frame/2 s. (Data taken from Shuai and Jung, 2003a).

distributed (Shuai and Jung, 2003b). This effect is demonstrated in Fig. 6. The left upper panel shows the cell-averaged Ca^{2+} signal in response to a sub-threshold IP_3 concentration, when 14,400 channels were placed homogeneously at a distance of 0.5 μm over an area of 60 μm × 60 μm. The upper panel of the second column in Fig. 6 shows the cell-averaged Ca^{2+} response at cluster distance of 3 μm with a corresponding cluster size of 36 IP_3Rs. The cell-averaged Ca^{2+} signal has now a large amplitude and almost perfect phase coherence, although no parameters other than the geometric distribution of the channels have been changed. The small IP_3 signal is now well encoded in the Ca^{2+} signal. Increasing the cluster distance further to 5 μm with a corresponding cluster size of 100 channels, the cell-averaged Ca^{2+} signal is almost constant with some small fluctuations. Thus the IP_3 signal is not encoded.

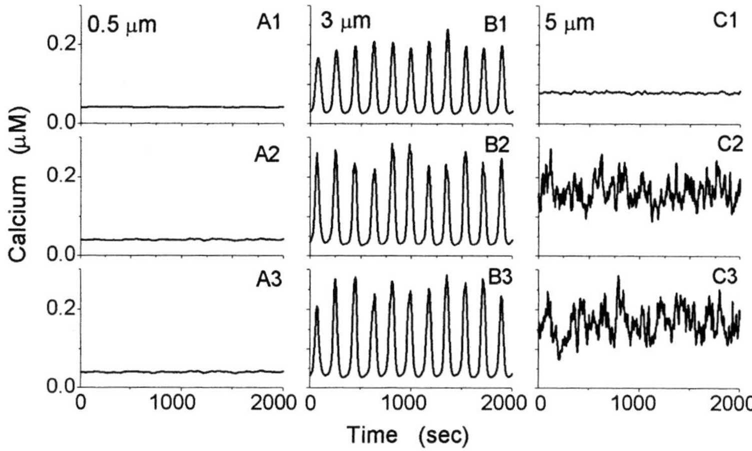

Fig. 6. Ca^{2+} concentrations at two neighboring active sites 0.5 μm apart (A2, A3), 3 μm apart (B2, B3) and 5 μm apart (C2, C3) and the corresponding cell-averaged Ca^{2+} concentrations (A1, B1, C1). The Ca^{2+} diffusion coefficient in the cytosol, $D = 20$ μm^2/s and $[IP_3] = 0.21$ μM (Shuai and Jung, 2003b).

Although in most computational studies the clusters of the IP_3Rs are assumed to be organized on a regular grid this is not so in cells. Exceptions are the two recent publications by Falcke (2003a,b) on nucleation of calcium waves and the effects of slow buffers. The clusters are typically close to the cell membrane and are clustered themselves forming 'hot spots'. It remains to be shown whether optimal clustering is robust for these more realistic scenarios.

3. Modeling of intercellular Ca^{2+} signaling (IRCW) in astrocytes

Calcium signals can travel through the cell membrane and propagate through many cells. Intercellular Ca^{2+} waves have been observed in astrocyte cultures (Cornell-Bell et al., 1990; Giaume and Venance, 1998; Charles, 1998), hippocampal slice cultures of mice (Harris-White et al., 1998), cultured glioma cells (Charles et al., 1992), neurons (Charles et al., 1996), and hepatocytes (Combettes et al., 1994; Nathanson and Burgstahler, 1995; Robb-Gaspers and Thomas, 1995; Patel et al., 1999). Although the mechanism for intercellular Ca^{2+} waves (IRCW) is controversial, it is clear that it is different from the mechanism for intracellular Ca^{2+} waves (IACW). There exists a substantial amount of indirect evidence that gap junctions connecting neighboring astrocytes are important for the propagation of IRCW, in that they provide permeability for the intracellular messenger IP_3. If an increased concentration of IP_3 is generated in an astrocyte, it may not only aid in providing intracellular Ca^{2+} via calcium-induced calcium release in the cell in question, but it also may diffuse through gap junction to neighboring astrocytes and contribute to the generation of an intracellular Ca^{2+} response. Such a mechanism has been modeled by Sneyd et al. (1995a,b) and Hofer et al. (2002), with numerous simplifications. The source density of IP_3 has to be supplemented by the divergence of the flux density through the gap-junctions at the cell-boundaries

$$\nabla \mathbf{j}_{\text{gap}}(\mathbf{x})|_{\text{cell-boundary}}. \tag{13}$$

Assuming Fick's law for the flux through the gap junctions (driven by a gradient in IP_3 concentration), i.e.

$$\mathbf{j}_{\text{gap}} = -D_{\text{gap}} \nabla c_{\text{IP}}, \tag{14}$$

and discretizing the Laplacian in the equation for IP_3 (last term on the right-hand side of Eq. (12)), the additional terms due to the gap junctions are

$$-\frac{D_{\text{gap}}}{\Delta x^2}(c_{\text{IP}}^{n+1} - 2c_{\text{IP}}^{n} + c_{\text{IP}}^{n-1}), \tag{15}$$

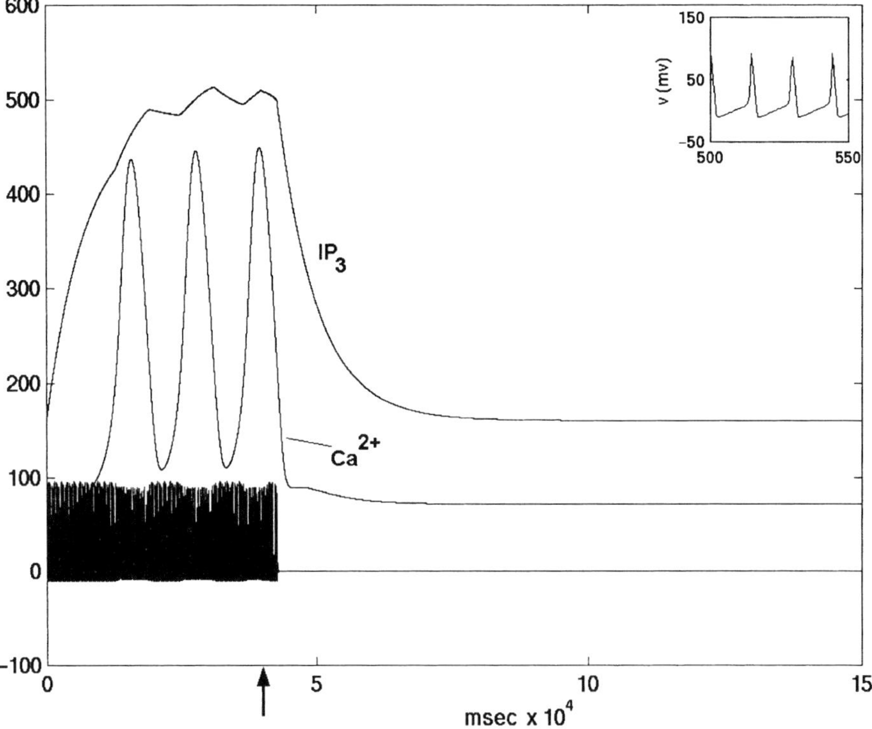

Fig. 7. A DC stimulus is applied to the neuron during the time interval [0–40 s] (end of stimulation interval is indicated by an arrow) at an IP_3 production rate of $\alpha = 0.5$ in the astrocyte. The lowest (dense) curve depicts the neuronal oscillations and an expanded section of it is shown in the inset. As the neuron fires, the concentration of IP_3 in the astrocyte (upper curve) is increasing. If the IP_3 concentration is high enough, intracellular Ca^{2+} oscillations occur. When the stimulus to the neuron is stopped, IP_3 degradation overwhelms the positive feedback into the neuron, and the Ca^{2+} oscillations and the neuronal oscillations disappear.

if the cell-boundary is perpendicular to this spatial direction. Assuming the cell membranes, separating the two cells is located between x_n and x_{n+1}, and a steep gradient of c_{IP} between the cells in comparison to the one within the cells, i.e.

$$c_{\mathrm{IP}}^{n-1} \approx c_{\mathrm{IP}}^{n}, \tag{16}$$

one finds

$$-\frac{D_{\mathrm{gap}}}{\Delta x^2}(c_{\mathrm{IP}}^{n+1} - 2c_{\mathrm{IP}}^{n} + c_{\mathrm{IP}}^{n-1}) \approx \frac{D_{\mathrm{gap}}}{\Delta x^2}(c_{\mathrm{IP}}^{n+1} - c_{\mathrm{IP}}^{n}) \equiv P_{\mathrm{gap}}(c_{\mathrm{IP}}^{n+1} - c_{\mathrm{IP}}^{n})$$

with the gap-junction permeability P_{gap}. Gap junction permeability has not been available in the literature and thus its value needs to be inferred indirectly, e.g., by the propagation distance of a wave. Such a diffusive coupling mechanism leads to a diffusive type of IRCW with a speed that would decrease as it spreads. This prediction is in agreement with

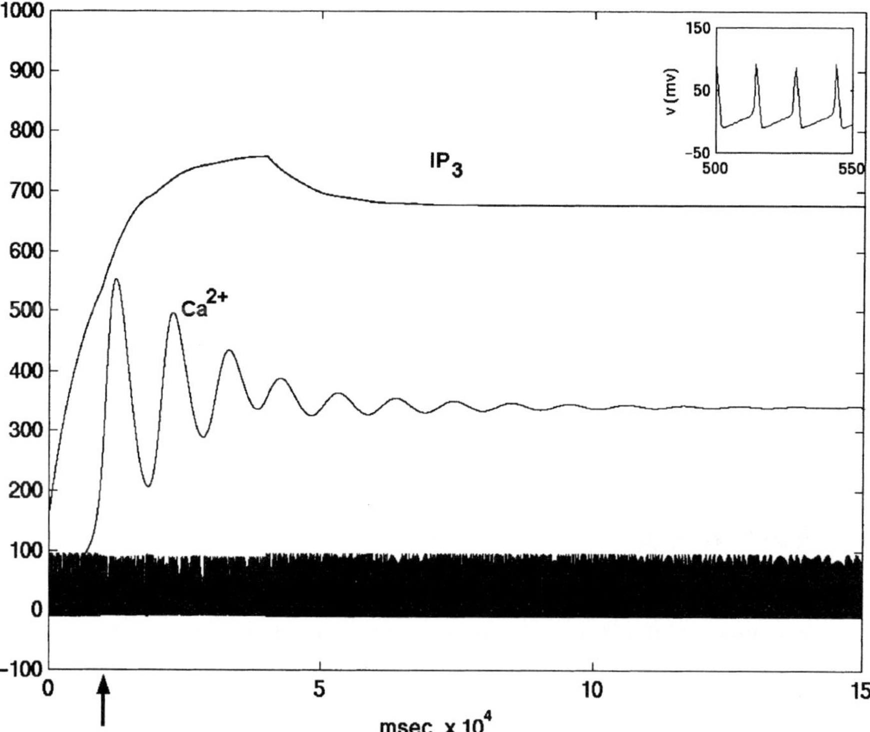

Fig. 8. DC stimulus is applied to the neuron during the time interval [0–40 s] (end of stimulation interval is indicated by an arrow) at the larger IP_3 production rate $\alpha = 0.8$ in the astrocyte. The lowest (dense) curve depicts the neuronal oscillations and an expanded section of it is shown in the inset. As the neuron fires, the concentration of IP_3 in the astrocyte (upper curve) is increasing. If the IP_3 concentration is high enough, intracellular Ca^{2+} oscillations occur. When the stimulus to the neuron is stopped, the production rate of IP_3 (larger than in Fig. 7) overwhelms the degradation and the Ca^{2+} oscillations and neuronal oscillations continue indefinitely.

observations in astrocytes (Sneyd et al., 1995a,b; Harris-White et al., 1998; Jung et al., 1998). It has, however, also been suggested that the diffusion range of IP_3 is not large enough to account for the distance an IRCW is propagating (Giaume and Venance, 1998).

While there is general agreement regarding the role of IP_3 as a messenger (see, however, also chapter by Scemes and Spray), the role of additional extracellular messengers is controversial, and it is a current topic of research. It has been observed that cultured mouse hippocampal astrocytes that were not coupled to other astrocytes by gap junctions, also participate in the IRCW (Hassinger et al., 1996). Using a novel technology to image extracellular ATP it has been reported (Wang et al., 2000) that Ca^{2+} waves triggered by mechanical stimulation are synchronized with a spreading extracellular ATP signal. Based on the propagation range of this signal it has been concluded by Wang et al. (2000) that the signal is regenerative, although it spreads only across a finite distance, i.e., that cells receiving the external ATP signal (by binding of ATP to purinergic receptors), also generate extracellular ATP in response. It has also been speculated that IP_3 traveling through gap junctions may not be the mediator of intercellular Ca^{2+} waves (Wang et al., 2000). This hypothesis is supported by the report by Bushong et al. (2002) that the astrocytes in most of the brain form isolated domains (see chapter by Scemes and Spray) so that only a diffusible extracellular messenger could facilitate direct signals between the cells. However, Arcuino et al. (2002) challenge the hypothesis of a regenerative wave-like ATP signal, concluding that ATP released in response to stimulation from a group of astrocytes diffuses passively through the extracellular space, and Giaume and Venance (1998) find no evidence that extracellular ATP should be involved in the intercellular Ca^{2+} wave. Mathematical modeling that includes extracellular messengers has not been carried through according to our knowledge.

4. Bidirectional coupling between neurons and astrocytes

Recently our group (Nadkarni and Jung, in preparation) has proposed a model for neuronal dynamics, that takes into account the coupling between neurons and astrocytes. The key elements of the model are coupling of Ca^{2+} dynamics in the synaptic astrocytes to neuronal dynamics, described by a conductance-based model. An increased Ca^{2+} concentration in synaptic astrocytes causes an additional inward current in neurons (Parpura and Haydon, 2000), probably due to generation and release of extracellular glutamate from astrocytes. The quantitative relationship between Ca^{2+} concentration in the astrocyte and the additional inward current in the neuron has been fitted by a curve and added to the ionic conductance model. When the neuron fires, it releases glutamate that binds to the metabotropic glutamate receptors on the astrocytes and causes the generation in the astrocyte of IP_3, that in turn contributes to calcium-induced calcium release and an intracellular Ca^{2+} signal. An equation has been set up to simulate the IP_3 concentration (see Eq. (12)), based upon its rates of generation, α, and degradation, $1/\tau_{IP}$ (Wang et al., 1995). The rate of generation of IP_3 in the astrocyte in response to neuronal firing of an action potential is a free parameter. While there is no experimental value available for this rate, we know that it has to be proportional to the density of metabotropic glutamate receptors on the synaptic astrocytes. One of the important conclusions of this study is that the critical value of the injected current into the neuron (sum of all synaptic inputs) to generate repetitive (periodic) neuronal firing is reduced in the presence of the astrocyte feedback-loop (see Figs. 7 and 8). The IP_3 production rate α, proportional to the density of mGluR is 0.5 s^{-1} in Fig. 7 and 0.8 s^{-1} in Fig. 8 where spontaneous oscillations remain even after the neuronal stimulation is terminated.

As a matter of fact, if the density of metabotropic glutamate receptors on the synaptic astrocytes is large enough (see Fig. 8), oscillations can set off in the absence of external stimulus. In this context it is interesting to note that astrocytes in epileptic foci are known to over-express metabotropic glutamate receptors (Ulas et al., 2000; Aronica et al., 2000; Tang and Lee, 2001).

5. Concluding remarks

We reviewed the recent literature on modeling of intracellular and intercellular calcium waves with focus on brain tissue. We have pointed out the importance of stochastic modeling for intracellular calcium signaling in view of the discreteness of the elementary release events even for cellular signaling capability. Mathematical modeling of intracellular Ca^{2+} signaling of the Ca^{2+} release channel results in enhanced cellular signaling capability in response to weak stimuli with optimal clustering. While intracellular modeling is advanced and capable of reproducing experimental results, modeling of intercellular signals is still patchy and incomplete. Progress in this field requires a better and more complete understanding of the underlying mechanisms that are currently only poorly understood. Bidirectional interaction of astrocytes with neurons can alter the behavior of the neurons. We have reported on a recent study where it has been shown that the feedback from synaptic astrocytes into the same synapse can have the effect of generating spontaneous neuronal oscillations.

Acknowledgements

This material is based upon work supported by the National Science Foundation under Grant No. IBN-0078055.

References

Arcuino, G., Lin, J.H.C., Takano, T., Liu, C., Jiang, L., Gao, Q., Kang, J., Nedergaard, M., 2002. Intercellular calcium signaling mediated by point-source release of ATP. Proc. Natl Acad. Sci. USA 99, 9840–9845.

Aronica, E., van Vliet, E.A., Mayboroda, O.A., Troost, D., da Silva, F.H.L., Gorter, J.A., 2000. Upregulation of metabotrobic glutamate receptor subtype mGluR3 and mGluR5 in reactive astrocytes in a rat model of mesial temporal lobe epilepsy. Eur. J. Neurosci. 12, 2333–2344.

Atri, A., Amundson, J., Clapham, D., Sneyd, J., 1993. A single-pool model for intracellular calcium oscillations and waves in the *Xenopus laevis* oocyte. Biophys. J. 65, 1727–1739.

Bär, M., Falcke, M., Levine, H., Tsimring, L.S., 2000. Discrete stochastic modeling of calcium channel dynamics. Phys. Rev. Lett. 84, 5664–5667.

Bezprovanny, I., Watras, J., Ehrlich, B., 1991. Bell-shaped calcium response curves of Ins(1,4,5) P3- and calcium-gated channels from endoplasmic reticulum of cerebellum. Nature 351, 751–754.

Bootman, M., Niggli, E., Berridge, M., Lipp, P., 1997. Imaging the hierarchical Ca^{2+} signaling system in HeLa cells. J. Physiol. 499, 307–314.

Bordeay, A., Sontheimer, H., 1998a. Electrophysiological properties of human astrocytic tumor cells in situ: enigma of spriking glial cells. J. Neurophysiol. 79, 2782–2793.

Bordeay, A., Sontheimer, H., 1998b. Properties of human glial cells associated with epileptic seizure foci. Epilepsy Res. 32, 286–303.

Bushong, E.A., Martone, M.E., Jones, Y.Z., Ellisman, M.H., 2002. Protoplasmic astrocytes in CA1 stratum radiatum occupy separate anatomical domains. J.Neurosci. 22, 183–192.

Callamaras, N.J., Marchant, S., Sun, X., Parker, I., 1998. Activation and co-ordination of $InsP_3$ mediated elementary Ca^{2+} events during global Ca^{2+} signals in *Xenopus* oocytes. J. Physiol. 509, 81–91.

Charles, A., 1998. Intercellular calcium waves in glia. Glia 24, 39–49.

Charles, A.C., Kodali, S.K., Tyndale, R.F., 1996. Intercellular calcium waves in neurons. Mol. Cell. Neurosci. 7, 337–353.

Charles, A.C., Naus, C.C.G., Zhu, D., Kidder, G.M., Dirksen, E.R., Sanderson, M.J., 1992. Intercellular calcium signaling via gap junctions in glioma cells. J. Cell Biol. 118, 195–201.

Cheng, H., Song, L., Shirokova, N., Gonzalez, A., Lakatta, E.G., Rios, E., Stern, M.D., 1999. Amplitude distribution of calcium sparks in confocal images: theory and studies with an automatic detection method. Biophys. J. 76, 606–617.

Combettes, L., Trans, D., Tordjmann, T., Laurent, M., Berthon, B., Claret, M., 1994. Ca^{2+} mobilizing hormones induce sequentially ordered Ca^{2+} signals in multicellular systems of rat hepathocytes. Biochem. J. 304, 585–594.

Cornell-Bell, A.H., Finkbeiner, S.M., Copper, M.S., Smith, S.J., 1990. Glutamate induces calcium waves in cultured astrocytes: long-range glial signaling. Science 247, 470–473.

Dawson, S.P., Keizer, J., Pearson, J.E., 1999. Fire-diffuse-fire model of dynamics of intracellular calcium waves. Proc. Natl Acad. Sci. USA 96, 6060–6063.

De Young, G.W., Keizer, J., 1992. A single-pool inositol 1,4,5-trisphosphate-receptor-based model for agonist-stimulated oscillations in Ca^{2+} concentration. Proc. Natl Acad. Sci. USA 89, 9895–9899.

Dupont, G., Goldbeter, A., 1993. One-pool model for Ca^{2+} oscillations involving Ca^{2+} and inositol 1,4,5 triphosphate as co-agonist for Ca^{2+} release. Cell Calcium 14, 311–322.

Dupont, G., Goldbeter, A., 1994. Properties of intracellular Ca^{2+} waves generated by a model based on Ca^{2+} induced Ca^{2+} release. Biophys. J. 67, 2191–2204.

Falcke, M., 2003. On the role of stochastic channel behavior in intracellular Ca^{2+} dynamics. Biophys. J. 84, 42–56.

Falcke, M., 2003. Buffers and oscillations in intracellular Ca^{2+} dynamics. Biophys. J. 84, 28–41.

Falcke, M., Hudson, J.L., Camacho, P., Lechleiter, J.D., 1999. Impact of mitochondrial Ca^{2+} cycling on pattern formation and stability. Biophys. J. 77, 37–44.

Falcke, M., Tsimring, L., Levine, H., 2000. Stochastic spreading of intracellular Ca^{2+} release. Phys. Rev. E62, 2636–2643.

Giaume, C., Venance, L., 1998. Intercellular calcium signaling and gap junctionional communication in astrocytes. Glia 24, 50–64.

Goldbeter, A., Dupont, G., Berridge, M.J., 1990. Minimal model for signal-induced Ca^{2+} oscillations and for their frequency encoding through protein phosphorylation. Proc. Natl Acad. Sci. USA 87, 1461–1465.

Gonzalez, A., Kirsch, W.G., Shirokova, N., Pizarro, G., Brum, G., Pessah, I.N., Stern, M.D., Cheng, H., Rios, E., 2000. Involvement of multiple intracellular release channels in calcium sparks of skeletal muscle. Proc. Natl Acad. Sci. USA 97, 4380–4385.

Harris-White, M.E., Zanotti, S.A., Fruatchy, S.A., Charles, A.C., 1998. Spiral intercellular calcium waves in hippocampal slice cultures. J. Neurophysiol. 79, 1045–1052.

Hassinger, T.D., Guthrie, P.B., Atkinson, P.B., Bennet, M.V.L., Kater, S.B., 1996. An external signaling component in propagation of astrocytic calcium waves. Proc. Natl Acad. Sci. USA 93, 13268–13273.

Hofer, T., Venance, L., Giaume, C., 2002. Control and plasticity of intercellular calcium waves in astrocytes: A modeling approach. J. Neurosci., 22, 4850–4859.

Jung, P., Cornell-Bell, A., Madden, K.S., Moss, F., 1998. Noise-induced spiral waves in astrocyte syncytia show evidence of self-organized criticality. J. Neurophysiol. 79, 1098–1101.

Jung, P., Mayer-Kress, G., 1995. Spatiotemporal stochastic resonance in excitable media. Phys. Rev. Lett. 74, 2130–2133.

Keener, J., Sneyd, J., 1998. Mathematical Physiology. Springer, Berlin.

Keizer, J., Smith, G.D., Ponce-Dawson, S., Pearson J.E., 1998a. Saltatory propagation of Ca^{2+} waves by Ca^{2+} sparks. Biophys. J. 75, 595–600.

Keizer, J., Smith, G.D., 1998b. Spark-to-wave transition: saltatory transmission of calcium waves in cardiac myocytes. Biophys. Chem. 72, 87–100.

Lechleiter, J., Clapham, D., 1992. Molecular mechanism on intracellular calcium excitability in *Xenopus laevis* oocytes. Cell 69, 283–294.

Lechleiter, J., Girard, S., Clapham, D., Peralta, E., 1991. Subcellular patterns of calcium release determined by G-protein specific residues of muscarinic receptors. Nature 350, 505–508.

Li, Y., Rinzel, J., 1994. Equations for $InsP_3$ receptor-mediated Ca^{2+} oscillations derived from a detailed kinetic model: a Hodgkin–Huxley like formalism. J. Theor. Biol. 166, 461–473.

Lipp, P., Niggli, E., 1998. Fundamental calcium release events revealed by two-photon excitation photolysis of caged calcium in guinea-pig cardiac myocytes. J. Physiol. 508, 801–809.

Magnus, G., Keizer, J., 1997. Minimal model of β-cell mitochondrial Ca^{2+} handling. Am. J. Physiol. 273, C717–C733.

Magnus, G., Keizer, J., 1998. Model of β-cell mitochondrial calcium handling and electrical activity. I. Cytoplasmic variables. Am. J. Physiol. 274, C1158–C1173.

Magnus, G., Keizer, J., 1998. Model of β-cell mitochondrial calcium handling and electrical activity. II. Mitochondrial variables. Am. J. Physiol. 274, C1174–C1184.

Mahrl, M., Hzberichter, Th., Brumen, M., Heinrich, R., 2000. Complex calcium oscillations and the role of mitochondria and cytosolic proteins. Biosystems 57, 75–86.

Mahrl, M., Schuster, S., Brumen, M., Heinrich, R., 1997. Modeling the interrelation between calcium oscillations and ER membrane potential oscillations. Biophys. Chem. 63, 221–239.

Melamed-Book, N., Kachalsky, S.G., Kaiserman, I., Rahamimoff, R., 1999. Neuronal calcium sparks and intracellular calcium noise. Proc. Natl Acad. Sci. USA 26, 15217–15221.

Nadkarni, S., Jung, P, submitted.

Nathanson, M.H., Burgstahler, A.D., 1995. Coordination of hormone-induced calcium signals in isolated rat hepatocytes couplets; demonstration with confocal microscopy. Mol. Biol. Cell 3, 113–121.

Parpura, V., Haydon, P., 2000. Physiological astrocytic calcium levels stimulate glutamate release to modulate adjacent neurons. PNAS 97, 8629–8634.

Patel, S., Robb-Gaspers, L.D., Stellato, K.A., Shon, M., Thomas, A.P., 1999. Coordination of calcium signaling by endothelia derived nitric oxide in the intact liver. Nat. Cell Biol. 1, 467–471.

Porter, J.T., McCarthy, K.D., 1996. Hippocampal astrocytes in situ respond to glutamate released from synaptic terminals. J. Neurosci. 16, 5073–5081.

Robb-Gaspers, L.D., Thomas, A.P., 1995. Coordination of Ca^{2+} signaling by intercellular propagation of Ca^{2+} waves in the intact liver. J. Biol. Chem. 270, 8102–8107.

Schuster, S., Marhl, M., Hofer, T., 2002. Modelling of simple and complex calcium oscillations—from single-cell responses to intercellular signaling. Eur. J. Biochem. 269, 1333–1355.

Shuai, J.W., Jung, P., 2002a. Optimal intracellular calcium signaling. Phys. Rev. Lett. 88, 068102.

Shuai, J.W., Jung, P., 2002b. Stochastic properties of Ca^{2+} release of inositol 1,4,5-trisphosphate receptor clusters. Biophys. J. 83, 87–97.

Shuai, J.W., Jung, P., 2003. Selection of intracellular calcium patterns in a model with clustered Ca^{2+} release channels. Phys. Rev. E 67, article #031905.

Shuai, J.W., Jung, P., 2003a. Phys. Rev. E (in press).

Shuai, J.W., Jung, P., 2003b. Optimal ion channel clustering for intracellular calcium signaling. Proc. Natl Acad. Sci. USA 100, 506–510.

Smith, G.D., Keizer, J.E., Stern, M.D., Lederer, W.J., Cheng, H., 1998. A simple numerical model of calcium spark formation and detection in cardiac myocytes. Biophys. J. 75, 15–32.

Sneyd, J., Girard, S., Clapham, D., 1993. Calcium wave propagation by calcium-induced calcium release: an unusual excitable system. Bull. Math. Biol. 55, 315–344.

Sneyd, J., Keizer, J., Sanderson, M.J., 1995. Mechanism of calcium oscillations and waves: a quantitative analysis. FASEB J. 9, 1463–1472.

Sneyd, J., Wetton, B.T.R., Charles, A.C., 1995. Intercellular calcium waves mediated by diffusion of inositol thiphosphate: a two dimensional model. Am. J. Physiol. 268, C1537–C1545.

Somorgyi, R., Stucki, J.W., 2000. Hormone-induced calcium oscillations in liver cells can be explained by a simple one pool model. J. Biol. Chem. 266, 11068–11077.

Sun, X.N., Callamaras, N.J., Marchant, J.S., Parker, I., 1998. A continuum of $InsP^3$-mediated elementary Ca^{2+} signaling events in *Xenopus* oocytes. J. Physiol. 509, 67–80.

Tang, F.R., Lee, W.L., 2001. Expression of the group II and III metabotropic glutatmate receptors in the hippocampus of patients with mesial temporal lobe epilepsy. J. Neurocytol. 30, 137–143.

Ulas, J., Satou, T., Ivins, J.P., Kesslak, C.W., Balazs, R., 2000. Expression of metabotropic gluatamet receptor 5 Is increased in astrocytes after kainate-induced epileptic seizures. Glia 30, 352–361.

Wang, S.S.H., Alousi, A.A., Thompson, S.H., 1995. The life time of inositol 1,4,5-triphosphate in single cells. J. Gen. Physiol. 105, 149–171.

Wang, Z., Haydin, P.G., Yeung, E.S., 2000. Direct observation of calcium-independent intercellular ATP signaling in astrocytes. Anal. Chem. 72, 2001–2007.

Advances in
Molecular and
Cell Biology

pH regulation in non-neuronal brain cells and interstitial fluid

Suzanne D. McAlear and Mark O. Bevensee*

Department of Physiology and Biophysics, University of Alabama at Birmingham, 1918 University Blvd, 846 MCLM, Birmingham, AL 35294-0005, USA
**Correspondence address: Tel.: +1-205-975-9084; fax: +1-205-975-7679. E-mail: bevensee@physiology.uab.edu*

Contents

1. Introduction
2. Basic principles of pH_i physiology
   2.1. Chronic acid loading
   2.2. pH_i-regulating mechanisms
   2.3. Steady-state pH_i
   2.4. Consequences of pH_i regulation in brain
3. Importance of pH regulation in brain
   3.1. General cellular activity
   3.2. Effects of pH on neuronal activity
   3.3. Effects of neuronal activity on pH_{ECF}
4. pH regulation in glial cells
   4.1. Acid-loading conductances
   4.2. Acid-loading transporters: HCO_3^--independent
   4.3. Acid-loading transporters: HCO_3^--dependent
   4.4. Acid-extruding transporters: HCO_3^--independent
   4.5. Acid-extruding transporters: HCO_3^--dependent
5. pH regulation of the CSF
   5.1. Overview
   5.2. pH regulation by choroid epithelial cells
   5.3. pH regulation by endothelial cells of brain capillaries
6. Concluding remarks

pH regulation in brain is important because changes in brain pH can influence neuronal activity, synaptic transmission, and possibly memory and learning. The pH of the cerebrospinal fluid bathing the brain is determined by mechanisms that regulate intracellular pH (pH_i) in the choroid plexus epithelium and capillary endothelium.

Advances in Molecular and Cell Biology, Vol. 31, pages 707–745

ISBN: 0-444-51451-1

In addition, the pH of the extracellular fluid (pH_{ECF}) surrounding individual brain cells is determined by pH_i-regulating mechanisms of neurons and glia. Changes in pH_{ECF} and pH_i are therefore intertwined because H^+ and HCO_3^- shuttle between the extra- and intracellular compartments. By modulating both pH_i and pH_{ECF}, these mechanisms directly influence pH-sensitive neuronal activity. Glial cells play a fundamental role in controlling both pH_{ECF} and pH_i with the use of powerful acid–base transporters including the Na–H exchanger, the Na/HCO_3 cotransporter, and the Cl–HCO_3 exchanger. Such transporters are also found in the choroid plexus epithelium and the capillary endothelium.

1. Introduction

The regulation of the ionic and chemical environment of brain cells is necessary for normal neuronal activity and brain function. In addition, the regulation of brain pH is important because changes in pH can influence many cellular processes including the activity of cellular enzymes, ion channels, and transporters. Decreases in brain pH associated with pathological conditions such as ischemia, hypoxia, and epileptic events lead to neuronal necrosis and overall brain damage (Katsura and Siesjo, 1998). One of the many functions of glial cells is to regulate the acid–base status of the environment surrounding brain cells. This environment is comprised of two types of fluid: the cerebrospinal fluid (CSF) and the interstitial or extracellular fluid (ECF). The CSF, which comprises the macroenvironment of brain cells, is produced by the choroid plexus and bathes both the internal and external surfaces of the brain. Endothelial cells of the blood–brain barrier also contribute to the composition of the CSF by filtering ions and molecules in the fluid that passes from brain capillaries into the CSF.

The second general fluid is the ECF that occupies the space between cells throughout the brain. ECF is in direct contact with brain cells, and therefore comprises their microenvironment. Although the ECF is ~20% of total brain volume, the volume between adjacent cells is very small because cells are in close apposition to one another (~20 nm apart) (see Ransom, 1992). Thus, the movement of ions (including acid–base equivalents) and other substances across cell membranes can substantially affect extracellular concentrations. For example, a single action potential can be estimated to elevate K^+ in the periaxonal space by as much as 1 mM (Adelman and Fitzhugh, 1975; also see chapter by Walz). The regulation of pH_i by brain cells can also lead to large changes in pH_{ECF}.

The focus of the present chapter is mechanisms by which non-neuronal cells such as glia regulate both pH_i and pH_{ECF}. We will begin by examining some of the basic principles of cellular pH regulation. Subsequently, we will evaluate the relationship between pH and neuronal activity, and the involvement of glial cells. We will then describe the cellular and molecular physiology of the specific acid–base transport mechanisms identified in glia. These mechanisms will include three of the most powerful acid–base transporters glia use to regulate pH_i and pH_{ECF}: the Na-H exchanger, the Na/HCO_3 cotransporter, and the Cl–HCO_3 exchanger. In Section 5 of this chapter, we will examine how some of the acid–base transporters found in the choroid plexus and the capillary endothelium contribute to the pH of the CSF.

2. Basic principles of pH_i physiology

2.1. Chronic acid loading

For a typical cell with a resting membrane potential of − 60 mV and bathed in a pH-7.3 solution, an electrochemical gradient favors the influx of H^+ (or protonated weak acids) and the efflux of HCO_3^- (or deprotonated weak bases) (see Bevensee and Boron, 1998b). If the plasma membrane of our typical cell is permeable to such acid–base equivalents, then their passive movement will tend to acidify the cell to a pH_i of 6.3. In addition, any acid produced by metabolism will tend to remain in the cell. Both the passive movement of acid–base equivalents and metabolic-acid production are regarded as passive acid-loading mechanisms, which subject most cells to a chronic acid load. Nearly all cells have pH_i values well above that predicted for H^+ to be in electrochemical equilibrium. To maintain pH_i above equilibrium pH_i, cells subjected to passive acid-loading mechanisms must therefore expend energy to extrude intracellular acid.

2.2. pH_i-regulating mechanisms

Nearly all cells use a system of acid–base transporters in their plasma membranes to regulate pH_i. A detailed description of the major types of acid-base transporters can be found in Bevensee et al., 2000a. As organized in this review, these transporters can be categorized as acid loaders or acid extruders. Acid loaders move H^+ into cells or bases such as HCO_3^- out of cells, whereas acid extruders move H^+ and bases in the opposite directions. In general, acid loaders contribute to pH_i recoveries following acute intracellular alkali loads, whereas acid extruders contribute to pH_i recoveries following acute intracellular acid loads. As further organized in this review, acid loaders and acid extruders can also be categorized as HCO_3^--independent or HCO_3^--dependent. Acid–extruding transporters are primary, secondary, or tertiary active transporters. Primary active transporters use free energy released from ATP hydrolysis to transport substrate against an electrochemical gradient. An example is the vacuolar-type H^+ pump. Secondary active transporters use free energy released from one substrate moving down an electrochemical gradient to transport another substrate against an electrochemical gradient. An example is the Na–H exchanger. Tertiary active transporters use free energy released from a secondary active transporter. For instance, the transport of monocarboxylates across the basolateral membrane of the kidney proximal tubule is mediated by a H-monocarboxylate cotransporter; the monocarboxylate gradient is established by the Na-monocarboxylate cotransporter in the apical membrane.

2.3. Steady-state pH_i

The steady-state pH_i of a cell is defined by the balance between acid-loading mechanisms (e.g., metabolic-acid production, passive movement of acid–base equivalents, and acid-loading transporters such as the Cl–HCO_3 exchanger) and acid-extruding mechanisms (e.g., acid-extruding transporters such as the Na–H exchanger). If there is an imbalance between the two, then pH_i will change. As reviewed in detail by Bevensee et al. (2000a), the rate of such pH_i change is described by the equation

$$\frac{\mathrm{d}pH_i}{\mathrm{d}t} = \frac{\rho(J_E - J_L)}{\beta_T}$$

where J_E is total acid efflux from all acid-extruding mechanisms, J_L is total acid influx from all acid-loading mechanisms, β_T is the total proton buffering power of the cell, and ρ is the cell's surface area-to-volume ratio. When $J_E > J_L$, pH_i will increase at a rate that is proportional to the magnitude of $J_E - J_L$. When $J_E < J_L$, pH_i will decrease at a rate that is proportional to the magnitude of $J_E - J_L$. The rate of the pH_i change is inversely proportional to the ability of the cell to buffer intracellularly introduced acids or bases (β_T). Also, $\mathrm{d}pH_i/\mathrm{d}t$ is proportional to ρ. When J_E and J_L are equal, $\mathrm{d}pH_i/\mathrm{d}t$ is zero and the cell is at a steady-state pH_i.

A range of steady-state pH_i values for glial cells has been reported. Steady-state pH_i can depend not only on the cell type, but also on other factors such as external pH (pH_o), temperature, and the presence vs. absence of CO_2/HCO_3^-. In the leech neuropile glial cell for instance, steady-state pH_i is 6.9–7.0 in the nominal absence of CO_2/HCO_3^- (Deitmer and Schlue, 1987; Deitmer, 1992), and approximately 0.3 units higher in the presence of the physiological buffer (Deitmer, 1998). For vertebrate glial cells, reported steady-state pH_i values range from 6.7 to 7.3 in the nominal absence of CO_2/HCO_3^-, and 6.9–7.6 in the presence of CO_2/HCO_3^- (pH_o 7.3–7.5) (see Rose and Ransom, 1998). Compared to the steady-state pH_i values of cultured astrocytes from mouse and rat, the values are higher for rat C6 glioma cells in both the presence and absence of CO_2/HCO_3^-.

2.4. Consequences of pH$_i$ regulation in brain

pH_i regulation by cells in close proximity has two important consequences. First, the movement of acid–base equivalents across the plasma membrane of a cell will not only change that cell's pH in one direction, but will also change pH_o in the opposite direction. As an example, acid extrusion from an astrocyte will increase pH_i of the astrocyte and simultaneously lower pH_o. The second important consequence is that the change in pH_o will also influence the activity of acid–base transporters that are pH_o-sensitive in adjacent cells. Thus, in our example of acid extrusion from an astrocyte, the lower pH_o will likely stimulate acid loaders and inhibit acid extruders in nearby neurons and glia.

3. Importance of pH regulation in brain

3.1. General cellular activity

Many cellular processes are sensitive to changes in pH (see Roos and Boron, 1981). For instance, changes in pH can alter enzyme activity. A classic example is the glycolytic enzyme phosphofructokinase, which is inhibited ~90% in vitro by a pH decrease of 0.1

(Trivedi and Danforth, 1966). A decrease in pH_i will also inhibit the phosphorylation-induced conversion of inactive phosphorylase b to active phorphorylase a, which catalyzes the breakdown of glycogen in tissues such as liver and muscle (Danforth, 1965). Other cellular processes that are sensitive to pH changes include growth-factor stimulation (Ganz et al., 1990; Pouysségur et al., 1984, 1985), microtubule-dependent changes in cell structure (Parton et al., 1991), and electrical coupling between cells (Spray et al., 1981; O'Beirne et al., 1987). As described in more detail below, changes in pH can alter the activity of ion channels.

3.2. Effects of pH on neuronal activity

The relationship between neuronal activity and changes in brain pH_{ECF} is complex because they influence one another. As reviewed by Chesler and Kaila (1992) and Ransom (2000), decreases in pH_{ECF} generally inhibit neuronal activity, whereas increases generally stimulate activity. In the rat hippocampus, for example, raising pH_{ECF} elicits an increase in the amplitude of population spikes evoked by a given stimulus (Balestrino and Somjen, 1988). In addition, as shown in both clinical and animal studies, the generation of epileptiform activity can be stimulated by alkalosis and inhibited by acidosis (Cohen and Kassirer, 1982; Woodbury et al., 1984; Aram and Lodge, 1987; Somjen et al., 1987; Lee et al., 1996).

pH-mediated effects on neuronal firing are likely due to their influence on many voltage- and ligand-gated channels (Tombaugh and Somjen, 1998; Traynelis, 1998), and neurotransmitter transporters that are sensitive to shifts in pH_i and/or pH_{ECF}. The NMDA-activated, ionotropic glutamate receptor has a pK in the physiological range and is inhibited by decreases in pH_o (Traynelis and Cull-Candy, 1990; Tang et al., 1990). In the hippocampus, an increase in pH_{ECF} can augment NMDA-mediated synaptic responses to evoked stimuli (Gottfried and Chesler, 1994; Taira et al., 1993). pH-induced changes in the activity of H^+-activated channels may also influence synaptic transmission, and possibly plasticity and memory. In trying to elucidate the role of H^+-activated cation currents in the CNS, Wemmie et al. (2002) (see also Cooke and Lilley, 2002) generated a mouse with a targeted knockout of the gene encoding ASIC1, a H^+-gated cation channel. ASIC1-mediated currents are typically observed at pH_o less than 6.9 (see Waldmann et al., 1999). Cultured hippocampal neurons from the ASIC1 knockout mice—in contrast to wild-type neurons—failed to elicit fast depolarizations when stimulated by decreasing pH_o (Wemmie et al., 2002). In addition, long-term potentiation was impaired in hippocampal slices from the knockout mice. The authors also found that the knockout mice displayed defects in both hippocampal-dependent memory and hippocampal-independent learning. Therefore, the synaptically expressed channel appears to be involved in both memory and learning.

Neuronal activity can be directly influenced by HCO_3^- currents per se, which have the additional effect of changing pH_i/pH_{ECF}. In the hippocampus, GABA-mediated inhibition of pyramidal-neuron activity is caused by an outwardly directed $GABA_A$-activated Cl^- current. However, $GABA_A$ receptors also conduct HCO_3^-, albeit to a lesser extent (Kaila and Voipio, 1987; Kaila et al., 1990). As mentioned above, the electrochemical gradient would favor HCO_3^- efflux (a depolarizing response) in a cell with a typical membrane potential. As discussed by Sun and Alkon (2002), significant HCO_3^- efflux in hippocampal neurons can convert an inhibitory GABA response into a stimulatory one. Such HCO_3^- conductances may underlie the observation that the enzyme carbonic anhydrase (CA) can influence synaptic activity and memory. CA catalyzes the general reaction $CO_2 + H_2O \leftrightarrow HCO_3^- + H^+$ and facilitates the formation of intracellular HCO_3^-. The specific reaction catalyzed by CA is $CO_2 + OH^- \leftrightarrow HCO_3^-$; the OH^- arising from deprotonation of H_2O bound to the enzyme (Liljas et al., 1994). Activators of CA (e.g., many amines and amino acids such as phenylalanine) substantially enhance synaptic efficacy, spatial learning, and memory in rat (Sun et al., 2001b). In contrast, inhibitors of CA (e.g., sulfonamides such as acetazolamide) impair spatial learning in rat (Sun et al., 2001a).

3.3. *Effects of neuronal activity on* pH_{ECF}

It is well established that neuronal firing elicits changes in pH_{ECF} that are often biphasic: an alkaline shift followed by an acid shift (Chesler and Kaila, 1992). The magnitudes of these shifts vary in different brain regions. For example, initial alkaline shifts are prominent in regions of gray matter (e.g., hippocampus and cerebellum), whereas acid shifts are more prevalent in regions of white matter (e.g., optic nerve). With excessive firing, an extended acid shift dominates in all regions, probably due to the metabolic production of acid and/or CO_2, and subsequent release into the extracellular space. Alkaline shifts in some cases are caused by activation of a neuronal Ca^{2+} pump that exchanges intracellular Ca^{2+} for extracellular H^+ following activity-induced increases in intracellular Ca^{2+} (Schwiening et al., 1993; Paalasmaa et al., 1994; Smith et al., 1994; Paalasmaa and Kaila, 1996; Trapp et al., 1996). HCO_3^- efflux through $GABA_A$-activated channels will also contribute to increases in pH_{ECF} (Chen and Chesler, 1990, 1992a; Kaila et al., 1992). An additional component of the alkaline shift could in principle be mediated by H^+-coupled uptake of glutamate by the glutamate transporter. The acid shift, which can either attenuate or immediately follow the alkaline shift, is due to activation of acid extruders in neurons and glia, particularly astrocytes (Chesler, 1990). Two prominent transporters in astrocytes that contribute to acid shifts are the electrogenic Na/HCO_3 cotransporter and the H-Lactate cotransporter.

It should be noted that extracellular acid and alkaline shifts are associated with changes in the interstitial partial pressure of CO_2 (P_{CO2}). Using CO_2/H^+-sensitive microelectrodes, Voipio and Kaila (1993) demonstrated that substantial P_{CO2} and pH_o gradients exist between the CA1 layer of the rat hippocampal slice and the bath solution. At the cell layer, pH_o was reported to be lower (range: 7.24–7.37) and P_{CO2} higher (range: 50–37 mm Hg) compared to values in the perfusion solution ($pH = 7.4$; $P_{CO2} = 24.3$ mm Hg). Such gradients are likely the result of continual metabolic production of intracellular acid that is titrated by HCO_3^- to form CO_2, which then diffuses from the cell layer to the bath solution. The authors found that repetitive stimulation in the hippocampal slice elicited an extracellular alkaline shift and decrease in P_{CO2}, followed by a pronounced acid shift and increase in P_{CO2}. Although the alkaline shift was accompanied by an increase in extracellular HCO_3^-, the acid shift occurred at a constant HCO_3^-. The acid shift, but not the increase in P_{CO2}, was blocked by applying the extracellular CA inhibitor benzolamide. The increase in P_{CO2} associated with the acid shift is consistent with activity-evoked increases in metabolism that generate intracellular CO_2, which then diffuses into the bath.

As mentioned above, the activity of acid-extruding mechanisms can account for either an attenuation of the alkaline shift or the generation of the acid shift near the onset of stimulation. Interestingly, HCO_3^--dependent acid extruders such as the electrogenic Na/HCO_3 cotransporter in astrocytes are predicted to cause an increase in extracellular P_{CO2} during the acid shift if they transport CO_3^{2-}. CO_3^{2-} transport has been described for the Na/HCO_3 cotransporter in gliotic slices (Grichtchenko and Chesler, 1994a), and for the cloned electrogenic Na/HCO_3 cotransporter NBCe1, as well as the cloned electroneutral Na-driven Cl-HCO_3 exchanger NDCBE expressed in *Xenopus* oocytes (Grichtchenko and Boron, 2002a,b). The P_{CO2} increase would arise from two sequential processes. First, CO_3^{2-} transport into cells causes a decrease in extracellular CO_3^{2-}, which then drives the deprotonation of HCO_3^- and formation of more extracellular CO_3^{2-} and H^+ (p$K \sim 10.3$). See Voipio (1998) for a description of pH changes caused by CO_3^{2-}. Second, because the increase in H^+ is disproportionately larger than the decrease in HCO_3^-, the CO_2/HCO_3^- equilibrium favors the formation of CO_2 and H_2O. The opposite events would occur inside the cell: the transporter-mediated influx of CO_3^{2-} rapidly equilibrates with H^+ to form HCO_3^-. Because the decrease in H^+ is disproportionately greater than the increase in HCO_3^-, the CO_2/HCO_3^- equilibrium favors the hydration of CO_2. Thus, intracellular P_{CO2} decreases. In all likelihood, the non-CO_2/HCO_3^- buffering power is lower outside than inside cells, and the increase in extracellular H^+ will be larger than the decrease in intracellular H^+ at similar pH values. Consequently, the increase in extracellular P_{CO2} will be larger than the decrease in intracellular P_{CO2}. In principle, the transporter-generated transmembrane P_{CO2} gradient could lead to transient CO_2 diffusion from the extracellular space into the cell, before net CO_2 diffusion out of the tissue.

4. pH regulation in glial cells

In this section, we will examine the main acid-loading and acid-extruding mechanisms identified in glial cells. These mechanisms contribute to steady-state pH_i, as well as pH_i recoveries from acid–base perturbations such as acute intracellular acid and alkali loads. The majority of the studies involve invertebrate glia and mammalian astrocytes. Investigators have also examined the pH_i physiology of additional vertebrate glia, including oligodendrocytes, microglia, Müller cells of the retina, and Schwann cells of the peripheral nervous system. Investigators have used several techniques to measure pH_i or pH_o. For example, the pH_i physiology of giant neuropile glial cells of the leech *Hirudo medicinalis* has been particularly well characterized with the use of double-barreled, pH-sensitive microelectrodes. Advantages of this glial preparation are that it is well established (Kuffler and Potter, 1964), and the cells are large enough to tolerate impalement by multiple microelectrodes. Microelectrodes have also been used to examine extracellular pH changes in mammalian brain-slice preparations. In studies on smaller mammalian glia (e.g., astrocytes in culture), most investigators have studied

pH_i-regulating transporters with either radioactive tracers (e.g., $^{22}Na^+$) or pH-sensitive dyes (e.g., 2′,7′-bis(carboxethyl)-5,6-carboxyfluorescein, or BCECF) and ratiometric fluorescence techniques. Below, we will review both functional and molecular studies on acid–base transporters identified in glial cells.

4.1. Acid-loading conductances

As described previously, the passive movement of acid–base equivalents can produce a chronic intracellular acid load. However, such movement through channels can also elicit acute decreases in pH_i and increases in pH_{ECF}. Channels that conduct acid–base equivalents can therefore be described as acid loaders. In the crayfish muscle fiber for instance, HCO_3^- exits through $GABA_A$-stimulated Cl^- channels and causes pH_i to decrease (Kaila and Voipio, 1987; Kaila et al., 1990). Through a similar mechanism, GABA stimulates HCO_3^- efflux from cells in the turtle cerebellum (Chen and Chesler, 1990) and the hippocampal slice (Chen and Chesler, 1992a; Kaila et al., 1992), thereby eliciting a transient extracellular alkaline shift. $GABA_A$ receptors have been reported in several types of glial cells, although at lower levels than in neurons (see Riquelme et al., 2002). Functional $GABA_A$ receptors have been found in cultured astrocytes (Blankenfeld and Kettenmann, 1992; Fraser et al., 1994; Bovolin et al., 1992; Bormann and Kettenmann, 1988; Blankenfeld et al., 1991), where their activation causes a depolarization, rather than a hyperpolarization, due to the higher intracellular Cl^- concentration than found in neurons (see chapter by Hansson and Rönnbäck). These receptors have also been demonstrated in cultured oligodendrocytes (Blankenfeld et al., 1991), and in both astrocytes and Bergmann glial cells in brain slices (Kang et al., 1998; Muller et al., 1994; Riquelme et al., 2002). Furthermore, activation of the $GABA_A$ receptor in cultured rat astrocytes does elicit a HCO_3^--dependent decrease in pH_i (Kaila et al., 1991).

Another ion conductance that mediates pH_i/pH_o changes is the voltage-dependent H^+ conductance first described in snail neurons (Thomas and Meech, 1982). In this preparation, the conductance is activated at depolarized membrane potentials close to ~ 0 mV, thereby favoring H^+ efflux. Voltage-activated H^+ currents with similar properties have been reported in many other cell types including phagocytes, microglia, skeletal muscle, and alveolar epithelia (DeCoursey and Cherny, 1994; Eder and DeCoursey, 2001). In microglia, decreases in pH_i and increases in pH_o lower the depolarization threshold for channel activation. H^+ movement only occurs in the outward direction when there is a large pH_o/pH_i gradient. In phagocytes such as neutrophils, these H^+ channels become activated during respiratory bursts when O_2 is converted into superoxide anion during phagocytosis (Henderson et al., 1987, 1988a,b). In a similar fashion, these channels might also be activated in phagocytosing microglia, particularly if there are accompanying increases in pH_{ECF}. Consequent changes in microglia pH may influence cell migration, ion-channel function, and microglia activation during pathological events (Faff and Nolte, 2000; Eder and DeCoursey, 2001).

4.2. Acid-loading transporters: HCO_3^--independent

4.2.1. Plasma membrane Ca^{2+} pump

Functional studies. The plasma membrane Ca^{2+}-ATPase (PMCA) is an ATP-driven transporter that exchanges intracellular Ca^{2+} for extracellular H^+ (*A*, Table 1). A vanadate-sensitive PMCA elicits increases in cell-surface pH of snail neurons following

Table 1
Acid-loading and acid-extruding transporters in glia

	Acid loaders	Net charge	Inhibitors
A	Plasma membrane Ca^{2+} pump (PMCA)	–	Vanadate
B	Glutamate transporter (EAAT)	+1 in	DL-threo-β-hydroxyasparatate (putative)
C	Cl-HCO_3 exchanger (AE)	–	Stilbene derivatives, Oxonol dyes
D	1:3 Na/HCO_3 cotransporter (NBC)	−2 out	Stilbene derivatives

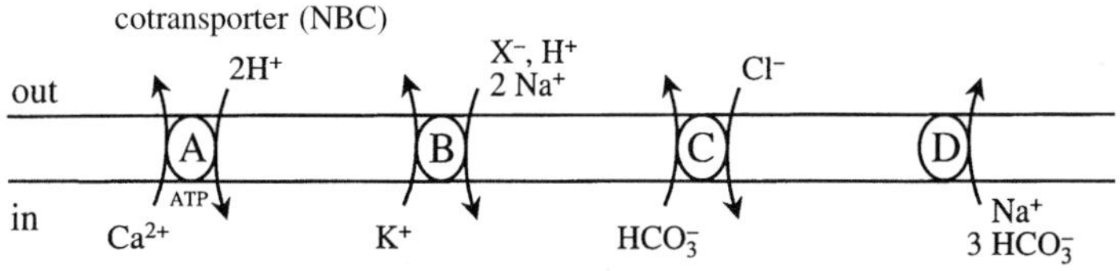

out
A: $2H^+$ out; Ca^{2+} in; ATP
B: X^-, H^+, 2 Na^+ out; K^+ in
C: Cl^- out; HCO_3^- in
D: Na^+, 3 HCO_3^- in
in

	Acid extruders	Active transport	Net charge	Inhibitors
F	Na-H exchanger (NHE)	2°	–	Amiloride and analogues, benzoylguanidines
G	Vacuolar-type H^+ pump	1°	+1 out	NEM, bafilomycin A_1, dicyclohexyl-carbodiimide
H	H-K pump	1°	–	Omeprazole, some Schering compounds
I	H/monocarboxylate cotransporter (MCT)	3°	–	CHC, *p*CMBS
J	1:2 Na/HCO_3 cotransporter (NBC)	2°	−1 in	Stilbene derivatives
K	Na-driven Cl-HCO_3 exchanger (NDCBE)	2°	–	Stilbene derivatives

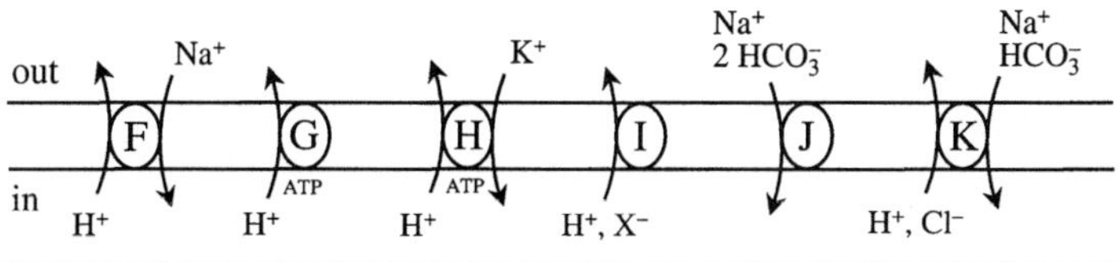

out
F: Na^+ out; H^+ in
G: H^+ in; ATP
H: K^+ out; H^+ in; ATP
I: H^+, X^- in
J: Na^+, 2 HCO_3^-
K: Na^+, HCO_3^- out; H^+, Cl^- in
in

elevations of intracellular Ca^{2+} (Schwiening et al., 1993). A similar Ca^{2+} pump has been described in mammalian hippocampal neurons (Paalasmaa et al., 1994; Smith et al., 1994; Paalasmaa and Kaila, 1996; Trapp et al., 1996), where it is responsible for both an extracellular alkaline shift and an intracellular acid shift following activity-evoked increases of $[Ca^{2+}]_i$.

PMCAs may also couple Ca_i^{2+} and pH_i changes in glia. Kawai et al. (1989) used an ultracytochemical technique to identify Ca^{2+}-pump activity in reactive and non-reactive astrocytes. Applying dry ice to the exposed rat scalp induced astrocyte activation and repair. The reactive astrocytes surrounding the lesion exhibited increased Ca^{2+}-pump activity compared to basal-level activity of the non-activated astrocytes. Activity of a vanadate-sensitive PMCA has also been described in the leech giant glial cell with the use of the Ca^{2+}-sensitive dye Fura-2 (Nett and Deitmer, 1998). The authors found that the pump contributes to low basal $[Ca^{2+}]_i$ levels, and is the primary means by which the cells recover from depolarization-induced elevations of intracellular Ca^{2+}. The functional coupling of changes in $[Ca^{2+}]_i$ and pH_i by PMCAs in glia has yet to be explored.

Molecular studies. cDNAs encoding four PMCAs (PMCA-1 through PMCA-4) have been identified (see Żylińska and Soszyński, 2000). According to both mRNA and protein localization studies, PMCA-1 and PMCA-4 are ubiquitously expressed, whereas PMCA-2 and PMCA-3 are predominantly expressed in brain and heart (Guerini, 1998; Stauffer et al., 1995). Using immunoblotting, immunohistochemical, and RT-PCR approaches, Fresu et al. (1999) identified PMCA-1, PMCA-2, and PMCA-4 in primary cultures of rat cortical astrocytes, and PMCA-1 and PMCA-4 in rat C6 glioma cells.

4.2.2. Glutamate transporter

Functional studies. Glutamate transporters are electrogenic, secondary active transporters that exchange extracellular glutamate, $2Na^+$, and H^+ for intracellular K^+ (*B*, Table 1). Glutamate and aspartate are the major excitatory neurotransmitters in the brain, and their extracellular concentrations therefore have to be exquisitely regulated. A prolonged elevation of extracellular glutamate in the brain leads to excitatory neurotoxicity and subsequent cell death due to glutamate-induced cellular entry of Ca^{2+} (Choi, 1988). As described in detail in the chapter by Schousboe and Waagepetersen, neurons and glia use plasma-membrane glutamate transporters to remove released glutamate from chemical synapses following transmitter release. These transporters in astrocytes are particularly important for neuronal function and survival, because they can help terminate glutamatergic transmission (Danbolt, 2001). From the standpoint of pH_i physiology, glutamate transporters are acid loaders because glutamate is cotransported with H^+ usually into the cell.

Several studies have documented pH_i changes elicited by glutamate transporters in invertebrate glia and mammalian astrocytes. For example, Bouvier et al. (1992) examined glutamate transport in voltage-clamped salamander retinal glia using either pH-sensitive microelectrodes to measure pH_o or intracellular BCECF to record pH_i. This preparation is ideal for these studies because the glia lack glutamate-gated ion channels that may conduct H^+. In one series of experiments with the microelectrodes, activating glutamate uptake by hyperpolarizing the cells elicited an increase in pH_o that was dependent on external glutamate and Na^+. In another series of experiments with BCECF, activating glutamate uptake generated a decrease in pH_i that was also Na^+-dependent. In an invertebrate preparation, Deitmer and Schneider (1997) used microelectrodes to record pH_i decreases and depolarizations elicited by applying glutamate to leech giant glial cells. These pH_i decreases appear to be mediated by the glutamate transporter for the following four reasons. First, similar results were obtained with D-aspartate, which has a low affinity for glutamate receptors but is transported by the glutamate-uptake system. Second, the glutamate/aspartate-induced pH_i decreases were unaffected by glutamate-receptor blockers. Third, the responses were Na^+-dependent. Fourth, the responses were partially inhibited by several putative glutamate-uptake inhibitors including DL-threo-β-hydroxy-aspartate.

Glutamate has been reported to elicit a decrease in the pH_i of astrocytes from several brain regions including mouse cerebrum (Brookes and Turner, 1993), rat cerebellum (Brune and Deitmer, 1995), and rat hippocampus (Rose and Ransom, 1996). In cultured rat cerebellar astrocytes for example, glutamate and aspartate caused decreases in pH_i that were independent of any increases in intracellular Ca^{2+} (Brune and Deitmer, 1995). Working on rat hippocampal astrocytes, Rose and Ransom (1996) demonstrated that such glutamate/aspartate-induced decreases in pH_i were paralleled by increases in $[Na^+]_i$. These changes were observed even when non-NMDA glutamate receptors, which might conduct H^+, were inactivated. These data are consistent with glutamate transporters mediating glutamate-induced pH_i decreases in astrocytes. Similar results have been obtained from hippocampal cells in the slice preparation (Amato et al., 1994).

Molecular studies. There is a considerable body of literature on the identification and localization of the five cloned glutamate transporters (EAAT1-EAAT5). As reviewed by Danbolt (2001), EAAT1 (GLAST) and EAAT2 (GLT) are predominantly found in glia of the normal adult mammalian CNS, whereas EAAT3 (EAAC) is found in several types of neurons. EAAT4 is principally expressed in Purkinje cells of the cerebellum, and EAAT5 appears to be found mainly in retinal cells, including neurons and Müller cells.

4.3. Acid-loading transporters: HCO_3^--dependent

4.3.1. Cl–HCO_3 exchanger

Functional studies. Cl–HCO_3 exchangers (also known as anion exchangers, AEs) are acid loaders that normally exchange extracellular Cl^- for intracellular HCO_3^- (*C*, Table 1). AEs are electroneutral, Cl^--dependent, Na^+-independent, and sensitive to a class of compounds called stilbene derivatives (e.g., 4,4′-diisothiocyanatostilbene-2,2′-disulfonic acid or DIDS), as well as oxonol dyes (Knauf et al., 1995; Alper et al., 1998).

Investigators have used several experimental approaches to identify AE activity in glia. Using double-barreled, pH-sensitive microelectrodes, Szatkowski and Schlue (1994) identified a Cl–HCO_3 exchange mechanism in connective glial cells of the leech. Recovery from an alkali load induced by removing acetate in the presence of CO_2/HCO_3^- was partially dependent on both extracellular Cl^- and HCO_3^-, and was inhibited by the stilbene derivative SITS (4-acetamido-4′-isothiocyanostilbene-2,2-disulfonic acid). Furthermore, removing external Cl^- elicited a DIDS-sensitive alkalization that was Na^+-independent yet HCO_3^--dependent.

Functional AEs have also been documented in mammalian astrocytes. Based on SITS-sensitive $^{36}Cl^-$ uptake and efflux measurements, Kimelberg et al. (1979) suggested the presence of a Cl–Cl or Cl–HCO_3 exchange mechanism in primary astroglial cultures from neonatal rat brain. The following two groups subsequently identified Cl–HCO_3 exchange in rat astrocytes using fluorometric techniques with the pH-sensitive dye BCECF. Mellergård et al. (1993) observed that the pH_i of primary cultures of rat cortical astrocytes recovered rapidly from alkali loads elicited by decreasing bath CO_2/HCO_3^- at constant pH_o. Applying DIDS or removing external Cl^- reduced the rate of recovery. Shrode and Putnam (1994) alkali loaded both C6 cells and primary cultures of rat cortical astrocytes with an acute exposure to the weak acid 5,5-dimethyl-2,4-oxazolidinedione (DMO). Beforehand, cells were incubated in a Cl^--free, CO_2/HCO_3^- solution. Returning the cells to a Cl^--containing solution following the alkali load elicited a DIDS-sensitive pH_i decrease that was Na^+-independent.

AE activity has been examined in at least two other glial preparations. Primary cultures of Schwann cells from the rat sciatic nerve exhibited a pH_i increase in response to removing external Cl^- (Nakhoul et al., 1994). This pH_i increase occurred in the absence of external Na^+, and was substantially slower in the nominal absence of CO_2/HCO_3^-. However, the pH_i increase was unaffected by 100 μM DIDS. A similar pH_i increase elicited by removing external Cl^- has also been documented in oligodendrocyte progenitors (Boussouf and Gaillard, 2000). As with the Schwann cells, the pH_i increase in these progenitor cells required HCO_3^- and was unaffected by removing external Na^+. In contrast to the Schwann cells however, the pH_i increase was blocked by DIDS (500 μM). The authors demonstrated the lack of AE activity in the more differentiated pro-oligodendrocytes, as well as the mature oligodendrocytes. Therefore, Cl–HCO_3 exchange activity in oligodendrocytes appears to be developmentally regulated.

Molecular studies. cDNAs encoding four Cl–HCO_3 exchangers are known: AE1, AE2, AE3, and AE4. The AEs in conjunction with the Na/HCO_3 cotransporters (NBCs) and the Na-driven Cl–HCO_3 exchangers (NDCBEs) to be described shortly, are members of a superfamily of HCO_3^- transporters. AE4 has been found primarily in kidney collecting duct, gastrointestinal tract, and submandibular gland (Tsuganezawa et al., 2001; Ko et al., 2002) and will therefore not be discussed further. AE1 (or the 'band-3 protein') was the first anion-exchanger cloned (Kopito and Lodish, 1985), and is found in vertebrate erythrocytes. AE2 is found on the basolateral membrane of mouse choroid plexus based on in situ hybridization and immunohistochemistry studies (Lindsey et al., 1990). AE3 mRNA is predominantly expressed in brain and heart (Kopito et al., 1989). According to in situ hybridization data, AE3 is present throughout rat brain and is found in nearly all types of neurons. In contrast, AE3 is not detected in many regions with high levels of glial-cell bodies (e.g., corpus callosum and cerebellar peduncles). Both the brain and cardiac isoforms of AE3 are expressed in the rat retina (Kobayashi et al., 1994). Interestingly, the brain isoform is expressed in Müller cells, whereas the cardiac isoform is found in horizontal cells. The authors found that expression of the brain isoform gradually increases from postnatal day 3 (P3) to P15. In contrast, expression of the cardiac isoform is not apparent until P15; after which time its rapid expression coincides with the onset of retinal function. The highest expression of the brain isoform in Müller cells is seen in basal endfoot processes that contact the vitreous humor and nearby blood vessels. This AE3 isoform may therefore facilitate CO_2 removal from the retina to blood vessels near the vitreous humor (Kobayashi et al., 1994).

4.3.2. Na/HCO$_3$ cotransporter

Functional studies. Na/bicarbonate cotransporters (NBCs) mediate the cotransport of one Na^+ ion and one or more HCO_3^- ions (*D* and *J*, Table 1). These transporters have been identified with one of three Na^+:HCO_3^- stoichiometries: 1:1, 1:2, and 1:3. The cloned electrogenic NBCe1 appears to transport CO_3^{2-} based on an elegant series of experiments on the transporter expressed in *Xenopus* oocytes (Grichtchenko and Boron, 2002a). Similar to the AEs, NBCs are inhibited by stilbene derivatives. The electrogenic NBC in the proximal tubule of salamander kidney was the first NBC to be functionally characterized (Boron and Boulpaep, 1983). In the kidney, this transporter has a 1:3 Na^+:HCO_3^- stoichiometry (Soleimani et al., 1987) and is responsible for ~80% of HCO_3^- reabsorption in the proximal tubule. By moving Na^+ and HCO_3^- across the basolateral membrane from cell to blood, the transporter functions as an acid loader. NBCs with Na^+:HCO_3^- stoichiometries of 1:1 (electroneutral) and 1:2 (electrogenic) found in other cells usually function as acid extruders by transporting Na^+ and HCO_3^- into cells (see Boron et al., 1997). As described below under 'acid-extruding transporters: HCO_3^--dependent', the 1:2 NBC has been identified in many glial cells. In the retinal Müller cell however, the transporter has a 1:3 Na^+:HCO_3^- stoichiometry.

Using the whole-cell, voltage-clamp technique on freshly dissociated Müller cells, Newman and Astion (Newman and Astion, 1991; Newman, 1991) demonstrated that HCO_3^--induced outward currents required the presence of external Na^+, but not external Cl^-. These currents were blocked by either stilbene derivatives (e.g., 4,4′-dinitro stilbene-2,2′-disulfonic acid, DNDS) or harmaline. By examining the reversal potentials of the DNDS-sensitive HCO_3^- currents at different transmembrane Na^+ gradients, the authors computed a Na^+:HCO_3^- stoichiometry of 1:3. Interestingly, the NBC-mediated current was an order of magnitude larger at the endfoot near the vitreous humor than at the distal end of the cell. The authors speculated that this polarized distribution of NBC activity may either facilitate CO_2 removal from active photoreceptors to the vitreous humor, or contribute to pH-mediated dilation of blood vessels near the endfeet.

Molecular studies. The molecular physiology of these transporters will be described below in the NBC section of 'acid-extruding transporters: HCO_3^--dependent' (p. 726).

4.4. Acid-extruding transporters: HCO_3^--independent

4.4.1. Na–H exchanger

Functional studies. Na–H exchangers (NHEs) are ubiquitous acid–base transporters that have been found in nearly all eukaryotic cell types studied to date (see Counillon and Pouysségur, 2000). These electroneutral, secondary active transporters exchange extracellular Na^+ for intracellular H^+ (*F*, Table 1). Pharmacologically, the isoforms have varied sensitivities to amiloride, amiloride analogs such as ethylisopropyl amiloride (EIPA), and benzoylguanidines including HOE642 (cariporide) and HOE694 (Putney et al., 2002). Cariporide is a particularly potent competitive inhibitor of NHE-1 (Counillon et al., 1993). NHEs have been found in both neuronal and glial cells (Bevensee and Boron, 1998a; Deitmer, 1998; Rose and Ransom, 1998). Na–H exchange activity generally elicits a pH_i recovery from an intracellular acid load (usually studied in the nominal absence of CO_2/HCO_3^-) that is dependent on external Na^+ and sensitive to amiloride or its analogs. In addition, a functional NHE at steady-state pH_i is evident by a decrease in pH_i elicited by applying amiloride and unmasking background acid-loading mechanisms (Boyarsky et al., 1990; Sjaastad et al., 1992). In glia, these transporters are potent regulators of pH_i during recoveries from acute intracellular acid loads.

Using double-barreled, pH-sensitive microelectrodes on leech neuropile glial cells, Deitmer and Schlue (1987) determined that the pH_i recovery from an acute acid load in the nominal absence of CO_2/HCO_3^- was inhibited ~50% by 2–3 mM amiloride and required the presence of external Na^+. These data are consistent with the presence of an amiloride-sensitive NHE. Using the same approach, Szatkowski and Schlue (1992) identified a similar NHE in leech connective glial cells.

Na–H exchange activity has been reported in mammalian preparations such as rat C6 glioma cells, as well as mammalian astrocytes. Measuring $^{22}Na^+$ uptake in C6 glioma cells, Benos and Sapirstein (1983) noted that upon serum starvation for at least 4 h, an amiloride-sensitive Na^+ transporter was expressed de novo. Jean et al. (1986) used the $^{22}Na^+$-uptake technique and the pH-sensitive dye BCECF in both the C6 rat glioma cell line and NN hamster astrocytes to identify a similar transporter that was inhibited by EIPA, amiloride, and benzamil. For example, the pH_i recovery from an acid load in BCECF-loaded C6 cells was dependent on external Na^+ (Fig. 1A) and blocked by 50 μM EIPA (Fig. 1B). NHE activity has been examined further in glioma cell lines including rat C6 (Shrode and Putnam, 1994; McLean et al., 2000), as well as human U-118, U-87, and U-251 cell lines (McLean et al., 2000). As we mentioned previously, gliomas have a higher steady-state pH_i than normal astrocytes. McLean et al. (2000) reported that increased NHE-1 activity in malignant rat and human gliomas is responsible for their higher steady-state pH_i compared to non-transformed rat astrocytes.

NHEs contribute to the pH physiology of rodent astrocytes cultured from multiple brain regions including the hippocampus (Pappas and Ransom, 1993; Pizzonia et al., 1996; Bevensee et al., 1997b), cortex (Chow et al., 1992; Mellergård et al., 1993; Shrode and Putnam, 1994; McLean et al., 2000), forebrain (Boyarsky et al., 1993), and cerebellum (Brune et al., 1994). It is worth noting that the NHE in cultured astrocytes from rat forebrain (Boyarsky et al., 1993) is insensitive to the amiloride analog EIPA. Thus, NHEs in astrocytes from different brain regions may not be the same, at least pharmacologically. In a similar pattern, some hippocampal neurons have been reported to exhibit no sensitivity to amiloride compounds (see Bevensee and Boron, 1998a).

In addition to regulating pH_i, NHEs also contribute to cell-volume regulation and are activated by cell shrinkage (see reviews by Hallows and Knauf, 1994; Hoffmann and Simonsen, 1989; Lang et al., 1995). The regulation of astrocyte volume is important

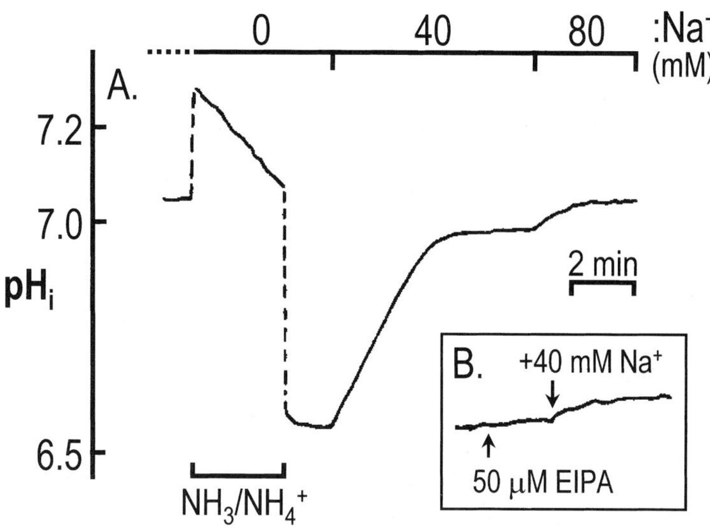

Fig. 1. EIPA-sensitive NHE in rat C6 glioma cells bathed in the nominal absence of CO_2/HCO_3^-. Modified from Jean et al. (1986) with permission from European Journal of Biochemistry. Copyright 1986, Federation of European Biochemical Societies. (A) pH_i was measured in a cell population using the pH-sensitive dye BCECF and a commercial spectrometer. Prior to the beginning of the experiment, the cells were incubated in a Na^+-free solution. In the continued absence of external Na^+, the cells had an average steady-state pH_i of ~7.05 at the beginning of the experiment. The cells were subsequently acid loaded using the NH_4^+-prepulse technique (Boron and De Weer, 1976a). Applying 30 mM NH_3/NH_4^+ elicited an initial increase in pH_i as NH_3 entered the cell and combined with H^+ to form NH_4^+. Subsequently, the pH_i decreased due to influx of NH_4^+ or activation of acid-loading processes. Removing external NH_3/NH_4^+ elicited a pronounced decrease in pH_i because accumulated intracellular NH_4^+ dissociated into H^+ (which remained trapped in the cell) and NH_3 (which diffused out of the cell). As shown in the figure, the pH_i does not recover in the absence of external Na^+. Adding 40 mM Na^+ to the bath solution elicited a rapid recovery of pH_i to ~7.0. Increasing external Na^+ to 80 mM caused a further increase of pH_i to the initial steady-state pH_i. (B) In a separate experiment in which cells were subjected to the same acid-loading protocol, pH_i failed to recover in the absence of external Na^+. The pH_i scale in part (A) also applies to part (B). Applying 50 μM EIPA had little effect on the pH_i recovery in the absence of Na^+. Furthermore, adding 40 mM external Na^+ in the continued presence of EIPA produced no appreciable increase in the pH_i recovery. Thus, the Na^+-dependent pH_i recovery was entirely EIPA sensitive at the lowest pH_i after the acid load.

because the brain is confined within a rigid skull, and neurological defects are associated with brain-cell swelling that undoubtedly alters cell–cell contacts and reduces the extracellular volume (Andrew, 1991). Swollen astrocytes can also release excessive amounts of excitatory amino acids that cause neurotoxicity (Kimelberg, 1995). Shrinkage-induced activation of NHE activity has been reported in the rat C6 glioma cell line (Jean et al., 1986; Shrode et al., 1995, 1997), and appears to involve phosphorylation of myosin light chain in confluent cell monolayers (Shrode et al., 1995, 1997).

There is evidence for Na–H exchange in mammalian oligodendrocytes, microglia, and Schwann cells. Kettenmann and Schlue (1988) used H^+-sensitive microelectrodes on cultured oligodendrocytes from mouse spinal cord to identify a HCO_3^--independent pH_i recovery from an acid load that was completely blocked by removing external Na^+ or applying 1 mM amiloride. In studies involving the use of BCECF, an amiloride-sensitive

NHE was found in both immature oligodendrocyte progenitor cells (Boussouf and Gaillard, 2000) and mature oligodendrocytes (Boussouf et al., 1997) from rat. Finally, in both microglia cultured from mouse (Faff et al., 1996) and Schwann cells cultured from the rat sciatic nerve (Nakhoul et al., 1994), pH_i recoveries from acid loads in the nominal absence of CO_2/HCO_3^- required external Na^+ and were inhibited by amiloride compounds.

Molecular studies. cDNAs encoding seven Na–H exchangers (NHE-1 through NHE-7) have been identified to date. We will focus our attention on NHE-1 through NHE-5; all of which appear to mediate Na–H exchange at the plasma membrane (Counillon and Pouysségur, 2000; Baird et al., 1999; Attaphitaya et al., 1999). NHE-6 is found in recycling endosomes (Brett et al., 2002), whereas NHE-7 is targeted to the *trans*-Golgi network (Numata and Orlowski, 2001). Ma and Haddad (1997) performed both Northern blot analysis and in situ hybridization to determine mRNA expression of NHE-1 through NHE-4 in rat brain. The authors concluded that NHE expression depends on transporter subtype, brain region, and animal age. NHE-1 is the most abundantly and ubiquitously expressed, with high levels found in the hippocampus, cerebellum, and the second/third layers of the periamygdaloid cortex. NHE-2 and NHE-4 are expressed in low levels mainly in the cerebral cortex and brainstem-diencephalon. Low levels of NHE-4 are seen in glial cells within the cerebellar white matter. Finally, NHE-3 is predominantly found in Purkinje and glial cells of the cerebellum. Regarding developmental profiles, the levels of mRNA for NHE-1, -2 and -4 increase in the cortex, but decrease in the cerebellum as animals age from postnatal day 0 (P0) to P30. In contrast, mRNA encoding NHE-3 increases in the cerebellum from P0 to P30. Based on other mRNA studies, NHE-5 is present in multiple brain regions (Baird et al., 1999), and has been localized to neurons in the dentate gyrus of rat hippocampus (Attaphitaya et al., 1999). Expression of NHE-5 mRNA in glial cells has not been reported.

Immunochemical data on NHEs are consistent with the aforementioned mRNA localization. In an animal-development study, Douglas et al. (2001) used immunoblotting with NHE isoform-specific antibodies to examine NHE-1, -2, and -4 expression in cortex, cerebellum, and brainstem-diencephalon. NHE-1 expression levels in all three regions gradually increase during animal development from embryonic day 16 to postnatal day 105. In contrast, NHE-2 and NHE-4 expression levels tend to peak at 3–4 weeks postnatal and then decline. These data may reflect different functional roles of NHE isoforms during development. Antibodies have been used to examine glial-cell expression of NHEs. Pizzonia et al. (1996) used immunoblotting to identify NHE-1 expression in astrocytes cultured from the rat hippocampus. Different expression profiles of the NHE isoforms may contribute to differences in the pH_i physiology of brain cells.

4.4.2. *Vacuolar-type H^+ pump*

Functional studies. Vacuolar-type (or V-type) H^+-ATPases are present in vesicles and other organellar membranes where they establish the low intraorganellar pH necessary for many organellar functions including enzyme activation, ligand–receptor uncoupling, and

H^+-coupled carrier transport. V-type H^+ pumps are also found in the plasma membrane of many cells (e.g., macrophages, neutrophils, and osteoclasts), where they are involved in pH_i regulation (see Bevensee et al., 2000a). On the apical membrane of the proximal and distal nephron of the kidney, they contribute to acid secretion into the lumen. These electrogenic pumps are primary active transporters that use the free energy released from ATP hydrolysis to extrude H^+ from the cell (*G*, Table 1). They are inhibited by *N*-ethylmaleimide (NEM), dicyclohexylcarbodiimide, and bafilomycin A_1.

A V-type H^+ pump has been found in both cultured astrocytes and gliomas. Pappas and Ransom (1993) used BCECF in cultured astrocytes from the rat hippocampus to monitor pH_i recoveries from acute acid loads in the nominal absence of CO_2/HCO_3^-. The authors identified a Na^+-independent component of the pH_i recovery that was inhibited by either bafilomycin A_1 or NEM. Furthermore, depolarizing the cell by raising external K^+ increased the rate of the pH_i recovery—consistent with activation of an electrogenic H^+ pump. There is electrophysiological evidence for a V-type H^+ pump in the C6 glioma cells, as well as in DI TNC_1 cells—an immortalized cell line derived from primary cultures of rat diencephalon astrocytes. In voltage-clamp experiments on these cells, bafilomycin A_1 inhibited a residual hyperpolarizing current that was unmasked after blocking ion channels (Philippe et al., 2002).

Molecular studies. The V-type H^+ pump is composed of a V_0 and a V_1 domain (see review by Kawasaki-Nishi et al., 2003). The V_0 domain, with five different subunits (a,b,c,c′,c″), is involved in H^+ translocation. The V_1 domain, with eight different subunits (A–H), is involved in ATP hydrolysis. Using RT-PCR and immunoblotting techniques with primary astrocyte cultures from rat striata and cortex, as well as cultures of C6 and DI TNC1 cells, Philippe et al. (2002) identified the presence of the a and c subunits (V_0 domain) and the A subunit (V_1 domain) of the V-type H^+ pump. Mouse genes encoding the G1 and G2 isoforms of the G subunit have recently been identified (Murata et al., 2002). The authors found at the RNA and protein levels that the G1 isoform is ubiquitously expressed in multiple tissues (including brain), and the G2 isoform is predominantly expressed in neurons in the CNS. Based on immunohistochemical studies, the G1 isoform is expressed in both astrocytes and oligodendrocytes cultured from mouse hippocampus.

4.4.3. H–K pump

Functional studies. H–K ATPases are primary active transporters that exchange extracellular K^+ and intracellular H^+ (*H*, Table 1). The gastric H–K pump in parietal cells is responsible for the acidity of gastric secretions (Okamoto and Forte, 2001), and both gastric and non-gastric H–K pumps in the collecting duct of the kidney contribute to K^+ reabsorption (Silver and Soleimani, 1999; Jaisser and Beggah, 1999). These pumps are inhibited by omeprazole and certain Schering compounds.

Shirihai et al. (1998) have used an elegant approach to identify an H–K pump in microglia. With ion-selective electrodes in self-referencing mode (generated by rapidly moving between two positions), the authors measured an external K^+ gradient within 10 μm of the cell surface. This gradient appeared to be generated by an H–K pump because it was enhanced by an increase in extracellular K^+, and was dissipated by omeprazole or the Schering compound 28080. Unexpectedly, lowering pH_o increased the K^+ gradient, presumably by stimulating the H–K pump. This pump may be particularly active in microglia following brain injury when $[K^+]_o$ rises and pH_{ECF} falls.

4.4.4. *Monocarboxylate transporter*

Functional studies. Monocarboxylate transporters (MCTs) are tertiary active transporters that mediate the electroneutral cotransport of H^+ and a monocarboxylate (e.g., lactate or pyruvate) (*I*, Table 1). Transport inhibitors include α-cyano-4-hydroxycinnamate (CHC) and *p*-chloromercuribenzenesulfonate (*p*CMBS). MCTs play an important role in metabolic coupling between glia and neurons because glycolytically produced lactate in astrocytes might be used as an energy source by neurons (however, see chapter by Roberts and Chih). Perhaps more importantly, MCTs are also essential for the transport of pyruvate into mitochondria. We will focus here on the influence of MCTs on the pH_i physiology of glial cells.

Using an enzyme assay, Walz and Mukerji (1988) measured extracellular lactate efflux from both neurons and astrocytes cultured from rat cortex. Lactate efflux from both cell types was inhibited by pyruvate and the non-transportable analog DL-*p*-hydrophenyl-lactate, but not by CHC. Although MCTs normally mediate transport out of glial cells, many investigators have characterized their activity operating in the opposite direction. For example, Nedergaard and Goldman (1993) used BCECF to measure the pH_i of astrocytes cultured from the forebrain of embryonic rat. The authors concluded that a lactate transport system is responsible for intracellular acidification rates that are a saturable function of externally applied lactic acid. This transport system has a low K_M for lactate (0.4 mM), and is insensitive to CHC and *p*CMBS. However, in radioactive tracer studies on primary cultures of rat astrocytes, Tildon et al. (1993) reported two lactate-transport systems: both high-affinity ($K_M = 0.5$ mM) and low-affinity ($K_M = 11$ mM) ones. Uptake was stimulated by low external pH, and inhibited by high external pH and CHC. The low-affinity transporter has a similar K_M to a CHC-sensitive lactate transporter ($K_M = 7.7$ mM) identified in rat astroglial cells (Bröer et al., 1997). According to another report, lactate transport in rat C6 glioma cells and cortical rat astrocytes is insensitive to CHC, but sensitive to quercetin (Volk et al., 1997). Dringen et al. (1995) on the other hand found two lactate-uptake mechanisms in the C6 cells: a CHC-insensitive, non-saturable one, and a CHC-sensitive saturable one. According to the same study, primary cultures of rat astroglial cells only exhibit a CHC-insensitive, non-saturable mechanism. Based on the aforementioned studies, at least two lactate transporters with different lactate affinities have been identified in astrocytes and C6 cells.

Molecular studies. cDNA sequences of eight mammalian monocarboxylate transporters (MCT1-MCT8) have been identified. While MCT1 is the predominant isoform found in astrocytes, MCT2 with a higher substrate affinity (at least when expressed in oocytes) is particularly prevalent in the astrocytic foot processes that abut blood vessels (see review by Halestrap and Price, 1999). The MCT2 expression profile may optimize efficient transport of metabolic substrates from the endothelium into astrocytes. MCT1-MCT4 have all been documented in retina.

4.5. Acid-extruding transporters: HCO_3^--dependent

4.5.1. Na/HCO_3 cotransporter

Functional studies. As described previously, the NBC with a 1:3 Na^+:HCO_3^- stoichiometry normally functions as an acid loader (e.g., in the proximal tubule of the kidney), whereas those with 1:1 and 1:2 stoichiometries normally function as secondary-active acid extruders. Based on many studies, the predominant acid-extruding HCO_3^- transporter in glial cells is an electrogenic NBC with a 1:2 Na^+:HCO_3^- stoichiometry that moves net-negative charge in the direction of ion transport (*J*, Table 1). Characteristics of an acid-extruding NBC include HCO_3^--dependent increases in pH_i (e.g., following intracellular acid loads) that are Na^+-dependent, Cl^--independent, and usually stilbene-sensitive.

Electrogenic NBC activity in glia was first reported in the giant neuropile glial cell of the leech. Using pH-sensitive microelectrodes, Deitmer and Schlue (1987) identified a Na^+- and HCO_3^--dependent pH_i recovery from an acid load that was inhibited by the stilbene derivative SITS. In subsequent studies, the authors used ion-sensitive microelectrodes to monitor pH_i, intracellular Na^+ activity (aNa_i^+), and V_m in the same preparation (Deitmer and Schlue, 1989; Deitmer, 1992). Exposing the cells to a CO_2/HCO_3^--containing solution elicited an increase in steady-state pH_i and a hyperpolarization; both of which were Na^+-dependent and DIDS-sensitive (Deitmer and Schlue, 1989). CO_2/HCO_3^- also induced a DIDS-sensitive increase in aNa_i^+. The authors estimated a 1:2 Na^+:HCO_3^- stoichiometry from the Na^+:HCO_3^- coupling ratio. Electrogenic NBC activity was particularly evident when Deitmer (1992) exposed the cells to a Na^+-free, HCO_3^--containing solution. Removing external Na^+ had three HCO_3^--dependent effects that occurred simultaneously: the cell depolarized, the pH_i decreased markedly, and aNa_i^+ decreased at a faster rate than observed in the absence of HCO_3^- (Fig. 2). These observations are consistent with NBC operating in the reverse direction and mediating the efflux of Na^+, HCO_3^-, and net-negative charge during the Na^+-removal protocol.

Electrogenic NBCs were subsequently identified in other glial preparations including connective glial cells from leech (Szatkowski and Schlue, 1992) and astrocytes from the optic nerve of the mudpuppy *Necturus* (Astion and Orkand, 1988). The transporter was also further characterized in the leech glia. Using the two-electrode, voltage-clamp technique, Munsch and Deitmer (1994) confirmed the stoichiometry of the NBC in the neuropile leech glial cells by evaluating DIDS-sensitive current-voltage relationships. In addition, the authors demonstrated the reversibility of the transporter, which has a reversal potential near the resting V_m of the cell (~ -75 mV). Normal inward transport can be reversed with only small changes in either V_m, or the transmembrane Na^+ and HCO_3^- gradients. In other words, the NBC can readily function as either an acid extruder or acid loader depending on the prevailing electrical and chemical gradients.

NBC activity has been examined in considerable detail in mammalian astrocytes. The transporter has been studied in astrocytes from rat forebrain (Boyarsky et al., 1993), rat hippocampus (Pappas and Ransom, 1994; O'Connor et al., 1994; Bevensee et al., 1997a,b), rat cerebellum (Brune et al., 1994), rat cortex (Shrode and Putnam, 1994), and

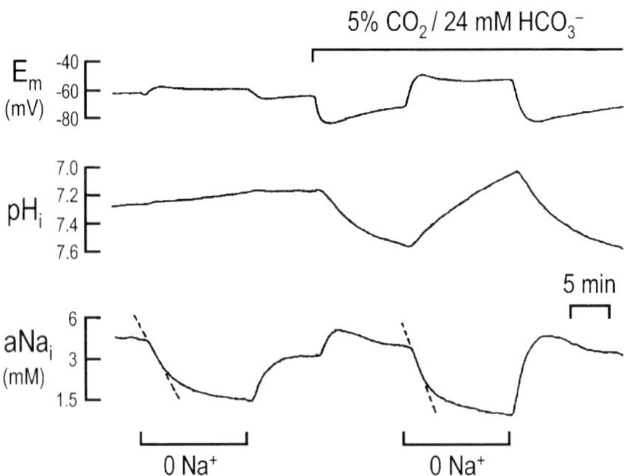

Fig. 2. Electrogenic NBC activity in leech giant neuropile glia. Modified from Deitmer (1992) with permission from Pflügers Archives. Copyright 1992, Springer-Verlag. Ion-sensitive microelectrodes were used to measure membrane potential (E_m), intracellular pH (pH_i), and activity of intracellular Na^+ (aNa_i) simultaneously. The experimental protocol involved removing external Na^+ from the bath solution either in the nominal absence or presence of CO_2/HCO_3^- (constant pH_o of 7.4). Compared to results in the absence of HCO_3^-, the following three observations were made for the cell in the presence of the physiological buffer: (i) The cell depolarized due to net-negative charge leaving the cell; (ii) pH_i decreased at a considerably faster rate due to Na-coupled HCO_3^- efflux; and (iii) aNa_i^+ decreased at a faster rate in the absence of external Na^+ (compare dotted lines) due to HCO_3^--coupled Na^+ efflux. These data are consistent with electrogenic NBC transporting Na^+, HCO_3^-, and net-negative charge out of the cell during the Na^+-removal protocol.

mouse cortex (Brookes and Turner, 1994; Chow et al., 1991). In most of these studies, pH_i was measured with BCECF, and transporter activity was evident from one or more of the following observations. The Na^+- and HCO_3^--dependent transporter contributes to pH_i recoveries following intracellular acid loads. The transporter is responsible for increases in steady-state pH_i when astrocytes are exposed to CO_2/HCO_3^--containing solutions. Finally, electrogenic NBC activity appears to cause an increase in steady-state pH_i when cells bathed in CO_2/HCO_3^- are depolarized by high $[K^+]_o$ (termed a depolarization-induced alkalization or DIA). A depolarization increases the electrical driving force for $Na^+/2HCO_3^-$ cotransport into cells. DIA was first described by Siebens and Boron (1989a,b) working on the kidney proximal tubule. In this preparation, DIA is mediated by both a SITS-sensitive, Na^+-dependent transporter and a monocarboxylate transport system involving Na-Lactate and H-Lactate cotransporters.

Conclusively identifying an electrogenic NBC requires examining two key characteristics of transport: Cl^- independence and electrogenicity. Ruling out the involvement of Cl^- is important to distinguish the transporter from a related one—the Na-driven Cl–HCO_3 exchanger. In rat hippocampal astrocytes, transporter-mediated intracellular alkalinizations still occurred after reducing intracellular Cl^- to near zero (Bevensee et al., 1997a). In addition, the transporter could still operate in the reverse direction in the absence of external Cl^-. Both observations are consistent with the transporter being an NBC and not a Cl^--dependent process. Examining the electrogenicity is important to distinguish the transporter from electroneutral NBCs. The electrogenicity of the transporter in mammalian astrocytes has been examined using patch-clamp techniques (O'Connor et al., 1994; Brune et al., 1994; Bevensee et al., 1997a). In rat hippocampal astrocytes for instance, removing external Na^+ elicited a DIDS-sensitive, HCO_3^--dependent depolarization (Bevensee et al., 1997a). The magnitude of the mean depolarization compared to the mean HCO_3^- flux in parallel pH experiments is consistent with the transporter having a 1:2 Na^+:HCO_3^- stoichiometry.

A functional electrogenic Na/HCO_3 cotransporter has also been characterized in the gliotic hippocampal slice. Using pH-sensitive microelectrodes, Grichtchenko and Chesler (1994a,b) found that depolarizing astrocytes by elevating $[K^+]_o$ induced both an extracellular acidification and an astrocyte alkalinization that were both partly due to a Na^+- and HCO_3^--dependent transporter. These data agree well with the aforementioned NBC studies on cultured astrocytes. There is one discrepancy however: the NBC in gliotic tissue is insensitive to stilbenes.

Less is known about NBC activity in oligodendrocytes. Kettenmann and Schlue (1988) identified a Na^+- and HCO_3^--dependent acid–base transporter in cultured oligodendrocytes from embryonic mouse spinal cord. The transporter does not appear to require Cl^-, and therefore may be an NBC. As in the gliotic astrocytes described above, the transporter is not sensitive to stilbene derivatives. It is not clear if the transporter is electrogenic or not. In a more recent study, Boussouf et al. (1997) reported similar DIDS-insensitive transporter activity in cultured mature oligodendrocytes from the rat cerebellum. The transporter appears to be electrogenic because membrane depolarizations (generated by raising $[K^+]_o$) elicited Na^+- and HCO_3^--dependent alkalizations. In a later study, the authors obtained similar results on oligodendrocyte progenitors and more differentiated pro-oligodendrocytes (Boussouf and Gaillard, 2000). Interestingly, the authors found that the HCO_3^- transporter in progenitor cells is DIDS-sensitive, whereas the one in mature cells is not. Thus, the DIDS-sensitivity of the HCO_3^- transporter in oligodendrocytes appears to decrease with development. Further identification of an electrogenic NBC in these cells will require a direct measure of transporter-mediated currents or voltage changes.

Activity during neuronal firing. Although many acid–base transporters contribute to pH_i/pH_{ECF} regulation in the brain, the electrogenic NBC occupies an unusual niche. When active, the transporter alters both pH_i and pH_{ECF} (by moving HCO_3^-) and V_m (by moving net-negative charge). In addition, the V_m influences the direction and activity of the transporter (Munsch and Deitmer, 1994). Therefore, this transporter may serve as a link between neuronal activity and pH changes in the brain. According to the model shown in Fig. 3, and originally proposed by Chesler (1990) and Ransom (1992), an electrogenic NBC elicits pH changes in response to nerve activity. Neuronal excitability causes an increase in $[K^+]_{ECF}$, which depolarizes adjacent astrocytes with a high K^+ conductance (see chapter by Walz). The depolarization stimulates the activity of an electrogenic NBC that transports HCO_3^- from the extracellular space into the astrocyte. The pH_i of the astrocyte increases due to the CA-catalyzed general reaction: $HCO_3^- + H^+ \rightarrow CO_2 + H_2O$, which consumes H^+. Although somewhat controversial, CA II is found in at least some astrocytes (see review by Ridderstrale and Winstrand, 1998). Simultaneous transport of HCO_3^- out of the extracellular space elicits a decrease in pH_{ECF} due to the reaction $CO_2 + H_2O \rightarrow H^+ + HCO_3^-$, which generates H^+. This reaction is catalyzed by

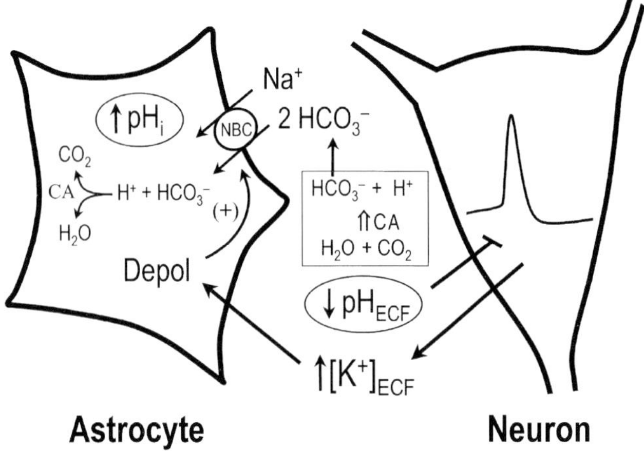

Fig. 3. A model of changes in pH_i and pH_{ECF} mediated by an NBC (Na/HCO_3 cotransporter) during electrical activity (see model from Ransom, 2000). An action potential in a neuron elicits an increase in extracellular K^+ ($[K^+]_{ECF}$) that depolarizes adjacent astrocytes with a high K^+ conductance. Depolarization stimulates an electrogenic NBC to transport Na^+, $2HCO_3^-$, and net-negative charge into the cell. The pH_i of the astrocyte increases due to the CA-catalyzed general reaction: $HCO_3^- + H^+ \rightarrow CO_2 + H_2O$, which consumes H^+. The opposite reaction in the ECF will cause pH_{ECF} to decrease. Through several potential mechanisms discussed in the text, a decrease in pH_{ECF} will inhibit further neuronal activity. The action potential shown in the model neuron is an actual V_m recording from a spontaneously firing cultured rat hippocampal neuron that was patch clamped under current-clamp conditions (McNicholas-Bevensee C.M. and Bevensee M.O., unpublished).

extracellular CA, which has been found in the brain slice (Chen and Chesler, 1992b; Kaila et al., 1992) and on the surface of both neurons and astrocytes (Svichar and Chesler, 2003). As described previously, the pH_{ECF} decrease can inhibit further neuronal excitability by inhibiting many voltage- and ligand-gated channels. This negative-feedback model would be neuroprotective in that excessive neuronal activity (e.g., during epileptic events) would elicit a greater decrease in pH_{ECF}, and consequently a greater inhibition of further neuronal activity (Ransom, 2000).

There is considerable evidence to support the model presented in Fig. 3. As already mentioned, DIAs have been documented in several glial preparations including the leech neuropile glial cell (see Deitmer, 1998), astrocytes cultured from several brain regions (Boyarsky et al., 1993; Pappas and Ransom, 1994; Shrode and Putnam, 1994; Brookes and Turner, 1994; Chow et al., 1991), and the gliotic hippocampal slice (Grichtchenko and Chesler, 1994b). Depolarizations in the gliotic hippocampal slice also cause decreases in pH_{ECF} (Grichtchenko and Chesler, 1994a). Using a very elegant approach, Newman (1996) used BCECF attached to the extracellular surface surrounding retinal Müller cells to demonstrate that cell depolarizations elicit pH_o decreases near the endfeet where NBC is predominantly expressed. Thus, depolarization of isolated glial cells and activation of NBC can elicit decreases in pH_o, in addition to increases in pH_i.

There is developmental evidence highlighting the importance of astrocytes in mediating activity-evoked extracellular acid shifts. In the rat spinal cord, activity-evoked changes in pH_o and $[K^+]_o$ are dependent on the age of the animal (Syková, 1998).

Compared to neonatal animals, older animals display smaller $[K^+]_o$ increases and larger acid shifts with spinal-cord stimulation (Jendelová and Syková, 1991). The larger acid shifts seem to coincide with gliogenesis during postnatal development as the number of mature glial cells increases from 31% at P1–P3 to 77% at P13–P15 (Chvátal et al., 1995). Also, 'early' postnatal X-irradiation, which selectively blocks gliogenesis, prevents the development of both larger acid shifts and smaller $[K^+]_o$ increases elicited by stimulation (Syková, 1998).

Molecular studies. Within the last several years, considerable advances have been made in the molecular identification of cation-coupled HCO_3^- transporters (e.g., NBCs and Na-driven anion exchangers). The first cDNA encoding an NBC was cloned by expression from the kidney of the salamander *Ambystoma* by Romero et al. (1997). Subsequently, numerous other NBC-related clones were identified from several tissues and species. Electrogenic NBCs have been cloned from human kidney (hkNBC) (Burnham et al., 1997), rat kidney (rkNBC) (Romero et al., 1998), human pancreas/heart (hpNBC, hhNBC) (Abuladze et al., 1998; Choi et al., 1999), rat pancreas (rpNBC) (Thévenod et al., 1999), and rat brain (rb1NBC, rb2NBC) (Bevensee et al., 2000b; Giffard et al., 2000). Based on differences at the amino and carboxy termini, these variants can be categorized into one of three groups: NBCe1-A (kidney clones), NBCe1-B (pancreas/heart and rb1NBC clones), and NBCe1-C (rb2NBC clone). Another group of NBCs (NBC4) distinct from the NBCe1s has been identified from human testis and heart (Pushkin et al., 2000a,b; Sassani et al., 2002) and a mixture of human tissue cDNAs (Virkki et al., 2002). At least one of the variants (NBC4c) functions as an electrogenic Na/HCO_3 cotransporter (NBCe2) (Sassani et al., 2002; Virkki et al., 2002). Electroneutral NBCs (NBCn1s) have been cloned and characterized from human skeletal muscle (Pushkin et al., 1999) and rat smooth muscle (Choi et al., 2000) after a clone was sequenced from human retina (Ishibashi et al., 1998).

The following antibodies were generated to distinguish between the NBCe1-B and NBCe1-C variants that were cloned from rat brain: αNBCe1-A/B and αNBCe1-C (Bevensee et al., 2000b). αNBCe1-A/B is expected to recognize both the A and B variants. Based on immunoblot data, αNBCe1-A/B and αNBCe1-C both recognize protein from rat brain, as well as protein from neurons and astrocytes cultured from rat cortex. Localization studies on brain slices with polynucleotide probes (Giffard et al., 2000; Schmitt et al., 2000) and polyclonal antibodies (Schmitt et al., 2000) have confirmed astrocytic, as well as some neuronal expression of NBCs. NBCe1 is present in both astrocytes and neurons in several regions of the CNS including the cortex and hippocampus. In a preliminary report, αNBCe1-A/B and αNBCe1-C labeled neurons in the hippocampus and cerebellum (Risso Bradley et al., 2001). According to a recent animal-development study using immunoblotting techniques, NBCe1 is expressed more abundantly in the cerebellum and brainstem-diencephalon than in cortex of P33 rats (Douglas et al., 2001). In all three regions, expression increases gradually from embryonic day 16 (when it is 25–40% of the adult level) to P105.

According to Northern blot and RT-PCR analyses, NBCn's (Pushkin et al., 1999; Choi et al., 2000) and NBCe2 (Pushkin et al., 2000b; Sassani et al., 2002) are also present in brain. NBCn1 protein has been localized to neuronal fibers within the hippocampus (Risso Bradley et al., 2001).

4.5.2. Na-driven Cl–HCO_3 exchanger

Functional studies. The Na-driven Cl–bicarbonate exchanger (NDCBE) is an electroneutral secondary active transporter that exchanges extracellular Na^+ for intracellular Cl^- (*K*, Table 1). In the transport process, two acid equivalents are extruded from the cell, probably by transport of CO_3^{2-} into the cell (Grichtchenko and Boron, 2002b). A stilbene-sensitive Na-driven Cl–HCO_3 exchanger was first identified in two invertebrate neuronal preparations: the squid giant axon (Boron and De Weer, 1976a,b; Russell and Boron, 1976; Boron and Russell, 1983) and the snail neuron (Thomas, 1976a, b, 1977). Subsequently, the transporter has been documented in vertebrate cells including fibroblasts (L'Allemain et al., 1985), kidney mesangial cells (Boyarsky et al., 1988a,b), and Ehrlich mouse ascites tumor cells (Kramhoft et al., 1994). The transporter is the predominant HCO_3^--dependent acid extruder in rat hippocampal CA1 neurons (Schwiening and Boron, 1994), and likely other mammallian neurons as well (Bevensee and Boron, 1998a).

Although an NBC is the primary HCO_3^--dependent acid extruder in astrocytes, NDCBE activity has also been reported. In rat cortical astrocytes studied with BCECF, the pH_i recovery from an acid load was inhibited by applying DIDS, or removing either external Na^+ or Cl^- (Mellergård et al., 1993). Shrode and Putnam (1994) obtained similar results with rat cortical astrocytes: CO_2/HCO_3^--induced alkalinizations that are DIDS-sensitive and HCO_3^--dependent can be inhibited by pre-incubating the cells in 0 Cl^- for 2 h. Na-driven Cl–HCO_3 exchange has also been reported in rat cerebellar astrocytes (Ko et al., 1999).

A HCO_3^--dependent transporter, which might be an NDCBE, also contributes to pH_i regulation of mouse cortical microglia (Faff et al., 1996). The authors identified a DIDS-sensitive, Na^+- and HCO_3^--dependent pH_i recovery from an acid load, which may be Cl^--dependent.

Molecular studies. NDCBEs are members of the HCO_3^--transporter superfamily that also includes the NBCs and AEs. Both a Na-driven anion exchanger from *Drosophila* (Romero et al., 2000), and an electroneutral Na-driven Cl–HCO_3 exchanger (NDCBE1) from human brain (Grichtchenko et al., 2001) have been cloned and functionally characterized. A partial NDCBE sequence was previously obtained from human NT-2 cells (Amlal et al., 1999), and a full-length one was more recently cloned from mouse kidney cells (Wang et al., 2001). Surprisingly, the corresponding full-length proteins in these two studies are reported to be likely Cl^--independent. According to mRNA analyses in the above studies, NDCBE and related transporters are all strongly present in the CNS. In a preliminary immunohistochemical study, NDCBE1 was found in the soma and dendrites of cerebellar Purkinje cells of rat (Risso Bradley et al., 2001). Further localization studies are required to assess NDCBE expression in glia, and to support the aforementioned functional studies.

A reported Na-driven Cl-HCO_3 exchanger (NCBE) has been cloned from an insulin-secreting cell line, and the mRNA is abundantly expressed in cerebellum and cerebrum of rat (Wang et al., 2000). Further functional studies however are required to confirm the Cl-independence and evaluate the electroneutrality of NCBE.

5. pH regulation of the CSF

5.1. Overview

The choroid plexus is an epithelial structure with a rich network of capillaries that secretes CSF into the brain ventricles (see Johanson, 2003; and chapter by Weaver et al.). The CSF circulates in the ventricles and subarachnoid space before being reabsorbed into the venous sinuses by the arachnoid villi. The CSF has three principle functions: (i) to cushion the brain, (ii) to serve as a fluid reservoir for controlling brain volume, and (iii) to serve as a nutritional source for brain cells. Although nascent CSF is produced by the choroid plexus, approximately one-third of the total CSF in the brain originates from secretion of capillary endothelial cells in brain parenchyma (Pollay and Curl, 1967). Compared to blood plasma, the CSF contains very little protein and significantly less glucose and amino acids. Both epithelial cells in the choroid plexus (the blood–CSF barrier) and endothelial cells in the capillaries in brain parenchyma (the blood–brain barrier) are responsible for the composition of CSF. There are two general reasons why the CSF is not simply an ultrafiltrate of blood plasma. First, the choroid epithelial cells and the capillary endothelial cells possess tight junctions that restrict the movement of ions, small polar molecules, and macromolecules from blood to CSF. Second, these polarized cells have apical and basolateral transporters and channels that regulate the composition of the secreted fluid.

5.2. pH regulation by choroid epithelial cells

Mechanisms that regulate the pH_i of the choroid epithelial cell will contribute to the pH of the CSF produced by the choroid plexus. According to both in vitro and in vivo studies, the choroid epithelial cell from rat has a pH_i of $\sim$7.0 (Johanson, 1978; Johanson et al., 1985). The following two main acid–base transporters are present in the choroid epithelial cell: Na-H exchanger (NHE) and Cl–HCO_3 exchanger (AE). Recent molecular evidence is consistent with the presence of a Na/HCO_3 cotransporter (NBC) as well. Below, we will review the evidence for each of these transporters in the choroid plexus.

NHE contributes to intracellular acid extrusion and Na^+ loading on the basolateral side of the epithelium (see Fig. 4). Consequently, NHE facilitates Na^+ transport from blood (or interstitial fluid) to the CSF. Using radioactive tracers to measure pH_i and $[Na^+]_i$ of rat choroid plexus epithelium in vivo, Murphy and Johanson (1990) observed NHE activity that elicited $[Na^+]$ decreases in both the epithelial cells and CSF of acid-loaded rats. In contrast, $[Na^+]$ increases were observed in the cells of alkali-loaded rats. The sidedness of amiloride inhibition is consistent with the transporter being on the basolateral (blood) side of the epithelium. In a subsequent study on rats, the authors found that systemic acetazolamide treatment decreased Na_i^+ and increased pH_i of choroid epithelial cells in vivo, and reduced Na^+ movement from plasma to CSF (Johanson and Murphy, 1990). The data are consistent with acetazolamide-induced inhibition of CA in the choroid plexus epithelial cells causing an increase in pH_i, which then inhibits basolateral NHE activity.

NHE activity has also been reported in vitro. For example, $^{22}Na^+$ uptake is inhibited by amiloride in rat choroid plexus epithelium in vitro (Murphy and Johanson, 1989b)—a

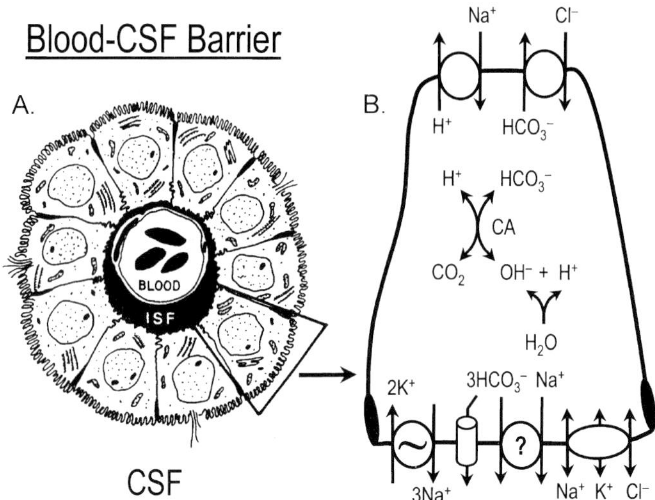

Fig. 4. Transporters in the choroid plexus epithelium. (A) The epithelial layer of the choroid plexus surrounding the interstitial fluid (ISF) that bathes a capillary. The tight junctions on the apical membranes define the blood–CSF barrier. The basolateral membranes face the ISF. Modified from Johanson et al. (1985) with permission from American Journal of Physiology. Copyright 1985, American Physiological Society. (B) A single epithelial cell of the choroid plexus. NHE and AE2 reside on the basolateral membrane, whereas a Na^+ pump and a Na/K/Cl cotransporter are present on the apical membrane. As mentioned in the text, the Na/K/Cl cotransporter has been reported to transport in either direction. An NBC and/or a HCO_3^--conducting anion channel on the apical membrane could mediate HCO_3^- movement into the CSF.

result that nicely complements the aforementioned in vivo NHE studies. In cultures of choroid epithelia from rabbit, the pH_i recovery from an acid load was Na^+-dependent and inhibited by amiloride compounds at low external Na^+ (Mayer and Sanders-Bush, 1993). At higher external Na^+, the transport process was amiloride-insensitive. NHE in the lateral ventricle choroid plexus from several mammals including pig and human has been reported based on radiolabeled amiloride binding (Kalaria et al., 1998). In addition, the authors used RT-PCR techniques to identify the presence of NHE-1 in these preparations. However, Alper et al. (1994) were unsuccessful in identifying NHE-1 expression in the lateral ventricle choroid plexus of human by immunoblotting and immunohistochemistry. Functional studies are presently the strongest evidence for a basolateral NHE in choroid epithelia.

An AE is the other main acid–base transporter functionally studied in the choroid epithelium. Johanson et al. (1985) used radiolabeled DMO to perform in vitro measurements of pH_i in adult rat choroid plexus where they found that epithelium $[HCO_3^-]$ decreased in proportion to synthetic CSF $[HCO_3^-]$. The presence of AE was reflected by a decrease in $[HCO_3^-]_i$ elicited by applying SITS, and an increase in $[HCO_3^-]_i$ caused by removing Cl^- from the synthetic CSF. These results confirm the suggestion by others that a $Cl–HCO_3$ exchange mechanism is responsible for decreases in CSF $[Cl^-]$ elicited by applying systemic DIDS (Frankel and Kazemi, 1983; Deng and Johanson, 1989).

Molecular work on AE expression in the choroid plexus complements the aforementioned functional studies. Lindsey et al. (1990) used in situ hybridization and immunocytochemistry to reveal AE2 expression in the rat choroid plexus epithelia, and more specifically on the basolateral membrane (see Fig. 4). Similar expression in the lateral and fourth ventricle choroid plexus of rat was later shown by both Alper et al. (1994) and Wu et al. (1998).

In support of earlier proposals (Johanson, 1984; Johanson and Murphy, 1990), there is evidence for the presence of an NBC. Based on short-circuit current measurements of frog chororid plexus, HCO_3^- movement from blood to CSF across the epithelium is dependent on Na^+ and sensitive to ouabain, acetazolamide, furosemide, and stilbene derivatives (Saito and Wright, 1983). Transepithelial HCO_3^- movement could involve a HCO_3^--conductance on the apical membrane (as suggested by the authors), and/or Na^+-coupled HCO_3^- transporters on either the apical or basolateral membranes. Using both in situ hybridization and immunohistochemical techniques, Schmitt et al. (2000) demonstrated the expression of NBCe1 in epithelial cells of choroid plexus, ependyma, and meninges of rat. The probes cannot distinguish between the three similar NBCe1 variants (see above). Labeling may therefore reflect the presence of any one (or more) of the variants. As shown in Fig. 4, an acid-loading NBC on the apical membrane may secrete HCO_3^- into the CSF. Based on reported transmembrane Na^+ and HCO_3^- gradients on the apical membrane of the epithelium, we estimate that an NBC with a 1:3 Na^+:HCO_3^- stoichiometry would have a reversal potential of ~ -50 mV. Because the cells have approximately the same V_m, it is not unreasonable to suggest that this NBC could mediate HCO_3^- secretion into the CSF. On the other hand, an acid-extruding NBC with a 1:2 stoichiometry—or a 1:1 stoichiometry for that matter—on the basolateral membrane could also contribute to transepithelial HCO_3^- movement into the CSF. On the basolateral membrane, such an NBC would transport HCO_3^- into the epithelial cell. Subsequently, the HCO_3^- could exit on the apical membrane through HCO_3^--conducting anion channels (Saito and Wright, 1984; also see Speake et al., 2001). Clearly, further functional and molecular studies are required to determine the stoichiometry and location of NBCs and related proteins in these cells.

The following is one model of pH_{CSF} regulation by the choroid epithelium (see Speake et al., 2001; Nattie, 1998; Johanson, 1984; Johanson et al., 1985). The pH of the secreted CSF is regulated by the combined activities of NHE and AE on the basolateral membrane and possibly NBC on the apical membrane (Fig. 4). CA catalyzes the hydration of CO_2 to H^+ and HCO_3^-. NHE on the basolateral membrane exchanges the newly formed intracellular H^+ for blood Na^+. The Na^+ gradient that drives NHE is established by the Na^+ pump on the apical membrane. AE on the basolateral membrane exchanges the newly formed intracellular HCO_3^- for blood Cl^-. However, HCO_3^- might have another fate: secretion into the CSF by an NBC on the apical membrane that cotransports Na^+ and HCO_3^-. Alternatively, HCO_3^--conducting anion channels on the apical membrane could allow HCO_3^- entry into the CSF. Working in concert therefore, NHE and AE on the basolateral membrane move NaCl from blood to cell. Subsequently, the Na/K/Cl cotransporter, NBC, and probably Cl^- channels on the apical membrane move the NaCl from cell to CSF. The transepithelial movement of ions into the CSF drives the osmotic flow of water, probably through aquaporin 1 in the apical and maybe basolateral membranes (see Speake et al., 2001). The direction of transport by the Na/K/Cl cotransporter can vary. Keep et al. (1994) reported transport out of cells bathed in CO_2/HCO_3^--buffered solutions, whereas Wu et al. (1998) reported opposite transport for cells bathed in the nominal absence of the physiological buffer. HCO_3^- appears to reduce one or more of the ion gradients (e.g., Cl^- or Na^+) necessary for inward transport (Speake et al., 2001).

According to development studies, the choroid plexus of the neonatal rat (1 week old) displays poor secretory ability compared to that of the adult rat. In addition, the CSF pH and [HCO_3^-] is more susceptible to changes in arterial pH elicited by acid–base disturbances (Johanson et al., 1976, 1988). As described above, acetazolamide elicits an increase in the pH and [HCO_3^-] of choroid epithelial cells from adult rats. In contrast, the CA inhibitor has no effect on the pH of the cells from neonatal animals (Johanson et al., 1992). The neonatal data could be explained by low levels of either CA activity or specific acid–base transport mechanisms necessary for HCO_3^- secretion into the CSF. Examining the developmental profiles of acid–base transporters in the choroid plexus at both the functional and molecular levels will contribute to our identification and understanding of the specific processes involved in the regulation of CSF pH and [HCO_3^-].

5.3. pH regulation by endothelial cells of brain capillaries

Similar to the choroid plexus epithelia, capillary endothelial cells are polarized and express acid–base transporters on their basolateral (blood) and apical (ECF) membranes. The predominant acid–base transporter identified in these endothelial cells is the NHE. There is also some evidence for the presence of an AE.

An NHE appears responsible for saturable $^{22}Na^+$ uptake into rat brain following an intracarotid bolus injection, as shown by Betz (1983b). The uptake is amiloride-sensitive. In an accompanying study, the same investigator studied isolated microvessels from rat brain and reported amiloride-sensitive $^{22}Na^+$ uptake that was stimulated by low pH_i, and inhibited by extracellular Na^+, H^+, Li^+ and NH_4^+ (Betz, 1983a). The author concluded that $^{22}Na^+$ uptake by NHE probably occurs across the apical membrane because the collapsed lumen of the microvessels restricts access to the basolateral membrane. In an in vivo study documenting NHE activity, $^{22}Na^+$ uptake across rat cerebral capillaries into the parietal cortex, pons-medulla, and CSF was reduced by either systemic acidosis or amiloride (Murphy and Johanson, 1989a). The combined results from these three studies are consistent with the presence of NHE on both the apical and basolateral membranes. Cultured cerebral endothelial cells from pig cells, with a steady-state pH_i of ~ 7.2 in CO_2/HCO_3^-, also displayed amiloride-sensitive NHE activity as measured with BCECF (Hsu et al., 1996). According to amiloride-binding and RT-PCR studies, the Na–H exchanger in cerebral microvessels of the rat, pig and human is NHE-1 (Kalaria et al., 1998).

A Cl–HCO_3 exchanger may also contribute to the pH_i physiology of cerebrovascular endothelial cells. For example, Smith and Rapoport (1984) performed in vivo studies on rats to examine $^{36}Cl^-$ movement into the cerebrovascular endothelium following intravenous injection. The authors found that $^{36}Cl^-$ uptake displays Michaelis–Menten saturation kinetics, consistent with a carrier-mediated process.

A model of pH_{ECF} regulation by capillary endothelial cells is similar in some respects to that described above for the choroid plexus epithelium (see Nattie, 1998). For example, NHE and possibly AE are present on the basolateral membrane. The NHE functions as an acid extruder and the AE functions as an acid loader. The Na^+ gradient that fuels the NHE is established by a Na^+ pump. An NHE is also present on the apical membrane where it probably also functions as an acid extruder. As of yet, there is no evidence for an NBC in capillary endothelial cells.

6. Concluding remarks

As we have reviewed in this chapter, the regulation of pH_i and pH_{ECF} is paramount for proper brain function. Glial cells express an array of acid–base transporters that maintain pH_i and contribute to pH_{ECF} regulation. In addition, cells of the choroid plexus epithelium and capillary endothelium use some of these same transporters to regulate both their own pH_i and subsequently pH_{CSF} and pH_{ECF}. As detailed above, there is a rich literature that highlights many of the functional and molecular properties of non-neuronal acid–base transporters. Additional functional and molecular studies will undoubtedly contribute to our understanding of the importance of individual acid–base transporters in the brain. Future functional studies include further characterizing the pH_i physiology of microglia and NBC activity in oligodendrocytes, as well as identifying NBCs in the choroid plexus where they may mediate HCO_3^- secretion into the CSF. Future molecular studies include examining the expression profiles of NBCs and NDCBEs throughout the CNS. As found with the NHEs and AEs, NBC and NDCBE isoforms probably exhibit different expression profiles that depend on cell type, brain region, and stage of animal development. Cation-coupled HCO_3^- transporters can markedly influence neuronal physiology, as exemplified by the link between NBC activity and neuronal activity. The molecular information obtained on cation-coupled HCO_3^- transporters provides us the opportunity to use genetic approaches to characterize the function of these transporters in the CNS further. It would be particularly exciting, for example, to examine cellular pH physiology and neuronal firing patterns in an NBC or NDCBE knock-out mouse. The combination of cellular, molecular, and genetic approaches will be necessary to understand the complete physiology of individual acid–base transporters in the context of a functioning nervous system.

Acknowledgements

We thank Dr Carmel M. McNicholas-Bevensee and Ms Jennifer B. Williams for reading the manuscript carefully and making suggestions. In addition, we thank Drs Mitchell Chesler and Conrad E. Johanson for helpful information and valuable comments. Dr Carmel M. McNicholas-Bevensee kindly obtained the neuronal action potential shown in Fig. 3.

References

Abuladze, N., Lee, I., Newman, D., Hwang, J., Boorer, K., Pushkin, A., Kurtz, I., 1998. Molecular cloning, chromosomal localization, tissue distribution, and functional expression of the human pancreatic sodium bicarbonate cotransporter. J. Biol. Chem. 273, 17689–17695.

Adelman, W.J., Fitzhugh, R., 1975. Solutions of the Hodgkin–Huxley equations modified for potassium accumulation in periaxonal spaces. Fed. Proc. 34, 1322–1329.

Alper, S.L., Chernova, M.N., Williams, J., Zasloff, M., Law, F.-Y., Knauf, P.A., 1998. Differential inhibition of AE1 and AE2 anion exchangers by oxonol dyes and by novel polyaminosterol analogs of the shark antibiotic, squalamine. Biochem. Cell Biol. 76, 799–806.

Alper, S.L., Stuart-Tilley, A., Simmons, C.F., Brown, D., Drenckhahn, D., 1994. The fodrin–ankyrin cytoskeleton of choroid plexus preferentially colocalizes with apical Na^+,K^+-ATPase rather than with basolateral anion exchanger AE2. J. Clin. Invest. 93, 1430–1438.

Amato, A., Ballerini, L., Attwell, D., 1994. Intracellular pH changes produced by glutamate uptake in rat hippocampal slices. J. Neurophysiol. 72, 1686–1696.

Amlal, H., Burnham, C.E., Soleimani, M., 1999. Characterization of Na^+/HCO_3^- cotransporter isoform NBC-3. Am J. Physiol. Renal Physiol. 276: F903–F913.

Andrew, R.D., 1991. Seizure and acute osmotic change: clinical and neurophysiological aspects. J. Neurol. Sci. 101, 7–18.

Aram, J.A., Lodge, D., 1987. Epileptiform activity induced by alkalosis in rat neocortical slices: block by antagonists of *N*-methyl-D-aspartate. Neurosci. Lett. 83, 345–350.

Astion, M.L., Orkand, R.K., 1988. Electrogenic Na^+/HCO_3^- cotransport in neuroglia. Glia 1, 355–357.

Attaphitaya, S., Park, K., Melvin, J.E., 1999. Molecular cloning and functional expression of a rat Na^+/H^+ exchanger (NHE5) highly expressed in brain. J. Biol. Chem. 274, 4383–4388.

Baird, N.R., Orlowski, J., Szabo, E.Z., Zaun, H.C., Schultheis, P.J., Menon, A.G., Schull, G.E., 1999. Molecular cloning, genomic organization, and functional expression of Na^+/H^+ exchanger isoform 5 (NHE5) from human brain. J. Biol. Chem. 274, 4377–4382.

Balestrino, M., Somjen, G.G., 1988. Concentration of carbon dioxide, interstitial pH and synaptic transmission in hippocampal formation of the rat. J. Physiol. 396, 247–266.

Benos, D.J., Sapirstein, V.S., 1983. Characteristics of an amiloride-sensitive sodium entry pathway in cultured rodent glial and neuroblastoma cells. J. Cell. Physiol. 116, 213–220.

Betz, A.L., 1983a. Sodium transport in capillaries isolated from rat brain. J. Neurochem. 41, 1150–1157.

Betz, A.L., 1983b. Sodium transport from blood to brain: inhibition by furosemide and amiloride. J. Neurochem. 41, 1158–1164.

Bevensee, M.O., Apkon, M., Boron, W.F., 1997a. Intracellular pH regulation in cultured astrocytes from rat hippocampus. II. Electrogenic Na/HCO_3 cotransport. J. Gen. Physiol. 110, 467–483.

Bevensee, M.O., Alper, S.L., Aronson, P.S., Boron, W.F., 2000a. Control of intracellular pH. In: Seldin, D.W., Giebisch, G. (Eds.), The Kidney: Physiology and Pathophysiology. 3rd ed. Lippincott Williams & Wilkins, Philadelphia, pp. 391–442.

Bevensee, M.O., Boron, W.F., 1998a. pH regulation in mammalian neurons. In: Kaila, K., Ransom, B.R. (Eds.), pH and Brain Function. Wiley-Liss, New York, pp. 211–231.

Bevensee, M.O., Boron, W.F., 1998b. Thermodynamics and physiology of cellular pH regulation. In: Kaila, K., Ransom, B.R. (Eds.), pH and Brain Function. Wiley-Liss, New York, pp. 173–194.

Bevensee, M.O., Schmitt, B.M., Choi, I., Romero, M.F., Boron, W.F., 2000b. An electrogenic $Na^+–HCO_3^-$ cotransporter (NBC) with a novel COOH-terminus, cloned from rat brain. Am. J. Physiol. Cell Physiol. 278, C1200–C1211.

Bevensee, M.O., Weed, R.A., Boron, W.F., 1997b. Intracellular pH regulation in cultured astrocytes from rat hippocampus. I. Role of HCO_3^-. J. Gen. Physiol. 110, 453–465.

Blankenfeld, G.V., Kettenmann, H., 1992. Glutamate GABA receptors in vertebrate glial cells. Mol. Neurobiol. 1, 31–43.

Blankenfeld, G., Trotter, J., Kettenmann, H., 1991. Expression and developmental regulation of a $GABA_A$ receptor in cultured murine cells of the oligodendrocyte lineage. Eur. J. Neurosci. 3, 310–316.

Bormann, J., Kettenmann, H., 1988. Patch-clamp study of gamma-aminobutyric acid receptor Cl^- channels in cultured astrocytes. Proc. Natl Acad. Sci. 85, 9336–9640.

Boron, W.F., Boulpaep, E.L., 1983. Intracellular pH regulation in the renal proximal tubule of the salamander: basolateral HCO_3^- transport. J. Gen. Physiol. 81, 53–94.

Boron, W.F., De Weer, P., 1976a. Intracellular pH transients in squid giant axons caused by CO_2, NH_3 and metabolic inhibitors. J. Gen. Physiol. 67, 91–112.

Boron, W.F., De Weer, P., 1976b. Active proton transport stimulated by CO_2/HCO_3^- blocked by cyanide. Nature 259, 240–241.

Boron, W.F., Fong, P., Hediger, M.A., Boulpaep, E.L., Romero, M.F., 1997. The electrogenic Na/HCO_3 cotransporter. Weiner Klin. Wochenschr. 109, 445–456.

Boron, W.F., Russell, J.M., 1983. Stoichiometry and ion dependencies of the intracellular-pH-regulating mechanism in squid giant axons. J. Gen. Physiol. 81, 373–399.

Boussouf, A., Gaillard, S., 2000. Intracellular pH changes during oligodendrocyte differentiation in primary culture. J. Neurosci. Res. 59, 731–739.

Boussouf, A., Lambert, R.C., Gaillard, S., 1997. Voltage-dependent $Na^+–HCO_3^-$ cotransporter and Na^+/H^+ exchanger are involved in intracellular pH regulation of cultured mature rat cerebellar oligodendrocytes. Glia 19, 74–84.

Bouvier, M., Szatkowski, M., Amato, A., Attwell, D., 1992. The glial cell glutamate uptake carrier countertransports pH-changing anions. Nature 360, 471–474.

Bovolin, P., Santi, M.R., Puia, G., Costa, E., Grayson, D., 1992. Expression patterns of gamma-aminobutyric acid type A receptor subunit mRNAs in primary cultures of granule neurons and astrocytes from neonatal rat cerebella. Proc. Natl. Acad. Sci. 89, 9344–9348.

Boyarsky, G., Ganz, M.B., Sterzel, B., Boron, W.F., 1988a. pH regulation in single glomerular mesangial cells. II. Na-dependent and -independent $Cl–HCO_3$ exchangers. Am. J. Physiol. Cell Physiol. 255, C857–C869.

Boyarsky, G., Ganz, M.B., Sterzel, B., Boron, W.F., 1988b. pH regulation in single glomerular mesangial cells. I. Acid extrusion in absence and presence of HCO_3^-. Am. J. Physiol. Cell Physiol. 255, C844–C856.

Boyarsky, G., Ganz, M.B., Cragoe, E.J. Jr., Boron, W.F., 1990. Intracellular-pH dependence of Na–H exchange and acid loading in quiescent and arginine vasopressin-activated mesangial cells. Proc. Natl. Acad. Sci. 87, 5921–5924.

Boyarsky, G., Ransom, B., Schlue, W.-R., Davis, M.B.E., Boron, W.F., 1993. Intracellular pH regulation in single cultured astrocytes from rat forebrain. Glia 8, 241–248.

Brett, C.L., Wei, Y., Donowitz, M., Rao, R., 2002. Human Na^+/H^+ exchanger isoform 6 is found in recycling endosomes of cells, not in mitochondria. Am. J. Physiol. Cell Physiol. 282, C1031–C1041.

Bröer, S., Rahman, B., Pellegri, G., Pellerin, L., Martin, J.L., Verleysdonk, S., Hamprecht, B., Magistretti, P.J., 1997. Comparison of lactate transport in astroglial cells and monocarboxylate transporter 1 (MCT 1) expressing *Xenopus laevis* oocytes. Expression of two different monocarboxylate transporters in astroglial cells and neurons. J. Biol. Chem. 272, 30096–30102.

Brookes, N., Turner, R.J., 1993. Extracellular potassium regulates the glutamine content of astrocytes: mediation by intracellular pH. Neurosci. Lett. 160, 73–76.

Brookes, N., Turner, R.J., 1994. K^+-induced alkalinization in mouse cerebral astrocytes mediated by reversal of electrogenic $Na^+–HCO_3^-$ cotransport. Am. J. Physiol. Cell Physiol. 36, C1633–C1640.

Brune, T., Deitmer, J.W., 1995. Intracellular acidification and Ca^{2+} transients in cultured rat cerebellar astrocytes evoked by glutamate agonists and noradrenaline. Glia 14, 153–161.

Brune, T., Fetzer, S., Backus, K.H., Deitmer, J.W., 1994. Evidence for electrogenic $Na–HCO_3$ cotransport in cultured rat cerebellar astrocytes. Pflügers Arch. 429, 64–71.

Burnham, C.E., Amlal, H., Wang, Z., Shull, G.E., Soleimani, M., 1997. Cloning and functional expression of a human kidney $Na^+:HCO_3^-$ cotransporter. J. Biol. Chem. 272, 19111–19114.

Chen, J.C.T., Chesler, M., 1990. A bicarbonate-dependent increase in extracellular pH mediated by $GABA_A$ receptors in turtle cerebellum. Neurosci. Lett. 116, 130–135.

Chen, J.C.T., Chesler, M., 1992a. Extracellular alkaline shifts in rat hippocampal slice are mediated by NMDA and non-NMDA receptors. J. Neurophysiol. 68, 342–344.

Chen, J.C.T., Chesler, M., 1992b. pH transients evoked by excitatory synaptic transmission are increased by inhibition of extracellular carbonic anhydrase. Proc. Natl. Acad. Sci. 89, 7786–7790.

Chesler, M., 1990. The regulation and modulation of pH in the nervous system. Prog. Neurobiol. 34, 401–427.

Chesler, M., Kaila, K., 1992. Modulation of pH by neuronal activity. Trends Neurosci. 15, 396–402.

Choi, D.W., 1988. Calcium-mediated neurotoxicity: relationship to specific channel types and role in ischemic damage. Trends Neurosci. 11, 465–469.

Choi, I., Aalkjaer, C., Boulpaep, E.L., Boron, W.F., 2000. An electroneutral sodium/bicarbonate cotransporter NBCn1 and associated sodium channel. Nature 405, 571–575.

Choi, I., Romero, M.F., Khandoudi, N., Bril, A., Boron, W.F., 1999. Cloning and characterization of a human electrogenic Na^+–HCO_3^- cotransporter isoform (hhNBC). Am. J. Physiol. Cell Physiol. 276, C576–C584.

Chow, S.Y., Yen-Chow, Y.C., White, H.S., Woodbury, D.M., 1991. pH regulation after acid load in primary cultures of mouse astrocytes. Dev. Brain Res. 60, 69–78.

Chow, S.Y., Yen-Chow, Y.C., Woodbury, D.M., 1992. Studies on pH regulatory mechanisms in cultured astrocytes of DBA and C57 mice. Epilepsia 33, 775–784.

Chvátal, A., Pastor, A., Mauch, M., Syková, E., Kettenmann, H., 1995. Distinct populations of identified glial cells in the developing rat spinal cord slice: ion channel properties and cell morphology. Eur. J. Neurosci. 7, 129–142.

Cohen, J.J., Kassirer, J.P., 1982. Acid–Base. Little Brown, Boston.

Cooke, S.F., Lilley, S.J., 2002. Acidity at the synapse: pH influences plasticity and memory. Trends Neurosci. 25, 446–447.

Counillon, L., Pouysségur, J., 2000. The members of the Na^+/H^+ exchanger gene family: their structure, function, expression and regulation. In: Seldin, D.W., Giebisch, G. (Eds.), The Kidney: Physiology and Pathophysiology. 3rd ed. Lippincott Williams & Wilkins, Philadelphia, pp. 223–234.

Counillon, L., Scholz, W., Lang, H.J., Pouysségur, J., 1993. Pharmacological characterization of stably transfected Na^+/H^+ antiporter isoforms using amiloride analogs and a new inhibitor exhibiting anti-ischemic properties. Mol. Pharmacol. 44, 1041–1045.

Danbolt, N.C., 2001. Glutamate uptake. Prog. Neurobiol. 65, 1–105.

Danforth, W.H., 1965. Activation of glycolytic pathway in muscle. In: Chance, B., Estabrook, R.W., Williamson, J.B. (Eds.), Control of Energy Metabolism. Academic Press, New York, pp. 287–298.

DeCoursey, T.E., Cherny, V.V., 1994. Voltage-activated hydrogen ion currents. J. Membr. Biol. 141, 203–223.

Deitmer, J.W., 1992. Bicarbonate-dependent changes of intracellular sodium and pH in identified leech glial cells. Pflügers Arch. 420, 584–589.

Deitmer, J.W., 1998. pH regulation in invertebrate glia. In: Kaila, K., Ransom, B.R. (Eds.), pH and Brain Function. Wiley-Liss, New York, pp. 233–252.

Deitmer, J.W., Schlue, W.-R., 1987. The regulation of intracellular pH by identified glial cells and neurones in the central nervous system of the leech. J. Physiol. 388, 261–283.

Deitmer, J.W., Schlue, W.-R., 1989. An inwardly directed electrogenic sodium-bicarbonate cotransport in leech glial cells. J. Physiol. 411, 179–194.

Deitmer, J.W., Schneider, H.P., 1997. Intracellular acidification of the leech giant glial cell evoked by glutamate and aspartate. Glia 19, 111–122.

Deng, Q.S., Johanson, C.E., 1989. Stilbenes inhibit exchange of chloride between blood, choroid plexus and cerebrospinal fluid. Brain Res. 501, 183–187.

Douglas, R.M., Schmitt, B.M., Xia, Y., Bevensee, M.O., Biemesderfer, D., Boron, W.F., Haddad, G.G., 2001. Na/H exchangers and Na/HCO_3 cotransporter: ontogeny of protein expression in the rat brain. Neuroscience 102, 217–228.

Dringen, R., Peters, H., Wiesinger, H., Hamprecht, B., 1995. Lactate transport in cultured glial cells. Dev. Neurosci. 17, 63–69.

Eder, C., DeCoursey, T.E., 2001. Voltage-gated proton channels in microglia. Prog. Neurobiol. 64, 277–305.

Faff, L., Nolte, C., 2000. Extracellular acidification decreases the basal motility of cultured mouse microglia via the rearrangement of the actin cytoskeleton. Brain Res. 853, 22–31.

Faff, L., Ohlemeyer, C., Kettenmann, H., 1996. Intracellular pH regulation in cultured microglial cells from mouse brain. J. Neurosci. Res. 46, 294–304.

Frankel, H., Kazemi, H., 1983. Regulation of CSF composition—blocking chloride-bicarbonate exchange. J. Appl. Physiol. 55, 177–182.

Fraser, D.D., Mudrick-Donnon, L.A., MacVicar, B.A., 1994. Astrocytic GABA receptors. Glia 11, 83–93.

Fresu, L., Dehpour, A., Genazzani, A.A., Carafoli, E., Guerini, D., 1999. Plasma membrane calcium ATPase isoforms in astrocytes. Glia 28, 150–155.

Ganz, M.B., Perfetto, M.C., Boron, W.F., 1990. Effects of mitogens and other agents on rat mesangial cell proliferation, pH, and Ca^{2+}. Am. J. Physiol. Renal Physiol. 259, F269–F278.

Giffard, R.G., Papadopoulos, M.C., van Hooft, J.A., Xu, L., Giuffrida, R., Monyer, H., 2000. The electrogenic sodium bicarbonate cotransporter: Developmental expression in rat brain and possible role in acid vulnerability. J. Neurosci. 20, 1001–1008.

Gottfried, J.A., Chesler, M., 1994. Endogenous H^+ modulation of NMDA receptor-mediated EPSCs revealed by carbonic anhydrase inhibition in rat hippocampus. J. Physiol. 478, 373–378.

Grichtchenko, I., Boron, W.F., 2002a. Evidence for CO_3^{2-} transport by NBCe1, based on surface-pH measurements in voltage-clamped *Xenopus* oocytes co-expressing NBCe1 and CAIV. FASEB J. 16, A795.

Grichtchenko, I., Boron, W.F., 2002b. Surface-pH gradient measurements in *Xenopus* oocytes co-expressing the Na^+-driven Cl–HCO_3 exchanger (NDCBE1) and CAIV: evidence for CO_3^{2-} transport. FASEB J. 16, A797.

Grichtchenko, I.I., Chesler, M., 1994a. Depolarization-induced acid secretion in gliotic hippocampal slices. Neuroscience 62, 1057–1070.

Grichtchenko, I.I., Chesler, M., 1994b. Depolarization-induced alkalinization of astrocytes in gliotic hippocampal slices. Neuroscience 62, 1071–1078.

Grichtchenko, I.I., Choi, I., Zhong, X., Bray-Ward, P., Russell, J.M., Boron, W.F., 2001. Cloning, characterization, and chromosomal mapping of a human electroneutral Na^+-driven Cl–HCO_3 exchanger. J. Biol. Chem. 276, 8358–8363.

Guerini, D., 1998. The significance of the isoforms of plasma membrane calcium ATPase. Cell Tissue Res. 292, 191–197.

Halestrap, A.P., Price, N.T., 1999. The proton-linked monocarboxylate transporter (MCT) family: structure, function and regulation. Biochem. J. 343, 281–299.

Hallows, K.R., Knauf, P.A., 1994. Principles of cell volume regulation. In: Strange, K. (Ed.), Cellular and Molecular Physiology of Cell Volume. Regulation. CRC Press, Boca Raton, pp. 3–29.

Henderson, L.M., Chappell, J.B., Jones, O.T., 1987. The superoxide-generating NADPH oxidase of human neutrophils is electrogenic and associated with an H^+ channel. Biochem. J. 246, 325–329.

Henderson, L.M., Chappell, J.B., Jones, O.T., 1988a. Internal pH changes associated with the activity of NADPH oxidase of human neutrophils. Further evidence for the presence of an H^+ conducting channel. Biochem. J. 251, 563–567.

Henderson, L.M., Chappell, J.B., Jones, O.T., 1988b. Superoxide generation by the electrogenic NADPH oxidase of human neutrophils is limited by the movement of a compensating charge. Biochem. J. 255, 285–290.

Hoffmann, E.K., Simonsen, L.O., 1989. Membrane mechanisms in volume and pH regulation in vertebrate cells. Physiol. Rev. 69, 315–382.

Hsu, P., Haffner, J., Albuquerque, M.C., Leffler, C.W., 1996. pH_i in piglet cerebral microvascular endothelial cells: recovery from an acid load. Proc. Soc. Exp. Biol. Med. 212, 256–262.

Ishibashi, K., Sasaki, S., Marumo, G., 1998. Molecular cloning of a new sodium bicarbonate cotransporter cDNA from human retina. Biochem. Biophys. Res. Commun. 246, 535–538.

Jaisser, F., Beggah, A.T., 1999. The nongastric H^+/K^+-ATPases: molecular and functional properties. Am. J. Physiol. Renal Physiol. 276, F812–F824.

Jean, T., Frelin, C., Vigne, P., Lazdunski, M., 1986. The Na^+/H^+ exchange system in glial cell lines. Properties and activation by a hyperosmotic shock. Eur. J. Biochem. 160, 211–219.

Jendelová, P., Syková, E., 1991. Role of glia in K^+ and pH homeostasis in the neonatal rat spinal cord. Glia 4, 56–63.

Johanson, C.E., 1978. Choroid epithelial cell pH. Life Sci. 23, 861–868.

Johanson, C.E., 1984. Differential effects of acetazolamide, benzolamide and systemic acidosis on hydrogen and bicarbonate gradients across the apical and basolateral membranes of the choroid plexus. J. Pharmacol. Exp. Ther. 231, 502–510.

Johanson, C.E., 2003. The choroid plexus–cerebrospinal fluid nexus. Gateway to the brain. In: Conn, P.M. (Ed.), Neuroscience in Medicine. Humana Press, Totowa, NJ, pp. 165–195.

Johanson, C.E., Allen, J., Withrow, C.D., 1988. Regulation of pH and HCO_3 in brain and CSF of the developing mammalian central nervous system. Brain Res. 466, 255–264.

Johanson, C.E., Murphy, V.A., 1990. Acetazolamide and insulin alter choroid plexus epithelial cell [Na^+], pH, and volume. Am. J. Physiol. Renal Physiol. 258, F1538–F1546.

Johanson, C.E., Parandoosh, Z., Dyas, M.L., 1992. Maturational differences in acetazolamide-altered pH and HCO_3 of choroid plexus, cerebrospinal fluid, and brain. Am. J. Physiol. Regul. Integr. Comp. Physiol. 262, R909–R914.

Johanson, C.E., Parandoosh, Z., Smith, Q.R., 1985. Cl–HCO_3 exchange in choroid plexus: analysis by the DMO method for cell pH. Am. J. Physiol. Renal Physiol. 249, F478–F484.

Johanson, C.E., Woodbury, D.M., Withrow, C.D., 1976. Distribution of bicarbonate between blood and cerebrospinal fluid in the neonatal rat in metabolic acidosis and alkalosis. Life Sci. 19, 691–700.

Kaila, K., Paalasmaa, P., Taira, T., Voipio, J., 1992. pH transients due to monosynaptic activation of $GABA_A$ receptors in rat hippocampal slices. Neuroreport 3, 105–108.

Kaila, K., Panula, P., Karhunen, T., Heinonen, E., 1991. Fall in intracellular pH mediated by $GABA_A$ receptors in cultured rat astrocytes. Neurosci. Lett. 126, 9–12.

Kaila, K., Saarikoski, J., Voipio, J., 1990. Mechanism of action of GABA on intracellular pH and on surface pH in crayfish muscle fibres. J. Physiol. 427, 241–260.

Kaila, K., Voipio, J., 1987. Postsynaptic fall in intracellular pH induced by GABA-activated bicarbonate conductance. Nature 330, 163–165.

Kalaria, R.N., Premkumar, D.R., Lin, C.W., Kroon, S.N., Bae, J.Y., Sayre, L.M., LaManna, J.C., 1998. Identification and expression of the Na^+/H^+ exchanger in mammalian cerbrovascular and choroidal tissues: characterization by amiloride-sensitive [^{3}H]MIA binding and RT-PCR analysis. Mol. Brain Res. 58, 178–187.

Kang, J., Jiang, L., Goldman, S.A., Nedergaard, M., 1998. Astrocyte-mediated potentiation of inhibitory synaptic transmission. Nat. Neurosci. 1, 683–692.

Katsura, K., Siesjö, B.K., 1998. Acid–base metabolism in ischemia. In: Kaila, K., Ransom, B.R. (Eds.), pH and Brain Function. Wiley-Liss, New York, pp. 563–582.

Kawai, K., Takahashi, H., Wakabayashi, K., Ikuta, F., 1989. Ultracytochemical localization of Ca^{2+}-ATPase activity in reactive astrocytes. Acta Neuropathol. 78, 449–454.

Kawasaki-Nishi, S., Nishi, T., Forgac, M., 2003. Proton translocation driven by ATP hydrolysis in V-ATPases. FEBS Lett. 545, 76–85.

Keep, R.F., Xiang, J., Betz, A.L., 1994. Potassium cotransport at the rat choroid plexus. Am. J. Physiol. Cell Physiol. 267, C1616–C1622.

Kettenmann, H., Schlue, W.-R., 1988. Intracellular pH regulation in cultured mouse oligodendrocytes. J. Physiol. 406, 147–162.

Kimelberg, H.K., 1995. Current concepts of brain edema. Review of laboratory investigations. J. Neurosurg. 83, 1051–1059.

Kimelberg, H.K., Bowman, C., Biddlecome, S., Bourke, R.S., 1979. Cation transport and membrane potential properties of primary astroglial cultures from neonatal rat brains. Brain Res. 177, 533–550.

Knauf, P.A., Law, F.Y., Hahn, K., 1995. An oxonol dye is the most potent known inhibitor of band 3-mediated anion exchange. Am. J. Physiol. Cell Physiol. 269, C1073–C1077.

Ko, S.B.H., Luo, X., Hager, H., Rojek, A., Choi, J.Y., Licht, C., Suzuki, M., Muallem, S., Neilsen, S., Ishibashi, K., 2002. AE4 is a DIDS-sensitive Cl^-/HCO_3^- exchanger in the basolateral membrane of the renal CCD and the SMG duct. Am. J. Physiol. Cell Physiol. 283, C1206–C1218.

Ko, Y.P., Lang, H.J., Loh, S.H., Chu, K.C., Wu, M.L., 1999. Cl^--dependent and Cl^--independent Na^+/HCO_3^- acid extrusion in cultured rat cerebellar astrocytes. Chin. J. Physiol. 42, 237–248.

Kobayashi, S., Morgans, C.W., Casey, J.R., Kopito, R.R., 1994. AE3 anion exchanger isoforms in the vertebrate retina: developmental regulation and differential expression in neurons and glia. J. Neurosci. 14, 6266–6279.

Kopito, R.R., Lee, B.S., Simmons, D.M., Lindsey, A.E., Morgans, C.W., Schneider, K., 1989. Regulation of intracellular pH by a neuronal homolog of the erythrocyte anion exchanger. Cell 59, 927–937.

Kopito, R.R., Lodish, H.F., 1985. Primary structure and transmembrane orientation of the murine anion exchange protein. Nature 316, 234–238.

Kramhoft, B., Hoffmann, E.K., Simonsen, L.O., 1994. pH_i regulation in Ehrlich mouse ascites tumor cells: role of sodium-dependent and sodium-independent chloride–bicarbonate exchange. J. Membr. Biol. 138, 121–132.

Kuffler, S.W., Potter, D.D., 1964. Glia in the leech central nervous system: physiological properties and neuron–glia relationship. J. Neurophysiol. 27, 290–320.

L'Allemain, G., Paris, S., Pouysségur, J., 1985. Role of a Na^+-dependent Cl^-/HCO_3^- exchange in regulation of intracellular pH_i in fibroblasts. J. Biol. Chem. 260, 4877–4883.

Lang, F., Busch, G.L., Volkl, H., Haussinger, D., 1995. Cell volume: a second message in regulation of cell function. News Physiol. Sci. 10, 18–22.

Lee, J., Taira, T., Pihlaja, P., Ransom, B.R., Kaila, K., 1996. Effects of CO_2 on excitatory transmission apparently caused by changes in intracellular pH in the rat hippocampal slice. Brain Res. 706, 210–216.

Liljas, A., Hakansson, K., Jonsson, B.H., Xue, Y., 1994. Inhibition and catalysis of carbonic anhydrase. Eur. J. Biochem. 219, 1–10.

Lindsey, A.E., Schneider, K., Simmons, D.M., Baron, R., Lee, B.S., Kopito, R.R., 1990. Functional expression and subcellular localization of an anion exchanger cloned from choroid plexus. Proc. Natl. Acad. Sci. 87, 5278–5282.

Ma, E., Haddad, G.G., 1997. Expression and localization of Na^+/H^+ exchangers in rat central nervous system. Neuroscience 79, 591–603.

Mayer, S.E., Sanders-Bush, E., 1993. Sodium-dependent antiporters in choroid plexus epithelial cultures from rabbit. J. Neurochem. 60, 1308–1316.

McLean, L., Roscoe, J., Jorgensen, N.K., Gorin, F.A., Cala, P.M., 2000. Malignant gliomas display altered pH regulation by NHE1 compared with nontransformed astrocytes. Am. J. Physiol. Cell Physiol. 278, C676–C688.

Mellergård, P., Ouyang, Y.B., Siesjö, B.K., 1993. Intracellular pH regulation in cultured rat astrocytes in CO_2/HCO_3^--containing media. Exp. Brain Res. 95, 371–380.

Muller, T., Fritschy, J.M., Grosche, J., Pratt, G.D., Mohler, H., Kettenmann, H., 1994. Developmental regulation of voltage-gated K^+ channel and $GABA_A$ receptor expression in Bergmann glial cells. J. Neurosci. 14, 2503–2514.

Munsch, T., Deitmer, J.W., 1994. Sodium-bicarbonate cotransport current in identified leech glial cells. J. Physiol. 474, 43–53.

Murata, Y., Sun-Wada, G., Yoshimizu, T., Yamamoto, A., Wada, Y., Futai, M., 2002. Differential localization of the vacuolar H^+ pump with G subunit isoforms (G1 and G2) in mouse neurons. J. Biol. Chem. 277, 36296–36303.

Murphy, V.A., Johanson, C.E., 1989a. Acidosis, acetazolamide, and amiloride: effects on ^{22}Na transfer across the blood–brain and blood–CSF barriers. J. Neurochem. 52, 1058–1063.

Murphy, V.A., Johanson, C.E., 1989b. Alteration of sodium transport by the choroid plexus with amiloride. Biochim. Biophys. Acta 979, 187–192.

Murphy, V.A., Johanson, C.E., 1990. Na^+–H^+ exchange in choroid plexus and CSF in acute metabolic acidosis or alkalosis. Am. J. Physiol. Renal Physiol. 258, F1528–F1537.

Nakhoul, N.L., Abdulnour-Nakhoul, S., Khuri, R.N., Lieberman, E.M., Hargittai, P.T., 1994. Intracellular pH regulation in rat Schwann cells. Glia 10, 155–164.

Nattie, E., 1998. Control and disturbances of cerebrospinal fluid pH. In: Kaila, K., Ransom, B.R. (Eds.), pH and Brain Function. Wiley-Liss, New York, pp. 629–650.

Nedergaard, M., Goldman, S.A., 1993. Carrier-mediated transport of lactic acid in cultured neurons and astrocytes. Am. J. Physiol. Regul. Integr. Comp. Physiol. 265, R282–R289.

Nett, W., Deitmer, J.W., 1998. Intracellular Ca^{2+} regulation by the leech giant glial cell. J. Physiol. 507, 147–162.

Newman, E.A., 1991. Sodium-bicarbonate cotransport in retinal Müller (Glial) cells of the salamander. J. Neurosci. 11, 3972–3983.

Newman, E.A., 1996. Acid efflux from retinal glial cells generated by sodium bicarbonate cotransport. J. Neurosci. 16, 159–168.

Newman, E.A., Astion, M.L., 1991. Localization and stoichiometry of electrogenic sodium bicarbonate cotransport in retinal glial cells. Glia 4, 424–428.

Numata, M., Orlowski, J., 2001. Molecular cloning and characterization of a novel $(Na^+,K^+)/H^+$ exchanger localized to the *trans*-Golgi network. J. Biol. Chem. 276, 17387–17394.

O'Beirne, M., Bulloch, A.G.M., MacVicar, B.A., 1987. Dye and electrotonic coupling between cultured hippocampal neurons. Neurosci. Lett. 78, 265–270.

O'Connor, E.R., Sontheimer, H., Ransom, B.R., 1994. Rat hippocampal astrocytes exhibit elecrogenic sodium-bicarbonate co-transport. J. Neurophysiol. 72, 2580–2589.

Okamoto, C.T., Forte, J.G., 2001. Vesicular trafficking machinery, the actin cytoskeleton, and H^+–K^+-ATPase recycling in the gastric parietal cell. J. Physiol. 532, 287–296.

Paalasmaa, P., Kaila, K., 1996. Role of voltage-gated calcium channels in the generation of activity-induced extracellular pH transients in the rat hippocampal slice. J. Neurophysiol. 75, 2354–2360.

Paalasmaa, P., Taira, T., Voipio, J., Kaila, K., 1994. Extracellular alkaline transients mediated by glutamate receptors in the rat hippocampal slice are not due to a proton conductance. J. Neurophysiol. 72, 2031–2033.

Pappas, C.A., Ransom, B.R., 1993. A depolarization-stimulated, bafilomycin-inhibitable H^+ pump in hippocampal astrocytes. Glia 9, 280–291.

Pappas, C.A., Ransom, B.R., 1994. Depolarization-induced alkalinization (DIA) in rat hippocampal astrocytes. J. Neurophysiol. 72, 2816–2826.

Parton, R.G., Dotti, C.G., Bacallao, R., Kurtz, I., Simons, K., Prydz, K., 1991. pH-induced microtubule-dependent redistribution of late endosomes in neuronal and epithelial cells. J. Cell. Biol. 113, 261–274.

Philippe, J.M., Dubois, J.M., Rouzaire-Dubois, B., Cartron, P.F., Vallette, F., Morel, N., 2002. Functional expression of V-ATPases in the plasma membrane of glial cells. Glia 37, 365–373.

Pizzonia, J.H., Ransom, B.R., Pappas, C.R., 1996. Characterization of Na^+/H^+ exchange activity in cultured rat hippocampal astrocytes. J. Neurosci. Res. 44, 191–198.

Pollay, M., Curl, F., 1967. Secretion of cerebrospinal fluid by the ventricular ependyma of the rabbit. Am. J. Physiol. 213, 1031–1038.

Pouysségur, J., Franchi, A., L'Allemain, G., Paris, S., 1985. Cytoplasmic pH, a key determinant of growth factor-induced DNA synthesis in quiescent fibroblasts. FEBS Lett. 190, 115–119.

Pouysségur, J., Sardet, C., Franchi, A., L'Allemain, G., Paris, S., 1984. A specific mutation abolishing Na^+/H^+ antiport activity in hamster fibroblasts precludes growth at neutral and acidic pH. Proc. Natl. Acad. Sci. USA 81, 4833–4837.

Pushkin, A., Abuladze, N., Lee, I., Newman, D., Hwang, J., Kurtz, I., 1999. Cloning, tissue distribution, genomic organization, and functional characterization of NBC3, a new member of the sodium bicarbonate cotransporter family. J. Biol. Chem. 274, 16569–16575.

Pushkin, A., Abuladze, N., Newman, D., Lee, I., Xu, G., Kurtz, I., 2000a. Two C-terminal variants of NBC4, a new member of the sodium bicarbonate cotransporter family: cloning, characterization, and localization. IUBMB Life 50, 13–19.

Pushkin, A., Abuladze, N., Newman, D., Lee, I., Xu, G., Kurtz, I., 2000b. Cloning, characterization and chromosomal assignment of NBC4, a new member of the sodium bicarbonate cotransporter family. Biochim. Biophys. Acta 1493, 215–218.

Putney, L.K., Denker, S.P., Barber, D.L., 2002. The changing face of the Na^+/H^+ exchanger, NHE1: structure, regulation, and cellular actions. Annu. Rev. Pharmacol. Toxicol. 42, 527–552.

Ransom, B.R., 1992. Glial modulation of neural excitability mediated by extracellular pH: a hypothesis. Prog. Brain Res. 94, 37–46.

Ransom, B.R., 2000. Glial modulation of neural excitability mediated by extracellular pH: a hypothesis revisited. Prog. Brain Res. 125, 217–228.

Ridderstrale, Y., Winstrand, P.J., 1998. Carbonic anhydrase isoforms in the mammalian nervous system. In: Kaila, K., Ransom, B.R. (Eds.), pH and Brain Function. Wiley-Liss, New York, pp. 21–43.

Riquelme, R., Miralles, C.P., De Blas, A.L., 2002. Bergmann glia $GABA_A$ receptors concentrate on the glial processes that wrap inhibitory synapses. J. Neurosci. 22, 10720–10730.

Risso Bradley, S., Richerson, G.B., Rojas, J.D., Bouyer, P., Boron, W.F., 2001. Distribution of Na^+/HCO_3^- cotransporters in rodent brain with subtype-specific antibodies. Abstr. Soc. Neurosci., 27, Prog. No. 527.11.

Romero, M.F., Hediger, M.A., Boulpaep, E.L., Boron, W.F., 1997. Expression cloning and characterization of a renal electrogenic Na^+/HCO_3^- cotransporter. Nature 387, 409–413.

Romero, M.F., Fong, P., Berger, U.V., Hediger, M.A., Boron, W.F., 1998. Cloning and functional expression of rNBC, an electrogenic Na^+–HCO_3^- cotransporter from rat kidney. Am. J. Physiol. Renal. Physiol. 274, F425–F432.

Romero, M.F., Henry, D., Nelson, S., Harte, P.J., Dillon, A.K., Sciortino, C.M., 2000. Cloning and characterization of a Na^+-driven anion exchanger (NDAE1). A new bicarbonate transporter. J. Biol. Chem. 275, 24552–24559.

Roos, A., Boron, W.F., 1981. Intracellular pH. Physiol. Rev. 61, 296–434.

Rose, C.R., Ransom, B.R., 1996. Mechanisms of H^+ and Na^+ changes induced by glutamate, kainate, and D-aspartate in rat hippocampal astrocytes. J. Neurosci. 16, 5393–5404.

Rose, C.R., Ransom, B.R., 1998. pH regulation in mammalian glia. In: Kaila, K., Ransom, B.R. (Eds.), pH and Brain Function. Wiley-Liss, New York, pp. 253–275.

Russell, J.M., Boron, W.F., 1976. Role of chloride transport in regulation of intracellular pH. Nature 264, 73–74.

Saito, Y., Wright., E.M., 1983. Bicarbonate transport across the frog choroid plexus and its control by cyclic nucleotides. J. Physiol. 336, 635–648.

Saito, Y., Wright, E.M., 1984. Regulation of bicarbonate transport across the brush border membrane of the bull–frog choroid plexus. J. Physiol. 350, 327–342.

Sassani, P., Pushkin, A., Gross, E., Gomer, A., Abuladze, N., Dukkipati, R., Carpenito, G., Kurtz, I., 2002. Functional characterization of NBC4: a new electrogenic sodium-bicarbonate cotransporter. Am. J. Physiol. Cell Physiol. 282, C408–C416.

Schmitt, B.M., Berger, U.V., Douglas, R., Bevensee, M.O., Hediger, M.A., Haddad, G.G., Boron, W.F., 2000. Na/HCO_3 cotransporters in rat brain: expression in glia, neurons and choroid plexus. J. Neurosci. 20, 6839–6848.

Schwiening, C.J., Boron, W.F., 1994. Regulation of intracellular pH in pyramidal neurons from the rat hippocampus by Na^+-dependent Cl^-–HCO_3^- exchange. J. Physiol. 475, 59–67.

Schwiening, C.J., Kennedy, H.J., Thomas, R.C., 1993. Calcium–hydrogen exchange by the plasma membrane Ca-ATPase of voltage-clamped snail neurones. Proc. R. Soc. Lond. 253, 285–289.

Shirihai, O., Smith, P., Dagan, D., 1998. Microglia generate external proton and potassium ion gradients utilizing a member of the H/K ATPase family. Glia 23, 339–348.

Shrode, L.D., Klein, J.D., Douglas, P.B., O'Neill, W.C., Putnam, R.W., 1997. Shrinkage-induced activation of Na^+/H^+ exchange: role of cell density and myosin light chain phosphorylation. Am. J. Physiol. Cell Physiol. 272, C1968–C1979.

Shrode, L.D., Klein, J.D., O'Neill, W.C., Putnam, R.W., 1995. Shrinkage-induced activation of Na^+/H^+ exchange in primary rat astrocytes: role of myosin light-chain kinase. Am. J. Physiol. Cell Physiol. 269, C257–C266.

Shrode, L.D., Putnam, R.W., 1994. Intracellular pH regulation in primary rat astrocytes and C6 glioma cells. Glia 12, 196–210.

Siebens, A.W., Boron, W.F., 1989a. Depolarization-induced alkalinization in proximal tubules. I. Characteristics and dependence on Na^+. Am. J. Physiol. Renal Physiol. 25, F342–F353.

Siebens, A.W., Boron, W.F., 1989b. Depolarization-induced alkalinization in proximal tubules. II. Effects of lactate and SITS. Am. J. Physiol. Renal Physiol. 25, F354–F365.

Silver, R.B., Soleimani, M., 1999. H^+–K^+-ATPases: regulation and role in pathophysiological states. Am. J. Physiol. Renal Physiol. 276, F799–F811.

Sjaastad, M.D., Wenzl, E., Machen, T.E., 1992. pH_i dependence of Na–H exchange and H delivery in IEC-6 cells. Am. J. Physiol. Cell Physiol. 262, C164–C170.

Smith, S.E., Gottfried, J.A., Chen, J.C.T., Chesler, M., 1994. Calcium dependence of glutamate receptor-evoked alkaline shifts in hippocampus. Neuroreport 5, 2441–2445.

Smith, Q.R., Rapoport, S.I., 1984. Carrier-mediated transport of chloride across the blood–brain barrier. J. Neurochem. 42, 754–763.

Soleimani, M., Grassl, S.M., Aronson, P.S., 1987. Stoichiometry of Na^+–HCO_3^- cotransport in basolateral membrane vesicles isolated from rabbit renal cortex. J. Clin. Invest. 79, 1276–1280.

Somjen, G.G., Allen, B.W., Balestrino, M., Aitken, P.G., 1987. Pathophysiology of pH and Ca^{2+} in bloodstream and brain. Can. J. Physiol. Pharmacol. 65, 1078–1085.

Speake, T., Whitwell, C., Kajita, H., Majid, A., Brown, P.D., 2001. Mechanisms of CSF secretion by the choroid plexus. Microsc. Res. Tech. 52, 49–59.

Spray, D.C., Harris, A.L., Bennett, M.V.L., 1981. Gap junctional conductance is a simple and sensitive function of intracellular pH. Science 211, 712–715.

Stauffer, T.P., Guerini, D., Carafoli, E., 1995. Tissue distribution of the four gene products of the plasma membrane Ca^{2+} pump. A study using specific antibodies. J. Biol. Chem. 270, 12184–12190.
Sun, M.-K., Alkon, D.L., 2002. Carbonic anhydrase gating of attention: memory therapy and enhancement. Trends Pharmacol. Sci. 23, 83–89.
Sun, M.-K., Dahl, D., Alkon, D.L., 2001a. Heterosynaptic transformation of GABAergic gating in the hippocampus and effects of carbonic anhydrase inhibition. J. Pharmacol. Exp. Ther. 296, 811–817.
Sun, M.-K., Zhao, W.-Q., Nelson, T.J., Alkon, D.L., 2001b. Theta rhythm of hippocampal CA1 neuron activity: gating by GABAergic synaptic depolarization. J. Neurophysiol. 85, 269–279.
Svichar, N., Chesler, M., 2003. Surface carbonic anhydrase activity on astrocytes and neurons facilitates lactate transport. Glia 41, 415–419.
Syková, E., 1998. Extracellular pH and ionic shifts associated with electrical activity and pathological states in the spinal cord. In: Kaila, K., Ransom, B.R. (Eds.), pH and Brain Function. Wiley-Liss, New York, pp. 339–357.
Szatkowski, M., Schlue, W.-R., 1992. Mechanisms of pH recovery from intracellular acid loads in the leech connective glial cell. Glia 5, 193–200.
Szatkowski, M.S., Schlue, W.-R., 1994. Chloride-dependent pH regulation in connective glial cells of the leech nervous system. Brain Res. 665, 1–4.
Taira, T., Smirnov, S., Voipio, J., Kaila, K., 1993. Intrinsic proton modulation of excitatory transmission in rat hippocampal slices. Neuroreport 4, 93–96.
Tang, C.-M., Dichter, M., Morad, M., 1990. Modulation of the *N*-methyl-D-aspartate channel by extracellular H^+. Proc. Natl. Acad. Sci. USA 87, 6445–6449.
Thévenod, F., Roussa, E., Schmitt, B.M., Romero, M.F., 1999. Cloning and immunolocalization of rat pancreatic Na^+ bicarbonate cotransporter. Biochem. Biophys. Res. Commun. 264, 291–298.
Thomas, R.C., 1976a. The effect of carbon dioxide on the intracellular pH and buffering power of snail neurones. J. Physiol. 255, 715–735.
Thomas, R.C., 1976b. Ionic mechanism of the H^+ pump in a snail neurone. Nature 262, 54–55.
Thomas, R.C., 1977. The role of bicarbonate, chloride and sodium ions in the regulation of intracellular pH in snail neurones. J. Physiol. 273, 317–338.
Thomas, R.C., Meech, R.W., 1982. Hydrogen ion currents and intracellular pH in depolarized voltage-clamped snail neurones. Nature 299, 826–828.
Tildon, F.T., McKenna, M.C., Stevenson, J., Couto, R., 1993. Transport of L-lactate by cultured rat brain astrocytes. Neurochem. Res. 18, 177–184.
Tombaugh, G.C., Somjen, G.G., 1998. pH modulation of voltage-gated ion channels. In: Kaila, K., Ransom, B.R. (Eds.), pH and Brain Function. Wiley-Liss, New York, pp. 395–416.
Trapp, S., Lückermann, M., Kaila, K., Ballanyi, K., 1996. Acidosis of hippocampal neurones mediated by a plasmalemmal Ca^{2+}/H^+ pump. Neuroreport 7, 2000–2004.
Traynelis, S.F., 1998. pH modulation of ligand-gated ion channels. In: Kaila, K., Ransom, B.R. (Eds.), pH and Brain Function. Wiley-Liss, New York, pp. 417–446.
Traynelis, S.F., Cull-Candy, S.G., 1990. Proton inhibition of *N*-methyl-D-aspartate receptors in cerebellar neurons. Nature 345, 347–350.
Trivedi, B., Danforth, W.H., 1966. Effect of pH on the kinetics of frog muscle phosphofructokinase. J. Biol. Chem. 241, 4110–4112.
Tsuganezawa, H., Kobayashi, K., Iyori, M., Araki, T., Koizumi, A., Watanabe, S.-I., Kaneko, A., Fukao, T., Monkawa, T., Yoshida, T., Kim, D.K., Kanai, Y., Endou, H., Hayashi, M., Saruta, T., 2001. A new member of the HCO_3^- transporter superfamily is an apical anion exchanger of β-intercalated cells in the kidney. J. Biol. Chem. 276, 8180–8189.
Virkki, L.V., Wilson, D.A., Vaughan-Jones, R.D., Boron, W.F., 2002. Functional characterization of human NBC4 as an electrogenic Na^+–HCO_3^- cotransporter (NBCe2). Am. J. Physiol. Cell Physiol. 282, C1278–C1289.
Voipio, J., 1998. Diffusion and buffering aspects of H^+, HCO_3^-, and CO_2 movements in brain tissue. In: Kaila, K., Ransom, B.R. (Eds.), pH and Brain Function. Wiley-Liss, New York, pp. 45–65.
Voipio, J., Kaila, K., 1993. Interstitial P_{CO2} and pH in rat hippocampal slices measured by means of a novel fast CO_2/H^+-sensitive microelectrode based on a PVC-gelled membrane. Pflügers Arch. 423, 193–201.

Volk, C., Kempski, B., Kempski, O.S., 1997. Inhibition of lactate export by quercetin acidifies rat glial cells in vitro. Neurosci. Lett. 223, 121–124.

Waldmann, R., Champigny, G., Lingueglia, E., De Weille, J.R., Heurteaux, C., Lazdunski, M., 1999. H^+-gated cation channels. Ann. N. Y. Acad. Sci. 868, 67–76.

Walz, W., Mukerji, S., 1988. Lactate release from cultured astrocytes and neurons: a comparison. Glia 1, 366–370.

Wang, C.-Z., Yano, H., Nagashima, K., Seino, S., 2000. The Na^+-driven Cl^-/HCO_3^- exchanger. Cloning, tissue distribution, and functional characterization. J. Biol. Chem. 275, 35486–35490.

Wang, Z., Conforti, L., Petrovic, S., Amlal, H., Burnham, C.E., Soleimani, M., 2001. Mouse $Na^+:HCO_3^-$ cotransporter isoform NBC-3 (kNBC-3): cloning, expression, and renal distribution. Kidney Int. 59, 1405–1414.

Wemmie, J.A., Chen, J., Askwith, C.C., Hruska-Hageman, A.M., Price, M.P., Nolan, B.C., Yoder, P.G., Lamani, E., Hoshi, T., Freeman, J.H. Jr., Welsh, M.J., 2002. The acid-activated ion channel ASIC contributes to synaptic plasticity, learning, and memory. Neuron 34, 463–477.

Woodbury, D.M., Egstrom, F.L., White, H.S., Chen, C.F., Kemp, J.W., Chow, S.Y., 1984. Ionic and acid–base regulation of neurons and glia during seizures. Ann. Neurol. 16, S135–S144.

Wu, Q., Delpire, E., Hebert, S.C., Strange, K., 1998. Functional demonstration of $Na^+-K^+-2Cl^-$ cotransporter activity in isolated, polarized choroid plexus cells. Am. J. Physiol. Cell Physiol. 275, C1565–C1572.

Żylińska, L., Soszyński, M., 2000. Plasma membrane Ca^{2+}-ATPase in excitable and nonexcitable cells. Acta Biochim. Pol. 47, 529–539.

Advances in
Molecular and
Cell Biology

AVP effects and water channels in non-neuronal CNS cells

Ye Chen* and Maria Spatz[a]

*Correspondence address: Department of Operational and Undersea Medicine, Naval Medical Research Center, 503 Robert Grant Ave., Silver Spring, MD 20910, USA
E-mail: chenye503@yahoo.com
[a]Stroke Branch, NINDS, National Institutes of Health, 36 Convent Drive, MSC 4128, Bethesda, MD 20892-4128, USA

Contents

1. Introduction
2. Ion and water homeostasis in non-neuronal parenchymal cells
   2.1. Cerebrocortical astrocytes
   2.2. The hypothalamo-hypophysial system
3. Brain barrier functions in ion and water homeostasis
   3.1. Capillary endothelium
   3.2. Choroid plexus
   3.3. Ependyma and tanycytes
4. Concluding remarks

Arginine vasopressin (AVP) is a nona-peptide, which is generally considered as an antidiuretic and vasoconstrictive hormone. It is mainly synthesized and released from the hypothalamo-neurohypophysial system, but it has also been recognized to play a crucial role in the regulation of water and ion homeostasis in the brain. The responses to AVP by cerebral non-neuronal cells have been characterized as follows, according to cell type, function and mechanism. (1) Astrocytes and specialized astrocytes: In cerebrocortical astrocytes AVP regulates water content and cell volume changes induced by hydro-osmotic challenges through V_{1b}/V_3 receptors by increasing water permeability. Circumventricular astrocytes and pituicytes regulate AVP secretion and release by altering the spatial relationship between neurons and their adjacent astrocytes. In addition, attention has been drawn to the role of putative modulators (specially endothelins), which could be involved in modulation of AVP functions in neuronal and non-neuronal cells. (2) Capillary endothelium and choroid plexus epithelium: These cells are the main cellular components of the blood–brain barrier (BBB) and blood–cerebrospinal fluid (CSF) barrier. AVP regulates ion movements between blood and brain across BBB to control

Advances in Molecular and Cell Biology, Vol. 31, pages 747–771

ISBN: 0-444-51451-1

extracellular potassium concentration and water content in brain. This process includes activation of $Na^+,K^+,2Cl^-$ cotransporter activity at the luminal membrane of capillary endothelial cells and opening of K^+ channels at their abluminal membrane, secondary to an AVP-induced stimulation of endothelin (ET) release. AVP also regulates ion movement across the blood–CSF barrier to control of CSF formation by decreasing Cl^- secretion in the epithelial cells of the choroid plexus. Moreover, AVP regulates water movement between CSF and brain by altering the water permeability in ependymal cells. Although it is known that water channel proteins, aquaporins, are widely distributed in non-neuronal cells, the capacity of AVP to affect aquaporins is yet unknown.

1. Introduction

Water and ion homeostasis of the central nervous system (CNS) is of profound importance, both physiologically and under pathological conditions, when brain edema rapidly may become life-threatening, because the brain is encaged within the rigid skull with no possibility for expansion. The presence of a vasopressinergic neuroendocrine system, regulating water transport in the CNS was first proposed by Raichle (1981), and over the past two decades it has become evident that vasopressin (AVP) does have such a role in addition to its well-known regulation of whole-body water homeostasis (Chen et al., 1992; Hertz et al., 2000a; Niermann et al., 2001). AVP is synthesized in the hypothalamus, primarily by magnocellular neurons (see chapters by Mercier and Hatton and by Salm et al.), located in the supraoptic nucleus (SON) and paraventricular nuclei (PVN). It is released into the systemic circulation, where it exerts multiple hormonal effects, most notably regulating water reabsorption by the kidney. AVP released into the systemic circulation does not easily cross the blood–brain barrier, but physiologically significant amounts of AVP are also released into the brain and into the cerebrospinal fluid (CSF) (Ludwig, 1995; de Vries and Miller, 1998). Centrally released AVP is involved in control of brain water homeostasis (Raichle and Grubb, 1978), brain edema (Doczi et al., 1988), CSF formation (Davson and Segal, 1970), and secretion of pituitary peptides (Segal and Zlokovic, 1990).

Among non-neuronal cells, astrocytes, capillary endothelium, and epithelial cells of the choroid plexus, as primary targets of AVP, play important roles in maintenance of physiological brain volume by changing water permeability and by modifying ion transport (Hertz et al., 2000a). Moreover, the specialized astrocytes in circumventricular organs (CVOs), including pituicytes play important roles in regulating AVP release into the general circulation. In this chapter, the current status of knowledge regarding effects of AVP on cellular water content and ion transport system in non-neuronal cells of the CNS will be discussed together with their response to changes of extracellular tonicity. Attention will be drawn to the role of some neuromodulators and/or neurotransmitters in the regulation of AVP release by specialized astrocytes. The possibility that AVP acts on water channels, aquaporins (AQP), in non-neuronal cells will also be considered, although the existence of such a functional relationship is still uncertain.

2. Ion and water homeostasis in non-neuronal parenchymal cells

2.1. Cerebrocortical astrocytes

2.1.1. Ion uptake

Neuronal activity triggers net ion fluxes between different cellular and extracellular compartments. An increase of extracellular K^+ concentration ($[K^+]_e$) occurs in the brain physiologically as a response to neuronal activity and to an even larger degree pathologically during seizures, brain ischemia, and other insults (Sykova, 1983; Walz and Hertz, 1983; Hertz et al., 2000b—see also chapter by Walz). To ensure an appropriate neuronal environment, increased $[K^+]_e$ must be removed by redistribution between cells or by uptake into cells. Astrocytes play a crucial role in clearing $[K^+]_e$ by passive and/or active mechanisms (Hertz, 1990). Except over short distances, or long times, the diffusion through the extracellular space is of minor importance. The so-called 'spatial buffering', a rapid redistribution of a local increase in $[K^+]_e$ through an astrocytic syncytium or along individual Müller cells in the retina, taking up K^+ through K^+ channels in regions of high K^+ and liberating it in low K^+ regions, has been considered important for K^+ redistribution in the brain. It exerts such a function in the retina (see chapter by Bringmann et al.), but it probably plays a limited role in redistributing elevated $[K^+]_e$ in the CNS under normal conditions (see chapter by Walz), and it cannot increase K^+ inside the cells, as equal amounts of K^+ enter and leave the syncytium (Amedee et al., 1997). Nevertheless, it can contribute to changes in the magnitude of the extracellular space secondary to ion redistribution (Dietzel et al., 1980; Witte et al., 2001).

Two mechanisms exist by which the intracellular K^+ concentration in astrocytes increases. One is passive uptake of KCl through ion channels. The driving force for such an uptake is Donnan forces, and the uptake is crucially dependent upon a high membrane permeability for not only K^+ but also chloride ions (Cl^-). This mechanism plays probably also at most a minor role in the normal CNS, whereas it is important in reactive astrocytes (Walz, 2002—see also chapter by Walz). In addition, astrocytes actively take up K^+ by activation of the Na^+,K^+-ATPase and the $Na^+,K^+,2Cl^-$ cotransporter.

The Na^+,K^+-ATPase is known to be stimulated by K^+ at its extracellular site and by Na^+ at its intracellular site (Sweadner, 1995). The affinity of the astrocytic Na^+,K^+-ATPase is low (high K_m), so that its activity is stimulated by increases in $[K^+]_e$ within the physiological and pathological range (Grisar et al., 1979; Mercado and Hernandez, 1992; Hajek et al., 1996). In contrast, the neuronal Na^+-K^+-ATPase has a high affinity for extracellular K^+, so that the K^+-sensitive site is saturated at normal values of $[K^+]_e$ and not affected by an increase (Grisar et al., 1979; Sweadner, 1995; Hajek et al., 1996). However, Na^+,K^+-ATPase-mediated K^+ uptake in neurons can be activated by intraneuronal increase in sodium ions (Na^+) resulting from neuronal excitation. Moreover, at low $[K^+]_e$, e.g., during the so-called undershoot following neuronal excitation (Heinemann and Lux, 1975), the neuronal Na^+,K^+-ATPase might also be stimulated at its K^+-sensitive site, and due to its higher affinity it may favor neuronal uptake at lower $[K^+]_e$, enabling a gradual return to neurons of K^+ that initially was accumulated by astrocytes and subsequently released to the extracellular space (Hajek et al., 1996; Hertz et al., 2000a).

A substantial part of the K^+ clearance from the extracellular space is not directly dependent upon the Na^+,K^+-ATPase, but occurs via a $Na^+,K^+,2Cl^-$ cotransporter system (Walz and Hertz, 1984; Tas et al., 1987; Su et al., 2002a,b). In vivo expression of the $Na^+,K^+,2Cl^-$ cotransporter protein in rat astrocytes has recently been reported (Yan et al., 2001). The $Na^+,K^+,2Cl^-$ cotransporter, which operates in the inward direction in astrocytes, provides not only K^+ uptake, but also uptake of Na^+, which in turn stimulates the Na^+,K^+-ATPase at its intracellular Na^+-sensitive site, triggering efflux of accumulated Na^+ and stimulation of energy metabolism (see chapter by Hertz, Peng et al.) The joint operation of the Na^+,K^+-ATPase and the cotransporter makes the K^+ uptake very effective and essentially mediates combined uptake of K^+ and of Cl^-. The cotransporter is inhibited by loop diuretics, such as furosemide and ethacrynic acid, and more selectively by bumetanide, and it is stimulated by elevated $[K^+]_e$ (Walz and Hertz, 1984; Su et al., 2002b). Joint stimulation of the Na^+,K^+-ATPase/cotransporter in astrocytes and of the Na^+,K^+-ATPase in neurons following neuronal excitation may contribute to the undershoot in $[K^+]_e$ following neuronal excitation (Hertz et al., 2000a).

K^+ uptake into astrocytes is not stimulated by AVP (Chen et al., 1992). A stimulation of Cl^- uptake by AVP is inhibited by bumetanide (Katay et al., 1998), indicating that it is secondary to uptake by the cotransporter. However, this stimulation does not become apparent immediately, suggesting that it is not a direct effect on cotransporter activity but rather secondary to osmotically induced cell swelling.

2.1.2. Fluid spaces

Uptake of Na^+, K^+, and Cl^- in cells leads to an osmotically driven uptake of water, i.e., swelling of the cells. Astrocytes exposed to high $[K^+]_e$ with concomitant reduction of the extracellular Na^+ concentration showed a significant increase in water content (Del Bigio and Fedoroff, 1990; Chen et al., 1992; Su et al., 2002b), which can be abolished either by the cotransporter inhibitor bumetanide or by cotransporter deletion ($Na^+,K^+,2Cl^-$ cotransporter $-/-$ mice) (Su et al., 2002a,b). The increase in water content reflects uptake of Na^+, K^+, and Cl^- together with osmotically obliged water. In astrocytes, but not in neurons, the K^+-induced increase in water content is greatly enhanced (50% of non-stimulation water content) by exposure to AVP (Fig. 1). Since the AVP-induced swelling is not caused by further stimulation of K^+ uptake (see above), it was concluded that it was due to a facilitation of water uptake (Chen et al., 1992), as has been directly demonstrated by Latzkovits et al. (1993). No corresponding effect was found at normal $[K^+]_e$, indicating that AVP facilitates rapid vectorial water movement driven by osmotic gradients (Fig. 2). The effect of AVP is potent, with a distinct enhancement of the K^+-stimulated swelling at 10^{-12} M AVP.

AVP also increases free cytosolic Ca^{2+} concentration ($[Ca^{2+}]_i$) in primary cultures of cerebrocortical astrocytes (Chen et al., 1992; Torday et al., 1997), identifying the receptor as a phospholipase C-linked V_1 receptor. This effect is independent of extracellular Ca^{2+} (Chen et al., 2000), suggesting that it may be evoked by stimulation of a V_{1b}/V_3 AVP receptor, rather than of the V_{1a} receptor that depends upon Ca^{2+} influx (Thibonnier et al., 1997). An AVP-mediated increase in $[Ca^{2+}]_i$ in cortical astrocytes was 'rediscovered' by Zhao and Brinton (2002), stating that theirs was the first description of this phenomenon,

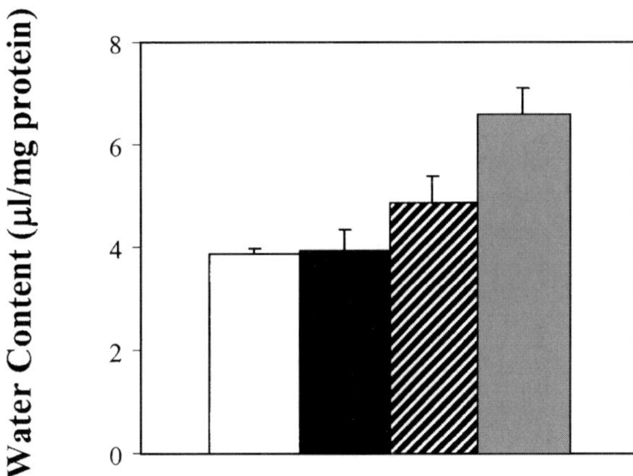

Fig. 1. Water content in primary cultures of mouse astrocytes, measured by aid of [^{14}C]urea as described by Walz (1987), after 30 min of incubation in tissue culture medium at 37 °C under control conditions (black column), exposure to 60 mM K^+ in the incubation medium (hatched column) and exposure to vasopressin (AVP) under control conditions (white column) and in the presence of the elevated K^+ concentration (gray column). The water content of 4 μL/mg protein under control conditions (exposure to 5 mM K^+ in the medium) is unaltered in the presence of 10^{-12} M vasopressin. After incubation at the elevated K^+ concentration (with concomitant reduction in the Na^+ concentration) the water content is slightly, but significantly, increased. Incubation in the joint presence of elevated K^+ and 10^{-12} M AVP leads to a large and significant additional increase in the water content, which reaches almost 7 μL/mg protein. No corresponding effect of AVP was seen in neuronal cultures (Chen et al., 1992).

and adding the new information that nuclear Ca^{2+} is also increased by AVP. However, these authors concluded that the receptor was a V_{1a} receptor, a difference from the results by Chen et al. (2000), which may reflect that the astrocytes they studied were relatively immature.

The effect of AVP on water permeability in astrocytes has been further studied during exposure of cultured astrocytes to severe hypotonic conditions, a non-physiological condition, which leads to rapid swelling, followed by a regulatory volume decrease (Sarfaraz and Fraser, 1999). In the presence of AVP, astrocytes exposed to hypotonic medium increase water space more than 50% at 5 min, at a time when cells in the absence of AVP had achieved complete regulatory volume decrease. This effect was inhibited by a subtype non-specific V_1 antagonist, whereas a V_2 agonist had no effect. Accordingly, although the effect of AVP on water permeability in the kidney is mediated by a V_2-induced increase in cAMP, all authors agree that AVP's volume-regulatory effect in cortical astrocytes is triggered by one of the two V_1 receptors (V_{1a} and V_{1b}/V_3) that are linked to the phosphatidyl inositide second messenger system, increasing $[Ca^{2+}]_i$ (Chen et al., 1992; Sarfaraz and Fraser, 1999; Zhao and Brinton, 2002). Moreover, the AVP mediated increase in Cl^- uptake mentioned above is obviously secondary to AVP-induced swelling, and it was also inhibited by a V_1 antagonist (Katay et al., 1998).

Recently the effect of AVP on activity-dependent water flux has been studied in cerebrocortical brain slices (Niermann et al., 2001). The slices were incubated in low-chloride incubation medium in order to prevent participation of the Na^+,K^+,Cl^-

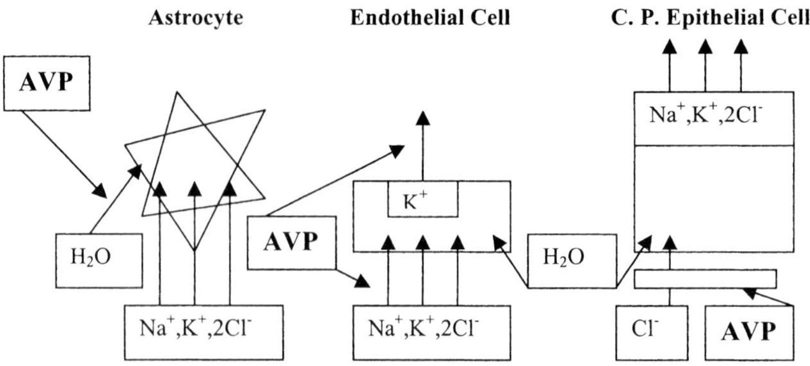

Fig. 2. Effects of AVP on Na^+,K^+,$2Cl^-$ cotransporter activity and/or vectorial H_2O movement created by the ion transport (as osmotically obliged H_2O) in astrocytes, brain capillary endothelial cells, and choroid plexus epithelial cells. In astrocytes, K^+-stimulated cotransporter-mediated ion uptake is independent of AVP, but movement of obliged water is enhanced by AVP; K^+-induced cell swelling is accordingly increased by AVP. In endothelial cells, H_2O permeability is unaffected by AVP, but cotransporter-mediated ion uptake at the luminal surface is stimulated by AVP, creating a demand for movement of osmotically obliged water; ion and water release across the abluminal surface is mediated by opening of ion channels, as indicated for K^+, which in order to maintain electroneutrality is likely to travel with an anion [probably Cl^-]; accordingly uptake of water and ions from the systemic circulation into brain is enhanced by AVP. In choroid plexus epithelial cells, H_2O permeability and cotransporter activity are unaffected by AVP, but channel-mediated uptake of Cl^- (in exchange with HCO_3^-) from the systemic circulation at the basal surface is decreased, reducing intracellular availability of Cl^- for transport by the cotransporter at the apical surface, which in these cells normally functions in the outward direction; accordingly secretion of Cl^- and of H_2O is reduced. If uptake of H_2O (and ions) in all three cell types is stimulated simultaneously, there will be an increase in water and ion contents in brain parenchyma, which is largely intracellular (although the accumulated ions and water originate from the systemic circulation), and this increase will be compensated for by a decrease of ion contents and amount of fluid in CSF, secondary to a decreased secretion by the choroids plexus.

cotransporter, and under these conditions evoked neuronal activity generates a rapid radial water flux in the slices, which at least partly may be due to the operation of the spatial buffer (Witte et al., 2001). AVP and V_{1a} receptor agonists facilitated the water flux, whereas it was reduced by a V_1 antagonists even in the absence of added agonist, suggesting a tonic vasopressinergic input. Due to the rapidity and high capacity of the flux it was concluded that it in all probably occurred as a result of modulation by vasopressin of aquaporin-mediated water flux through astrocytic membranes.

Astrocytes express the water channels, AQP4 and 9 (Nielsen et al., 1997; Elkjaer et al., 2000; Baduat et al., 2001). Highly polarized AQP4 expression is found in astrocytic foot processes near, or in direct contact with capillaries. AQP4 expression is also very pronounced in glia limitans (Nielsen et al., 1997). Although AQP4 deletion has no affect on general behavior, neuromuscular co-ordination, or response to sensory stimulation in AQP-null mice (Ma et al., 1997), the deletion is associated with greatly reduced cerebral edema in response to water intoxication and stroke (Manley et al., 2000), implying a key role of AQP4 in the development of brain edema. It has been shown that AQP4 is not AVP-sensitive in the kidney collecting duct (Dibas et al., 1998), but since the AVP receptor activating water flux in the kidney is an adenylate cyclase-associated V_2, and

AVP effects on astrocytes are exerted on the phospholipase C-associated V_1 receptor, this does not necessarily mean that the astrocytic AQP4 could not be stimulated by AVP. It is consistent with this point of view that the channel is regulated by protein kinase C (PKC) (Han et al., 1998; Nakahama et al., 1999). Because the promoter activity of AQP4 has not been analyzed, the precise regulatory mechanisms remain unknown (Venero et al., 2001). However, AQP4 is known to contain three putative phosphorylation sites, and it was demonstrated by Niermann et al. (2001), that AVP lost its ability to facilitate water flux in brain cortex slices after treatment with either an inhibitor of PKC or thapsigargin, which by inhibiting refill of intracellular Ca^{2+} stores (Norup et al., 1986) prevents the increase in $[Ca^{2+}]_i$ evoked by stimulation of phospholipase-C associated receptors such as the V_1 receptor. It was concluded that both PKC and Ca^{2+} are involved in the facilitation of water redistribution, and that PKC may act either by phosphorylation of the channel itself or by phosphorylating other components of the signal transduction pathway (Niermann et al., 2001). The latter possibility may appear more likely, since phosphorylation of AQP4 may lead to a decrease in channel activity (Han et al., 1998).

AQP4 is tethered to perivascular astrocytic endfeet by binding of its C terminus to the PDZ domain of synthrophin, a component of the dystrophin complex, and its expression in astrocytic membranes facing the neuropil is considerably lower (Neely et al., 2001). Its total expression in brain is normal in dystrophin null mice, but the normal cellular polarity is disturbed, with AQP expression being markedly reduced in perivascular astrocyte endfeet, but present at higher than normal levels in astrocyte membranes facing the neuropil. Another aquaporin, AQP9, is found on astrocytic processes and cell bodies. Similarities of distribution pattern within the brain between AQP4 and 9 in rodents suggest that the two proteins mediate common functions, and that they can act in synergy (Badaut et al., 2002). Recent studies, determining gene expression by reverse transcription polymerase chain reaction, have demonstrated that astrocytes also express mRNA for AQP3, 5, and 8 (Yamamoto et al., 2001). The physiologic role of these proteins remains to be determined.

2.2. *The hypothalamo-hypophysial system*

2.2.1. *Astrocytes in the hypothalamo-hypophysial system*

As described elsewhere (see chapters by Mercier and Hatton and by Salm et al.) AVP and oxytocin are produced in magnocellular and parvocellular neuronal somata, respectively, of hypothalamic nuclei like the SON, the PVN, both of which contain mainly AVP-secreting neurons, and the arcuate nucleus, containing oxytocin-secreting neurons. The fibers terminate in the posterior, neural lobe of the pituitary (the neurohypophysis). In response to appropriate stimuli, astrocytes undergo significant morphological changes, possibly triggered by stimulation of β-adrenergic receptors expressed by SON astrocytes (Lafarga et al., 1992), as will be discussed below. Astrocytic membranes covering neuronal somata and dendrites withdraw and neuronal surfaces become directly apposed to each other (Tweedle and Hatton, 1977; Theodosis et al., 1981; Beagley and Hatton, 1994). The resulting cessation of astrocytic removal of neuronally released K^+ and glutamate as well as the absence of astrocytical release of the inhibitory transmitter taurine enhance neuronal stimulation (van den Pol et al., 1990; Mecker et al., 1993; see also chapters by Mercier and Hatton and by Salm et al.). When hormone demand returns to basal levels, astrocytic processes are once again observed between the neuronal elements (Hatton et al., 1984). In the neurohypophysis, specialized astrocytes, the pituicytes, envelop axons and terminals of vasopressin- and oxytocin-secreting neurons and their processes occupy portions of the basal lamina, constituting a barrier between neuronal terminals and capillaries (Wittkowski, 1986). During stimulation of hormonal secretion, release of engulfed axon and retraction of pituicyte endfeet from the vascular surface favor release of AVP and of oxytocin into the general circulation.

AVP is not only released from axonal terminals but also from the dendrites of magnocellular neurons in the SON during its activation (Ludwig, 1995). Cultured astrocytes from SON express AVP receptors, which appear to be of V_{1b}/V_3 subtype (Hatton, 1997). Pituicytes respond to AVP with an increase in $[Ca^{2+}]_i$ in the absence of extracellular Ca^{2+} (Hatton et al., 1992), which identifies also this receptor as being of the V_{1b} subtype. This is consistent with an immunocytochemical study, which showed that pituicytes express V_{1b} receptors (Hernando et al., 2001). AVP induced $[Ca^{2+}]_i$ elevation in pituicytes may serve as an inhibitory mechanism of peptide release in neural lobe terminals (Hatton, 1999) by spread of Ca^{2+} waves from glia to neurons (see chapter by Cornell-Bell et al.). Induction of increased $[Ca^{2+}]_i$ in the terminals by the closely apposed astrocytes might inactivate Ca^{2+} channels and inhibit further release.

In contrast to most brain areas, where astrocytic AQP4 protein localization is highly polarized, AQP4 in SON and PVN is evenly distributed over the membrane of astrocytes, notably in lamellae in direct apposition with neuronal elements (Nielsen et al., 1997). This distribution suggests a particular functional significance of regulation of water movement through astrocytes. Considering that SON astrocytes reside at an interface between extraparenchymal fluid (CSF and blood) and magnocellular neurons, and that astrocytes undergo morphological changes in response to osmotic stimulations, it is possible that astrocytes in SON may sense the surrounding extracellular tonicity and act as 'vesicular osmometers' (Wells, 1998). If AVP increases water permeability in these cells as in other astrocytes, this might facilitate such a function.

2.2.2. *Astrocytes in afferent nuclei*

Two CVOs, the subfornical organ (SFO) and the organum vasculosum lamina terminalis (OVLT) send direct projections to SON and PVN. Since CVOs are devoid of a blood–brain barrier and are in direct contact with both plasma and CSF, cells in CVOs are ideally placed to readily sense any change in extracellular fluid composition. Astrocytes in SFO and OVLT express AVP receptors and the receptor subtype proved to be V_{1b}, because AVP induced $[Ca^{2+}]_i$ increase in cultured astrocytes from SFO and OVLT is Ca^{2+}-independent (Jurzak et al., 1995). The presence of functional AVP receptors in CVOs may reflect CNS control of peripheral AVP level. AQP4 is densely expresses in SFO (Venero et al., 1999). Astrocytes from SFO and OVLT also express receptors for ETs, and binding sites for both the ET_A and the ET_B receptor were demonstrated by autoradiography. In addition ET-3 induced an intracellular Ca^{2+} transient through ET_A

receptors, whereas ET-1-stimulated Ca^{2+} mobilization was mediated by ET_A and ET_B-receptors (Gebke et al., 2000). ETs, whether produced locally or circulating, may participate with AVP to maintain extracellular fluid balance, as will be discussed below.

2.2.3. Putative modulators

The likelihood of neuronal–glial interaction has been substantiated by the observed formation of synaptic contact between axonal terminals containing various neurohormones/neurotransmitters (i.e., AVP, oxytocin, enkephalin, adrenaline, serotonin, and GABA) and their respective receptors expressed on the pituicytes in the neurohypophysis. Some of these substances are co-localized or localized in adjacent neuronal and non-neuronal tissues. Moreover, circulating substances derived from peripheral organs may reach the AVP and oxytocin releasing brain areas due to the absence of blood–brain barrier (pituitary) or localization in close proximity to a large vascular bed (SON glia), as illustrated in the chapter by Mercier and Hatton. Based on recent in vivo and in vitro studies, evidence has been accumulated indicating that catecholamines, ATP (endogenous and exogenous) and its metabolite adenosine, as well as cardiovascular peptides may play a role in regulating the secretion of AVP and oxytocin and their control of water homeostasis (Ritz et al., 1992; Wall and Ferguson, 1992; Yamamoto et al., 1992; Lange et al., 1994; Rossi et al., 1997).

Catecholamines. SON receives a rich catecholaminergic innervation (McNiel and Sladek, 1980) and (as mentioned above) the SON astrocytes express β-adrenergic receptors (Lafarga et al., 1992). The involvement of a catecholaminergic system in the modulation of SON and pituitary astrocytes was suggested by osmotically stimulated increase in number of these receptors that was associated with a change in astrocyte morphology (Beagley and Hatton, 1994). The absence of osmotically induced astrocytic retraction in adrenalectomized animals indicated that catecholaminergic signals from the adrenal gland could also directly or indirectly modulate the function of the hypophysio-hypothalamic system. In support of this contention, morphological changes were observed in cultured pituicytes exposed to adrenergic agonist, which altered their shape from flat polygonal to stellate. This event was mimicked with 8-bromo cAMP, a permeable cAMP analog or forskolin (an activator of adenylate cyclase). The effect of this treatment of the pituicytes suggests an involvement of cAMP-mediated process in modulation of pituicyte morphology. Thus, cAMP signal transduction induced by catecholamine or other neurotransmitters/neurohormones in pituicytes may influence the neuronal release of AVP and/or oxytocin from the neurohypophysis (Miyata et al., 1999).

ATP and adenosine. ATP is co-stored with neuropeptides in the secretory granules of the neurohypophysial nerve endings, and an ectoATPase terminating the extracellular action of ATP is localized in the same organ. The ATP, which is released along with AVP and oxytocin by electrical stimulation of the neurohypophysis, may have a role in regulation of AVP secretion via P_{2X2}-purine receptor. The effects of ATP on AVP secretion have been shown to be both stimulatory and inhibitory (Troadec et al., 1998; Sperlagh et al., 1999). Lemos and Wang have suggested that endogenous and exogenous ATP exert opposite effects on AVP secretion in neurohypophysial terminals (Lemos and Wang, 2000). Alternatively, the conflicting results of ATP could be explained by assuming that ATP, co-released with AVP, initially stimulates further release of AVP, whereas adenosine, which is rapidly formed by hydrolysis of ATP, could then act through A1 receptor to decrease the release of AVP (Wang et al., 2002). As mentioned above, the morphologic plasticity of astrocytes and pituicytes, demonstrated in vitro studies, mirror the in vivo depicted changes in these cells during hypothalmamo-hypophysial hormonal secretion. Therefore, the latest reported studies focused on the responses and mechanisms responsible for the morphologic changes in the pituicytes induced by either ATP or adenosine (Rosso et al., 2002a,b). ATP was demonstrated to elicit stellation of pituicytes (Fig. 3, lower left part). This effect is due to the ATP metabolite, adenosine, and it is mediated by A1-receptors and independent of intracellular Ca^{2+} and mitogen protein kinase pathway. It involves the adenosine-induced down-regulation of Rho A activity associated with F-actin depolymerization and reorganization of microtubular filaments in

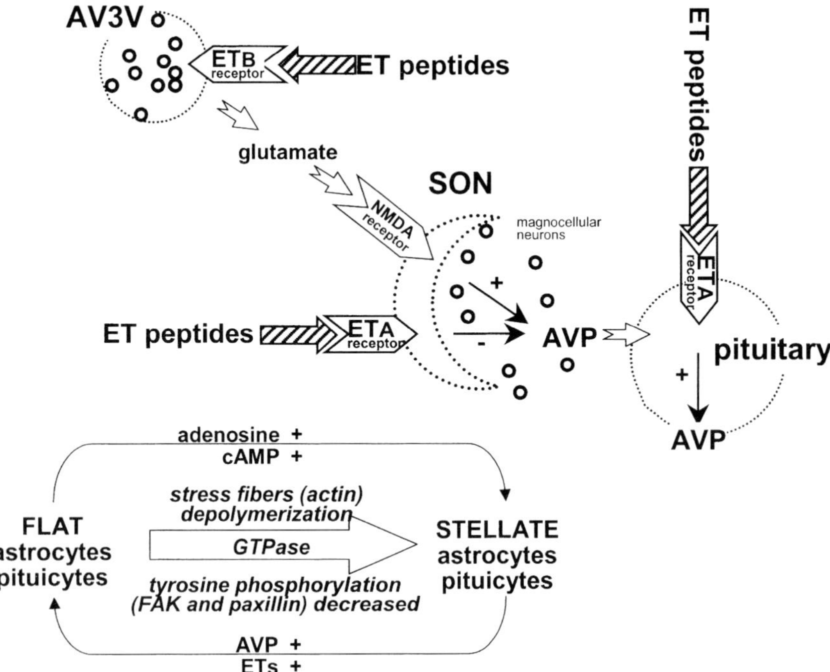

Fig. 3. Upper part: Schematic illustration of proposed synergistic and antagonistic effects between AVP and ET_A and ET_B receptor stimulation on AVP secretion from the neurohypophysis and on astrocyte stellation. AVP secretion from magnocellular neurons in the SON is inhibited by stimulation of ET_A receptors on these cells, but it is enhanced by stimulation of ET_B receptors on AV3V cells, leading to efferent glutamatergic stimulation of NMDA receptors on the magnocellular neurons. In addition, AVP release from the neurohypophysis is enhanced by stimulation of pituitary ET_A receptors. Lower left part: Mechanisms involved in adenosine- and cAMP-mediated stellation of astrocytes (stress fiber (actin) depolymerization and decreased tyrosine phosphorylation of the focal adhesion associated proteins FAK and paxillin). Oppositely directed reactions (enhanced polymerization of actin and inhibited tyrosine dephosphorylation) are presumably involved in the conversion of stellate astrocytes to flat, polygonal astrocytes by endothelins (ETs) and AVP.

the pituicytes. Rho, a member of small GTPase proteins, was shown to control the basal (flat—nonstellate) state of cultured pituicytes and glial cells. In addition, it was found that AVP and oxytocin prevented the adenosine-induced stellation of pituicytes (Fig. 3, lower left part). This effect was mediated by a V_{1a}-receptor mechanism and entailed Ca^{2+}-dependent activation of Cdc42, another small GTPase protein linked to alterations of the cytoskeleton (Rosso et al., 2002b). It appears therefore that the signal transduction pathway taking part in altering the adenosine-elicited structural features of pituicytes differs from that described to prevent it.

The AVP and oxytocin prevention of adenosine-induced stellation of pituicytes is indicative of an existing potential negative feedback mechanism provided by these substances. It is still unknown whether morphologic responses of hypothalamic astrocytes to adenosine alone or in the presence of AVP and oxytocin is similar to those of pituicytes. Nevertheless, the morphologic observations in vitro support the concept of purinergic regulation of functional plasticity in pituicytes, which may be involved in modulation of hormonal (AVP and oxytocin) release occurring during physiological stimulation.

Endothelin. The ET family of peptides is known to exert various and diverse biological effects on many organs, tissues and cell types through activation of ET_A and ET_B receptors (Rubanyi and Polokoff, 1994). Therefore it is conceivable that ET-elicited action and interaction with other peptides may play a crucial regulatory role in the function of a given organ. In the brain, ETs were shown to affect neuronal AVP secretion irrespective of whether they had been produced centrally or locally. However, little is known about the effects of ETs on hypothalamo-hypophysial astrocytes. Therefore the information presented below pertains to localization of ETs and their involvement in regulating neuronal release of AVP, which may have ramifications for the functional participation of astrocytes and pituicytes in these endeavors.

Endothelin-1 (ET-1), the most potent vasoconstrictor, is a member of 21-amino acid family (ET-1,2,3) originally isolated from porcine aortic endothelium. Subsequently, ET-1 and ET-3 were localized in various organs including CNS (Takahashi et al., 1991; Nakamura et al., 1993; Rubanyi and Polokoff, 1994; Gajkowska and Viron, 1997; Lange et al., 2002). In the brain, ET peptides, ET mRNA, ET1-converting enzymes, and ET-1 receptors have also been identified in non-vascular tissues, and they may be derived from neurons as well as non-neuronal cells (astrocytes, endothelial cells). Particularly, ET-1 and ET-3 were localized within the region involved in the secretion of AVP and in the control of body fluid homeostasis (i.e., neurohypophysis, SON and anteroventrical (AV3V) neurons, projecting to the SON (Wall and Ferguson, 1992; Yamamoto et al., 1995). An increased AVP plasma content was reported after intravenous or intracerebroventricular administration of ET-1 or ET-3 (Beagley and Hatton, 1994). These observations suggested activation of SON and paraventricular neurons by ETs, reaching these areas and/or AV3V neurons from the systemic circulation. A reported reversible ET-3 induced inhibition of water reabsorption in rats and the attenuation of the ET-3 induced increase in AVP plasma levels by brain natriuretic peptides is indicative of peptidergic interactive effects (Makino et al., 1992; Yamamoto et al., 1995). These findings also suggest that endogenously released ETs may play a role in the central control of fluid and electrolyte homeostasis.

A potentiating effect of ET-1 and ET-3 on AVP secretion, observed in K^+-depolarized isolated nerve endings, implicated involvement of the ETs as an autocrine regulator.

However the reports concerned with mechanism and mediation involved in the AVP release are somewhat contradictory, dependent on the specific region studied in the compartmentalized hypothalamo-neurohypophysial explants. Thus, the accumulated experimental data indicate that the ET-1 inducible modulation of AVP secretion depends on the region and the subtype of ET-receptor activation. The ET-stimulated AVP release in neurohypophysis is mediated by ET_A receptor and depolarization (Rossi, 1995), but the role of Ca^{2+} is still debatable. The demonstrated ET-effects on the hypothalamo-hypophysial region are rather complex (Rossi and Chen, 2002), as indicated in Fig. 3.

Previous electrophysiologic studies demonstrated that ET directly inhibited the phasic firing of AVP neurons within SON. However, it stimulated the neuronal activity within the AV3V region, which contains projections to SON, and therefore excited the magnocellular neurons (Yamamoto et al., 1993). In hypothalamo-neurohypophysial explants, which included the AV3V region, Rossi and Chen (2002) lately demonstrated an increased AVP secretion from the neurohypophysis through pharmacological activation of ET_B receptors in the hypothalamic area, probably in the AV3V region, whereas the stimulation of ET_A-receptors on the vasopressinergic neurons of the SON inhibited the AVP release. Moreover, the stimulatory effect on AVP release induced by activation of the ET_B receptor was shown to be mediated by hypothalamic NMDA-receptors, mediating the afferent stimulation from AV3V (Fig. 3). In addition, it was suggested that NMDA-activation may be associated with local release of GABA which in turn could decrease the AVP secretory response to ET_B activation.

A recent report by Miyata et al. (1999) indicated that ET-1 or ET-3 reversed the adenosine or cAMP-induced stellation of pituicytes in vitro (lower left part of Fig. 3). This observation implies that ETs known to be directly involved in the regulation of neuronal AVP release may also influence the AVP release through pituicyte activities. As far as the mechanism is concerned, the studies on pituicytes suggested participation of tyrosine phosphorylation in the pituicyte shape conversion, as it has been described in astrocytes. Until now, no other reports exist regarding the mechanism of ETs partaking in reversal of pituicyte stellation. Therefore, it is noteworthy to mention some of the factors entailed in altering astrocytic shape that might be relevant to those in pituicytes. In general, it appears that the mechanism involved in this event depends to a certain extent on the substances used for induction of cellular stellation (Goldman and Abramson, 1990; Koyama and Baba, 1994; Koyama and Baba, 1996; Padmanabhan et al., 1999). The observed common denominator for stellation is the reorganization of actin filaments due to its depolymerization, caused by various agents (i.e., cytochalasin, a disruptor of actin fibers, cAMP- or cAMP-mediated hormones). The association of tyrosine phosphorylation with focal adhesion molecules and involvement of tyrosine kinase was shown to be responsible for cAMP-induced changes in astrocytes. The stellation was manifested by decreased tyrosine phosphorylation of focal adhesion kinase (FAK) and paxillin (focal adhesion associated proteins). This process seems to be downstream to Rho protein activity. On the other hand, ET-evoked prevention of astrocytic process formation was linked to induction of stress fibers (actin), ET_B receptor-mediated tyrosine phosphorylation, and activation of Rho proteins (Miyata et al., 1999). It is therefore possible the same mechanism is involved in the reported ET-induced reversion of pituicyte stellation by ETs. Provided this is the case, and the same scenario occurs in vivo during a given physiological stimulus, it can be envisioned that retraction and recovery of pituicytes as well as of astrocytes, induced by AVP release involves a co-ordination of AVP and ET signals between neurons and glia in the hypothalamo-hypophysial region of the brain.

2.2.4. Adrenocorticotrops

The anterior pituitary releases several hormones in response to hypohysiotropic hormones produced in the hypothalamus. Among these, the release of adrenocorticotropic hormone (ACTH) has repeatedly been found to be stimulated by application of AVP to preparations of isolated hormone-secreting cells (Jard et al., 1986; Tse and Lee, 1998; Livesey et al., 2000). The stimulation is associated with the evoked increase in $[Ca^{2+}]_i$, which is temporally correlated with a burst of exocytosis and can be replaced by release of Ca^{2+} via flash photolysis of caged IP_3 (Tse and Lee, 1998). The stimulatory effect is independent of extracellular Ca^{2+}, identifying the receptor as being of the V_{1b} subtype, and the V_{1b} subtype of the AVP receptor was first identified in pituitary corticotrops (Jard et al., 1986). In addition, systemic ET-1-induced stimulation of the hypothalamo-hypophysial–adrenal axis is mediated at the early stages by AVP and subsequently by AVP and corticotropin-releasing hormone, supporting the notion of interaction between ET-1 and AVP in a variety of endocrine systems (Malendowicz et al., 1998).

3. Brain barrier functions in ion and water homeostasis

3.1. Capillary endothelium

Cerebral capillaries and microvessels have a very limited permeability to most plasma constituents, including Na^+ and K^+, due to tight junctions between the endothelial cells and specialized membrane properties (see chapter by Couraud et al.). This 'blood–brain barrier' is extremely important for neuronal function by preventing systemic changes in K^+ concentration from altering $[K^+]_e$ in the brain. Thus, during hypo- and hyperkalemia, despite the change in plasma, K^+ concentration in brain is maintained constant by the regulation of flux of K^+ between blood and brain (Bradbury, 1979; Jones and Keep, 1988—see also chapter by Walz). A recent cytochemical study reveals localization of Na^+,K^+-ATPase on both luminal and abluminal membranes (Manoonkitiwongsa et al., 1998). Consistent with this observation, a biochemical study of luminal and abluminal membrane vesicles derived from bovine brain endothelial cells after fractionation in a discontinuous Ficoll gradient led to the conclusions that although Na^+,K^+-ATPase activity is primarily located on the abluminal membrane, approximately 25% of the activity is of luminal origin, and that different isoforms of the enzyme may be found at the two surfaces (Sanchez del Pino et al., 1995). During chronic dietary hyperkalemia, a significant decrease in the amount of the $\alpha 3$ subunit of Na^+,K^+-ATPase may reflect a down-regulation of the luminal enzyme to reduce transportation of K^+ from blood to brain (Keep et al., 1999). In contrast to this evidence suggesting Na^+,K^+-ATPase activity both luminally and abluminally, several studies have shown a distinct abluminal polarity of the enzyme, which possibly may be explained by a decrease of luminal, but not of abluminal, activity when certain fixation procedures are used (Manoonkitiwongsa et al., 2000).

Similar to what has been found in astrocytes, increases in $[K^+]_e$ above the normal resting level stimulate Na^+,K^+-ATPase activity in microvessels (Schielke et al., 1990).

Cultured brain microvessels and capillary endothelial cells express Na^+,K^+,$2Cl^-$ cotransporter proteins (Yerby et al., 1997) and exhibit Na^+,K^+,$2Cl^-$ cotransporter activity (Keep et al., 1994; Vigne et al., 1994; Sun et al., 1995; Kawai et al., 1995a,b). Although there is substantial evidence suggesting a luminal localization of Na^+,K^+,$2Cl^-$ cotransporters (Keep et al., 1993; Kawai et al., 1995b), in vivo direct cytochemistry for its existence on the luminal membrane is still lacking. The cotransporter mediates uptake of all three ions, and Ca^{2+}-dependent K^+ channels on the abluminal surface may mediate K^+ entry into the CNS (Fig. 2).

Brain vascular cells express mRNA for the V_{1a} receptor. AVP induced increase in $[Ca^{2+}]_i$ is dependent upon extracellular Ca^{2+}, and therefore by definition acting on a V_{1a} subtype (Hess et al., 1991). AVP receptors have been identified at the luminal membrane of brain endothelial cells, and systemic administration of AVP antagonists of V_1 subtype reduce vasogenic brain edema and Na^+ accumulation (Rosenberg et al., 1990; Nagao et al., 1994). However, a reduction in blood-to-brain Na^+ flux in AVP-deficient Brattleboro rat during cerebral ischemia is abolished by intraventricularly administrated AVP, but not by systemic treatment with AVP, suggesting that AVP receptors are also expressed abluminally (Dickinson and Betz, 1992). Along similar lines, intracereboventricular injection of AVP exacerbates acute ischemic brain edema (Liu et al., 1991).

AVP enhances release of ET-1 in cultured endothelial cells (Spatz et al., 1993; Kawai et al., 1995a, 1997), and ET-1, in turn, stimulates Na^+,K^+,$2Cl^-$ cotransporter activity (Vigne et al., 1994; Spatz et al., 1994, 1997, 1998) and Na^+,K^+-ATPase activity (Kawai et al., 1995b), which may enhance uptake of ions and osmotically obliged water into endothelial cells, as illustrated in Fig. 2. The effects of ET-1 on the ionic transport systems occur through activation of ET_A receptors and phospholipase C (PLC), linked to intracellular Ca^{2+} and PKC. This may also be the reason why AVP-induced increase of Na^+,K^+,$2Cl^-$ cotransporter activity has been found to be secondary to a Ca^{2+}- and calmodulin-dependent phosphorylation of the cotransporter protein (O'Donnell et al., 1995). On the other hand, an ET-1 stimulated Na^+/H^+ exchange, coupled to ET_A and PLC activation, is PKC independent and is partially mediated by tyrosine kinase and Ca^{2+} calmodulin (Kawai et al., 1995b). At the abluminal side, AVP and ET separately or possibly together mediates opening of Ca^{2+}-dependent K^+ channels (Keep et al., 1993; Van Renterghem et al., 1995). It is quite interesting that AVP has been found to induce the ET-1 gene in endothelium, although this has not yet been demonstrated in brain capillaries (Imai et al., 1992), and that astrocyte-conditioned medium increases cotransporter activity in endothelial cells (O'Donnell et al., 1995). It is therefore possible that an AVP/ET circuit takes part in a co-ordination of astrocytic and capillary event in order to maintain the intracellular and extracellular volume and ionic composition in the brain.

AVP-activated Na^+,K^+,$2Cl^-$ cotransporter activity in endothelial cells may also be functionally co-ordinated with AVP-facilitated water uptake in astrocytes in general during normalization of resting $[K^+]_e$ in brain after physiological activity. In the absence of AVP, astrocytic removal of K^+ and Cl^- without osmotically obliged water uptake, together with Na^+-stimulated neuronal Na^+/K^+ exchange after neuronal activity, lead to a post-excitatory decrease in $[K^+]_e$, i.e., an undershoot. AVP-mediated uptake of blood-borne Na^+, K^+, and Cl^- in endothelial cells, together with the opening of Ca^{2+}-dependent K^+ channel on the abluminal membrane, may respond to this decrease by transporting K^+ from blood to brain and thereby normalize $[K^+]_e$ (Hertz et al., 2000a,b). Thanks to the concomitant AVP-induced increase in water flux across the astrocytic cell membrane, such a transendothelial uptake of K^+ may cause $[K^+]_e$ to rise above its normal level. In turn, the increase in $[K^+]_e$ may stimulate the abluminally located Na^+,K^+-ATPase, mediating uptake of K^+ across the abluminal endothelial membrane to normalize $[K^+]_e$.

There are no indications that AVP directly affects water permeability in cerebral endothelial cells, possibly reflecting an absence of AQP protein in endothelial cells in the brain in situ (Nielsen et al., 1997). However, as discussed above, AQP4 is abundant in perivascular astrocytic endfeet, and astrocytic processes cover >99% of the capillary surface in brain (see chapter by Wolff and Chao). This arrangement opens the possibility that at least part of the exchange of H_2O between the systemic circulation and the brain may involve both transendothelial and transastrocytic processes. Moreover, it might explain the detection of mRNA and proteins of AQP4 and 9 in isolated rat cerebral microvessels and in cultured cerebral microvascular endothelial cells (Sobue et al., 1999; Kobayashi et al., 2001, 2002). In any case there is substantial reduction in brain edema following water intoxication by intraperitoneal water infusion in AQP null mice (Manley et al., 2000). This infusion leads to serum hyponatremia, which creates an osmotic gradient driving water into the brain as a 'vasogenic brain edema' (Klatzo, 1987). Nevertheless, the loss of pronounced polarity of AQP expression to the astrocytic endfeet in dystrophin null mice is not accompanied by a lack of ability to develop brain edema or a reduced mortality after intraperitoneal injection of distilled water and AVP, but the development of the edema is slightly delayed (Vajda et al., 2002).

As summarized in legend of Fig. 2, simultaneous stimulation of astrocytes and of capillary endothelial cells in the brain will cause an increase in water and ion contents in brain parenchyma, which is largely intracellular, although the accumulated ions and water originate from the systemic circulation.

3.2. Choroid plexus

The choroid plexuses secrete CSF and consist of villi containing a connective tissue stroma, which is a richly vascularized extension of the subarachnoid space, protruding into the ventricles and covered by epithelial cells connected by tight junctions (see chapters by Mercier and Hatton and by Weaver et al.). The epithelial cells are the site of the blood–CSF barrier (see chapter by Couraud et al.). In isolated choroid plexuses, Na^+,K^+-ATPase and $Na^+,K^+,2Cl^-$ cotransporter are located on apical membrane of the choroid plexus epithelial cells (Masuzawa et al., 1984; Ernst et al., 1986; Keep et al., 1994; Klarr et al., 1997; Plotkin et al., 1997), as indicated in Fig. 2. The direction of the fluxes mediated by the cotransporter has been uncertain. Keep et al. (1994) suggested that they are involved in secretion of CSF, but Wu et al. (1998) found in vitro experiments that inhibition of this transporter by bumetanide caused cell shrinkage and therefore concluded that the cotransporter normally mediates inwards fluxes of the three ions. Recently, the pendulum has swung back to the concept that the cotransporter in vivo functions in the outward direction, partly accounted for by high inward concentrations of Na^+ and Cl^-, established by the operation on the basolateral membrane of H^+/Na^+ and bicarbonate/Cl^- exchangers, respectively (Murphy and Johanson, 1990; Garner and Brown, 1992), and driven by metabolic activity of the cells, producing CO_2, which is dissociated to H^+ and HCO_3^- (Speake et al., 2001).

Whereas the cotransporter and the Na^+,K^+-ATPase both mediate K^+ uptake in astrocytes and in endothelial cells, the outward direction of Na^+, K^+ and Cl^- transport by the cotransporter is contrasted by a 'normal' inwardly directed Na^+,K^+-ATPase, which may serve the purpose of regulating CSF K^+ concentration. In isolated preparations from adult animals this uptake has been found to increase in a stepwise fashion with each 2 mM increase in $[K^+]_e$, up to a 90% increase over control (3 mM $[K^+]_e$) with 9 mM $[K^+]_e$ (Parmelee et al., 1991). In contrast, choroid plexus from 3-day-old animals increased uptake in 5 mM $[K^+]_e$, but could not increase K^+ uptake further with higher $[K^+]_e$, and kinetic analysis of the ouabain-inhibitable component suggested that the immature tissues may express a different isoform of the α-subunit of the Na^+,K^+-ATPase. These data may explain an inability of neonatal rats to regulate CSF $[K^+]_e$, when serum $[K^+]_e$ is elevated, and they indicate that active K^+ transport by the choroid plexus epithelial cells plays an integral role in the regulation of $[K^+]_e$ in CSF. In contrast, in intact adult animals with increasing plasma $[K^+]_e$ during dietary hyperkalemia, the α_1 and β_1 subunits of the Na^+,K^+-ATPase were up-regulated as an indication of enhanced K^+ extrusion, whereas during hypokalemia, Na^+,K^+-ATPase activity was decreased (Klarr et al., 1997).

In addition to ion channels and carriers discussed above the choroid plexus epithelial cells are likely to express K^+ and HCO_3^- channels on their apical membrane (Speake et al., 2001), and on the basolateral membrane, a Na^+–Cl^- cotransporter has been identified, which may participate in K^+ exit into blood (Pearson et al., 2001).

The concentration of AVP is several times higher in CSF than in plasma (Vorherr et al., 1968; Luerssen and Robertson, 1980; Reppert et al., 1982). Synthesis of AVP by choroidal epithelium close to the apical membrane has been demonstrated in situ (Chodobski et al., 1997, 1998b), and AVP release can be regulated by cAMP-dependent signaling, although other second messenger systems may also be involved (Chodobski and Szmydynger-Chodobska, 2001). Therefore, it is conceivable that changes in blood-borne AVP do not affect CSF formation under normal physiological conditions (Chodobski et al., 1998a), although morphological changes in choroid plexus epithelial cells mimicking those observed after intraventricular administration of AVP (see below) have been reported after systemic administration of AVP (Schultz et al., 1977), and circulating AVP is a potent vasoconstrictor of choroidal arterioles (Segal et al., 1992), which might secondarily decrease CSF production.

Intracerebroventricular injection of AVP reduces blood flow in the choroid plexus and decreases CSF secretion (Faraci et al., 1988, 1990; Maktabi et al., 1993; Chodobski et al., 1998a) as well as transfer of ^{22}Na from blood to CSF (Davson and Segal, 1970). AVP binding sites are abundant in choroid plexus epithelial cells (van Leeuwen et al., 1987; Tribollet et al., 1999), and they are both of the V_{1a} and V_{1b} subtype (Ostrowski et al., 1994; Burbach et al., 1995). In addition, V_2 receptor mRNA was detected in the choroid plexus of newborn rats, but it was not detectable in the adult (Kato et al., 1995).

Hypernatremia in rats increases the expression of both AVP and $V_{1b}R$ mRNA expression in choroid plexus, indicating the involvement of the V_{1b} receptor in the regulation of ion and water homeostasis (Zemo and McCabe, 2001).

A study of the effect of AVP on Cl^- efflux from acutely isolated choroid plexuses after previous labeling with ^{36}Cl have shown that AVP potently reduces Cl^- efflux (Fig. 2) and increases the number of dehydrated epithelial cells by a V_1 receptor-mediated effect (Johanson et al., 1999). The inhibitory effect of AVP on Cl^- efflux is probably due to interference with Cl^-/HCO_3^- and Na^+/H^+ exchange at the basolateral membrane. As a result of slower uptake of Cl^- and Na^+, efflux of Cl^- efflux and thus of osmotically obliged fluid across the apical membrane into CSF decreases and many choroid plexus epithelial cells become dehydrated, which is consistent with the observed V_1-dependent inhibition of CSF formation. An alternative explanation, i.e., that cotransporter-mediated efflux of Cl^- together with Na^+ and K^+ should be inhibited is less likely, since it would lead to cell swelling rather than shrinkage.

Immunocytochemical studies have shown that the aquaporin 1 (AQP1) protein is expressed in the apical membrane of the rat choroid plexus epithelium (Nielsen et al., 1997; Wu et al., 1998; Speake et al., 2003), and it is likely that it plays an important role in mediating water transport across this membrane during CSF secretion. A recent study has shown that AQP4 protein is also expressed in the choroid plexus of the fourth and the lateral ventricles of rat brain (Speake et al., 2003), but is diffusely distributed throughout the cytoplasma. The route by which water crosses the basolateral membrane remains to be established. The application of AVP to *Xenopus* oocytes injected with AQP1 cRNA increased the membrane permeability to water, implying that AQP1 may be an AVP-regulated water channel (Patil et al., 1997). If this is the case, the AVP-mediated cell dehydration could reflect not only a reduced influx of Cl^- and Na^+, but perhaps also a more rapid flux of osmotically obliged water across the apical membrane together with the cotransporter-induced efflux of Na^+, K^+, and Cl^-.

3.3. Ependyma and tanycytes

Ependymal cells line the cerebral ventricles of the brain and form the interface that separates the CSF and the brain (see chapter by Wolff and Chao). They may have functional roles in regulating the transport of ions, small molecules, and water between cerebrospinal and interstitial fluids (Bruni, 1998). Although the morphology is similar to that of most epithelial membranes, ependyma possess electrophysiological characteristics of glia cells (Bruni, 1998). In hypothalamus, ependyma may take up and spatially buffer K^+ released from adjacent endocrine neurons, and tanycytes, a related cell type lining the ventricles at certain locations (see chapter by Wolff and Chao), shunt K^+ to the CSF or to capillaries, thereby influencing neuronal excitability (Jarvis and Andrew, 1988). AVP increases water permeability of ependymal cells as shown by an augmentation in water movement from CSF to blood across the ependymal and capillary interfaces in the presence of AVP (Rosenberg et al., 1986). AQP4 is abundant in ependymal cells (Jung et al., 1994; Nielsen et al., 1997).

As summarized in the legend of Fig. 2, the reduction of CSF formation during exposure to AVP is purposeful, because it will provide a partial compensation for the increase in water and ions within brain parenchyma, which will be produced if astrocytes and brain capillaries are exposed to AVP at the same time.

4. Concluding remarks

The role of AVP in the regulation of water and ion balances in brain is explored, and in particular, its important influences on non-neuronal brain cells. AVP affects non-neuronal cells in diverse manners, an effect which may be modulated by other neuronal hormone and/or neurotransmitter systems. The modulation by ETs may be specially important and takes place both in astrocytes and in endothelial cells. In cerebral astrocytes, AVP facilitates water permeability to regulate water content at the cellular level of brain parenchyma. This effect is marked in astrocytic endfeet covering microvessels, which may lead to an enhanced fluid uptake in brain, facilitating the development of brain edema. In endothelial cells, AVP enhances cotransporter activity and channel opening to control $[K^+]_e$ in brain and further enhance edema formation under pathological conditions. In ependymal cells, AVP increases water permeability to facilitate transport between CSF and CSF extracellular fluid. In choroid plexus, AVP decreases Cl^- efflux into CSF to reduce formation of CSF. Circumventricular astrocytes and pituicytes regulate systemic AVP secretion by morphological changes in their relationship with neurons, and they may react to local AVP release by reducing further AVP secretion. In the anterior lobe of the pituitary gland AVP facilitates release of ACTH. Many of these effects are associated with aquaporins, either by directly increasing water permeability or by mediating ion transport that drives vectorial water movement.

References

Amedee, T., Robert, A., Coles, J.A., 1997. Potassium homeostasis and glial energy metabolism. Glia 21, 46–55.

Baduat, J., Hirt, L., Granziera, C., Bogousslavsky, J., Magistretti, P.J., Regli, L., 2001. Astrocyte-specific expression of aquaporin-9 in mouse brain is increased after transient focal cerebral ischemia. J. Cereb. Blood Flow Metab. 21, 477–482.

Badaut, J., Lasbennes, F., Magistretti, P.J., Regli, L., 2002. Aquaporins in brain: distribution, physiology, and pathophysiology. J. Cereb. Blood Flow Metab. 22, 367–378.

Beagley, G.H., Hatton, G.I., 1994. Systemic signals contribute to induced morphological changes in the hypothalamo-neurohypophysial system. Brain Res. Bull. 33, 211–218.

Bradbury, M.W.B., 1979. The Concept of a Blood–Brain Barrier. Wiley, Chichester.

Bruni, J.E., 1998. Ependymal development, proliferation, and functions: a review. Microsc. Res. Tech. 41, 2–13.

Burbach, J.P.H., Adan, R.A., Lolait, S.J., van Leeuwen, F.W., Nezey, E., Palkovits, M., Barberis, C., 1995. Molecular neurobiology and pharmacology of the vasopressin/oxytocin receptor family. Cell Mol. Neurobiol. 15, 573–595.

Chen, Y., McNeill, J.R., Hajek, I., Hertz, L., 1992. Effect of vasopressin on brain swelling at the cellular level—do astrocytes exhibit a furosemide–vasopressin-sensitive mechanism for volume regulation? Can. J. Physiol. Pharmacol. 70, S367–S373.

Chen, Y., Zhao, Z., Hertz, L., 2000. Vasopressin increases $[Ca^{2+}]_i$ in differentiated astrocytes by activation of V_{1b}/V_3 receptors but has no effect in mature cortical neurons. J. Neurosci. Res. 60, 761–766.

Chodobski, A., Loh, Y.P., Corsetti, S., Szmydynger-Chobodska, J., Johanson, C.E., Lim, Y.P., Monfils, P.R., 1997. The presence of arginine vasopressin and its mRNA in rat choroid plexus epithelium. Mol. Brain Res. 48, 67–72.

Chodobski, A., Szmydynger-Chobodska, J., 2001. Choroid plexus: target for polypeptides and site of their synthesis. Microsc. Res. Tech. 52, 65–82.

Chodobski, A., Szmydynger-Chobodska, J., McKinley, M.J., 1998a. Cerebrospinal fluid formation and absorption in dehydrated sheep. Am. J. Physiol. 275, F235–F238.

Chodobski, A., Wojcik, B.E., Peng Loh, Y., Dodd, K.A., Szmydynger-Chobodska, J., Johanson, C.E., Demers, D.M., Chun, Z.G., Limthong, N.P., 1998b. Vasopressin gene expression in rat choroid plexus. Adv. Exp. Med. Biol. 449, 59–65.

Davson, H., Segal, M.B., 1970. The effects of some inhibitors and accelerators of sodium transport on turnover of ^{22}Na in the cerebrospinal fluid. J. Physiol. Lond. 209, 131–153.

Del Bigio, M.R., Fedoroff, S., 1990. Swelling of astroglia in vitro and the effect of arginine vasopressin and atrial natriuretic peptide. Acta Neurochir. Suppl. (Wien) 51, 14–16.

Dibas, A.I., Mia, A.J., Yorio, T., 1998. Aquaporins (water channels): role in vasopressin-activated water transport. Proc. Soc. Exp. Biol. Med. 219, 183–199.

Dickinson, L.D., Betz, A.L., 1992. Attenuated development of ischemic brain edema in vasopressin-deficient rats. J. Cereb. Blood Flow Metab. 12, 681–690.

Dietzel, I., Heinemann, U., Hofmeier, G., Lux, H.D., 1980. Transient changes in the size of the extracellular space in the sensorimotor cortex of cats in relation to stimulus-induced changes in potassium concentration. Exp. Brain Res. 40, 432–439.

Doczi, T., Joo, F., Szerdahelhy, P., Bodosi, M., 1988. Regulation of brain water and electrolyte contents: the opposite actions of central vasopressin and atrial natriuretic factor (ANF). Acta Neurochir. 43, 186–188.

Elkjaer, M., Vajda, Z., Nejsum, L.N., Kwon, T., Jensen, U.B., Amiry-Moghaddam, M., Frokiaer, J., Nielsen, S., 2000. Immunolocolazation of AQP9 in liver, epididymis, testis, spleen, and brain. Biochem. Biophys. Res. Commun. 276, 1118–1128.

Ernst, S.A., Palacios, J.R., Siegel, G.J., 1986. Immunocytochemical localization of Na^+,K^+-ATPase catalytic polypeptide in mouse choroid plexus. J. Histochem. Cytochem. 34, 189–195.

Faraci, F.M., Mayhan, W.G., Farrell, W.J., Heistad, D.D., 1988. Humoral regulation of blood flow to choroid plexus: role of arginine vasopressin. Circ. Res. 63, 373–379.

Faraci, F.M., Mayhan, W.G., Heistad, D.D., 1990. Effect of vasopressin on production of cerebrospinal fluid: possible role of vasopressin (V_1)-receptors. Am. J. Physiol. 258, R94–R98.

Gajkowska, B., Viron, A., 1997. Protracted elevation of endothelin immunoreactivity in hypothalamo-neurohypophysial system after ischemia. Folia Neuropathol. 35, 107–114.

Garner, C., Brown, P.D., 1992. Two types of chloride channel in the apical membrane of rat choroid plexus epithelial cells. Brain Res. 591, 137–145.

Gebke, E., Muller, A.R., Pehl, U., Gerstberger, R., 2000. Astrocytes in sensory circumventricular organs of the rat brain express functional binding sites for endothelin. Neuroscience 97, 371–381.

Goldman, J.E., Abramson, B., 1990. Cyclic AMP-induced shape changes of astrocytes are accompanied by rapid depolymerization of actin. Brain Res. 528, 189–196.

Grisar, T., Frere, J.M., Franck, G., 1979. Effects of K^+ ions on kinetic properties of the (Na^+,K^+)ATPase (EC 3.6.1.3.) of bulk isolated glia, perikarya and synaptosomes from rabbit brain cortex. Brain Res. 165, 87–103.

Hajek, I., Subbarao, K.V., Hertz, L., 1996. Stimulation of Na^+,K^+-ATPase activity in astrocytes and neurons by K^+ and/or noradrenaline. Neurochem. Int. 28, 335–342.

Han, Z., Wax, M.B., Patil, R.V., 1998. Regulation of aquaporin-4 water channels by phorbol ester-dependent protein phosphorylation. J. Biol. Chem. 273, 6001–6004.

Hatton, G.I., 1997. Function-related plasticity in hypothalamus. Annu. Rev. Neurosci. 20, 375–397.

Hatton, G.I., 1999. Astroglial modulation of neurotransmitter/peptide release from the neurohypophysis: present status. J. Chem. Neuroanat. 16, 203–222.

Hatton, G.I., Bicknell, R.J., Hoyland, J., Bunting, R., Mason, W.T., 1992. Arginine vasopressin mobilises intracellular calcium via V_1-receptor activation in astrocytes (pituicytes) cultured from adult rat neural lobes. Brain Res. 588, 75–83.

Hatton, G.I., Perlmutter, L.S., Salm, A.K., Tweedle, C.D., 1984. Dynamic neuronal-glial interactions in hypothalamus and pituitary: implications for control of hormone synthesis and release. Peptides 5(Suppl. 1), 121–138.

Heinemann, U., Lux, H.D., 1975. Undershoots following stimulus-induced rises of extracellular potassium concentration in the cerebral cortex of cat. Brain Res. 93, 63–76.

Hernando, F., Schoots, O., Lolait, S.J., Burbach, J.P.H., 2001. Immunohistochemical localization of the vasopressin V_{1b} receptor in the rat brain and pituitary gland: anatomical support for its involvement in the central effects of vasopressin. Endocrinology 142, 1659–1668.

Hertz, L., 1990. Regulation of potassium homeostasis by glial cells. In: Levi, G. (Ed.), Development and Function of Glial Cells. Wiley-Liss, New York, pp. 225–234.

Hertz, L., Chen, Y., Spatz, M., 2000a. Involvement of non-neuronal brain cells in AVP-mediated regulation of water space at the cellular, organ, and whole-body level. J. Neurosci. Res. 62, 480–490.

Hertz, L., Chen, Y., Spatz, M., 2000b. Effects of arginine vasopressin on water space in astrocytes and in whole-brain. Am. J. Physiol. 278, E1175–E1176.

Hess, J., Jensen, C.V., Diemer, N.H., 1991. The vasopressin receptor of the blood–brain barrier in the rat hippocampus is linked to calcium signaling. Neurosci. Lett. 132, 8–10.

Imai, T., Hirata, Y., Emori, T., Yanagisawa, M., Masaki, T., Marumo, F., 1992. Induction of endothelin-1 gene by angiotensin and vasopressin in endothelial cells. Hypertension 19, 753–757.

Jard, S., Gaillard, R.C., Guillon, G., Marie, J., Schoenenberg, P., Muller, A.F., Manning, M., Sawyer, W.H., 1986. Vasopressin antagonists allow demonstration of a novel type of vasopressin receptor in the rat adenohypophysis. Mol. Pharmacol. 30, 171–177.

Jarvis, C.R., Andrew, R.D., 1988. Correlated electrophysiology and morphology of the ependyma in rat hypothalamus. J. Neurosci. 8, 3691–3702.

Johanson, C.E., Preston, J.E., Chodobski, A., Stopa, E.G., Szmydynger-Chodobska, J., McMillan, P.N., 1999. AVP V_1 receptor-mediated decrease in Cl^- efflux and increase in dark cell number in choroid plexus epithelium. Am. J. Physiol. 276, C82–C90.

Jones, H.C., Keep, R.F., 1988. Brain fluid calcium concentration and response to acute hypercalcaemia during development in the rat. J. Physiol. 402, 579–593.

Jung, J.S., Bhat, R.V., Preston, G.M., Guggino, W.B., Baraban, J.M., Agre, P., 1994. Molecular characterization of an aquaporin cDNA from brain: candidate osmoreceptor and regulator of water balance. Proc. Natl Acad. Sci. USA 91, 13052–13056.

Jurzak, M., Muller, A.R., Gerstberger, R., 1995. Characterization of vasopressin receptors in cultured cells derived from the region of rat brain circumventricular organs. Neuroscience 65, 145–1159.

Katay, L., Latzkovits, L., Fonagy, A., Janka, Z., Lajtha, A., 1998. Effects of arginine vasopressin and atriopeptin on chloride uptake in cultured astroglia. Neurochem. Res. 23, 831–836.

Kato, Y., Igarashi, N., Hirasawa, A., Tsujimoto, G., Kobayashi, M., 1995. Distribution and developmental changes in vasopressin V_2 receptor mRNA in rat brain. Differentiation 59, 163–169.

Kawai, N., McCarron, R.M., Spatz, M., 1995a. Endothelins stimulate sodium uptake into rat brain capillary endothelial cells through endothelin A-like receptors. Neurosci. Lett. 190, 85–88.

Kawai, N., Yamamoto, T., Yamamoto, H., McCarron, R.M., Spatz, M., 1995b. Endothelin-1 stimulates $Na(^+)$, $K(^+)$-ATPase and $Na(^+)$–$K(^+)$-Cl^- cotransport through ET_A receptors and protein kinase C-dependent pathway in cerebral capillary endothelium. J. Neurochem. 65, 1588–1596.

Kawai, N., Yamamoto, T., Yamamoto, H., McCarron, R.M., Spatz, M., 1997. Functional characterization of endothelin receptors on cultured brain capillary endothelial cells of the rat. Neurochem. Int. 31, 597–605.

Keep, R.F., Ulanski, L.J. 2nd, Xiang, J., Ennis, S.R., Betz, L.A., 1999. Blood–brain barrier mechanisms involved in brain calcium and potassium homeostasis. Brain Res. 815, 200–205.

Keep, R.F., Xiang, J., Betz, A.L., 1993. Potassium transport at the blood–brain and blood–CSF barriers. Adv. Exp. Med. Biol. 331, 43–54.

Keep, R.F., Xiang, J., Betz, A.L., 1994. Potassium cotransport at the rat choroid plexus. Am. J. Physiol. 267, 1616–1622.

Klarr, A., Ulanski, L.J., Stummer, W., Xiang, J., Betz, A.L., Keep, R.F., 1997. Effects of hypo- and hyperkalemia on choroid plexus potassium transport. Brain Res. 758, 39–44.

Klatzo, I., 1987. Pathophysiological aspects of brain edema. Acta Neuropathol. (Berl) 72, 236–239.

Kobayashi, H., Minami, S., Itoh, S., Shiraishi, S., Yokoo, H., Yanagita, T., Uezono, Y., Mohri, M., Wada, A., 2001. Aquaporin subtypes in rat cerebral microvessels. Neurosci. Lett. 197, 163–166.

Kobayashi, H., Minami, S., Itoh, S., Yokoo, H., Yanagita, T., Sugano, T., Kis, B., Ueta, Y., Wada, A., 2002. Aquaporin subtypes expressed in rat cerebral microvessels: effect of hypoxia. Soc. Neurosci. (Abstract).

Koyama, Y., Baba, A., 1994. Endothelins are extracellular signals modulating cytoskeletal actin organization in rat cultured astrocytes. Neuroscience 61, 1007–1016.

Koyama, Y., Baba, A., 1996. Endothelin-induced cytoskeletal actin re-organization in cultured astrocytes: inhibition by C3 ADP-ribosyltransferase. Glia 16, 342–350.

Lafarga, M., Berciano, M.T., del Olmo, E., Andres, M.A., Pazos, A., 1992. Osmotic stimulation induces changes in the expression of beta-adrenergic receptors and nuclear volume of astrocytes in supraoptic nucleus of the rat. Brain Res. 588, 311–316.

Lange, M., Pagotto, U., Hopfner, U., Ehrenreich, H., Oeckler, R., Sinowatz, F., Stalla, GK. 1994. Endothelin expression in normal human anterior pituitaries and pituitary adenomas. J. Clin. Endocrinol. Metab. 79, 1864–1870.

Lange, M., Pagotto, U., Renner, U., Arzberger, T., Oeckler, R., Stalla, G.K., 2002. The role of endothelins in the regulation of pituitary function. Exp. Clin. Endocrinol. Diab. 110, 103–112.

Latzkovits, L., Cserr, H.F., Park, J.T., Patlak, C.S., Pettigrew, K.D., Rimanoczy, A., 1993. Effects of arginine vasopressin and atriopeptin on glial cell volume measured as 3-MG space. Am. J. Physiol. 264, C603–C608.

van Leeuwen, L.W., van der Beek, E.M., van Heerikhuize, J.J., Wolters, P., van der Meulen, G., Wan, Y.P., 1987. Quantitative light microscopic autoradiographic localization of binding sites labelled with [^{3}H]vasopressin antagonist d(CH_3)$_5$Tyr(Me)VP in the rat brain, pituitary and kidney. Neurosci. Lett. 80, 121–126.

Lemos, J.R., Wang, G., 2000. Excitatory versus inhibitory modulation by ATP of neurohypophysial terminal activity in the rat. Exp. Physiol. 85(Spec No), 67S–74S.

Liu, X.F., Shi, Y.M., Lin, B.C., 1991. Mechanism of action of arginine vasopressin on acute ischemic brain edema. Chin. Med. J. 104, 480–483.

Livesey, J.H., Evans, M.J., Mulligan, R., Donald, R.A., 2000. Interactions of CRH, AVP and cortisol in the secretion of ACTH from perifused equine anterior pituitary cells: "permissive" roles for cortisol and CRH. Endocr. Res. 26, 445–463.

Ludwig, M., 1995. Functional role of intrahypothalamic release of oxytocin and vasopressin: consequences and controversies. Am. J. Physiol. 268, E537–E545.

Luerssen, T.G., Robertson, G.L., 1980. Cerebrospinal fluid vasopressin and vasotocin in health and diseases. In: Wood, J.H. (Ed.), Neurobiology of Cerebrospinal Fluid. Plenum, New York, pp. 613–623.

Ma, T., Yang, B., Gillespie, A., Carlson, E.J., Epstein, C.J., Verkman, A.S., 1997. Generation and phenotype of a transgenic knockout mouse lacking the mercurial-insensitive water channel aquaporin-4. J. Clin. Invest. 100, 957–962.

Makino, S., Hashimoto, K., Hirasawa, R., Hattori, T., Ota, Z., 1992. Central interaction between endothelin and brain natriuretic peptide on vasopressin secretion. J. Hypertens. 10, 25–28.

Maktabi, M.A., Elbokl, F.F., Faraci, F.M., Todd, M.M., 1993. Halothane decreases the rate of production of cerebrospinal fluid. Possible role of vasopressin V_1 receptors. Anesthesiology 78, 72–82.

Malendowicz, L.K., Belloni, A.S., Nussdorfer, G.G., Hochol, A., Nowak, M., 1998. Arginine–vasopressin and corticotropin-releasing hormone are sequentially involved in the endothelin-1-induced acute stimulation of rat pituitary-adrenocortical axis. J. Steroid Biochem. Mol. Biol. 66, 45–49.

Manley, G.T., Fujimura, M., Ma, T., Noshita, N., Filiz, F., Bollen, A.W., Chan, P., Verkman, A.S., 2000. Aquaporin-4 deletion in mice reduces brain edema after acute water intoxication and ischemic stroke. Nat. Med. 6, 159–163.

Manoonkitiwongsa, P.S., Schultz, R.L., Wareesangtip, W., Whitter, E.F., Nava, P.B., McMillan, P.J., 2000. Luminal localization of blood–brain barrier sodium, potassium adenosine triphosphatase is dependent on fixation. J. Histochem. Cytochem. 48, 859–865.

Manoonkitiwongsa, P.S., Whitter, E.F., Schultz, R.L., 1998. An in situ cytochemical evaluation of blood–brain barrier sodium, potassium-activated adenosine triphosphatase polarity. Brain Res. 798, 261–270.

Masuzawa, T., Ohta, T., Kawamura, M., Nakahara, N., Sato, F., 1984. Immunohistochemical localization of Na^+, K^+-ATPase in the choroid plexus. Brain Res. 302, 357–362.

McNiel, T.H., Sladek, J.R., 1980. Simultaneous monoamine histofluorescence and neuropeptide immunocytochemistry: II. Correlative distribution of catecholamine varicosities and magnocellular neurosecretory neurons in the rat supraoptic and paraventricular nuclei. J. Comp. Neurol. 193, 1023–1033.

Mecker, R.B., Greenwood, R.S., Hayward, J.N., 1993. Glutamate is the major excitatory transmitter in the supraoptic nuclei. Ann. N. Y. Acad. Sci. 689, 636–639.

Mercado, R., Hernandez, J., 1992. Regulatory role of a neurotransmitter (5-HT) on glial Na^+/K^+-ATPase in the rat brain. Neurochem. Int. 21, 119–127.

Miyata, S., Furuya, K., Nakai, S., Bun, H., Kiyohara, T., 1999. Morphological plasticity and rearrangement of cytoskeletons in pituicytes cultured from adult rat neurohypophysis. Neurosci. Res. 33, 299–306.

Murphy, V.A., Johanson, C.E., 1990. $Na(^+)–H^+$ exchange in choroid plexus and CSF in acute metabolic acidosis or alkalosis. Am. J. Physiol. 258, F1528–F1537.

Nagao, S., Kagawa, M., Bemana, I., Kuniyoshi, T., Ogawa, T., Honma, Y., Kuyama, H., 1994. Treatment of vasogenic brain edema with arginine vasopressin receptor antagonist—an experimental study. Acta Neurochir. Suppl. 60, 502–504.

Nakahama, K., Nagano, M., Fujioka, A., Shinoda, K., Sasaki, H., 1999. Effect of TPA on aquaporin 4 mRNA expression in cultured rat astrocytes. Glia 25, 240–246.

Nakamura, S., Naruse, M., Naruse, K., Shioda, S., Nakai, Y., Uemura, H., 1993. Colocalization of immunoreactive endothelin-1 and neurohypophysial hormones in the axons of the neural lobe of the rat pituitary. Endocrinology 132, 530–533.

Neely, J.D., Amiry-Moghaddam, M., Ottersen, O.P., Froehner, S.C., Agre, P., Adams, M.E., 2001. Syntrophin-dependent expression and localization of Aquaporin-4 water channel protein. Proc. Natl Acad. Sci. USA 98, 14108–14113.

Nielsen, S., Nagelhus, E.A., Amiry-Moghaddam, M., Bourque, C., Agre, P., Ottersen, O.P., 1997. Specialized membrane domains for water transport in glial cells: high-resolution immunogold cytochemistry of aquaporin-4 in rat brain. J. Neurosci. 17, 171–180.

Niermann, H., Amiry-Moghaddam, M., Holthoff, K., Witte, O.W., Ottersen, O.P., 2001. A novel role of vasopressin in the brain: modulation of activity-dependent water flux in the neocortex. J. Neurosci. 21, 3045–3051.

Norup, E., Smitt, U.W., Christensen, S.B., 1986. The potencies of thapsigargin and analogues as activators of rat peritoneal mast cells. Planta Med. 4, 251–255.

O'Donnell, M.E., Martinez, A., Sun, D., 1995. Endothelial Na^+,K^+,Cl^- cotransport regulation by tonicity and hormones: phosphorylation of cotransport protein. Am. J. Physiol. 269, C1513–C1523.

Ostrowski, N.L., Lolait, S.J., Young, W.S. 3rd, 1994. Cellular localization of vasopressin V_{1a} receptor messenger ribonucleic acid in adult male rat brain, pineal and brain vasculature. Endocrinology 135, 1511–1528.

Padmanabhan, J., Clayton, D., Shelanski, M.L., 1999. Dibutyryl cyclic AMP-induced process formation in astrocytes is associated with a decrease in tyrosine phosphorylation of focal adhesion kinase and paxillin. J. Neurobiol. 39, 407–422.

Parmelee, J.T., Bairamian, D., Johanson, C.E., 1991. Response of infant and adult rat choroid plexus potassium transporters to increased extracellular potassium. Brain Res. Dev. Brain Res. 60, 229–233.

Patil, R.V., Han, Z., Wax, M.B., 1997. Regulation of water channel activity of aquaporin-1 by arginine vasopressin and atrial natriuretic peptide. Biochem. Biophys. Res. Commun. 204, 861–866.

Pearson, M.M., Lu, J., Mount, D.B., Delpire, E., 2001. Localization of the $K^+–Cl^-$ cotransporter, KCC3, in the central and peripheral nervous systems: expression in the choroid plexus, large neurons and white matter tracts. Neuroscience 103, 481–491.

Plotkin, M.D., Kaplan, M.R., Peterson, L.N., Gullans, S.R., Hebert, S.C., Delpire, E., 1997. Expression of the $Na^+–K^+–2Cl^-$ cotransporter BSC2 in the nervous system. Am. J. Physiol. 272, C173–C183.

van den Pol, A.N., Wuarin, J.P., Dudek, F.E., 1990. Glutamate, the dominant excitary transmitter in neuroendocrine regulation. Science 250, 1276–1278.

Raichle, M.E., 1981. Hypothesis: a central neuroendocrine system regulates brain ion homeostasis and volume. In: Martin, J.B., Reichlin, S., Bick, K.L., (Eds.), Neurosecretion and Brain Peptides. Raven Press, New York, pp. 329–336.

Raichle, M.E., Grubb, R.L. Jr., 1978. Regulation of brain water permeability by centrally-released vasopressin. Brain Res. 143, 191–194.

Reppert, S.M., Coleman, R.J., Heath, H.W., Keutmann, H.T., 1982. Circadian properties of vasopressin and melatonin rhythms in cat cerebrospinal fluid. Am. J. Physiol. 243, E489–E498.

Ritz, M.F., Stuenkel, E.L., Dayanithi, G., Jones, R., Nordmann, J.J., 1992. Endothelin regulation of neuropeptide release from nerve endings of the posterior pituitary. Proc. Natl Acad. Sci. USA 89, 8371–8375.

Rosenberg, G.A., Estrada, E., Kyner, W.T., 1990. Vasopressin-induced brain edema is mediated by the V_1 receptor. Adv. Neurol. 52, 149–154.

Rosenberg, G.A., Kyner, W.T., Fenstermacher, J.D., Patlak, C.S., 1986. Effect of vasopressin on ependymal and capillary permeability to tritiated water in cat. Am. J. Physiol. 251, F485–F489.

Rossi, N.F., 1995. Differential effects of ET_A and ET_B receptor activation at hypothalamic and posterior pituitary sites on basal vasopressin secretion in vitro. J. Am. Soc. Nephrol. 6, 745.

Rossi, N.F., Chen, H., 2002. Modulation of ET(B) receptor-induced arginine-vasopressin secretion by *N*-methyl-D-aspartate (NMDA) and gamma-aminobutyric acid (GABA)-dependent mechanisms in hypothalamo-neurohypophysial explants. Clin. Sci. (Lond.) 103(Suppl. 48), 162S–166S.

Rossi, N.F., O'Leary, D.S., Scislo, T.J., Caspers, M.L., Chen, H., 1997. Central endothelin 1 regulation of arterial pressure and arginine vasopressin secretion via the AV3V region. Kidney Int. 61, S22–S26.

Rosso, L., Peteri-Brunback, B., Vouret-Craviari, V., Deroanne, C., Troadec, J.D., Thirion, S., Van Obberghen-Schilling, E., Mienville, J.M., 2002a. RhoA inhibition is a key step in pituicyte stellation induced by A(1)-type adenosine receptor activation. Glia 38, 351–362.

Rosso, L., Peteri-Brunback, B., Vouret-Craviari, V., Deroanne, C., Van Obberghen-Schilling, E., Mienville, J.M., 2002b. Vasopressin and oxytocin reverse adenosine-induced pituicyte stellation via calcium-dependent activation of Cdc42. Eur. J. Neurosci. 16, 2324–2332.

Rubanyi, G.M., Polokoff, M.A., 1994. Endothelins: molecular biology, biochemistry, pharmacology, physiology, and pathophysiology. Pharmacol. Rev. 46, 325–415.

Sanchez del Pino, M.M., Hawkins, R.A., Peterson, D.R., 1995. Biochemical discrimination between luminal and abluminal enzyme and transport activities of the blood–brain barrier. J. Biol. Chem. 270, 14907–14912.

Sarfaraz, D., Fraser, C.L., 1999. Effects of arginine vasopressin on cell volume regulation in brain astrocytes in culture. Am. J. Physiol. 276, E596–E601.

Schielke, G.P., Moises, H.C., Betz, A.L., 1990. Potassium activation of the Na,K-pump in isolated brain microvessels and synaptosomes. Brain Res. 524, 291–296.

Schultz, W.J., Brownfield, M.S., Kozlowski, G.P., 1977. The hypothalamo-choridal tract. II. Ultrastructural response of the choroid plexus to vasopressin. Cell Tissue Res. 178, 129–141.

Segal, M.B., Chodobski, A., Szemydynger-chodobska, J., Cammish, H., 1992. Effect of arginine vasopressin on blood vessels of the perfused choroid plexus of the sheep. Prog. Brain Res. 91, 451–453.

Segal, M.B., Zlokovic, B.V., 1990. The Blood–Brain Barrier, Amino Acids and Peptides. Kluwer Academic, Lancaster, pp. 103–105.

Sobue, K., Yamamoto, N., Yoneda, K., Fujita, K., Miura, Y., Asai, Y., Tsuda, T., Katsuya, H., Kato, T., 1999. Molecular cloning of two bovine aquaporin-4 cDNA isoforms and their expression in brain endothelial cells. Biochim. Biophys. Acta 1489, 393–398.

Spatz, M., Kawai, N., Bembry, J., Lenz, F., McCarron, R.M., 1998. Human brain capillary endothelium: modulation of K^+ efflux and K^+, Ca^{2+} uptake by endothelin. Neurochem. Res. 23, 1125–1132.

Spatz, M., Kawai, N., Merkel, N., Bembry, J., McCarron, R.M., 1997. Functional properties of cultured endothelial cells derived from large microvessels of human brain. Am. J. Physiol. 272, C231–C239.

Spatz, M., Stanimirovic, D., Bacic, F., Uematsu, S., Bembry, J., McCarron, R.M., 1993. Peptidergic induction of endothelin 1 and prostanoid secretion in human cerebromicrovascular endothelium. In: Drewes, L.R., Betz, A.L. (Eds.), Frontiers in Cerebral Vascular Biology: Transport and its Regulation. Plenum Press, New York, pp. 165–169.

Spatz, M., Stanimirovic, D., Bacic, F., Uematsu, S., McCarron, R.M., 1994. Vasoconstrictive peptides induce endothelin-1 and prostanoids in cerebrovascular endothelium. Am. J. Physiol. 266, C654–C660.

Speake, T., Freeman, L.J., Brown, P.D., 2003. Expression of aquaporin 1 and aquaporin 4 water channels in rat choroid plexus. Biochim. Biophys. Acta 1609, 80–86.

Speake, T., Whitwell, C., Kajita, H., Majid, A., Brown, P.D., 2001. Mechanisms of CSF secretion by the choroid plexus. Microsc. Res. Tech. 52, 49–59.

Sperlagh, B., Mergl, Z., Juranyi, Z., Vizi, E.S., Makara, G.B., 1999. Local regulation of vasopressin and oxytocin secretion by extracellular ATP in the isolated posterior lobe of the rat hypophysis. J. Endocrinol. 160, 343–350.

Su, G., Kintner, D.B., Fagella, M., Shukk, G.E., Sun, D., 2002a. Astrocytes from Na^+,K^+,Cl^- cotransporter-null mice exhibit absence of swelling and decrease in EAA release. Am. J. Physiol. 282, C1147–C1160.

Su, G., Kintner, D.B., Sun, D., 2002b. Contribution of Na^+–K^+–Cl^- cotransporter to high-$[K^+]_o$-induced swelling and EAA release in astrocytes. Am. J. Physiol. 282, C1136–C1146.

Sun, D., Lytle, C., O'Donnell, M.E., 1995. Astroglial cell-induced expression of Na^+,K^+,Cl^- cotransporter in brain microvascular endothelial cells. Am. J. Physiol. 269, C1505–C1512.

Sweadner, K.J., 1995. Na,K-ATPase and its isoforms. In: Kettenmann, H., Ransom, B.R. (Eds.), Neuroglia. Oxford University Press, New York, pp. 259–272.

Sykova, E., 1983. Extracellular K^+ accumulation in the central nervous system. Prog. Biophys. Mol. Biol. 42, 135–189.

Takahashi, K., Ghatei, M.A., Jones, P.M., Murphy, J.K., Lam, H.C., O'Halloran, D.J., Bloom, S.R., 1991. Endothelin in human brain and pituitary gland: presence of immunoreactive endothelin, endothelin messenger ribonucleic acid, and endothelin receptors. J. Clin. Endocrinol. Metab. 72, 693–699.

Tas, P.W.L., Massa, P.T., Kress, H.G., Koschel, K., 1987. Characterization of an Na^+,K^+,Cl^- co-transport in primary cultures of astrocytes. Biochim. Biophys. Acta 903, 411–416.

Theodosis, D.T., Poulain, D.A., Vincent, J.D., 1981. Possible morphological bases for synchronisation of neuronal firing in the rat supraoptic nucleus during lactation. Neuroscience 6, 919–929.

Thibonnier, M., Preston, J.A., Dulin, N., Wilkins, P.L., Berti-Mattera, L.N., Mattera, R., 1997. The human V_3 pituitary vasopressin receptor: ligand binding profile and density-dependent signaling pathways. Endocrinology 138, 4109–4122.

Torday, C., Fonagy, A., Latzkovits, L., 1997. Intracellular free Ca^{2+} elevations in cultured astroglia induced by neuroligands playing a role in cerebral ischemia. Acta Chir. Hung. 36, 362–363.

Tribollet, E., Raufaste, D., Maffrand, J., Maffrand, J., 1999. Binding of the non-peptide vasopressin V_{1a} receptor antagonist SR-49059 in the rat brain: an in vitro and in vivo autoradiographic study. Neuroendocrinology 69, 113–120.

Troadec, J.D., Thirion, S., Nicaise, G., Lemos, J.R., Dayanithi, G., 1998. ATP-evoked increases in $[Ca^{2+}]_i$ and peptide release from rat isolated neurohypophysial terminals via a P_{2X2} purinoceptor. J. Physiol. 511, 89–103.

Tse, A., Lee, A.K., 1998. Arginine vasopressin triggers intracellular calcium release, a calcium-activated potassium current and exocytosis in identified rat corticotropes. Endocrinology 139, 2246–2252.

Tweedle, C.D., Hatton, G.I., 1977. Ultrastructural changes in rat hypothalamic neurosecretory cells and their associated glia during minimal dehydration and rehydration. Cell Tissue Res. 181, 59–72.

Vajda, Z., Pedersen, M., Fuchtbauer, E.M., Wertz, K., Stodkilde-Jorgensen, H., Sulyok, E., Doczi, T., Neely, J.D., Agre, P., Frokiaer, J., Nielsen, S., 2002. Delayed onset of brain edema and mislocalization of aquaporin-4 in dystrophin-null transgenic mice. Proc. Natl Acad. Sci. USA 99, 13131–13136.

Van Renterghem, C., Vigne, P., Frelin, C., 1995. A charybdotoxin-sensitive Ca(2 +)-activated K + channel with inward rectifying properties in brain microvascular endothelial cells: properties and activation by endothelins. J. Neurochem. 65, 1274–1281.

Venero, J.L., Vizuete, M.L., Machado, A., Cano, J., 2001. Aquaporins in the central nervous system. Prog. Neurobiol. 63, 321–336.

Venero, J.L., Vizuete, M.L., Ilundain, A.A., Machado, A., Echevarria, M., Cano, J., 1999. Detailed localization of aquaporin-4 messenger RNA in the CNS: preferential expression in periventricular organs. Neuroscience 94, 239–250.

Vigne, P., Lopez Farre, A., Frelin, C., 1994. Na^+–K^+Cl^- cotransporter of brain capillary endothelial cells. Properties and regulation by endothelins, hyperosmolar solutions, calyculin A, and interleukin-1. J. Biol. Chem. 269, 19925–19930.

Vorherr, H., Bradbury, M.W., Hoghoughi, M., Kleeman, C.R., 1968. Antidiuretic hormone in cerebrospinal fluid during endogenous and exogenous changes in its blood level. Endocrinology 83, 246–250.

de Vries, G.J., Miller, M.A., 1998. Anatomy and function of extrahypothalamic vasopressin systems in the brain. Prog. Brain Res. 119, 3–20.

Wall, K.M., Ferguson, A.V., 1992. Endothelin acts at the subfornical organ to influence the activity of putative vasopressin and oxytocin-secreting neurons. Brain Res. 586, 111–116.

Walz, W., 1987. Swelling and potassium uptake in cultured astrocytes. Can. J. Physiol. Pharmacol. 65, 1051–1057.

Walz, W., 2002. Chloride/anion channels in glial cell membranes. Glia 40, 1–10.

Walz, W., Hertz, L., 1983. Functional interactions between neurons and astrocytes. II. Potassium homeostasis at the cellular level. Prog. Neurobiol. 20, 133–183.

Walz, W., Hertz, L., 1984. Intense furosemide-sensitive potassium accumulation into astrocytes in the presence of pathologically high extracellular potassium levels. J. Cereb. Blood Flow Metab. 4, 301–304.

Wang, G., Dayanithi, G., Custer, E.E., Lemos, J.R., 2002. Adenosine inhibition via A(1) receptor of N-type Ca(2 +) current and peptide release from isolated neurohypophysial terminals of the rat. J. Physiol. 540, 791–802.

Wells, T., 1998. Vesicular osmometers, vasopressin secretion and aquaporin-4: a new mechanism for osmoregulation. Mol. Cell. Endocrinol. 136, 103–107.

Witte, O.W., Niermann, H., Holthoff, K., 2001. Cell swelling and ion redistribution assessed with intrinsic optical signals. An. Acad. Bras. Cienc. 73, 337–350.

Wittkowski, W., 1986. Pitiucytes. In: Fedoroff, S., Vernadakis, A. (Eds.), Astrocytes, vol. 1. Academic Press, New York, pp. 173–208.

Wu, Q., Delpire, E., Hebert, C.S., Strange, K., 1998. Functional demonstration of Na^+–K^+–$2Cl^-$ cotransporter activity in isolated, polarized choroid plexus cells. Am. J. Physiol. 275, C1565–C1572.

Yamamoto, S., Inenaga, K., Eto, S., Yamashita, H., 1995. Cardiovascular-related peptides influence hypothalamic neurons involved in control of body water homeostasis. Obes. Res. Suppl. 5, 789S–794S.

Yamamoto, S., Inenaga, K., Kannan, H., Eto, S., Yamashita, H., 1993. The actions of endothelin on single cells in the anteroventral third ventricular region and supraoptic nucleus in rat hypothalamic slices. J. Neuroendocrinol. 5, 427–434.

Yamamoto, S., Morimoto, I., Yamashita, H., Eto, S., 1992. Inhibitory effects of endothelin-3 on vasopressin release from rat supraoptic nucleus in vitro. Neurosci. Lett. 141, 147–150.

Yamamoto, N., Yoneda, K., Asai, K., Sobue, K., Tada, T., Fujita, Y., Katsuya, H., Fujita, M., Aihar, N., Mase, M., Yamada, K., Miura, Y., Kato, T., 2001. Alterations in the expression of the AQP family in cultured rat astrocytes during hypoxia and reoxygenation. Mol. Brain Res. 90, 26–38.

Yan, Y., Dempsey, R.J., Sun, D., 2001. Expression of Na^+–K^+–Cl^- cotransporter in rat brain during development and its localization in mature astrocytes. Brain Res. 911, 43–55.

Yerby, T.R., Vibat, C.R., Sun, D., Payne, J.A., D'Donnell, M.E., 1997. Molecular characterization of the Na^+, K^+,Cl^- cotransporter of bovine aortic endothelial cells. Am. J. Physiol. 273, C188–C197.

Zemo, D.A., McCabe, J.T., 2001. Salt-loading increases vasopressin and vasopressin 1b receptor mRNA in the hypothalamus and choroid plexus. Neuropeptides 35, 181–188.

Zhao, L., Brinton, R.D., 2002. Vasopressin-induced cytoplasmic and nuclear calcium signaling in cultured cortical astrocytes. Brain Res. 943, 117–131.